Student Manual

for

The Art of Electronics

Thomas C. Hayes
Paul Horowitz

Harvard University

CAMBRIDGE UNIVERSITY PRESS

Cambridge

New York Port Chester Melbourne Sydney

TO TURNER, TESSA AND DEBBIE; TO MISHA, JACOB AND CAROL

Published by the Press Syndicate of the University of Cambridge
The Pitt Building, Trumpington Street, Cambridge CB2 1RP
40 West 20th Street, New York, NY 10011, USA
10 Stamford Road, Oakleigh, Melbourne 3166, Australia

© Cambridge University Press 1989

Printed in the United States of America

ISBN 0-521-37709-9

Contents

ANALOG

Chapter 1: Foundations
Overview: Chapter 1 1
Class 1: DC Circuits 3
   Worked example: Resistors & instruments 15
Lab 1: DC circuits 24

Class 2: Capacitors & *RC* Circuits 32
   Worked example: *RC* circuits 46
   A Note on reading capacitor values 51
Lab 2: Capacitors 54

Class 3: Diode Circuits 61
   Worked example: Power supply 71
Lab 3: Diode circuits 75
   Wrap-up: Ch. 1: Review 80
                 Jargon and terms 81

Chapter 2: Transistors (bipolar)
Overview: Chapters 2 & 3 82
Class 4: Transistors I: First Model 84
   Worked example: Emitter follower 90
Lab 4: Transistors I 94

Class 5: Transistors II: Corrections to the first model: Ebers-Moll: r_e; applying this new view 100
   Worked example: Common-emitter amplifier (bypassed emitter) 115
Lab 5: Transistors II 118

Class 6: Transistors III: Differential amplifier; Miller effect 124
   Worked example: Differential amplifier 131
Lab 6: Transistors III 134
   Wrap-up Ch. 2: Review *139*
                 Jargon and terms 140

Chapter 3: Field Effect Transistors
Class 7: FETs I (linear applications) 142
   Worked example: Current source, source follower 153
Lab 7: FETs I (linear) 156
   (We return to FETs in Lab 11)

Chapter 4: Feedback and Operational Amplifiers
Overview: Feedback: Chapters 4, 5 & 6 163
Class 8: Op Amps I: Idealized view 166
   Worked Example: Inverting amplifier; summing circuit 175
Lab 8: Op amps I 175

Class 9: Op Amps II: Departures from ideal 184
   Worked example: Integrators; effects of op amp errors 196
Lab 9: Op amps II 200

Chapter 4 (continued); Chapter 5 Active Filters & Oscillators
Class 10: Op Amps III: Positive Feedback, Good and Bad: comparators, oscillators, and unstable circuits; a quantitative view of the effects of negative feedback 207
   Appendix: Notes on op amp frequency compensation 222
   Worked example: Effects of feedback (quantitative) 224
                Schmitt trigger 227
           Op Amp Innards: Annotated schematic of the LF411 232
Lab 10: Positive feedback, good and bad 233
   Wrap-up Chs. 4 & 5:   Review 242
                 Jargon and terms 243

Chapter 3: Field Effect Transistors (revisited)
Class 11: FETs II: Switches (power switching and analog switch applications) 244
   Worked example: Sample and hold 250
Lab 11: FET switches 255
   Wrap-up Ch. 3: Review 264
                 Jargon and terms 265

Chapter 6: Voltage Regulators and Power Circuits
Class 12: Voltage Regulators 267
Lab 12: Voltage regulators 274
   Wrap-up Ch. 6: Jargon and terms 280

(This course omits the Text's *Chapter 7: Precision Circuits & Low-Noise Techniques*)

DIGITAL

Chapter 8: Digital Electronics

Overview: Chapters 8 & 9 281
Class 13: Digital Gates; Combinational Logic 281
          Worked example: Multiplexers 283
                        Binary arithmetic 295
Lab 13: Digital gates 309

Class 14: Sequential Circuits: Flip-Flops 320
          Worked example: combinational logic 332
Lab 14: Flip-flops 334

Class 15: Counters 342
          Worked example: counter applications 351
Lab 15: Counters 362

Class 16: Memory; Buses; State Machines 375
          Worked example: state machines 384
Lab 16: Memory; State Machines 394
          Wrap-up Ch. 8: Review *403*
                        Jargon and terms 404

Chapter 9: Digital Meets Analog

Class 17: Analog <—> Digital;
    Phase-Locked Loop 406
Lab 17: Analog <—> Digital;
    Phase-Locked Loop 421
          Wrap-up Ch. 9: Review *430*
                        Jargon and terms 430

Chapters 10, 11: Microcomputers; Microprocessors

Overview: Chapters 10 & 11 431
Class 18: µ1: IBM PC and our lab
    microcomputer 433
          Worked example: minimal 68008 controller .. 441
Lab 18: Add CPU 443

Class 19: Assembly Language; Inside the CPU; I/O
    Decoding 455
          Supplementary notes: introduction to assembly
          language 467
Lab 19: µ2: I/O 471

Class 20: µ3: A/D <—> D/A Interfacing; Masks;
    Data tables 479
Lab 20: Subroutines; More I/O Programming 489

Class 21: µ4: More Assembly-Language
               Programming; 12-bit port 498
          Worked example: 10 tiny programs 503
          Worked Example: 12-bit frequency counter. 518
          Hand assembly: table of codes 521
Lab 21: A/D, D/A, Data Handling 525

Class 22: µ5: Interrupts & Other 'Exceptions' 535
          Debugging aid: *Register Check* in two
              forms 541
Lab 22: 'Storage scope'; Interrupts & other
    'exceptions' 548

Class 23: µ6: Wrap-up: Buying and Building 562
Lab 23: Applying your microcomputer ('Toy
    Catalog') 567
    Wrap-up Chapters 10, 11:
                    Review *586*
                    Jargon and terms 587

Appendix:
    A Equipment and Parts List 588
    B Selected data sheets
                2N5485 JFET 592
                DG403 analog switch 593
                74HC74 dual D FLIP-FLOP 595
                AD7569 8-bit A/D, D/A 597
                68008 execution times and timing
                    diagram 600
                25120 write-only memory 605
    C Big Picture: Schematic of lab
        microcomputer 606
    D Pinouts 608

Index 612

Laboratory Exercises *(a more detailed listing)*

PART I: ANALOG LABS

Lab 1. DC Circuits
Ohm's law; A Nonlinear device; The diode; Voltage divider; Thevenin model; Oscilloscope; AC voltage divider

Lab 2. Capacitors
RC circuit; Differentiator; Integrator; Low-pass filter; High-pass filter; Filter example I; Filter example II; Blocking capacitor; *LC* filter

Lab 3. Diodes
LC resonant circuit; Confirming Fourier series; Half-wave rectifier; Full-wave bridge rectifier; Ripple; Signal diodes; Diode clamp; Diode limiter; Impedances of test instruments

Lab 4. Transistors I
Transistor junctions are diodes; Emitter follower; Transistor current gain; Current source; Common emitter amplifier; Transistor switch

Lab 5. Transistors II
Dynamic diode curve tracer; Grounded emitter amplifier; Current mirror; Ebers-Moll equation; Biasing: good & bad; Push-Pull

Lab 6. Transistors III
Differential amplifier; Bootstrap; Miller effect; Darlington; Superbeta

Lab 7. Field Effect Transistors I
FET characteristics; FET current sources; Source follower; FET as Voltage-controlled resistance; Amplitude modulation; 'Radio broadcast'

Lab 8. Op Amps I
Op–amp open-loop gain; Inverting amplifier; Non- inverting amplifier; Follower; Current source; Current-to-voltage converter; Summing amplifier; Push-pull buffer

Lab 9. Op Amps II
Op–amp limitations; AC amplifier; Integrator; Differentiator; Active rectifier; Active clamp

Lab 10. Oscillators
Comparator; Schmitt trigger; IC relaxation oscillator; Sawtooth wave oscillator; ; Voltage-controlled oscillator; Wien bridge sine oscillator; Unwanted oscillations: discrete follower & op amp stability problems

Lab 11. Field Effect Transistors II
Analog switch characteristics; Applications: chopper circuit; sample-&-hold; switched-capacitor filters; negative voltage from positive

Lab 12. Power Supplies
The 723 regulator; Three- terminal fixed regulator; Three-terminal adjustable regulator;
Three-terminal regulator as current source; Voltage reference; 'Crowbar' clamp

PART II: DIGITAL LABS

Lab 13. Gates
Logic probe; IC gates: TTL & CMOS; Logic functions with NANDs; Gate innards: TTL; CMOS: CMOS NOT, NAND, 3-state

Lab 14. Flip-Flops
Latch; D flop; J-K flop; Ripple counter; Synchronous counter; Shift-register; Digitally-timed *one-shot*

Lab 15. Counters
8-bit counter; Cascading; Load from keypad; Programmable divide-by-n counter; Period meter; Capacitance meter

Lab 16. Memory; State Machines
RAM; Divide-by-3 (your design); Memory-based state machines: Single-loop; External control added

Lab 17. A/D; Phase-Locked Loop: Two Digital Feedback Machines:
D/A; A/D: Slow motion; Full speed; Displaying search tree; Speed limit; Latching output; Phase-Locked Loop: frequency multiplier.

Lab 18. µ1: Adding CPU
Clock; CPU preliminary test; Fixing *busgrant**; Memory enable logic; Memory write logic; Single-step; Test program; Full-speed: timing diagram

Lab 19. µ2: I/O: Output: First small programs
Battery backup; Power-fail detector; I/O decoder; Data displays; Timing program

Lab 20. µ3: Input; More small programs
Delay as subroutine; Improved delay routines; Input hardware: Data input hardware; Input/output program; Ready signal; I/O program with enter/ready function; Decimal arithmetic

Lab 21. µ3: A/D <—> D/A
A/D-D/A wiring details; Programs: confirming that D/A, A/D work; In & Out; Invert, rectify, low-pass;

Lab 22. µ4: 'Storage scope;' Interrupts & other 'Exceptions'
'Storage scope;' keyboard control;
Exceptions: A software exception: illegal; Interrupt: hardware to request interrupt; Program: main & service routine; NMI; Applying interrupts

Register-Check: a debugging aid (optional program; install if you choose to)

Lab 23. Applying Your Microcomputer (*'Toy Catalog'*)
X-Y scope displays; Light-pen; Voice output; Driving a stepper motor; Games; Sound sampling/generation

PREFACE

This manual is intended to be used along with The Art of Electronics by Horowitz and Hill (Cambridge University Press, New York, 2d ed. 1989) in an introductory electronics course. The manual includes three principal elements:

- laboratory exercises: 23 of these, each meant to occupy a 3-hour lab period; each set of laboratory notes except the last includes a reading assignment in the Text;
- explanatory notes: one for each laboratory exercise or class;
- worked examples: a total of 20: approximately one for each reading assignment.

In addition, we have included some reference materials:

- a glossary of frequently-used terms and jargon;
- review notes for each chapter, noting the most important circuits and topics;
- selected data sheets, analog and digital.

The students this course might suit

These notes arose out of a course at Harvard; they define what we try to teach in that busy term. The course does less than all of Horowitz and Hill, of course. We treat chapters 1-11, omitting Chapter 7, on *Precision Circuits...*, which is more specialized than the rest, and skimming Chapter 4 on *Active Filters and Oscillators*. Even this selection includes more information than we expect students to absorb fully on a first pass through the book. This Manual tries to guide students to the most important material.

The *typical* student that we see—if there really is a typical student—is an undergraduate majoring in Physics, and wanting to learn enough electronics to let him or her do useful work in a laboratory. But we do not assume such background in these notes. Students very different from that typical student thrive in our course. Graduate students in the sciences appear regularly; during the summer we see many high school students, and some of these do brilliantly; now and then a professor of Physics takes the course (and they do all right, too!). In the 'extension' version of the course, we see lots of programmers who want to know what's going on in their machines, and we see people who just happen to be curious about electronics. That curiosity, in fact, is the only prerequisite for this course, and suggests the only good rule to define who will enjoy it. Someone looking for an engineering course will find our treatment oddly informal, but a person eager to learn how to design useful circuits will like this course.

Laboratory Exercises

The laboratory exercises build upon a set of labs that were set out in the 1981 edition of the Laboratory Manual, by Horowitz and Robinson. The new exercises replace all of the original digital labs and substantially revise the analog labs on FET's and oscillators. In the digital section we have switched over from LSTTL to HCMOS, but the major change has been the enlarged role given to the microprocessor labs, and the shift from the Z80 processor programmed rather laboriously via a DIP switch to a 68008 processor programmed through a keypad. (A complete schematic is included. See Lab 15. Complete keypad units are available through the authors. See Parts list).

We have held to our intention that students should build their computer from the chip level, and that they should not be handed a ROM cleverly programmed by someone else.

We want our students to feel that they know their computer intimately, and that it is fully *their* product.

The digital half of the course now centers on the microcomputer: we meet simpler digital devices—gates, flip-flops, counters, memory—partly because we want to be able to build small digital circuits, but also partly in order to understand the full microcomputer circuit. To put this point another way, the final series of labs, in which the microcomputer gradually takes form, draws together every one of the several circuit elements met earlier: combinational logic networks, flip-flop circuits, counters, memory, and analog/digital conversion. The A-D conversion experiments have been expanded to include the effects of sampling-rate and of filters applied to input and output.

Notes

The notes that introduce each lab respond to two needs that students often voice:

- The notes *select* a few points from the much broader coverage of the Text; those selected points are, of course, those that we think most important to a student meeting practical electronics for the first time.
- The notes *explain* at length. They do this at a level more basic than the Text's, and they provide explanations in a step-by-step style that the Text cannot afford, given its need to cover far more material.

A suggestion: how to use the notes

Here's a proposal; you will, of course, find your own way to use Text, Notes and all the other course materials. But here is one way to begin.

- Start by reading the day's assignment in the Text. It will include some material that is subtler than what we expect you to pick up in a first course. You may want to hear some points restated in another way, or you may want to see an example worked. Primed with this specific sort of curiosity, you might then—
- Look at the day's Notes and Lab: scan, first, to see which circuits and which points are selected. Read the Notes on any points that puzzled you; if you still are puzzled, return to the Text for a second look at the topics you now know are most important.
- *Skip* topics in the Notes that you understand already. The Notes are meant to help you, *not* to burden you with additional reading: if you have read and understood the Text's discussion of a topic, you will miss nothing by omitting the corresponding section in the Notes.
- Try the day's worked example, at least in your head. If it looks easy, you may want to skip it. If it looks hard, probably you should try to do your own solution. If you find yourself heading into a lot of work—especially any involved calculations—probably you are doing unnecessary labor, and it is time to peek at our solution. We hope to teach you an approach to problems of circuit design, not just a set of particular rules. If there is a laborious way and a quick way to reach a good design, we want to push you firmly toward the quick way.

We expect that some of these notes will strike you as babyish, some as excessively dense: your reaction naturally reflects the uneven experience you have had with the topics the Text and Manual treat. Some of you are sophisticated programmers, and will sail through the assembly-language programming near the course's end; others will find it heavy going.

That's all right. The course out of which this Manual grew—and, earlier, the Text as well—has a reputation as fun, and not difficult in one sense, but difficult in another: the concepts are straightforward; abstractions are few. But we do pass a lot of information to our students in a short time; we do expect them to achieve literacy rather fast. This course is a lot like an introductory language course, and we hope to teach by the method sometimes called *immersion*. It is the laboratory exercises that do the best teaching; we hope the Text and this Manual will help to make those exercises instructive.

Why our figures and text look the way they do

You will discover very quickly that this manual is informal in language and layout. The figures all are hand-drawn. They are done by hand partly because we like the *look* of hand-drawn figures (when they are done right; not all our figures are pretty), and partly because we want to encourage students to do their own free-hand drafting of schematics. In some cases we did *draft* drawings on a computer, then drew the final versions by hand! The text was produced as camera-ready copy, put out by an ordinary PC word processor. So—as writers used to say, long ago—dear reader, look with sympathy, if you can, when you find a typo, or a figure drawn amiss. Don't blame the publisher for corporate sloth. Picture, instead, two fellows hunched over their keyboard and drawing board, late at night and beginning to get drowsy.

Who helped especially with this book

Two teaching fellows gave us good advice on uncounted occasions: Shahn Majid, a mathematical Physicist who taught with us for years in the Harvard College course, and Steve Morss, a digital engineer who once took the course and then returned to teach. Steve often would linger late into the night helping to try out a new circuit or analyze an old one. Both of these two could perfectly well have taught the course, and chose nevertheless to linger—Bodhissattva-like—giving their expert help in this quieter way.

A pair of our former students, Jeff Hobson and Wei-Jing Zhu, helped us first by drawing figures—and then gradually turned into this book's godparents, helping in all sorts of ways. Often they would arrive in the evening, at the end of a long day's work, and then would labor to help us organize, check, re-check—and also to make judgments on how to make our points clearly. Often the end of the workday was defined by the departure of the last bus, at 1:00 in the morning. Their devotion to the project was invaluable, and touching.

Finally, Debbie Mills deserves thanks for putting up with her husband Tom's strange, long hours, and then, toward the end, doing much more: providing essential help in organizing, checking, and correcting the growing stacks of printouts and drawings.

Tom Hayes
Paul Horowitz

Cambridge, Mass.
July 1989

CHAPTER 1
Overview

The title of this first chapter, "Foundations," describes its place pretty well: here you will learn techniques that will underlie circuitry that later produces impressive results. Chapter 1's circuits are humbler than what you will see later, and the devices you meet here are probably more familiar to you than, say, transistors, operational amplifiers—or microprocessors: Ohm's Law will surprise none of you; $I = C\, dV/dt$ probably sounds at least vaguely familiar.

But the circuit elements that this chapter treats—*passive* devices—appear over and over in later *active* circuits. So, if a student happens to tell us, 'I'm going to be away on the day you're doing Lab 2,' we tell him he will have to make up the lab somehow: that the second lab, on RC circuits, is the most important in the course. If you do not use that lab to cement your understanding of RC circuits—especially filters—then you will be haunted by muddled thinking for at least the remainder of *analog* part of the course.

Resistors will give you no trouble; diodes will seem simple enough, at least in the view that we settle for: they are one-way conductors. Capacitors and inductors behave more strangely. We will see very few circuits that use inductors, but a great many that use capacitors. You are likely to need a good deal of practice before you get comfortable with the central facts of capacitors' behavior—easy to state, hard to get an intuitive grip on: they pass AC, block DC, and *sometimes* cause large phase shifts.

We should also restate a word of reassurance offered by the Text (p. 29), but seldom believed by students: you can manage this course perfectly even if you cannot follow the mathematical arguments that begin in sec. 1.18 (use of complex quantities to represent voltage and current), and even if, after reading the spectacularly-dense *Math Review* in appendix B you feel that *you* must be spectacularly dense. This is the place in the Text and course where the squeamish usually begin to wonder if they ought to retreat to some slower-paced treatment of the subject. Do not give up at this point; hang on until you have seen transistors, at least. The mathematical arguments of 1.18 are not at all characteristic of this Text or of this course. To the contrary, one of the most striking qualities of this Text is its cheerful evasion of complexity whenever a simpler account can carry you to a good design. The treatment of transistors offers a good example, and you ought to stay with the course long enough to see that: the transistor chapter is difficult, but wonderfully simpler than most other treatments of the subject. You will begin designing useful transistor circuits on your first day with the subject.

It is also in the first three labs that you will get used to the lab instruments—and especially to the most important of these, the oscilloscope. It is a complex machine; only practice will teach you to use it well. Do not make the common mistake of thinking that the person next to you who is turning knobs so confidently, flipping switches and adjusting trigger level—all on the first day of the course—is smarter than you are. No, that person has done it before. In two weeks, you too will be making the scope do your bidding—assuming that you don't leave the work to that person next to you—who knew it all from the beginning.

The images on the scope screen make silent and invisible events visible, though strangely abstracted as well; these scope traces will become your mental images of what happens in your circuits. The scope will serve as a time microscope that will let you see events that last a handful of *nanoseconds*: the length of time light takes to get from you to the person sitting a little way down the lab bench. You may even find yourself reacting emotionally to shapes on the screen: feeling good when you see a smooth, handsome sine wave; disturbed when you see the peaks of the sine clipped, or its shape warped; annoyed when fuzz grows on your waveforms.

Anticipating some of these experiences, and to get you in the mood to enjoy the coming weeks in which small events will paint their self-portraits on your screen, we offer you a view of some scope traces that never quite occurred, and that nevertheless seem just about right: just what a scope *would* show if it could. This drawing has been posted on one of our doors for years, now, and students who happen by pause, peer, hesitate—evidently working a bit to put a mental frame around these not-quite-possible pictures; sometimes they ask if these *are* scope traces. They are not, of course; the leap beyond what a scope can show was the artist's: Saul Steinberg's. Graciously, he has allowed us to show his drawing here. We hope you enjoy it. Perhaps it will help you to look on your less exotic scope displays with a little of the respect and wonder with which we have to look on the traces below.

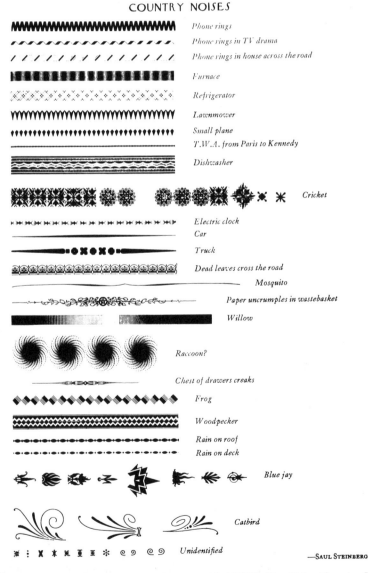

Figure IN1.1: Drawing by Saul Steinberg; copyright 1979 The New Yorker Magazine, Inc.

Class 1: DC Circuits

Topics:

- What this course treats: <u>Art?</u> of Electronics
  DC circuits
  Today we will look at circuits made up entirely of

    – DC voltage sources (things whose output voltage is constant over time; things like a battery, or a lab power supply);
    and
    – resistors.

Sounds simple, and it is. We will try to point out quick ways to handle these familiar circuit elements. We will concentrate on one circuit fragment, the voltage divider.

Preliminary: What is "the art of electronics?"

Not an art, perhaps, but a craft. Here's the Text's formulation:

> [E]lectronics is basically a simple art, a combination of some basic laws, rules of thumb, and a large bag of tricks. (P. 1)

As you may have gathered, if you have looked at the text, this course differs from an engineering electronics course in concentrating on the "rules of thumb" and the "large bag of tricks." You will learn to use rules of thumb and reliable "tricks" without apology. With their help you will be able to leave the calculator-bound novice engineer in the dust!

Two Laws

Text sec. 1.01

First, a glance at two *laws*: Ohm's Law, and Kirchhoff's Laws (V,I).

We rely on these rules continually, in electronics. Nevertheless, we rarely will mention Kirchhoff again. We use his observations *implicitly*. We will see and use Ohm's Law a lot, in contrast; no one has gotten around to doing what's demanded by the bumper sticker one sees around MIT: *Repeal Ohm's Law!*)

Ohm's Law: $E = IR$

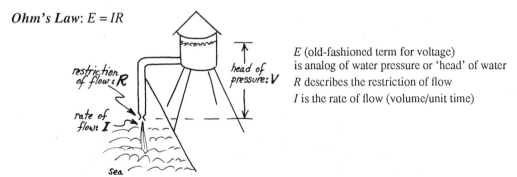

E (old-fashioned term for voltage) is analog of water pressure or 'head' of water
R describes the restriction of flow
I is the rate of flow (volume/unit time)

Figure N1.1: Hydraulic analogy: voltage as head of water, etc. Use it if it helps your intuition

The homely hydraulic analogy works pretty well, if you don't push it too far—and if you're not too proud to use such an aid to intuition.

Ohm's is a very useful rule; but it applies only to things that behave like *resistors*. What are these? They are things that obey Ohm's Law! (Sorry folks: that's as deeply as we'll look at this question, in this course[1].)

We begin almost at once to meet devices that do *not* obey Ohm's Law (see Lab 1: a lamp; a diode). Ohm's Law describes *one* possible relation between V and I in a component; but there are others.

As the text says,

> Crudely speaking, the name of the game is to make and use gadgets that have interesting and useful I vs V characteristics. (P. 3)

Kirchhoff's Laws (V,I)

These two 'laws' probably only codify what you think you know through common sense:

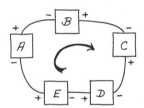

Sum of voltages around loop (circuit) is zero.

Sum of currents in & out of node is zero (algebraic sum, of course).

Figure N1.2: Kirchhoff's two laws

Applications of these laws: series and parallel circuits

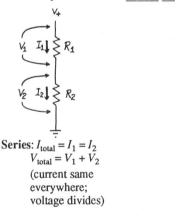
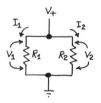

Series: $I_{total} = I_1 = I_2$
$V_{total} = V_1 + V_2$
(current same everywhere; voltage divides)

Parallel: $I_{total} = I_1 + I_2$
$V_{total} = V_1 = V_2$
(voltage same across all parts; current divides)

Figure N1.3: Applications of Kirchhoff's laws: Series and parallel circuits: a couple of truisms, probably familiar to you already

Query: Incidentally, where is the "loop" that Kirchhoff's law refers to?

This is *kind of boring*. So, let's hurry on to less abstract circuits: to applications—and tricks. First, some labor-saving tricks.

---

1. If this remark frustrates you, see an ordinary E & M book; for example, see the good discussion of the topic in E. M. Purcell, Electricity & Magnetism, cited in the Text (2d ed., 1985), or in S. Burns & P. Bond, Principles of Electronic Circuits (1987).

Parallel Resistances: calculating equivalent R

The *conductances* add:
conductance$_{total}$ = conductance$_1$ + conductance$_2$ = $1/R_1 + 1/R_2$

Figure N1.4: Parallel resistors: the *conductances* add; unfortunately, the *resistances* don't

This is the easy notion to remember, but not usually convenient to apply, for one rarely speaks of conductances. The notion "resistance" is so generally used that you will sometimes want to use the formula for the effective resistance of two parallel resistors:

$$R_{tot} = R_1 R_2 / (R_1 + R_2)$$

Believe it or not, even this formula is messier than what we like to ask you to work with in this course. So we proceed immediately to some tricks that let you do most work in your head.

Text sec. 1.02

Consider some easy cases:

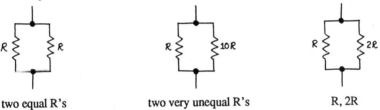

Figure N1.5: Parallel R's: Some easy cases

The first two cases are especially important, because they help one to *estimate* the effect of a circuit one can liken to either case. Labor-saving tricks that give you an estimate are not to be scorned: if you see an easy way to an estimate, you're likely to make the estimate. If you have to work to hard to get the answer, you may find yourself simply *not* making the estimate.

In this course we usually are content with answers good to 10%. So, if two parallel resistors differ by a factor of ten, then we can ignore the larger of the two.

Let's elevate this observation to a <u>rule of thumb</u> (our first). While we're at it, we can state the equivalent rule for resistors in series.

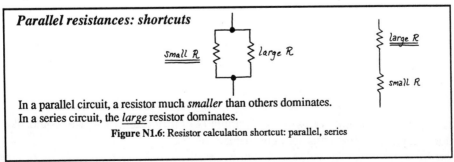

Parallel resistances: shortcuts

In a parallel circuit, a resistor much *smaller* than others dominates.
In a series circuit, the <u>large</u> resistor dominates.

Figure N1.6: Resistor calculation shortcut: parallel, series

Voltage Divider

Text sec. 1.03

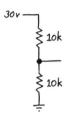

Figure N1.7: Voltage divider

At last we have reached a circuit that does something useful.

First, a *note on labeling*: we label the resistors "10k"; we omit "Ω." It goes without saying. The "k" means kilo- or 10^3, as you probably know.

One can calculate V_{out} in several ways. We will try to push you toward the way that makes it easy to get an answer in your head.

Three ways:

1. Calculate the current through the series resistance; use that to calculate the voltage in the lower leg of the divider.

$I = V_{in} / (R_1 + R_2)$
Here, that's 30v / 20kΩ = 1.5 mA
$V_{out} = I \cdot R_2$
Here, that's 1.5 mA · 10k = 15 v

Figure N1.8: Voltage divider: first method (too hard!): calculate current explicitly

That takes too long.

2. Rely on the fact that I is constant in top and bottom, but do that implicitly. If you want an algebraic argument, you might say,

$$V_2/(V_1 + V_2) = IR_2 / (I[R_1 + R_2]) = R_2/(R_1 + R_2)$$

or,

(1), $V_{out} = V_{in} \{R_2/(R_1 + R_2)\}$

In this case, that means

$$V_{out} = V_{in} (10k/20k) = V_{in}/2$$

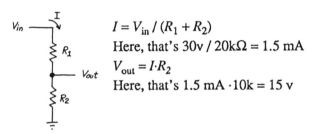

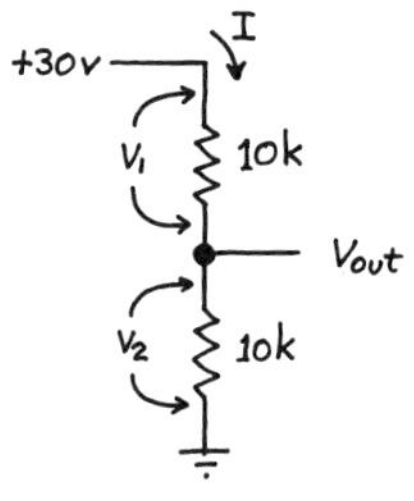

Figure N1.9: Voltage divider: second method: (a little better): current implicit

That's *much better*, and you will use formula *(1)* fairly often. But we would like to push you not to memorize that equation, but instead to—

3. Say to yourself in words how the divider works: something like,

> *Since the currents in top and bottom are equal, the voltage drops are proportional to the resistances* (later, *impedances*—a more general notion that covers devices other than resistors).

So, in this case, where the lower R makes up half the total resistance, it also will show half the total voltage.

For another example, if the lower leg is 10 times the upper leg, it will show about 90% of the input voltage (10/11, if you're fussy, but 90%, to our usual tolerances).

Loading, and "output impedance"
Text sec. 1.05,

Now—after you've calculated V_{out} for the divider—suppose someone comes along and puts in a third resistor:

Text exercise 1.9

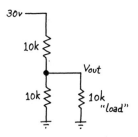

Figure N1.10: Voltage divider *loaded*

(*Query:* Are you entitled to be outraged? Is this no fair?) Again there is more than one way to make the new calculation—but one way is tidier than the other.

Two possible methods:

1. Tedious Method:

Text exercise 1.19

Model the two lower R's as one R; calculate V_{out} for this new voltage divider:

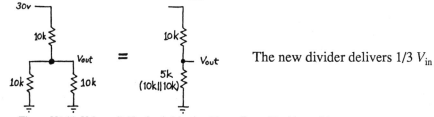

The new divider delivers $1/3\ V_{in}$

Figure N1.11: Voltage divider *loaded*: load and lower R combined in model

That's reasonable, but it requires you to draw a new model to describe each possible loading.

2. Better method: Thevenin's.

Text sec. 1.05

> ***Thevenin Model***
>
> **Thevenin's good idea:**
> Model the actual circuit (unloaded) with a simpler circuit—the *Thevenin model*—which is an idealized voltage source in series with a resistor. One can then see pretty readily how that simpler circuit will behave under various loads.
>
>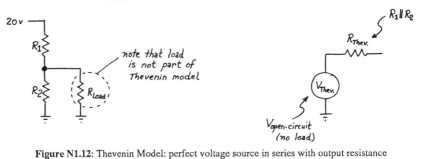
>
> **Figure N1.12**: Thevenin Model: perfect voltage source in series with output resistance

Here's how to calculate the two elements of the Thevenin model:

V_{Thevenin}:   Just $V_{\text{open circuit}}$: the voltage out when nothing is attached ("no load")

R_{Thevenin}:   *Defined* as the quotient of $V_{\text{Thevenin}} / I_{\text{short-circuit}}$, which is the current that flows from the circuit output to ground if you simply *short* the output to ground.

In practice, you are not likely to discover R_{Thev} by so brutal an experiment; and if you have a diagram of the circuit to look at, there is a much faster shortcut:

> ***Shortcut calculation of R_{Thev}***
> Given a circuit diagram, the fastest way to calculate R_{Thev} is to see it as *the parallel resistance of the several resistances viewed from the output*.
>
>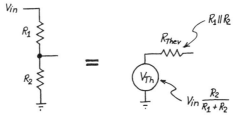
>
> **Figure N1.13**: $R_{Thev} = R_1$ parallel R_2
>
> (This formulation assumes that the voltage sources are ideal, incidentally; when they are not, we need to include their output resistance. For the moment, let's ignore this complication.)

You saw this result above, but this still may strike you as a little odd: why should R_1, going up to the positive supply, be treated as *parallel* to R_2? Well, suppose the positive supply were set at 0 volts. Then surely the two resistances would be in parallel, right?

Or suppose a different divider (chosen to make the numbers easy): twenty volts divided by two 1k resistors. To discover the *impedance* at the output, do the usual experiment (one that we will speak of again and again):

> ***A general definition and procedure to determine <u>impedance</u> at a point:***
> To discover the *impedance* at a point:
> apply a ΔV; find ΔI.
> The quotient is the impedance

This you will recognize as just a "small-signal" or "dynamic" version of Ohm's Law.

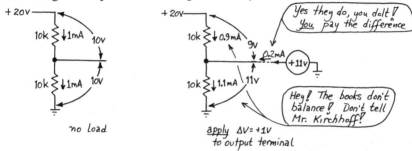

Figure N1.14: Hypothetical divider: current = 1 mA; apply a wiggle of voltage, ΔV; see what ΔI results

In this case 1 mA was flowing before the wiggle. After we force the output up by 1v, the currents in top and bottom resistors no longer match: upstairs: 0.9 mA; downstairs, 1.1 mA. The difference must come from you, the wiggler.

Result: impedance = $\Delta V / \Delta I$ = 1v / 0.2 mA = 5 k

And—happily—that is the parallel resistance of the two R's. Does that argument make the result easier to accept?

You may be wondering why this model is useful. Here is one way to put the answer, though probably you will remain skeptical until you have seen the model at work in several examples: Any non-ideal voltage source "droops" when loaded. How much it droops depends on its "output impedance". The Thevenin equivalent model, with its $R_{Thevenin}$, describes this property neatly in a single number.

Applying the Thevenin model

First, let's make sure Thevenin had it right: let's make sure his model behaves the way the original circuit does. We found that the 10k, 10k divider from 30 volts, which put out 15v when not loaded, drooped to 10V under a 10k load. Does the model do the same?

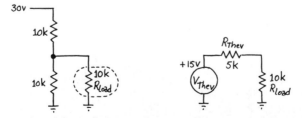

Figure N1.15: Thevenin model and load: droops as original circuit drooped

Yes, the model droops to the extent the original did: down to 10 v. What the model provides that the original circuit lacked is that single value, R_{Thev}, expressing how droopy/stiff the output is.

If someone changed the value of the load, the Thevenin model would help you to see what droop to expect; if, instead, you didn't use the model and had to put the two lower resistors in parallel again and recalculate their parallel resistance, you'd take longer to get each answer, and still you might not get a feel for the circuit's *output impedance*.

Let's try using the model on a set of voltage sources that differ *only* in R_{Thev}. At the same time we can see the effect of an instrument's *input impedance*.

Suppose we have a set of voltage dividers, dividing a 20v input by two. Let's assume that we use 1% resistors (value good to ±1%).

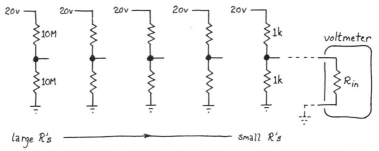

Figure N1.16: A set of similar voltage dividers: same V_{Th}, differing R_{Th}'s

V_{Thev} is obvious, and is the same in all cases. R_{Th} evidently varies from divider to divider.

Suppose now that we try to *measure* V_{out} at the output of each divider. If we measured with a *perfect* voltmeter, the answer in all cases would be 10v. (*Query*: is it 10.000v? 10.0v?)

But if we actually perform the measurement, we will encounter the R_{in} of our imperfect lab voltmeters. Let's try it with a VOM ("volt-ohm-meter," the conventional name for the old-fashioned "analog" meter, which gives its answers by deflecting its needle to a degree that forms an analog to the quantity measured), and then with a DVM ("digital voltmeter," a more recent invention, which usually can measure current and resistance as well as voltage, despite its name; both types sometimes are called simply "multimeters").

Suppose you poke the several divider outputs, beginning from the right side, where the resistors are 1kΩ. Here's a table showing what we find, at three of the dividers:

R values, divider	Measured V_{out}	Inference
1k	9.95	within R tolerance
10k	9.76	loading barely apparent
100k	8.05	loading obvious

The 8.05 v reading shows such obvious loading—and such a nice round number, if we treat it as "8 v"—that we can use this to calculate the meter's R_{in} without much effort:

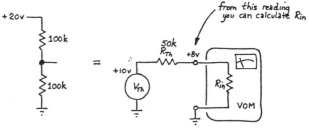

Figure N1.17: VOM reading departs from ideal; we can infer $R_{in\text{-VOM}}$

As usual, one has a choice now, whether to pull out a formula and calculator, or whether to try, instead, to do the calculation "back-of-the-envelope" style. Let's try the latter method.

First, we know that R_{Thev} is 100k parallel 100k: 50k. Now let's talk our way to a solution (an *approximate* solution: we'll treat the measured V_{out} as just "8 volts"):

> The meter shows us 8 parts in 10; across the divider's R_{Thev} (or call it "R_{out}") we must be dropping the other 2 parts in 10. The relative sizes of the two resistances are in proportion to these two voltage drops: 8 to 2, so $R_{\text{in-VOM}}$ must be $4 \cdot R_{\text{Thev}}$: 200k.

If we squint at the front of the VOM, we'll find a little notation,
$$20{,}000 \text{ ohms/volt}$$
That specification means that if we used the meter on its *1 V* scale (that is, if we set things so that an input of 1 volt would deflect the needle fully), then the meter would show an input resistance of 20k. In fact, it's showing us 200k. Does that make sense? It will when you've figured out what must be inside a VOM to allow it to change voltage ranges: a set of big series resistors. You'll understand this fully when you have done problem 1.8 in the text; for now, take our word for it: our answer, 200k, is correct when we have the meter set to the *10 V* scale, as we do for this measurement.

This is probably a good time to take a quick look at what's inside a multimeter—VOM or DVM:

How a meter works:
   Depends on type.
      —depends whether the basic works of the meter sense *current* or *voltage*.

- analog: senses current

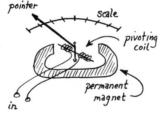

Figure N1.18: Analog meter senses current, in its guts

- digital: senses voltage

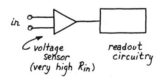

Figure N1.19: Digital meter senses voltage, in its innards

The VOM specification, 20,000 ohms/volt, describes the sensitivity of the meter *movement*—the guts of the instrument. This movement puts a fairly low ceiling on the VOM's input resistance at a given range setting.

Let's try the same experiment with a DVM, and let's suppose we get the following readings:

R values, divide	Measured V_{out}	Inference
100k	9.92	within R tolerance
1M	9.55	loading apparent
10M	6.69	loading obvious

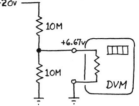

Figure N1.20: DVM reading departs from ideal; we can infer $R_{in\text{-}DVM}$

Again let's use the case where the droop is obvious; again let's talk our way to an answer:

This time R_{Th} is 5M; we're dropping 2/3 of the voltage across $R_{in\text{-}DVM}$, 1/3 across R_{Th}. So, $R_{in\text{-}DVM}$ must be $2 \cdot R_{Th}$, or 10M.

If we check the data sheet for this particular DVM we find that its R_{in} is specified to be "≥ 10M all ranges." Again our readings make sense.

VOM vs DVM: a conclusion?

Evidently, the DVM is a better voltmeter, at least in its R_{in}—as well as much easier to use. As a current meter, however, it is no better than the VOM: it drops 1/4 v full scale, as the VOM does: it measures current simply by letting it flow through a small resistor; the meter then measures the voltage across that resistor.

Digression on ground

The concept "ground" ("earth," in Britain) sounds solid enough. It turns out to be ambiguous. Try your understanding of the term by looking at some cases:

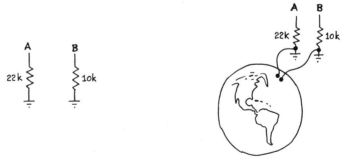

Figure N1.21: Ground in two senses

Query: what is the resistance between points A and B? (Easy, if you don't think about it too hard.) We know that the ground symbol means, in any event, that the bottom ends of the two resistors are electrically joined. Does it matter whether that point is also tied to the pretty planet we live on? It turns out that it does not.

And where is "ground" in this circuit:

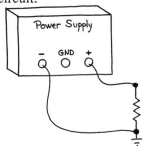

Figure N1.22: Ground in two senses, revisited

Local ground is what we care about: the common point in our circuit that we arbitrarily choose to call zero volts. Only rarely do we care whether or not that local reference is tied to a spike driven into the earth. But, *be warned*, sometimes you are confronted with lines that are tied to world ground—for example, the ground clip on a scope probe, and the "ground" of the breadboards that we use in the lab; then you must take care not to clip the scope ground to, say, +15 on the breadboard.

Generalizing what we've learned of R_{in} and R_{out}

The voltage dividers whose outputs we tried to measure introduced us to a problem we will see over and over again: some circuit tries to "drive" a load. To some extent, the load changes the output. We need to be able to predict and control this change. To do that, we need to understand, first, the characteristic we call R_{in} (this rarely troubles anyone) and, second, the one we have called $R_{Thevenin}$ (this one takes longer to get used to). Next time, when we meet frequency-dependent circuits, we will generalize both characteristics to "Z_{in}" and "Z_{out}."

Here we will work our way to another rule of thumb; one that will make your life as designer relatively easy. We start with a goal: *Design goal:* When circuit A drives circuit B: arrange things so that B loads A lightly enough to cause only insignificant attenuation of the signal. And this goal leads to the rule of thumb:

> **Design rule of thumb:**
>
> When circuit A drives circuit B:
> Let R_{out} for A be $\leq 1/10\ R_{in}$ for B
>
>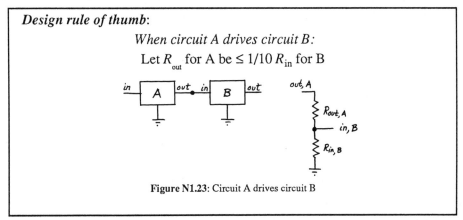
>
> Figure N1.23: Circuit A drives circuit B

How does this rule get us the desired result? Look at the problem as a familiar voltage divider question. If R_{outA} is much smaller than R_{inB}, then the divider delivers nearly all of the original signal. If the relation is 1 : 10, then the divider delivers 10/11 of the signal: attenuation is just under 10%, and that's good enough for our purposes.

We like this arrangement not just because we like big signals. (If that were the only concern, we could always boost the output signal, simply amplifying it.) We like this arrangement above all because it *allows us to design circuit-fragments independently*: we can design A, then design B, and so on. We need not consider A,B as a large single circuit. That's good: makes our work of design and analysis lots easier than it would be if we had to treat every large circuit as a unit.

An example, with numbers: What R_{Thev} for droop of < 10%? What R's, therefore?

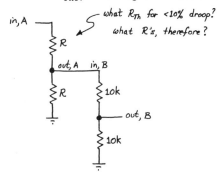

Figure N1.24: One divider driving another: a chance to apply our rule of thumb

The effects of this rule of thumb become more interesting if you extend this chain: from A and B, on to C.

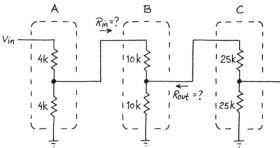

Figure N1.25: Extending the divider: testing the claim that our rule of thumb lets us consider one circuit fragment at a time

As we design C, what R_{Thev} should we use for B? Is it just 10K parallel 10K? That's the answer if we can consider B by itself, using the usual simplifying assumptions: source ideal ($R_{out} = 0$) and load ideal (R_{in} infinitely large).

But should we be more precise? Should we admit that the upper branch really looks like 10K + 2K: 12k? Oh dear! That's 20% different. Is our whole scheme crumbling? Are we going to have to look all the way back to the start of the chain, in order to design the next link? Must we, in other words, consider the whole circuit at once, not just the fragment B, as we had hoped?

No. Relax. That 20% error gets diluted to half its value: R_{Thev} for B is 10k *parallel 12k*, but that's a shade under 5.5k. So—fumes of relief!—we need not look beyond B. We can, indeed, consider the circuit in the way we had hoped: fragment by fragment.

If this argument has not exhausted you, you might give our claim a further test by looking in the other direction: does C alter B's input resistance appreciably (>10%)? You know the answer, but confirming it would let you check your understanding of our rule of thumb and its effects.

Chapter 1: Worked Examples: Resistors & Instruments

Five worked examples:

1. Design a voltmeter and ammeter from bare meter movement
2. Effects of instrument imperfections, in first lab (L1-1)
3. Thevenin models
4. R_{in}, R_{out}
5. Effect of loading

1. Design a Voltmeter, Current Meter

Text sec. 1.04, ex. 1.8, p. 10

> ***Problem: Modify a meter movement to form a voltmeter and ammeter***
>
> A 50µA meter movement has an internal resistance of 5k. What shunt resistance is needed to convert it to a 0-1 amp meter? What series resistance will convert it to a 0-10 volt meter?

This exercise gives you a useful insight into the instrument, of course, but it also will give you some practice in judging when to use approximations: how precise to make your calculations, to say this another way.

1 amp meter

"50µA meter movement" means that the needle deflects fully when 50µA flows through the movement (a coil that deflects in the magnetic field of a permanent magnet: see Class 1 notes for a sketch). The remaining current must bypass the movement; but the current through the movement must remain proportional to the whole current.

Such a long sentence makes the design sound complicated. In fact, as probably you have seen all along, the design is extremely simple: just add a resistance in parallel with the movement (this is the "shunt" mentioned in the problem):

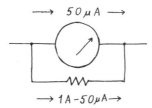

Figure X1.1: Shunt resistance allows sensitive meter movement to measure a total current of 1 A

What value?

Well, what else do we know? We know the *resistance* of the meter movement. That characteristic plus the full-scale current tell us the *full-scale voltage drop* across the movement: that's

$$V_{movement(full-scale)} = I_{full-scale} \cdot R_{movement} = 50 \, \mu A \cdot 5 k\Omega = 250 \, mV$$

Now we can choose R_{shunt}, since we know current and voltage that we want to see across the parallel combination. At this point we have a chance to work too hard, or instead to use a sensible approximation. The occasion comes up as we try to answer the question, 'How much current should pass through the shunt?'

One possible answer is '1A less 50μA, or 0.99995 A."

Another possible answer is '1A.'

Which do you like? If you're fresh from a set of Physics courses, you may be inclined toward the first answer. If we take that, then the resistance we need is

$$R = V_{\text{full-scale}}/I_{\text{full-scale}} = 250 \text{ mV}/0.99995\text{A} = 0.2500125\Omega$$

Now in some settings that would be a good answer. In this setting, it is not. It is a very silly answer. That resistor specification claims to be good to a few parts in a million. If that were possible at all, it would be a preposterous goal in an instrument that makes a needle move so we can squint at it.

So we should have chosen the second branch at the outset: seeing that the 50μA movement current is small relative to the the 1A total current, we should then ask ourselves, 'About how small (in fractional or percentage terms)?' The answer would be '50 parts in a million.' And that fraction is so small relative to reasonable resistor tolerances that we should conclude at once that we should neglect the 50μA.

Neglecting the movement current, we find the shunt resistance is just 250mV/1A = 250 mΩ. In short, the problem is very easy if we have the good sense to let it be easy. You will find this happening repeatedly in this course: if you find yourself churning through a lot of math, and especially if you are carrying lots of digits with you, you're probably overlooking an easy way to do the job. There is no sense carrying all those digits and then having to reach for a 5% resistor and 10% capacitor.

Voltmeter

Here we want to arrange things so that 10V applied to the circuit causes a full-scale deflection of the movement. Which way should we think of the cause of that deflection—'50μA flowing,' or '250 mV across the movement?'

Either is fine. Thinking in *voltage* terms probably helps one to see that most of the 10V must be dropped across some element we are to add, since only 0.25V will be dropped across the meter movement. That should help us sketch the solution:

← 9.75V → ← 0.25V →
Figure X1.2: Voltmeter: series resistance needed

What series resistance should we add? There are two equivalent ways to answer:

1. The resistance must drop 9.75 volts out of 10, when 50μA flows; so R = 9.75V/50μA = 195kΩ
2. Total resistance must be 10V /50μA = 200kΩ. The meter movement looks like 5k, we were told; so we need to add the difference, 195kΩ.

If you got stung on the first part of this problem, giving an answer like "0.2500125Ω," then you might be inclined to say, 'Oh, 50μA is very small; the meter is delicate, so I'll neglect it. I'll put in a 200k series resistor, and be just a little off.'

Well, just to keep you off-balance, we must now push you the other way: this time, "50μA," though a small current is not negligibly small, because it is not to be compared with some much larger current. On the contrary, it is the crucial characteristic we need to work with: it determines the value of the series resistor. And we should *not* say '200k is close enough,' though 195k is the exact answer. The difference is 2.5%: much less than what we ordinarily worry about in this course (because we need to get used to components with 5 and 10% tolerances); but in a meter it's surely worth a few pennies extra to get a 1% resistor: a 195k.

2. Lab 1-1 questions: working around imperfections of instruments

The very first lab exercise asks you to go through the chore of confirming Ohm's Law. But it also confronts you at once with the difficulty that you cannot quite do what the experiment asks you to do: measure *I* and *V* in the resistor simultaneously. Two placements of the DVM are suggested (one is drawn, the other hinted at):

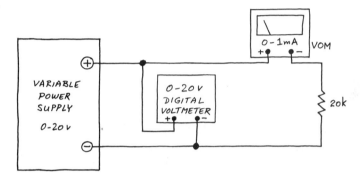

Figure X1.3: Lab 1-1 setup: DVM and VOM cannot both measure the relevant quantity

A Qualitative View

Just a few minutes' reflection will tell you that the voltage reading is off, in the circuit as drawn; moving the DVM solves that problem (above), but now makes the current reading inaccurate.

A Quantitative View

Here's the problem we want to spend a few minutes on:

Problem:

> ### Errors caused by the Meters
>
> If the analog meter movement is as described in the Text's problem 1.8, what percentage error in the *voltage* reading results, if the voltage probe is connected as shown in the figure for the first lab 1 experiment, when the measured resistor has the following values. Assume that you are applying 20 volts, and that you can find a meter setting that lets you get *full-scale deflection* in the current meter.
>
> - $R = 20k$ ohms.
> - $R = 200$ ohms.
> - $R = 2M$ ohms.
>
> Same question, but concerning *current* measurement error, if the voltmeter probe is moved to connect directly to the top of the resistor, for the same resistor values. Assume the DVM has an input resistance of 20 M ohms.

Errors in Voltage readings

The first question is easier than it may appear. The error we get results from the voltage drop across the current meter; but we know what that drop is, from problem 1.8: full-scale: 0.25V. So, the resistor values do not matter. Our voltage readings always are high by a quarter volt, if we can set the current meter to give full-scale deflection. The value of the resistor being measured does *not* matter.

When the DVM reads 20V, the true voltage (at the top of the resistor) is 19.75V. Our voltage reading is high by 0.25V/19.75V—about 0.25/20 or 1 part in 80: 1.25% (If we applied a lower voltage, the voltage error would be more important, assuming we still managed to get full-scale deflection from the current meter, as we might be able to by changing ranges).

Errors in Current readings

If we move the DVM to the top of the resistor, then the voltage reading becomes correct: we are measuring what we meant to measure. But now the current meter is reading a little high: it measures not only the resistor current but also the DVM current, which flows parallel to the current in R.

The size of *this* error depends directly on the size of R we are measuring. You don't even need a pencil and paper to calculate how large the errors are:

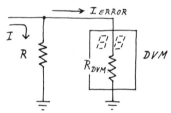

Figure X1.4: DVM causes current-reading error: how large? % error same as ratio of R to R_{DVM}

If R is 20kΩ	—and the DVM looks like 20M—then one part in a thousand of the total current flows through the DVM: the current reading will be high by 0.1%.
If R is 200Ω	then the current error is minute: 1 part in 100,000: 0.001%.
If R is 2MΩ	then the error is large: 1 part in ten.

Conclusion?

There is *no* general answer to the question, 'Which is the better way to place the DVM in this circuit?' The answer depends on R, on the applied voltage and on the consequent ammeter range setting.

And before we leave this question, let's notice the implication of that last phrase: the error depends on the VOM *range* setting. Why? Well, this is our first encounter with the concept we like to call Electronic Justice, or the principle that The Greedy Will Be Punished. No doubt these names mystify you, so we'll be specific: the thought is that if you want good resolution from the VOM, you will pay a price: the meter will alter results more than if you looked for less resolution:

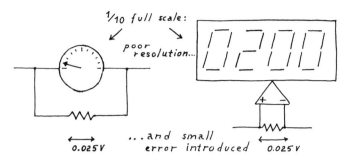

Figure X1.5: *Tradeoffs, or Electronic Justice I:* VOM or DVM as *ammeter:* the larger the reading, the larger the voltage error introduced; VOM as *voltmeter:* the larger the deflection at a given V_{in}, the lower the input impedance

If you want the current meter needle to swing nearly all the way across the dial (giving best resolution: small changes in current cause relatively large needle movement), then you'll get nearly the full-scale 1/4-volt drop across the ammeter. The same goes for the DVM as ammeter, if you understand that 'full scale' for the DVM means filling its digital range: "3 1/2 digits," as the jargon goes: the "half digit" is a character that takes only the values zero or one. So, if you set the DVM current range so that your reading looks like

.093

you have poor resolution: about one percent. If you are able to choose a setting that makes the same current look like 0.930, you've improved resolution 10X. But you have also increased the voltage drop across the meter by the same factor; for the DVM, like the analog VOM drops 1/4V full-scale, and proportionately less for smaller "deflection" (in the VOM) or smaller fractions of the full-scale range (for the DVM).

3. Thevenin Models

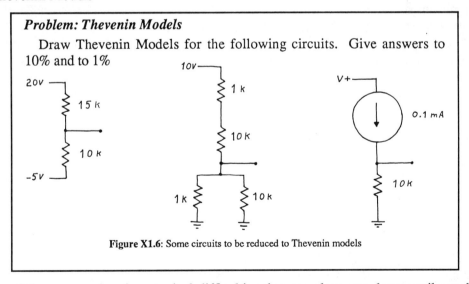

Problem: Thevenin Models

Draw Thevenin Models for the following circuits. Give answers to 10% and to 1%

Figure X1.6: Some circuits to be reduced to Thevenin models

Some of these examples show typical difficulties that can slow you down until you have done a lot of Thevenin models.

The leftmost circuit is most easily done by temporarily redefining *ground*. That trick puts the circuit into entirely familiar form:

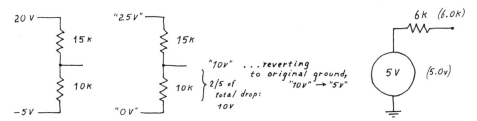

Figure X1.7: A slightly-novel problem reduced to a familiar one, by temporary redefinition of ground

The only difficulty that the middle circuit presents comes when we try to approximate. The 1% answer is easy, here. The 10% answer is tricky. If you have been paying attention to our exhortations to use 10% approximations, then you may be tempted to model each of the resistor blocks with the dominant R: the small one, in the parallel case, the big one in the series case:

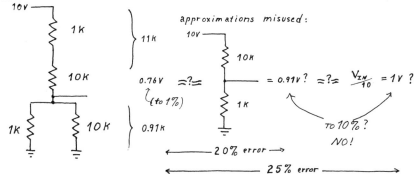

Figure X1.8: 10% approximations: errors can accumulate

Unfortunately, this is a rare case when the errors gang up on us; we are obliged to carry greater precision for the two elements that make up the divider.

This example is not meant to make you give up approximations. It makes the point that it's the *result* that should be good to the stated precision, not just the intermediate steps.

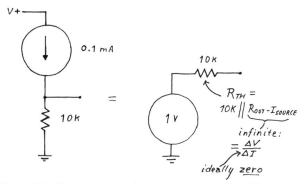

Figure X1.9: Current source feeding resistor, and equivalent Thevenin model

The *current source* shown here probably looks queer to you. But you needn't understand how to make one to see its effect; just take it on faith that it does what's claimed: sources (squirts) a fixed current, down toward the negative supply. The rest follows from Ohm's Law. (In Chapter 2 you will learn to design these things—and you will discover that some devices just do behave like current sources without being coaxed into it: transistors behave this way—both bipolar and FET.)

The point that the current source shows a very high output impedance helps to remind us of the definition of impedance: always the same: $\Delta V/\Delta I$. It is better to carry that general notion with you than to memorize a truth like 'Current sources show high output impedance.' Recalling that definition of impedance, you can always figure out the current source's approximate output importance (large versus small); soon you will know the particular result for a current source, just because you will have seen this case repeatedly.

4. 'Looking through' a circuit fragment, and R_{in}, R_{out}

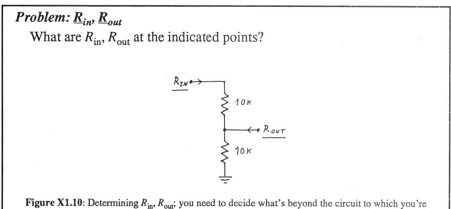

Problem: $\underline{R_{in}}$, $\underline{R_{out}}$
What are R_{in}, R_{out} at the indicated points?

Figure X1.10: Determining R_{in}, R_{out}: you need to decide what's beyond the circuit to which you're connecting

R_{in}

It's clear what R_{in} the divider should show: just $R_1 + R_2$. But when we say that are we answering the right question? Isn't the divider surely going to drive something down the line? If not, why was it constructed?

The answer is Yes, it *is* going to drive something else—the *load*. But that something else should show its own R_{in} high enough so that it does not appreciably alter the result you got when you ignored the load. If we follow our *10X* rule of thumb (see the end of Class 1 notes), you won't be far off this idealization: less than 10% off. To put this concisely, you might say simply that we assume an *ideal* load: a load with infinite input impedance.

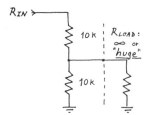

Figure X1.11: R_{in}: we need an assumption about the load that the circuit drives, if we are to determine R_{in}

R_{out}

Here the same problem arises—and we settle it in the same way: by assuming an ideal source. The difficulty, again, is that we need to make some assumption about what is on the far side of the divider if we are to determine R_{out}:

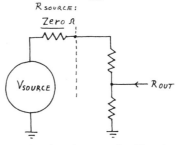

Figure X1.12: R_{out}: we need an assumption about the source that drives the circuit, if we are to determine R_{out}

4. Effects of loading

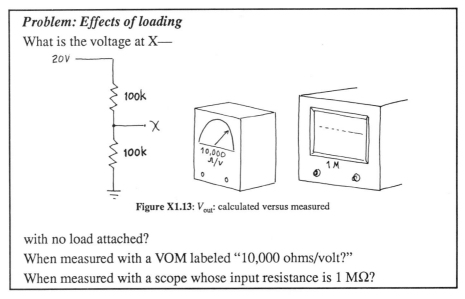

Problem: Effects of loading
What is the voltage at X—

Figure X1.13: V_{out}: calculated versus measured

with no load attached?
When measured with a VOM labeled "10,000 ohms/volt?"
When measured with a scope whose input resistance is 1 MΩ?

This example recapitulates a point made several times over in the the first day's class notes, as you recognize. *Reminder*: The "…ohms/volt" rating implies that on the 1-volt scale (that is, when 1V causes full deflection of the meter) the meter will present that input resistance. What resistance would the meter present when set to the 10 volt scale?

We start, as usual, by trying to reduce the circuit to familiar form. The Thevenin model does that for us. Then we add meter or scope as *load*, forming a voltage divider, and see what voltage results:

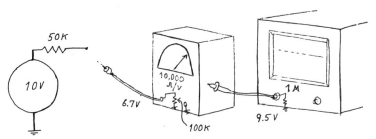

Figure X1.14: Thevenin model of the circuit under test; and showing the "load"—this time, a meter or scope

You will go through this general process again and again, in this course: reduce an unfamiliar circuit diagram to one that looks familiar. Sometimes you will do that by merely redrawing or rearranging the circuit; more often you will invoke a model, and often that model will be Thevenin's.

Lab 1: DC Circuits

Reading:	Chapter 1, secs 1.1 – 1.11.
	Appendix A (don't worry if there are things you don't understand)
	Appendix C
Problems:	Problems in text.
	Additional Exercises 1,2.

1-1. Ohm's Law

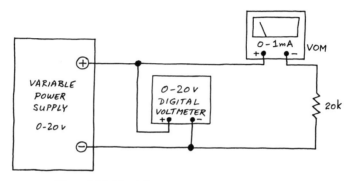

Figure L1.1: Circuit for measurement of resistor's **I** vs **V**

First, the pedestrian part of this exercise: *Verify that the resistor obeys Ohm's law*, by measuring **V** and **I** for a few voltages.

---

*A preliminary note on **procedure***

The principal challenge here is simply to get used to the breadboard and the way to connect instruments to it. We do not expect you to find Ohm's Law surprising. Try to build your circuit *on the breadboard*, not in the air. Novices often begin by suspending a resistor between the jaws of alligator clips that run to power supply and meters. Try to do better: plug the resistor into the plastic breadboard strip. Bring power supply and meters to the breadboard through jacks (banana jacks, if your breadboard has them); then plug a wire into the breadboard so as to join resistor, for example, to banana jack. Below is a sketch of the poor way and better way to build a circuit.

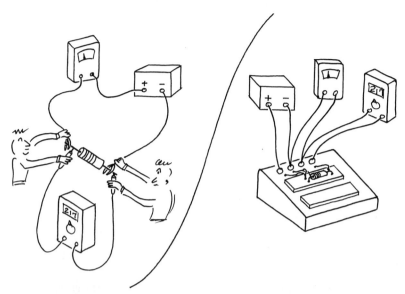

Figure L1.2: Bad and Good breadboarding technique: Left: labor intensive, mid-air method, in which many hands hold everything precariously in place; Right: tidy method: circuit wired in place

Have your instructor demonstrate which holes are connected to which, how to connect voltages and signals from the outside world, etc.

This is also the right time to begin to establish some conventions that will help you keep your circuits intelligible:

- Try to build your circuit so that it looks like its circuit diagram:

    - Let signal flow in from left, exit on right (in this case, the "signal" is just **V**; the "output" is just **I**, read on the ammeter);
    - Place *ground* on a horizontal breadboard *bus* strip *below* your circuit; place the positive supply on a similar bus *above* your circuit. When you reach circuits that include negative supply, place that on a bus strip *below* the ground bus.
    - Use **color coding** to help you follow your own wiring: use **black** for ground, **red** for the positive supply. Such color coding helps a little now, *a lot* later, when you begin to lay out more complicated *digital* circuits.

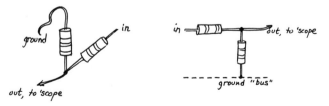

Figure L1.3: Bad and good breadboard layouts of a simple circuit

---

Use a variable regulated dc supply, and the hookup shown in the first figure, above, Fig. L1.1. Note that voltages are measured **between** points in the circuit, while currents are measured *through* a part of a circuit. Therefore you usually have to break the circuit to measure a current.

Measure a few values of **V** and **I** for the 20k resistor (*note:* you may well find *no 20k resistor in your kit*. Don't panic. Consider how to take advantage of some 10k's.) Next try a 10k resistor instead, and sketch the two curves that these resistors define on a plot of I vs V. You may be disinclined to draw these "curves," because you knew without doing this experiment what they would look like. Fair enough. But we encourage you to draw the plot for contrast with the devices you will meet next—which interest us just because their curves do *not* look like those of a resistor: just because these other devices *do not* obey Ohm's Law.

Effects of the instruments on your readings

Now that you have done what we called the pedestrian part of the experiment, consider a couple of practical questions that arise in even this simplest of "experiments."

A Qualitative View

The voltmeter is not measuring the voltage at the place you want, namely across the resistor. Does that matter? How can you fix the circuit so the voltmeter measures what you want? When you've done that, what about the accuracy of the current measurement? Can you summarize by saying what an ideal voltmeter (or ammeter) should do to the circuit under test? What does that say about its "internal resistance"?

A Quantitative View

How large is each error, given a 20k resistor. Which of the two alternative hookups is preferable, therefore? Would you have reached the same conclusion if the resistor had been 20MΩ?

(You will find this question pursued in one of the Worked Examples.)

Two Nonlinear Devices: *(Ohm's Law Defied!)*

1-2. An incandescent lamp

Now perform the same measurement (**I** vs **V**) for a #47 lamp. Use the 100mA and 500mA scales on your VOM. <u>Do not exceed 6.5 volts!</u> Again you need only a few readings. Again we suggest you plot your results on the drawing you used to show the *resistor's* behavior. Get enough points to show how the lamp *diverges* from resistor-like performance.

What is the "resistance" of the lamp? Is this a reasonable question? If the lamp's filament is made of a material fundamentally like the material used in the resistors you tested earlier, what accounts for the funny shape of the lamp's **V** vs **I** curve?

1-3. The Diode

Here is another device that does not obey Ohm's law: the **diode**. (We don't expect you to understand how the diode works yet; we just want you to meet it, to get some perspective on Ohm's Law devices: to see that they constitute an important but *special* case.

We need to modify the test setup here, because you can't just stick a voltage across a diode, as you did for the resistor and lamp above[1] You'll see why after you've measured the diode's **V** vs **I**. Do that by wiring up the circuit shown below.

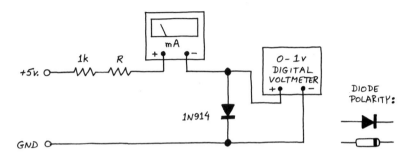

Figure L1.4: Diode VI measuring circuit

In this circuit you are applying a *current*, and noting the diode voltage that results; earlier, you applied a voltage and read resulting current. The 1k resistor limits the current to safe values. Vary **R** (use a 100k variable resistor (usually called a *potentiometer* or "pot" even when wired, as here, as a variable resistor), a resistor substitution box, or a selection of various fixed resistors), and look at **I** vs **V**, sketching the plot in two forms: linear and "semi-log."

First, get an impression of the shape of the linear plot; just four or five points should define the shape of the curve. Then draw the same points on a *semi-log* plot, which compresses one of the axes. (Evidently, it is the fast-growing current axis that needs compressing, in this case.) If you have some semi-log paper use it. If you don't have such paper, you can use the small version laid out below. The point is to see the pattern. You will see this shape again in Lab 5 when you let the scope plot diode and transistor characteristics for you.

---

[1]. Well, you *can*; but you can't do it twice with one diode!

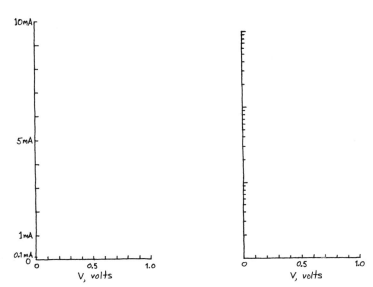

Figure L1.5: Diode I vs V: linear plot; semi-log plot

See what happens if you reverse the direction of the diode.

How would you summarize the **V** vs **I** behavior of a diode?

Now explain what would happen if you were to put 5 volts across the diode (**Don't try it!**). Look at a diode data sheet, if you're curious: see what the manufacturer thinks would happen. The data sheet won't say "Boom" or "Pfft," but that is what it will mean.

We'll do lots more with this important device; see, e.g., secs. 1.25-1.31 in the text, and Lab 3.

1-4. Voltage Divider

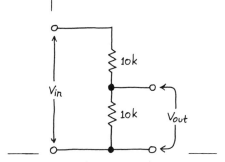

Figure L1.6: Voltage divider

Construct the voltage divider shown above (this is the circuit described in Exercise 1.9 (p. 10 of the text)). Apply V_{in} = 15 volts (use the dc voltages on the breadboard). Measure the (open circuit) output voltage. Then attach a 10k load and see what happens.

Now measure the short circuit current. (That means "short the output to ground, but make the current flow through your current meter. Don't let the scary word "short" throw you: the current in this case will be very modest. You may have grown up thinking "a short blows a fuse." That's a good generalization around the house, but it often does not hold in electronics.)

From $I_{Short\ Circuit}$ and $V_{Open\ Circuit}$ you can calculate the Thevenin equivalent circuit.

Now build the Thevenin equivalent circuit, using the variable regulated dc supply as the voltage source, and check that its open circuit voltage and short circuit current match those of the circuit that it models. Then attach a 10k load, just as you did with the original voltage divider, to see if it behaves identically.

> *A Note on Practical use of Thevenin Models*
>
> You will rarely do again what you just did: short the output of a circuit to ground in order to discover its **R**$_{Thevenin}$ (or "output impedance," as we soon will begin to call this characteristic). This method is too brutal in the lab, and too slow when you want to calculate R_{Th} on paper.
>
> In the lab, I_{SC} could be too large for the health of your circuit (as in your fuse-blowing experience). You will soon learn a gentler way to get the same information.
>
> On paper, if you are given the circuit diagram the fastest way to get R_{Th} for a divider is always to take the *parallel* resistance of the several resistances that make up the divider (again assuming R_{source} is ideal: zero Ω). So, in the case you just examined:
>
>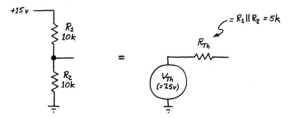
>
> **Figure L1.7**: R_{Th} = parallel resistances as seen from the circuit's output

1-5. Oscilloscope

We'll be using the oscilloscope ("scope") in virtually every lab from now on. If you run out of time today, you *can* learn to use the scope while doing the experiments of Lab 2. You will find Lab 2 easier, however, if today you can devote perhaps 20 minutes to meeting the scope.

Get familiar with scope and *function generator* (a box that puts out time-varying voltages: waveforms: things like sine waves, triangle waves and square waves) by generating a 1000 hertz (1kHz, 1000 cycles/sec) sine wave with the function generator and displaying it on the scope.

If both instruments seem to present you with a bewildering array of switches and knobs, don't blame yourself. These front-panels just *are* complicated. You will need several lab sessions to get fully used to them—and at term's end you may still not know all: it may be a long time before you find any occasion for use of the *holdoff* control, for example, or of *single-shot* triggering, if your scope offers this.

Play with the scope's sweep and trigger controls. Specifically, try the following:

- The vertical gain switch. This controls *"volts/div"*; note that "div" or "division" refers to the *centimeter* marks, not to the tiny 0.2 cm marks);
- The horizontal sweep speed selector: *time* per division.

  > On this knob as on the vertical gain knob, make sure the switch is in its *CAL* position, not *VAR* or "variable." Usually that means that you should turn a small center knob clockwise till you feel the switch *detent* click into place. If you don't do this, you can't trust *any* reading you take!)

- The trigger controls. Don't feel dumb if you have a hard time getting the scope to trigger properly. Triggering is by far the subtlest part of scope operation. When you think you have triggering under control, invite your partner to prove to you that you ***don't***: have your partner randomize some of the scope controls, then see if you can regain a sensible display (don't overdo it here!).

  > Beware the tempting so-called "normal" settings (usually labeled "**NORM**"). They rarely help, and instead cause much misery when misapplied. Think of "normal" here as short for ***abnormal***! Save it for the rare occasion when you know you need it. "**AUTO**" is almost always the better choice.

Switch the function generator to square waves and use the scope to measure the "risetime" of the square wave (defined as time to pass from 10% to 90% of its full amplitude).

At first you may be inclined to despair, saying "Risetime? The square wave rises instantaneously." The scope, properly applied, will show you this is not so.

> *A suggestion on triggering:*
> It's a good idea to *watch* the edge that *triggers* the scope, rather than trigger on one event and watch another. If you watch the *trigger event*, you will find that you can sweep the scope fast without losing the display off the right side of the screen.

What comes out of the function generator's **SYNC OUT** or **TTL** connector? Look at this on one channel while you watch a triangle or square wave on the other scope channel. To see how SYNC or TTL can be useful, try to trigger the scope on the peak of a sine wave *without* using these aids; then notice how entirely easy it is to trigger so when you *do* use SYNC or TTL to trigger the scope. (Triggering on a well-defined point in a waveform, such as peak or trough, is especially useful when you become interested in measuring a difference in *phase* between two waveforms; this you will do several times in the next lab.)

How about the terminal marked **CALIBRATOR** (or "CAL") on the scope's front panel? (We won't ask you to *use* this signal yet; not until Lab 3 do we explain how a scope probe works, and how you "calibrate" it with this signal. For now, just note that this signal is available to you). *Postpone* using scope probes until you understand what is within one of these gadgets. A "10X" scope probe is *not* just a piece of coaxial cable.

Put an "offset" onto the signal, if your function generator permits, then see what the **AC/DC** switch (located near the scope inputs) does.

> **Note on AC/DC switch:**
>
> Common sense may seem to invite you to use the *AC* position most of the time: after all, aren't these time-varying signals that you're looking at "AC"—alternating current (in some sense)? *Eschew* this plausible error. The *AC* setting on the scope puts a capacitor in series with the scope input, and this can produce startling distortions of waveforms if you forget it is there. (See what a 50 Hz square wave looks like on AC, if you need convincing.) Furthermore, the AC setting washes away DC information, as you have seen: it hides from you the fact that a sine wave is sitting on a DC offset, for example. You don't want to wash away information except when you choose to do so knowingly and purposefully. Once in a while you *will* want to look at a little sine with its DC level stripped away; but always you will want to *know* that this DC information has been made invisible.

Set the function generator to some frequency in the middle of its range, then try to make an accurate frequency measurement with the scope. (Directly, you are obliged to measure *period*, of course, not frequency.) You will do this operation hundreds of times during this course. Soon you will be good at it.

Trust the scope period readings; distrust the function generator frequency markings; these are useful only for very *approximate* guidance, on ordinary function generators.

Try looking at pulses, say 1μs wide, at 10kHz.

1-6. AC Voltage Divider

First spend a minute thinking about the following question: How would the analysis of the voltage divider be affected by an input voltage that changes with time (i.e., an input *signal*)? Now hook up the voltage divider from lab exercise 1-4, above, and see what it does to a 1kHz sine wave (use function generator and scope), comparing input and output signals.

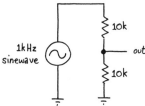

Figure L1.8: Voltage divider applied to a time-varying signal

Explain in detail, to your own satisfaction, why the divider must act as it does.

If this question seems silly to you, you know either too much or too little.

Class 2: Capacitors and *RC* Circuits

Topics:

- *old:*
    - Impedances: in, out: rule of thumb when A drives B: old rule in a new setting: filter drives filter.

- *new:*
    - Capacitors: dynamic description
    - *RC* circuits
        - Time domain:
            - step response
            - integrator, differentiator (approximate)
        - Frequency domain: filters

Capacitors

Today things get a little more complicated, and more interesting, as we meet frequency-dependent circuits, relying on the *capacitor* to implement this new trick. Capacitors let us build circuits that "remember" their recent history. That ability allows us to make timing circuits (circuits that let *this* happen a predetermined time after *that* occurs); the most important of such circuits are *oscillators*—circuits that do this timing operation over and over, endlessly, in order to set the frequency of an output waveform. The capacitor's memory also lets us make circuits that respond mostly to changes (*differentiators*) or mostly to averages (*integrators*), and—by far the most important—circuits that favor one frequency range over another (*filters*).

All of these circuit fragments will be useful within later, more complicated circuits. The filters, above all others, will be with us constantly as we meet other analog circuits. They are nearly as ubiquitous as the (resistive-) *voltage divider* that we met in the first class.

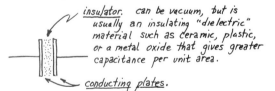

Figure N2.1: The simplest capacitor configuration: sandwich

This capacitor is drawn to look like a ham sandwich: metal plates are the bread, some dielectric is the ham (*ceramic* capacitors really are about as simple as this). More often, capacitors achieve large area (thus large capacitance) by doing something tricky, such as putting the dielectric between two thin layers of metal foil, then rolling the whole thing up like a roll of paper towel (*mylar* capacitors are built this way).

A *static* description of the way a capacitor behaves would say
$$Q = CV$$
where Q is total charge, C is the measure of how big the cap is (how much charge it can store at a given voltage: $C = Q/V$), and V is the voltage across the cap.

This statement just defines the notion of capacitance. It is a Physicist's way of describing how a cap behaves, and rarely will we use it again. Instead, we use a dynamic description—a statement of how things change with time:

$$I = C\, dV/dt$$

This is just the time derivative of the "static" description. C is constant with time; I is defined as the rate at which charge flows. This equation isn't hard to grasp: it says 'The bigger the current, the faster the cap's voltage changes.'

Again, flowing water helps intuition: think of the cap (with one end grounded) as a tub that can hold charge:

Figure N2.2: A cap with one end grounded works a lot like a tub of water

A tub of large diameter (cap) holds a lot of water (charge), for a given height (V). If you fill the tub through a thin straw (small I), the water level—V—will rise slowly; if you fill or drain through a fire hose (big I) the tub will fill ("charge") or drain ("discharge") quickly. A tub of large diameter (large capacitor) takes longer to fill or drain than a small tub. Self-evident, isn't it?

Time-domain Description

Text sec. 1.13

Now let's leave tubs of water, and anticipate what we will see when we watch the voltage on a cap change with time: when we look on a scope screen, as you will do in Lab 2.

*An easy case: constant **I***

Text sec. 1.15;
see Fig. 1.43

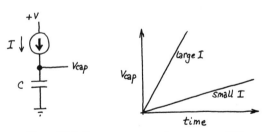

Figure N2.3: Easy case: constant $I \longrightarrow$ constant dV/dt

This tidy waveform, called a *ramp*, is useful, and you will come to recognize it as the signature of this circuit fragment: capacitor driven by constant current (or "current source").

This arrangement is used to generate a triangle waveform, for example:

Compare Text sec.1.15,
Fig. 1.42: ramp generator

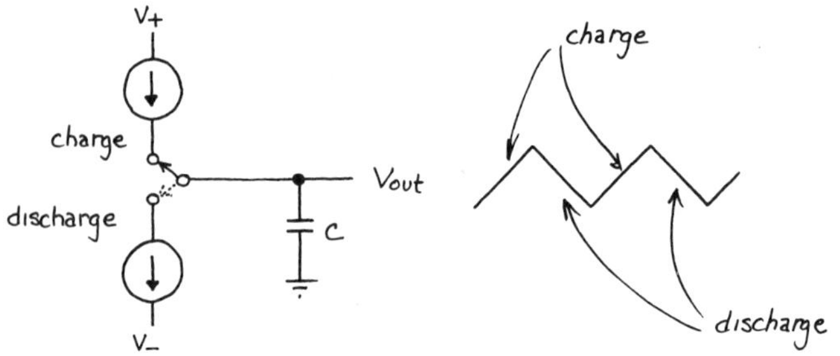

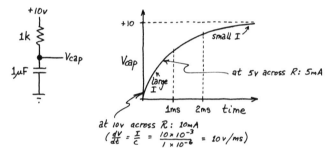

Figure N2.4: How to use a cap to generate a triangle waveform: ramp up, ramp down

But the ramp waveform is relatively rare, because current sources are relatively rare. Much more common is the next case.

A harder case but more common: **constant voltage** source in series with a resistor

Figure N2.5: The more usual case: cap charged and discharged from a *voltage* source, through a series resistor

Here, the voltage on the cap approaches the applied voltage—but at a rate that diminishes toward zero as V_{cap} approaches its destination. It starts out bravely, moving fast toward its V_{in} (charging at 10 mA, in the example above, thus at 10V/ms); but as it gets nearer to its goal, it loses its nerve. By the time it is 1 volt away, it has slowed to 1/10 its starting rate.

(The cap behaves a lot like the hare in Xeno's paradox: remember him? Xeno teased his fellow-Athenians by asking a question something like this: 'If a hare keeps going halfway to the wall, then again halfway to the wall, does he ever get there?' (Xeno really had a hare chase a tortoise; but the electronic analog to the tortoise escapes us, so we'll simplify his problem.) Hares do bump their noses; capacitors don't: V_{cap} never does reach $V_{applied}$, in an *RC* circuit. But it will come as close as you want.)

Here's a fuller account:

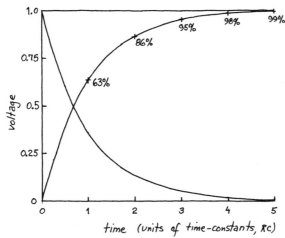

Figure N2.6: *RC* charge, discharge curves

Don't try to memorize these numbers, except two:

- in *one RC* (called "one time-constant") 63% of the way
- in *five RCs*, 99% of the way

If you need an exact solution to such a timing problem:
$$V_{cap} = V_{applied}(1 - e^{(-t/RC)})$$
In case you can't see at a glance what this equation is trying to tell you, look at
$$e^{(-t/RC)}$$
by itself:

- when $t = RC$, this expression is $1/e$, or 0.37.
- when t = very large ($\gg RC$), this expression is tiny, and $V_{cap} \approx V_{applied}$

A tip to help you calculate time-constants:
MΩ and µF give time-constant in seconds
kΩ and µF give time-constant in milliseconds

In the case above, *RC* is the product of 1k and 1µF: 1 ms.

Integrators and Differentiators

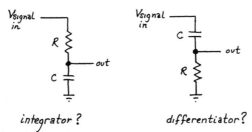

Figure N2.7: Can we exploit cap's $I = C\, dV/dt$ to make differentiator & integrator?

The very useful formula, $I = C dV/dt$ will let us figure out when these two circuits perform pretty well as differentiator and integrator, respectively.

Let's, first, consider this simpler circuit:

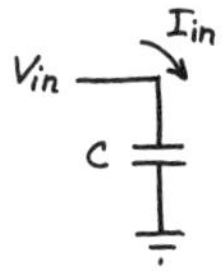

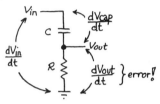

Figure N2.8: Useless "differentiator"?

The current that flows in the cap is proportional to dV_{in}/dt: the circuit differentiates the input signal. But the circuit is pretty evidently useless. It gives us no way to measure that current. If we could devise a way to measure the current, we would have a differentiator.

Here's our earlier proposal, again. Does it work?

differentiator ?

Figure N2.9: Differentiator? —again

Answer: Yes and No: Yes, to the extent that $V_{cap} = V_{in}$ (and thus $dV_{cap}/dt = dV_{in}/dt$), because the circuit responds to dv/dt across the cap, whereas what interests us is dV_{in}/dt—that is, relative to ground.

So, the circuit errs to the extent that the output moves away from ground; but of course it must move away from ground to give us an output. This differentiator is compromised. So is the *RC* integrator, it turns out. When we meet operational amplifiers, we will manage to make nearly-ideal integrators, and pretty good differentiators.

The text puts this point this way:

Text sec. 1.14

>for the differentiator:
>"…choose R and C small enough so that $dV/dt \ll dV_{in}/dt$…."

Text sec. 1.15

>for the integrator:
>"…[make sure that] $V_{out} \ll V_{in}$….$\omega RC \ll 1$."

We can put this simply—perhaps crudely: assume a sine wave input. Then,

>the *RC* differentiator (and integrator, too) works pretty well **if** it is **murdering** the signal (that is, attenuates it severely), so that V_{out} (and dV_{out}) is tiny: hardly moves away from ground.

It follows, along the way, that differentiator and integrator will impose a 90° phase shift on a sinusoidal input. This result, obvious here, should help you anticipate how *RC* circuits viewed as "filters" (below) will impose phase shifts.

Integrator

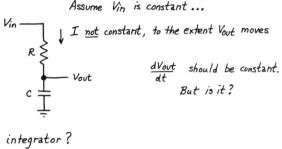

integrator ?

Figure N2.10: Integrator?—again

One can make a similar argument to explain the limitations of the *RC* integrator. To keep things simple, imagine that you apply a *step* input; ask what waveform out you would like to see out, and what, therefore, you would like the current to do.

RC Filters

These are the most important application of capacitors. These circuits are just voltage dividers, essentially like the resistive dividers you have met already. The resistive dividers treated DC and time-varying signals alike. To some of you that seemed obvious and inevitable (and maybe you felt insulted by the exercise at the end of Lab 1 that asked you to confirm that AC was like DC to the divider). It happens because resistors can't remember what happened even an instant ago. They're little existentialists, living in the present. (We're talking about ideal R's, of course.)

The impedance or *reactance* of a cap

A cap's impedance varies with frequency. ("Impedance" is the generalized form of what we called "resistance" for "resistors;" "reactance" is the term reserved for capacitors and inductors (the latter usually are coils of wire, often wound around an iron core)).

Compare Text sec. 1.12

It's obvious that a cap cannot conduct a DC current: just recall what the cap's insides look like: an insulator separating two plates. That takes care of the cap's "impedance" at DC: clearly it's infinite (or *huge*, anyway).

It is not obvious that a rapidly-varying voltage can pass "through" a capacitor, but that does happen. The Text explains this difficult notion at sec. 1.12, speaking of the *current* that passes through the cap. Here's a second attempt to explain how a *voltage* signal passes through a cap, in the high-pass configuration. If you're already happy with the result, skip this paragraph.

When we say the AC signal passes through, all we mean is that a wiggle on the left causes a wiggle of similar size on the right:

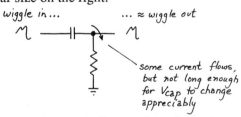

Figure N2.11: How a cap "passes" a signal

The wiggle makes it "across" the cap so long as there isn't time for the voltage on the cap to change much before the wiggle has ended—before the voltage heads in the other direction. In other words, quick wiggles pass; slow wiggles don't.

We can stop worrying about our intuition and state the expression for the cap's *reactance*:

$$X_C = -j/\omega C = -j/2\pi f C$$

And once we have an expression for the impedance of the cap—an expression that shows it varying continuously with frequency—we can see how capacitors will perform in voltage dividers.

RC Voltage dividers
text sec. 1.18

You know how a resistive divider works on a sine. How would you expect a divider made of capacitors to treat a time-varying signal?

Figure N2.12: Two dividers that deliver 1/2 of V_{in}

If this case worries you, good: you're probably worrying about phase shifts. Turns out they cause no trouble here: output is in phase with input. (If you can handle the complex notation, write $X_C = -j/\omega C$, and you'll see the j's wash out.)

But what happens in the combined case, where the divider is made up of both R and C? This turns out to be an extremely important case.

text sec. 1.19

This problem is harder, but still fits the voltage-divider model. Let's generalize that model a bit:

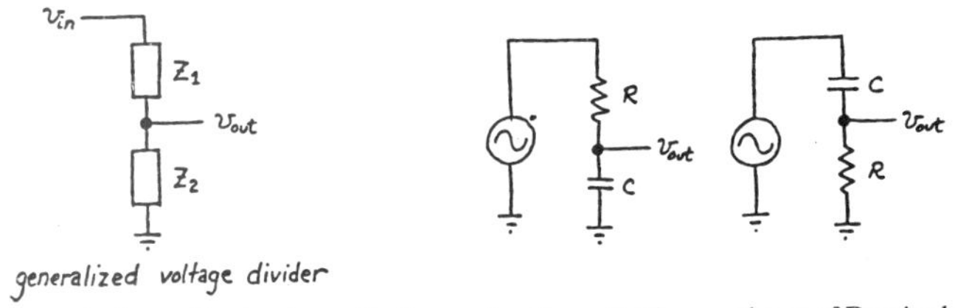

Figure N2.13: Generalized voltage divider; and voltage dividers made up of R paired with C

The behavior of these voltage dividers—which we call *filters* when we speak of them in frequency terms, because each favors either high or low frequencies— is easy to describe:

1. See what the filter does at the two frequency extremes. (This looking at extremes is a useful trick; you will soon use it again to find the filters' worst-case Z_{in} and Z_{out}.)

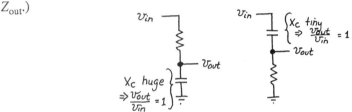

Figure N2.14: Establishing the endpoints of the filter's frequency response curve

At f = 0: what fraction out?   At f = very high: what fraction out?

2. Determine where the output "turns the corner" (corner defined arbitrarily[1]) as the frequency where the output is 3dB less than the input (always called just "the 3 dB point"; "minus" understood).

Knowing the endpoints, which tell us whether the filter is *high-pass* or *low-pass*, and knowing the 3dB point, we can draw the full frequency-response curve:

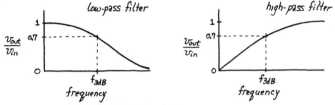

Figure N2.15: *RC* filter's frequency response curve

The "3dB point," the frequency where the filter "turns the corner" is

$$f_{3dB} = 1/(2\pi RC)$$

Beware the more elegant formulation that uses ω:

$$\omega_{3dB} = 1/RC.$$

That is tidy, but is very likely to give you answers off by about a factor of 6, since you will be measuring period and its inverse in the lab: frequency in hertz (or "cycles-per-second," as it used to be called), *not* in radians.

Two asides:

Caution!

Do not confuse these *frequency-domain* pictures with the earlier *RC* step-response picture, (which speaks in the *time-domain*).

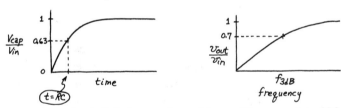

Figure N2.16: Deceptive similarity between shapes of time- and frequency- plots of *RC* circuits

Not only do the curves look vaguely similar. To make things worse, details here seem tailor-made to deceive you:

- *Step response*: in the *time RC* (time-constant), V_{cap} moves to about 0.6 of the applied step voltage (this is $1 - 1/e$).
- *Frequency domain*: at f_{3dB}, a frequency determined by RC, the filter's V_{out}/V_{in} is about 0.7 (this is $1/\sqrt{2}$)

Don't fall into this trap.

A note re Log Plots

You may wonder why the curves we have drawn, the curve in Fig. 1.59, and those you see on the scope screen (when you "sweep" the input frequency) don't look like the tidier curves shown in most books that treat frequency response, or like the curves in Chapter 5 of

---

[1]. Well, not quite arbitrarily: a signal reduced by 3dB delivers half its original power.

the Text. Our curves trickle off toward zero, in the low-pass configuration, whereas these other curves seem to fall smoothly and promptly to zero. This is an effect of the logarithmic compression of the axes on the usual graph. Our plots are linear; the usual plot ("Bode plot") is log-log:

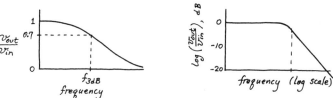

Figure N2.17: Linear versus log-log frequency-response plots contrasted

Input and output impedance of an *RC* Circuit

If filter A is to drive filter B—or anything else, in fact—it must conform to our *10X* rule of thumb, which we discussed last time, when we were considering only resistive circuits. The same reasons prevail, but here they are more urgent: if we don't follow this rule, not only will signals get attenuated; frequency response also is likely to be altered.

But to enforce our rule of thumb, we need to know Z_{in} and Z_{out} for the filters. At first glance, the problem looks nasty. What is Z_{out} for the low-pass filter, for example? A novice would give you an answer that's much more complicated than necessary. He might say,

$$Z_{out} = X_C \text{ parallel } R = -j/\omega C \cdot R / (-j/\omega C + R)$$

Yow! And then this expression doesn't really give an answer: it tells us that the answer is frequency-dependent.

We cheerfully sidestep this hard work, by considering only *worst case* values. We ask, 'How bad can things get?

We ask, 'How bad can Z_{in} get?' And that means, 'How *low* can it get?'

We ask, 'How bad can Z_{out} get?' And that means, 'How *high* can it get?'

This trick delivers a stunningly-easy answer: the answer is always just *R*! Here's the argument for a low-pass, for example:

worst Zin: cap looks like a short: Zin = R
(this happens at highest frequencies)

worst Zout: cap doesn't help at all; we look through to the source, and see only R: Zout = R (this happens at lowest frequencies)

Figure N2.18: Worst-case Z_{in} and Z_{out} for *RC* filter reduces to just *R*

Now you can string together *RC* circuits just as you could string together voltage dividers, without worrying about interaction among them.

Phase Shift

You already know roughly what to expect: the differentiator and integrator showed you phase shifts of 90°, and did this when they were severely attenuating a sine-wave. You need to *beware* the misconception that because a circuit has a cap in it, you should expect to see a 90° shift (or even just noticeable shift). That is *not so*. You need an intuitive sense of when phase shifting occurs, and of roughly its magnitude. You rarely will need to calculate the amount of shift.

Here is a start: a rough account of phase shift in *RC* circuits:

> If the amplitude *out* is close to amplitude *in*, you will see little or no phase shift. If the output is much attenuated, you will see considerable shift (90° is maximum)

And here are curves saying the same thing:

Text sec. 1.20,
fig. 1.60, p. 38

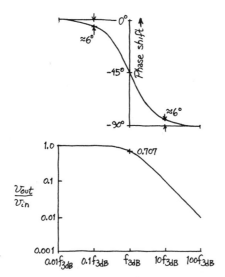

Figure N2.19: Attenuation and phase shift (log-log plot)

Why does this happen? Here's an attempt to rationalize the result:

- voltages in *R* and *C* are 90° out of phase, as you know.
- the sum of the voltages across *R* and *C* must equal, at every instant, V_{in}.
- as frequency changes, *R* and *C* share the total V_{in} differently, and this alters the phase of V_{out} relative to V_{in}:

Consider a low-pass, for example: if a lot of the total voltage, V_{in}, appears across the cap, then the phase of the input voltage (which appears across the <u>RC</u> series combination) will be close to the phase of the output voltage, which is taken across the cap alone. In other words, R plays a small part: V_{out} is about the same as V_{in}, in both amplitude and phase. Have we merely restated our earlier proposition? It almost seems so.

But let's try a drawing:

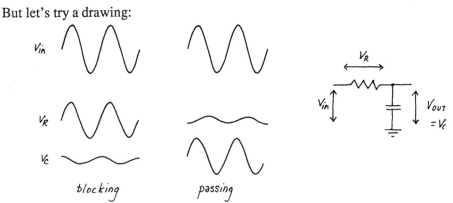

Figure N2.20: R and C sharing input voltage differently at two different frequencies

Now let's try another aid to an intuitive understanding of phase shift: phasors.

Phasor Diagrams

These diagrams let you compare phase and amplitude of input and output of circuits that shift phases (circuits including C's and L's). They make the performance graphic, and allow you to get approximate results by use of geometry rather than by explicit manipulation of complex quantities.

The diagram uses axes that represent resistor-like ("real") impedances on the horizontal axis, and capacitor or inductor-like impedances ("imaginary"—but don't let that strange name scare you; for our purposes it only means that voltages across such elements are 90° out of phase with voltages across the resistors). This plot is known by the extra-frightening name, "complex plane" (with nasty overtones, to the untrained ear, of 'too-complicated-for-you plane'!). But don't lose heart. It's all very easy to understand and use. Even better, *you don't need to understand phasors*, if you don't want to. We use them rarely in the course, and always could use direct manipulation of the complex quantities instead. Phasors are meant to make you feel better. If they don't, forget them.

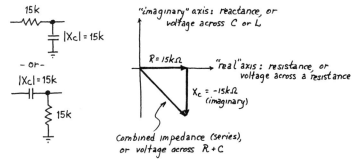

Figure N2.21: Phasor diagram: "complex plane;" showing an *RC* at f_{-3dB}

The diagram above shows an *RC* filter at its 3dB point, where, as you can see, the *magnitude* of the impedance of C is the same as that of R. The arrows, or vectors, show phase as well as amplitude (notice that this is the amplitude of the waveform: the peak value, not a voltage varying with time); they point at right angles so as to say that the voltages in R and C are 90° out of phase.

"Voltages?," you may be protesting, "but you said these arrows represent impedances." True. But in a series circuit the voltages are proportional to the impedances, so this use of the figure is fair.

The total impedance that R and C present to the signal source is *not* 2R, but is the vector sum: it's the length of the hypotenuse, R√2. And from this diagram we now can read two familiar truths about how an *RC* filter behaves at its 3dB point:

- the amplitude of the output is down 3dB: down to 1/√2: the length of either the R or the C vector, relative to the hypotenuse.
- the output is shifted 45° relative to the input: R or C vectors form an angle of 45° with the hypotenuse, which represents the phase of the input voltage.

So far, we're only restating what you know. But to get some additional information out of the diagram, try doubling the input frequency several times in succession, and watch what happens:

each time, the length of the X_C vector is cut to half what it was.

First doubling of frequency:

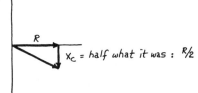

Figure N2.22: *RC* after a doubling of frequency, relative to the previous diagram

The first doubling also affects the length of the hypotenuse substantially, too, however; so the amplitude relative to input is not cut quite so much as 50% (6dB). You can see that the output is a good deal more attenuated, however, and also that phase shift has increased a good deal.

Second doubling of frequency

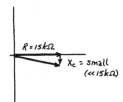

Figure N2.23: *RC* after a doubling of frequency, relative to the previous diagram

This time, the length of the hypotenuse is changed less, so the output shrinks nearly as the X_C vector shrinks: nearly 50%. Here, we are getting close to the *–6dB/octave* slope of the filter's rolloff curve. Meanwhile, the phase shift between output and input is increasing, too—approaching the limit of 90°.

We've been assuming a *low-pass*. If you switch assumptions, and ask what these diagrams show happening to the output of a *high-pass*, you find all the information is there for you to extract. No surprise, there; but perhaps satisfying to notice.

LC circuit on phasor diagram

Finally, let's look at an LC *trap* circuit on a phasor diagram.

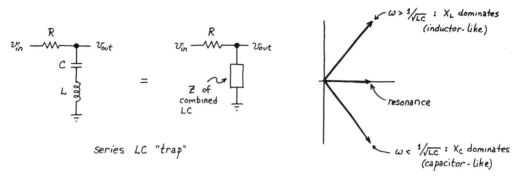

Figure N2.24: LC trap circuit, and its phasor diagram

This is less familiar, but pleasing because it reveals the curious fact—which you will see in Lab 3 when you watch a similar (parallel) LC circuit—that the LC combination looks sometimes like L, sometimes C, showing its characteristic phase shift—and at resonance, shows no phase shift at all. We'll talk about LC's next time; but for the moment, see if you can enjoy how concisely this phasor diagram describes this behavior of the circuit (actually a *trio* of diagrams appears here, representing what happens at *three* sample frequencies).

To check that these LC diagrams make sense, you may want to take a look at what the old voltage-divider equation tells you ought to happen:

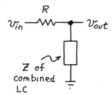

Figure N2.25: LC trap: just another voltage divider

Here's the expression for the output voltage as a fraction of input:

$$V_{out}/V_{in} = Z_{combination} / (Z_{combination} + R)$$

But

$$Z_{combination} = -j/\omega C + j\omega L.$$

And at some frequency—where the magnitudes of the expressions on the right side of that last equation are equal—the sum is zero, because of the opposite signs. Away from this magic frequency (the "resonant frequency"), either cap or inductor dominates. Can you see all this on the phasor diagram?

Better Filters

Having looked hard at *RC* filters, maybe we should remind of the point that the last exercise in Lab 2 means to make: *RC*'s make extremely useful filters, but if you need a better filter, you can make one, either with an LC combination, as in that circuit, or with operational amplifiers cleverly mimicking such an LC circuit (this topic is treated in Chapter 5; we will not build such a circuit), or with a clever circuit called a 'switched-capacitor' filter, a circuit that you will get a chance to try, in Lab 13, and again in Lab 21. Here is a sketch (based on a scope photo) comparing the output of an ordinary *RC* low-pass against a 5-pole Butterworth low-pass, like the one you will build at the end of Lab 2.

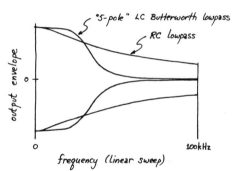

Figure N2.26: Simple *RC* low-pass contrasted with 5-pole Butterworth low-pass

("5-pole" is a fancy way to say that it something like, 'It rolls off the way 5 simple *RC*'s in series would roll off'—it's five times as steep as the plain *RC*. But this nice filter works a whole lot better than an actual string of 5 *RC*'s.)

Not only is the roll-off of the Butterworth much more abrupt than the simple *RC*'s, but also the "passband" looks much flatter: the fancy filter does a better job of passing. The poor old *RC* looks sickly next to the Butterworth, doesn't it?

Nevertheless, we will use *RC*'s nearly always. Nearly always, they are good enough. It would not be in the spirit of this course to pine after a more beautiful transfer function. We want circuits that work, and in most applications the plain old *RC* passes that test.

Chapter 1: Worked Examples: RC Circuits

Two worked exercises:

1. filter to keep "signal" and reject "noise"
2. bandpass filter

---

1. Filter to keep "signal" and reject "noise'

Problem:

> *Filter to remove fuzz:*
>
> Suppose you are faced with a signal that looks like this: a signal of moderate frequency, polluted with some fuzz
>
>
>
> **Figure X2.1**: Signal With Fuzz Added
>
> 1. Draw a *skeleton* circuit (no parts values, yet) that will keep most of the good signal, clearing away the fuzz.
> 2. Now choose some values:
>
>   a. If the *load* has value 100k, choose R for your circuit.
>   b. Choose f_{3dB}, explaining your choice briefly.
>   c. Choose C to achieve the f_{3dB} that you chose.
>   d. By about how much does your filter attenuate the noise "fuzz"?
>
>   What is the circuit's input impedance—
>
>   a. at very low frequencies?
>   b. at very high frequencies?
>   c. at f_{3dB}?
>
> 3. What happens to the circuit output if the load has resistance *10k* rather than 100k?

A Solution:

1. Skeleton Circuit

You need to decide whether you want a low-pass or high-pass, since the signal and noise are distinguishable by their frequencies (and are far enough apart so that you can hope to get one without the other, using the simple filters we have just learned about). Since we have called the lower frequency "good" or "signal," we need a *low-pass*:

Figure X2.2: Skeleton: just a low-pass filter

2. Choose R, given the load

This dependence of R upon load follows from the observation that R of an RC filter defines the *worst-case* input and output impedance of the filter (see Class 2 notes). We want that output impedance low relative to the load's impedance; our rule of thumb says that 'low' means low by a factor of 10. So, we want $R \leq R_{load}/10$. In this case, that means R should be $\leq$ 10k. Let's use 10k.

3. Choose f_{3dB}

This is the only part of the problem that is not entirely mechanical. We know we want to pass the low and attenuate the high, but does that mean put f_{3dB} halfway between good and bad? Does it mean put it close to good? ...Close to bad? Should both good and bad be on a steeply-falling slope of the filter's response curve?

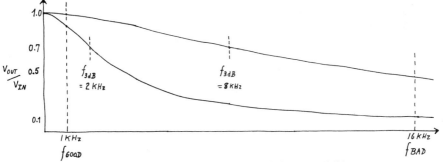

Figure X2.3: Where should we put f_{3dB} ? Some possibilities

Assuming that our goal is to achieve a large ratio of good to bad signal, then we should not put f_{3dB} close to the noise: if we did we would not do a good job of attenuating the bad. Halfway between is only a little better. Close to signal is the best idea: we will then attenuate the bad as much as possible while keeping the good, almost untouched.

An alert person might notice that the greatest relative preference for good over bad comes when both are on the steepest part of the curve showing frequency response: in other words, put f_{3dB} so low that *both* good and bad are attenuated. This is a clever answer—but wrong, in most settings.

The trouble with that answer is that it assumes that the signal is a single frequency. Ordinarily the signal includes a range of frequencies, and it would be very bad to choose f_{3dB} somewhere *within* that range: the filter would distort signals.

So, let's put f_{3dB} at $2 \cdot f_{signal}$: around 2kHz. This gives us 89% of the original signal amplitude (as you can confirm if you like with a phasor diagram or direct calculation). At the same time we should be able to attenuate the 16kHz noise a good deal (we'll see in a moment *how* much).

4...Choose C to achieve the f_{3dB} that you want

This is entirely mechanical: use the formula for the 3dB point:
$f_{3dB} = 1/(2\pi RC) \Longrightarrow C = 1/(2\pi f_{3dB} R) \approx 1/(6 \cdot 2 \cdot 10^3 \cdot 10 \cdot 10^3) = 1/(120 \cdot 10^{-6}) \approx 0.008 \cdot 10^{-6} F$
We might as well use a 0.01 µF cap ("cap" ≡ capacitor). It will put our f_{3dB} about 25% low—1.6Khz; but our choice was a rough estimate anyway.

5. By about how much does your filter attenuate the noise ("fuzz")?

The quick way to get this answer is to count octaves or decades between f_{3dB} and the noise. f_{3dB} is 2 kHz; the fuzz is around 16kHz: $8 \cdot f_{3dB}$.

Count octaves: we could also say that the frequency is doubled three times ($= 2^3$) between f_{3dB} and the noise frequency. *Roughly,* that means that the fuzz amplitude is cut in half the same number of times: down to $1/(2^3)$: 1/8.

This is only an approximate answer because a) near f_{3dB} the curve has not yet reached its terminal steepness of –6dB/octave; b) on the other hand, even *at* f_{3dB}, some attenuation occurs. But let's take our rough answer. (It happens that our rough answer is better than it deserves to be: we called V_{out}/V_{in} 0.125; the more exact answer is 0.124.)

6. What happens to the circuit output if the load has resistance 10k rather than 100k?

Here's a picture of such loading.

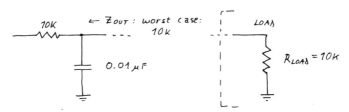

Figure X2.4: Overloaded filter

If you have gotten used to Thevenin models, then you can see how to make this circuit look more familiar:

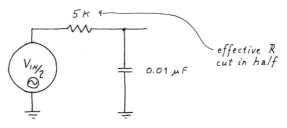

Figure X2.5: Loaded circuit, redrawn

The amplitude is down; but, worse, f3dB has changed: it has doubled. You will find a plot showing this effect at the start of the class notes for next time: Class 3.

7. What is the circuit's input impedance—

   1. at very low frequencies?
      Answer: *very large*: the cap shows a high impedance; the signal source sees only the load—which is assumed very high impedance (high enough so we can neglect it as we think about the filter's performance)

   2. at very high frequencies?
      Answer: *R*: The cap impedance falls toward zero—but *R* puts a lower limit on the input impedance.

   3. at f_{3dB}?

This is easy if you are willing to use phasors, a nuisance to calculate, otherwise. If you recall that the magnitude of $X_C = R$ at f_{3dB}, and if you accept the notion that the voltages across R and C are 90° out of phase, so that they can be drawn at right angles to each other on a phasor diagram, then you get the phasor diagram you saw in the class notes (fig. N2.21):

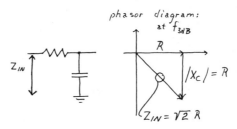

Figure X2.6: Phasor diagram showing an RC filter at its 3dB point

—then you can use a geometric argument to show that the hypotenuse—proportional to Z_{in}—is $R\sqrt{2}$.

2. Bandpass

Text exercise AE-6
(Chapter 1)

Problem:

> **Bandpass filter**
> Design a bandpass filter to pass signals between about 1.6 kHz and 8kHz (you may use these as 3dB frequencies)..
>   Assume that the *next* stage, which your bandpass filter is to drive, has an input impedance of 1 M ohms.

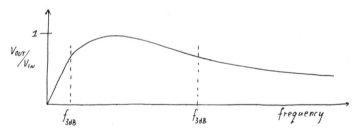

Figure X2.7: Bandpass frequency response

Once you recognize that to get this frequency response from the filter you need to put *high-pass* and *low-pass* in series, the task is mechanical. You can put the two filters in either order. Now we need to choose *R* values, because these will determine worst-case impedances for the two filter stages. The later filter must show Z_{out} low relative to the load, which is 1MΩ; the earlier filter must show Z_{out} low relative to Z_{in} of the second filter stage.

So, let the second-stage *R* = 100k; the first-stage *R* = 10k:

Choosing C's

Here the only hard part is to get the filter's right: it's hard to say to oneself, 'The high-pass filter has the lower f_{3dB};' but that is correct. Here are the calculations: notice that we try to keep things in *engineering notation*—writing '10•10³' rather than '10⁴.' This form looks clumsy, but rewards you by delivering answers in standard units. It also helps you scan for nonsense in your formulation of the problem: it is easier to see that '10•10³' is a good translation for '10k' than it is to see that, say, '10⁵' is *not* a good translation.

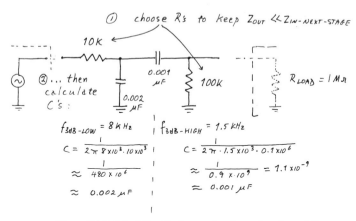

Figure X2.8: Calculating *C* values: a plug for engineering notation

A Note on Reading Capacitor Values

Why you may need this note

Most students learn pretty fast to read resistor values. They tend to have more trouble finding, say, a 100 pF capacitor.

That's not their fault. They have trouble, as you will agree when you have finished reading this note, because the cap manufacturers don't want them to be able to read cap values. ("Cap" is shorthand for "capacitor," as you probably know.) The cap markings have been designed by an international committee to be nearly unintelligible. With a few hints, however, you can learn to read cap markings, despite the manufacturers' efforts. Here are our hints:

Big Caps: electrolytics

These are easy to read, because there is room to write the value on the cap, including units. You need only have the common sense to assume that

Figure CAP.1: A big cap is labeled intelligibly

means 500 micro farads, not what it would mean if you took the capital M seriously.

All of these big caps are *polarized*, incidentally. That means the capacitor's innards are not symmetrical, and that you may destroy the cap if you apply the wrong polarity to the terminals: the terminal marked + must be at least as positive as the other terminal. (Sometimes, violating this rule will generate gas that makes the cap blow up; more often, you will find the cap internally shorted, after a while. Often, you could get away with violating this rule, at low voltages. But don't try.)

Smaller Caps

As the caps get smaller, the difficulty in reading their markings gets steadily worse.

Tantalum

These are the silver colored cylinders. They are polarized: a + mark and a metal nipple mark the positive end. Their markings may say something like

Figure CAP.2: Tantalum cap

That means pretty much what it says, if you know that the "R" marks the decimal place: it's a 4.7 µF cap.

The same cap is also marked,

Figure CAP.3: Tantalum cap: second marking scheme on same part

Here you meet your first challenge, but also the first appearance of an orderly scheme for labeling caps, a scheme that would be helpful if it were used more widely.

The challenge is to resist the plausible assumption that "k" means "kilo." It does not; it is not a unit marking, but a *tolerance* notation (it means ± 10%). (Wasn't that nasty of the labelers to choose "K?" Guess what's another favorite letter for tolerance. That's right: M. Pretty mean!)

The orderly labeling here mimics the resistor codes: 475 means
$$47 \times \text{ten to the fifth.}$$

What units?

10^5 what? 10^5 of something small. You will meet this question repeatedly, and you must resolve it by relying on a few observations:

1. The only units commonly used in this country are

   - microfarads: 10^{-6} Farad
   - picofarads: 10^{-12} Farad

   (You should, therefore, avoid using "mF" and "nF," yourself.)
   A Farad is a huge unit. The biggest cap you will use in this course is 500 µF. That cap is physically large. (We do keep a 1F cap around, but only for our freak show.) So, if you find a little cap labeled "680," you know it's 680 **pF**.
2. A picofarad is a tiny unit. You will not see a cap as small as 1 pF in this course. So, if you find a cap claiming that it is a fraction of some unstated unit—say,".01"—the unit is µF's: ".01" means 0.01 µF.
3. *Beware* the wrong assumption that a *picofarad* is only a bit smaller than a microfarad. A *pF* is *not* 10^{-9} F (10^{-3} µF); instead, it is 10^{-12}F: a *million* times smaller than a microfarad.

So, we conclude, this cap labeled "475" must be 4.7×10^6 *picofarads*. That, you will recognize, is a roundabout way to say
$$4.7 \times 10^{-6} \text{ F}$$
We knew that was the answer, before we started this last decoding effort. This way of labeling is indeed roundabout, but at least it is unambiguous. It would be nice to see it used more widely. You will see another example of this *exponential* labeling in the case of the CK05 ceramics, below.

Mylar

These are yellow cylinders, pretty clearly marked. .01M is just 0.01 µF, of course; and .1MFD is *not* a tenth of a megafarad. These caps are not polarized; the black band marks the outer end of the foil winding. We don't worry about that fine point. Orient them at random in your circuits.

Figure CAP.4: Mylar capacitor

Because they are long *coils* of metal foil (separated by a thing insulator/dielectric—the "mylar" that gives them their name), mylar caps must betray their coil-like construction at very high frequencies: that is, they begin to fail as capacitors, behaving instead like inductors, blocking the very high frequencies they ought to pass. Ceramics (below) do better in this respect, though they are poor in other characteristics.

Ceramic

These are little orange pancakes. Because of this shape (in contrast to the *coil* format hidden within the tubular shape of mylars) they act like capacitors even at high frequencies. The trick, in reading these, is to reject the markings that can't be units:

Disc

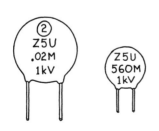

Z5U: Not a unit marking: cap type

.02M, 560M That's it: the M is a tolerance marking, as you know (± 20%); not a unit

Common sense tells you units:

".02?" microfarads. "560?" picofarads.

1kV Not a unit marking. Instead, this means—as you would guess—that the cap can stand 1000 volts.

Figure CAP.5: Disc capacitor markings

CK05

These are little boxes, with their leads 0.2" apart. They are handy, therefore, for insertion into a printed circuit.

101k: This is the neat resistor-like marking. This one is 100 pF.

Figure CAP.6: CK05 capacitor markings

Tolerance Codes

Just to be thorough—and because this information is hard to come by—let's list all the tolerance codes. These apply to both capacitors and resistors; the tight tolerances are relevant only to resistors; the strangely-asymmetric tolerance is used only for capacitors.

Tolerance Code	Meaning
Z	+80%, –20% (for big filter capacitors, where you are assumed to have asymmetric worries: too small a cap allows excessive "ripple;" more on this in Lab 3 and Notes 3)
M	± 20%
K	± 10%
J	± 5%
G	± 2%
F	± 1%
D	± 0.5%
C	± 0.25%
B	± 0.1%
A	± 0.05%
Z	± 0.025 (precision resistors; context will show the asymmetric cap tolerance "Z" makes no sense here)
N	± 0.02%

Figure CAP.7: Tolerance codes

Lab 2: Capacitors

> **Reading:** Chapter 1.12 – 1.20, pp 17-33;
> omit "Power in reactive circuits", pp 28-9.
> Appendix B (if you need it).
> **Warning:** This is by far the most mathematical portion of the course. Don't panic. Even if you don't understand the math, you'll be able to understand the rest of the book.
> Lab Manual note on reading capacitor values.
>
> **Problems:** Problems in text.
> Additional Exercises 3-6.

2-1 *RC* Circuit

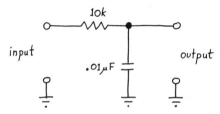

Figure L2.1: *RC* Circuit: step response

Verify that the *RC* circuit behaves in the time domain as described in Text sec. 1.13. In particular, construct the circuit above. Use a mylar capacitor (yellow tubular package, with one lead sticking out each end: "axial leads"). Drive the circuit with a 500Hz square wave, and look at the output. Be sure to use the scope's *DC* input setting. (Remember the warning about the AC setting, last time?)

Measure the time constant by determining the time for the output to drop to 37%. Does it equal the product ***RC***?

> *Suggestion:* The *percent* markings over at the left edge of the scope screen are made-to-order for this task: put the foot of the square wave on 0%, the top on 100%. Then crank up the sweep rate so that you use most of the screen for the fall from 100% to around 37%.

Measure the time to climb from 0% to 63%. Is it the same as the time to fall to 37%? (If not, something is amiss in your way of taking these readings!)

Try varying the frequency of the square wave.

2-2 Differentiator

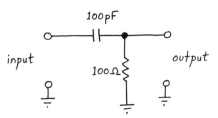

Figure L2.2: *RC* differentiator

Construct the *RC* differentiator shown above. Drive it with a square wave at 100kHz, using the function generator with its attenuator set to 20dB. Does the output make sense? Try a 100kHz triangle wave. Try a sine.

Input Impedance

Here's your first chance to try getting used to quick *worst-case* impedance calculations, rather than exact and frequency-dependent calculations (which often are almost useless).

- What is the impedance presented to the signal generator by the circuit (assume no load at the circuit's output) at f = 0?
- At infinite frequency?

Questions like this become important when the signal source is less ideal than the function generators you are using.

2-3 Integrator

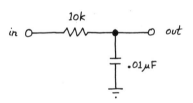

Figure L2.3: *RC* integrator

Construct the integrator shown above. Drive it with a 100kHz square wave at maximum output level (attenuator set at 0dB).

What is the input impedance at dc? At infinite frequency? Drive it with a triangle wave; what is the output waveform called? (Doesn't this circuit seem clever? Doesn't it remember its elementary calculus better than you do—or at least faster?)

To expose this as only an *approximate* or conditional integrator, try dropping the input frequency. Are we violating the stated condition (sec. 1.15):

$$v_{out} \ll v_{in}?$$

The differentiator is similarly approximate, and fails unless (sec. 1.14):

$$dV_{out}/dt \ll dV_{in}/dt$$

In a differentiator, *RC* too large tends to violate this restriction. If you are extra zealous you may want to look again at the differentiator of experiment 2-2, but this time increasing *RC* by a factor of, say, 1000. The "derivative" of the square wave gets ugly, and this will not surprise you; the derivative of the triangle looks odd in a less obvious way.

When we meet *operational amplifiers* in Chapter 3, we will see how to make "perfect" differentiators and integrators—those that let us lift the restrictions we have imposed on these *RC* versions.

2-4 Low-pass Filter

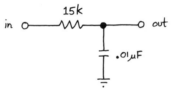

Figure L2.4: *RC* low-pass filter

Construct the low-pass filter shown above.

Aside: "Integrator" versus "Low-pass Filter"

'Wait a minute!,' you may be protesting, 'Didn't I just build this circuit?' Yes, you did. Then why do it again? We expect that you will gradually divine the answer to that question as you work your way through this experiment. One of the two experiments might be called a special case of the other. When you finish, try to determine which is which.)

What do you calculate to be the filter's –3dB frequency? Drive the circuit with a sine wave, sweeping over a large frequency range, to observe its low-pass property; the 1kHz and 10kHz ranges should be most useful.

Find f_{-3dB} *experimentally*: measure the frequency at which the filter attenuates by 3dB (v_{out} down to 70.7% of full amplitude).

Note: henceforth we will refer to "the 3dB point" and "f_{3dB}," henceforth, not to the *minus* 3dB point, or f_{-3dB}. This usage is confusing but conventional; you might as well start getting used to it.

What is the limiting phase shift, both at very low frequencies and at very high frequencies?

> *Suggestion:*
>
> As you measure phase shift, use the function generator's SYNC or TTL output to drive the scope's External Trigger. That will define the input phase cleanly. Then use the scope's *variable* sweep rate so as to make a full period of the input waveform use exactly 8 divisions (or centimeters). The output signal, viewed at the same time, should reveal its phase shift readily.

Check to see if the low-pass filter attenuates 6dB/octave for frequencies well above the –3dB point; in particular, measure the output at 10 and 20 times f_{3dB}. While you're at it, look at phase shift vs frequency: What is the phase shift for

$$f \ll f_{3dB},$$
$$f = f_{3dB},$$
$$f \gg f_{3dB}?$$

Finally, measure the attenuation at $f = 2f_{3dB}$ and write down the attenuation figures at $f = 2f_{3dB}$, $f = 4f_{3dB}$ and $f = 10f_{3dB}$ for later use: in section 2-9, below, we will compare this filter against one that shows a steeper *rolloff*.

> *Sweeping Frequencies*
>
> This circuit is a good one to look at with the function generator's *sweep* feature. This will let your scope draw you a plot of amplitude versus *frequency* instead of amplitude versus *time* as usual. If you have a little extra time, we recommend this exercise. If you feel pressed for time, save this task for next time, when the LC resonant circuit offers you another good target for sweeping.
>
> To generate such a display of v_{out} versus frequency, let the generator's *ramp* output drive the scope's horizontal deflection, with the scope in "X-Y" mode: in *X-Y*, the scope ignores its internal horizontal deflection ramp (or "timebase") and instead lets the input labeled "X" determine the spot's horizontal position.
>
> The function generator's **ramp** time control now will determine sweep rate. Keep the ramp *slow*: a slow ramp produces a scope image that is annoyingly intermittent, but gives the truest, prettiest picture, since the slow ramp allows more cycles in a given frequency range than are permitted by a faster ramp.

2-5 High-pass Filter

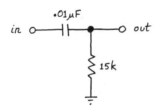

Figure L2.5: *RC* high-pass filter

Construct a high-pass filter with the components that you used for the low-pass. Where is this circuit's 3dB point? Check out how the circuit treats sine waves: Check to see if the output amplitude at low frequencies (well below the −3dB point) is proportional to frequency. What is the limiting phase shift, both at very low frequencies and at very high frequencies?

2-6 Filter Application I: Garbage Detector

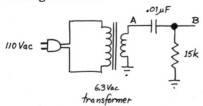

Figure L2.6: High-pass filter applied to the 60Hz ac power

The circuit above will let you see the "garbage" on the 110-volt power line. First look at the output of the transformer, at **A**. It should look more or less like a classical sine wave. (The transformer, incidentally, serves two purposes — it reduces the 110Vac to a more reasonable 6.3V, and it "isolates" the circuit we're working on from the potentially lethal power line voltage)

To see glitches and wiggles, look at **B**, the output of the high-pass filter. All kinds of interesting stuff should appear, some of it curiously time-dependent. What is the filter's attenuation at 60Hz? (No complex arithmetic necessary. *Hint:* count octaves, or use the fact—which you confirmed just above—that amplitude grows linearly with frequency, well below f_{3dB}.)

2-7 Filter Application II: Selecting signal from signal plus noise

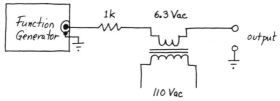

Figure L2.7: Composite signal, consisting of two sine waves. (The 1k resistor protects the function generator in case the composite output is accidentally shorted to ground)

Now we will try using high-pass and then low-pass filters to prefer one frequency range or the other in a composite signal, formed as shown in the figure above. The transformer adds a large 60Hz sine wave (peak value about 10 volts) to the output of the function generator.

Run this composite signal through the high-pass filter shown below.

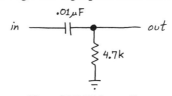

Figure L2.8: High-pass filter

Look at the resulting signal. Calculate the filter's 3dB point.

Is the attenuation of the 60Hz waveform about what you would expect? Note that this time the 60Hz is considered "noise." (In fact, as you will gather gradually, this is the most common and troublesome source of noise in the lab.)

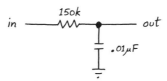

Figure L2.9: Low-pass filter

Now run the composite signal through the low-pass filter shown above, instead of running it through the high-pass. Look at the resulting signal. Calculate this circuit's 3dB point. Why were the 3dB frequencies chosen where they were?

2-8 Blocking Capacitor

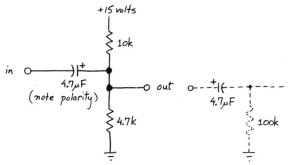

Figure L2.10: Blocking capacitor

Capacitors are used very often to "block" dc while coupling an ac signal. The circuit above does this. You can think of it as a high-pass filter, with all signals of interest well above the 3dB point. This way of describing what the circuit does is subtly different from the way one usually speaks of a *filter*: the filter's job is to prefer a range of good frequencies, while attenuating another range, the "noise." The blocking capacitor is doing something different; its mission is not to attenuate evil low frequencies; its mission, instead, is simply to block the irrelevant dc. (Once again the alert and skeptical student may be objecting, 'I've built this circuit *twice* before. You keep re-naming it and asking me to build it again: "differentiator," "high-pass," and now "blocking capacitor.' Again we must answer that it *is* the same circuit, but it is applied differently from the others.)

To see this application in action, wire up the circuitry labeled "A," above: the left side of the circuit.

Drive it with the function generator, and look at the output on the scope, *dc coupled*, as usual. The circuit lets the ac signal "ride" on +5 volts. Next, add the circuitry labeled "B," above (another blocking capacitor), and observe the signal back at ground. What is the low frequency limit for this blocking circuit?

This circuit fragment you have just built looks quite useless.

As it stands, it *is* useless, since it gets you back where you started, doing nothing useful between. Next time you will see applications for this *kind* of circuit. The difference will be that you will do something useful to the signal after getting it to ride on, say, +5 volts.

2-9 LC Filter

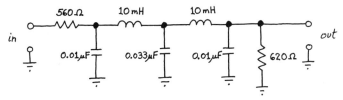

Figure L2.11: 5-pole Butterworth low-pass filter, designed using the procedure of Appendix H

It is possible to construct filters (high pass, low pass, etc.) with a frequency response that is far more abrupt than the response of the simple *RC* filters you have been building, by combining inductors with capacitors, or by using amplifiers or electronic switching (the latter two types are called "active filters"—see Chapter 5 in the text—or "switched capacitor" types—see sec. 5.11, p. 281ff).

To get a taste of what can be done, try the filter shown above. It should have a 3dB point of about 16kHz, and v_{out} should drop like a rock at higher frequencies. Measure its 3dB frequency, then measure its response at $f = 2f_{3dB}$ and at $f = 10f_{3dB}$—if you can; you may not be able to see any response beyond about $4 \times f_{3dB}$. Compare these measurements against the rather "soft" response of the *RC* low-pass filter you measured in section 2.4, and against the calculated response (i.e., ratio of output amplitude to input amplitude) of the two filters shown in the table below.

By far the best way to enjoy the performance of this filter is to *sweep* the frequency in, as suggested in section 2-4, above. You may, incidentally, find the theoretically-flat "passband" —the frequency range where the filter passes nearly all the signal— not quite flat; a slope or bump here results from our settling for standard values of R and C, and 10% capacitors, rather than using exactly the values called for.

Note: One must give a common-sense redefinition to "f_{3dB}" of this LC filter, because the filter attenuates to 1/2 amplitude even at DC. One must define the LC's 3dB point as the frequency at which the output amplitude is down 3dB *relative to its amplitude at DC*. This is not the last time you will need to modify the meaning of "3dB point" to give it a common sense reading. You will do so again when you meet amplifiers, for example, and discuss their frequency responses.

Rolloff of two filters: 5-pole vs simple *RC*

Frequency:	0	f_{3dB}	$2f_{3dB}$	$4f_{3dB}$	$10f_{3dB}$
RC	1.0	0.71	0.45	0.24	0.10
5-pole	1.0	0.71	0.031	0.001	0.00001

Figure L2.12: Amplitude out vs amplitude-at-DC: *RC* filter vs 5-pole LC filter

Optional: Contrast Ordinary RC against 5-Pole LC Filter

To appreciate how good the 5-pole filter is, it helps a lot to watch it against an ordinary *RC* filter. Here is an ordinary *RC* low-pass filter with f_{3dB} the same as the Butterworth's: about 16 kHz.

Figure L2.13: *RC* low pass with f_{3dB} matched to the LC filter's

Sweep the input frequency, and watch the outputs of the two filters simultaneously on the scope's two channels. This will challenge your scopesmanship: you no longer can use *X-Y* mode; instead you will need to use the function generator's ramp to *trigger* the scope.

Class 3: Diode Circuits

Topics:

- old:
    - Once again, the problem of A driving B; this time, A is frequency-dependent

    new:

    - scope probe; designing it; Fourier helps us check probe compensation
    - LC circuit: highly selective; fun to use as "Fourier analyzer"
    - Diode circuits

Old:

Remember our claim that our *10×* rule of thumb would let us design circuit fragments? Let's confirm it by watching what happens to a filter when we violate the rule.

Suppose we have a low-pass filter, designed to give us f_{3dB} a bit over 1kHz (this is a filter you built last time, you'll recall):

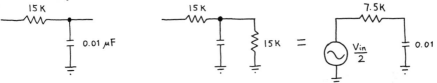

Figure N3.1: RC low-pass: not loaded versus loaded; redrawn to simplify

If R_{load} is around 150k or more, the load attenuates the signal only slightly, and f_{3dB} stays put. But what happens if we make R_{load} 15k? Attenuation is the lesser of the two bad effects. Look at what happens to f_{3dB}:

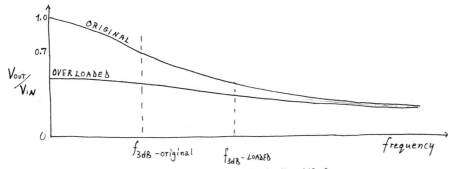

Figure N3.2: Excessive loading shifts f_{3dB}

New:

A. Scope Probe
Text exercise AE-8 (Ch 1)
Lab 3-8

That mishap leads nicely into the problem of how to design a scope probe.

Until now, we have fed the scope with BNC cables. They work, but at middling to high frequencies they don't work well. Their heavy capacitance (about 30 pF/foot) burdens the circuits you look at, and may make those circuits misbehave in strange ways—oscillating, for example. So, we nearly always use "10×" probes with a scope: that's a probe that makes the scope's input impedance 10× that of the bare scope. (The bare scope looks like 1MΩ parallel about 150pF—cable and scope.)

Here is a defective design for a *10×* probe:

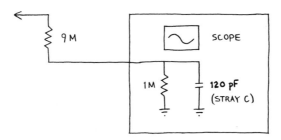

Figure N3.3: Crummy 10× probe

Do you see what's wrong? It works fine at DC. But try redrawing it as a Thevenin model driving a cap to ground, as in the example we did at the start of these notes. The flaw should appear. What is f_{3dB}?

Remedy

We need to make sure our probe does *not* have this low-pass effect: scope and probe should treat alike all frequencies of interest (the upper limit is set by the scope's maximum frequency: for most in our lab that is up to 50 or 60 MHz; a few are good to 100 MHz).

The trick is just to build two voltage dividers in parallel: one resistive, the other capacitive. At the two frequency extremes one or the other dominates (that is, passes most of the current); in between, they share. But if each delivers $V_{in}/10$, nothing complicated happens in this "in-between" range.

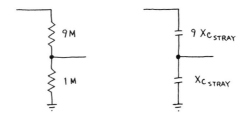

Figure N3.4: Two dividers to deliver $V_{in}/10$ to the scope

What happens if we simply *join* the outputs of the two dividers? Do we have to analyze the resulting composite circuit as one, fairly messy thing? No. No current flows along the line that joins the two dividers, so things remain utterly simple.

So, a good probe is just these two dividers joined:

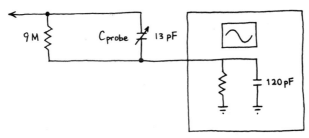

Figure N3.5: A good ×10 probe: one capacitor is trimmable, to allow use with scopes that differ in C_{in}

Practical probes make the probe's added C adjustable. This adjustment raises a question: how do you know if the probe is properly adjusted, so that it treats all frequencies alike?

Probe "compensation: Fourier again

One way to check the frequency response of probe and scope, together, would be to sweep frequencies from DC to the top of the scope's range, and watch the amplitude the scope showed. But that requires a good function generator, and would be a nuisance to set up each time you wanted to check a probe.

The easier way to do the same task is just to feed scope and probe a *square* wave, and then look to see whether it looks square on the scope screen. If it does, good: all frequencies are treated alike. If it does not look square, just adjust the trimmable C in the probe until the waveform *does* look square.

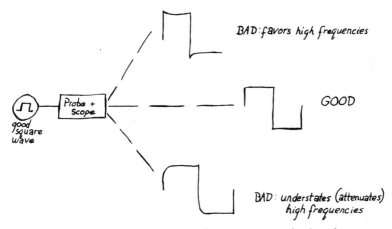

Figure N3.6: Using square wave to check frequency response of probe and scope

Neat? This is so clearly the efficient way to check probe *compensation* (as the adjustment of the probe's C is called) that every respectable scope offers a square wave on its front panel. It's labeled something like *probe comp* or *probe adjust*. It's a small square wave at around 1kHz.

B. Applying Lab 3's LC resonant circuit

Text sec. 1.22
fig. 1.63

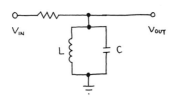
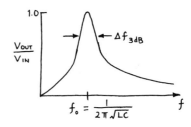

Figure N3.7: Lab 3's LC circuit, and its frequency response (generic)

As the lab notes point out, this circuit is highly selective, passing a narrow range of frequencies (the characteristic called "Q" describes how narrow: $Q \equiv f_0 / \Delta f_{3dB}$: f_0 is the resonant frequency—the favored frequency; Δf is the width at the amplitude that delivers half-power). It's entertaining to apply the circuit so as to confirm one of Fourier's claims. Here is an excerpt from those lab notes. Let's make sure we agree on what's proposed there.

Finding Fourier Components of a Square Wave

This resonant circuit can serve as a "Fourier Analyzer:" the circuit's response measures the amount of 16 kHz (approx.) present in an input waveform.

Try driving the circuit with a <u>square</u> wave at the resonant frequency; note the amplitude of the (sine wave) response. Now gradually lower the driving frequency until you get another peak response (it should occur at 1/3 the resonant frequency) and check the amplitude (it should be 1/3 the amplitude of the fundamental response). With some care you can verify the amplitude and frequency of the first five or six terms of the Fourier series. Can you think of a way to calculate *pi* with this circuit?

Here is a reminder of the Fourier series for a square wave:

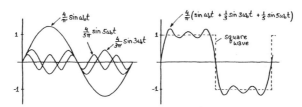

Figure N3.8: Fourier series for square wave

Classier: Frequency Spectrum Display

If you *sweep* the *square wave* input to your 16kHz-detector, you get a sort of inverse frequency spectrum: you should see a big bump at $f_{resonance}$, a smaller bump at 1/3 $f_{resonance}$, and so on.

Our detector-frequency is fixed. The square wave frequency changes. We get one or another of the several frequency components of the square wave. Make sense?

C. Diode Circuits

Diodes do a new and useful trick for us: they allow current to flow in one direction only:

The symbol looks like a one-way street sign, and that's handy: it's telling conventional current which way to go. For many applications, it is enough to think of the diode as a one-way current valve; you need to note, too, that when it conducts current it also shows a characteristic "diode drop": about 0.6 v. This you saw in Lab 1. Here's a reminder of the curve you drew on that first day:

Text sec. 1.25
Fig. 1.67

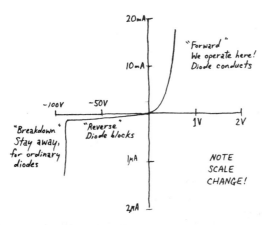

Figure N3.9: Diode *I-V* curve (reverse current is exaggerated by change of scale in this plot)

We often use diodes within voltage dividers, much the way we have used resistors, and then capacitors. Here is a set of such dividers: what should the outputs look like?

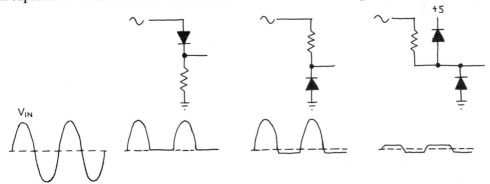

Figure N3.10: 3 dividers made with diodes: one rectifier, two clamps

The outputs for the first two circuits look strikingly similar. Yet only the rectifier is used to generate the DC voltage needed in a power supply. Why? What important difference exists—invisible to the scope display—between the "rectifier" and the "clamp?"

The bumpy output of the "rectifier" is still pretty far from what we need to make a power-supply—a circuit that converts the AC voltage that comes from the wall or "line" to a DC level. A capacitor will smooth the rectifier's output for us. But, first, let's look at a better version of the rectifier:

Full-wave bridge rectifier

This clever circuit gives a second bump out, on the *negative* swing of the input voltage. Clever, isn't it? Once you have seen this output, you can see why the simpler rectifier is called *half-wave*.

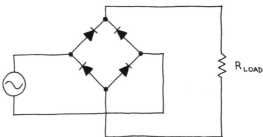

Figure N3.11: Full-wave bridge rectifier

Rarely is there an excuse for using anything other than a full-wave bridge in a power supply, these days. Once upon a time, "diode" meant a $5 vacuum tube, and then designers tried to limit the number of these that they used. Now you just buy the bridge: a little epoxy package with four diodes and four legs; a big one may cost you a dollar. In a minute, we'll proceed to looking at a full power supply circuit, which will include one of these bridges. But before we do, let's note one additional sort of diode: the zener.

Zener diodes
Text sec. 2.04[1]

Here is the *I-V* curve for a zener, a diode that breaks down at some low reverse voltage, and *likes to!* If you put it into a voltage divider "backwards"—that is, "back-biased:" with the voltage running the wrong way on the one-way street—you can form a circuit whose output voltage is pretty nearly constant, despite variation on the input, and despite variation in loading.

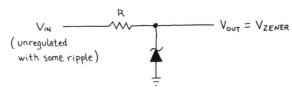

Figure N3.12: Zener voltage source (Text fig. 2.11)

Here is a chance to use a subtler view of the diode's behavior. A closer look at the diode's curve will show us, for example, why we need to follow the text's rule of thumb that says 'keep at least 10 mA flowing in the zener, even when the circuit is loaded.'

---

[1]. Yes, this is a *forward* reference; we'd just like you to think at one time about all the diodes that you'll meet.

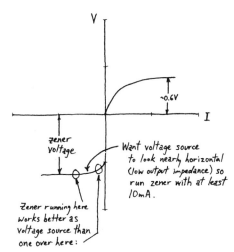

Figure N3.13: Zener diode *I-V* curve; turned on its side to reveal *impedances*

The curve turned on its side shows a slope that is $\Delta V/\Delta I$—the device's *dynamic resistance*. That value describes how good a voltage reference the zener is: how much the output voltage will vary as current varies. (Here, current varies because of two effects: variation of *input* voltage, and variation in *output current*, or loading.) Can you see how badly the diode would perform if you wandered up into the region of the curve where I_{zener} was very small?

We'll show you a worked example of such a zener reference (in another section of these notes)—and then we'll drop the subject for a while. A practical voltage source would never (well, hardy ever!) use a naked zener like the one we just showed you; it would always include a transistor or op-amp circuit after the zener, so as to limit the variation in *output current* (see Chapter 2, sec. 2.04, when you get there). The voltage regulators discussed in Chapter 6 used just such a scheme. At least you should understand those circuits the better for having glimpsed a zener today. Now on to the diode application you will see most often.

The Most Important Diode Application: Power Supply

Here is a standard unregulated power supply circuit (we'll learn later just what "unregulated" means: we can't understand fully until we meet regulated supplies):

Text sec. 6.11,
fig. 6.17,

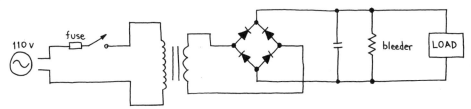

Figure N3.14: AC-"line"-to-DC Power Supply Circuit

Now we'll look at a way to choose component values. In these notes, we will do the job incompletely, as if we were just sketching a supply; in a *worked example* you will find a similar case done more thoroughly.

Assume we aim for the following specifications:

- V_{out}: about 12 volts
- Ripple: about 1 volt
- R_{load}: 120Ω

We must choose the following values:

- C size (μF)
- transformer voltage (V_{rms})
- fuse rating

We will postpone until the Worked Example two less fundamental tasks (because often a ball-park guess will do): specify:

- transformer current rating
- "bleeder" resistor value

Let's start, doing this in stages. Surely you should begin by drawing the circuit without component values, as we did above.

1. Transformer Voltage:
Compare Text sec. 6.12

This is just V_{out} plus the voltage lost across the rectifier bridge. The bridge always puts 2 diodes in the path. Specify as $_{rms}$ voltage: for a sine wave, that means $(1/\sqrt{2})(V_{peak})$.

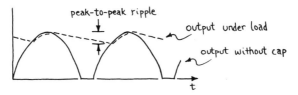

Figure N3.15: Transformer voltage, given $V_{out_{peak}}$

Here, that gives V_{peak} at the transformer of about $14V \approx 10V_{rms}$

2. Capacitor:
Compare Text sec. 6.13

This is the interesting part. This task could be hard, but we'll make it pretty easy by using two simplifying assumptions.

Here is what the "ripple" waveform will look like:

Text sec. 1.28, fig. 1.73, p. 46

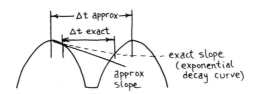

Figure N3.16: Ripple

We can figure the ripple (or choose a C for a specified ripple, as in this problem) by using the general equation

$$I = C\, dV/dt$$

"dV" or ΔV is ripple; "dt" or ΔT is the time during which the cap discharges; "I" is the current taken out of the supply. (You can, instead, memorize the Text's ripple formula; but that's probably a waste of memory space.)

To specify *exactly* the C that will allow a specified amount of ripple at the stated maximum load requires some thought. Specifically,

- What is I_{out}? If the load is resistive, the current out is not constant, but decays exponentially each half-cycle.
- What is ΔT? That is, for how long does the cap discharge (before the transformer voltage comes up again to charge it)?

We will, as usual, coolly sidestep these difficulties with some *worst-case* approximations:

- We assume I_{out} is constant, at its maximum value;
- We assume the cap is discharged for the full time between peaks.

Both these approximations tend to overstate the ripple we will get. Since ripple is not a good thing, we don't mind building a circuit that delivers a bit less ripple than called for.

Try those approximations here:

- I_{out} is just $I_{out\text{-}max}$: 100 mA
- ΔT is 1/2 the period of the 60 Hz input waveform: $1/120\text{Hz} \approx 8$ ms.

What cap size does this imply?

$$C = I\, \Delta T / \Delta V \approx 0.1 \cdot 8 \cdot 10^{-3} / 1\text{V} \approx 0.8 \cdot 10^{-3}\text{F} = 800\ \mu\text{F}.$$

That may sound big; it isn't, for a power supply filter.

3. Fuse Rating

The *current* in the primary is smaller than the current in the secondary, by about the same ratio as the primary *voltage* is larger. (This occurs because the transformer dissipates little power: $P_{in} \approx P_{out}$.)

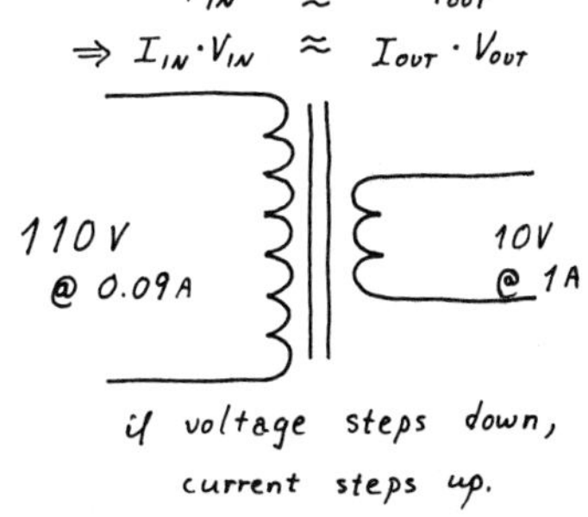

$$P_{IN} \approx P_{OUT}$$
$$\Rightarrow I_{IN} \cdot V_{IN} \approx I_{OUT} \cdot V_{OUT}$$

110 V @ 0.09 A   10V @ 1A

if voltage steps down, current steps up.

Figure N3.17: Transformer preserves power, roughly; so, a 'step down' transformer draws less current than it puts out

In the present case, where the *voltage* is stepped down about 11×, the primary current is lower than the secondary current by about the same factor. So, for 100 mA out, about 9mA must flow in the primary.

RMS Heating

*Text sec. 1.28,
p. 47; compare exercise 1.28
and figure 1.76*

In fact, the fuse feels its "9 mA" as somewhat more, because it comes in surges (see the Text at p.47, and Worked Example below); call it 20 mA. And we don't want the fuse to blow under the maximum-current load. So, use a fuse that blows at perhaps 4X the maximum I_{out}, adjusted for its heating effect: 80 mA. A 100 mA, or 0.1 A fuse is a pretty standard value, and we'd be content with that. The value is not critical, since the fuse is for emergencies, in which very large currents can be expected.

It's a good idea to use a *slow-blow* type, since on power-up a large initial current charges the filter capacitor, and we don't want the fuse to blow each time you turn on the supply.

Chapter 1: Worked Example: Power Supply

Another Power Supply

Here we will do a problem much like the one we did more sketchily in the Notes for Class 3. If you are comfortable with the design process, *skip* to sections 5 and 6, below, where we meet some new issues.

We are to design a standard *unregulated* power supply circuit (we'll learn later just what "unregulated" means when we meet voltage *regulators* in Chapter 6). In this example we will look a little more closely than we did in class, at the way to choose component values. Here's the particular problem:

> *Problem: Unregulated power supply*
>
> Design a power supply to convert 110VAC 'line' voltage to DC. Aim for the following specifications:
>
> - V_{out}: no less than 20 volts
> - Ripple: about 2 volts
> - R_{load}: 1A (maximum)
>
> Choose—
>
> - C size (µF)
> - transformer voltage (V_{rms})
> - fuse rating (I)
> - "bleeder" resistor value
> - transformer current rating
>
> *Questions:* What difference would you see in the circuit output—
>
> - If you took the circuit to Europe and plugged it into a wall outlet: there, the line voltage is 220V, 50Hz.
> - If one diode in the bridge rectifier burned out (open, not shorted)?

0. Skeleton Circuit

First, as usual, we would draw the circuit without component values; you will find that figure in the Class Notes (fig. N3.14). Fuse goes on primary side, to protect against as many mishaps as possible—including failures of the transformer and switch. Always use a *full-wave* rectifier (a bridge); most of the *half-wave rectifier* circuits you see in textbooks are relics of the days when "diode" meant an expensive vacuum tube. Now that diode means a little hunk of silicon, and they come as an integrated *bridge* package, there's rarely an excuse for anything but a bridge. A "bleeder" resistor is useful in a lab power supply, which might have no load: you want to be able to count on finding close to 0 volts a few seconds after you shut power off. The bleeder achieves that. Many power supplies are always loaded, at least by a regulator and perhaps by the circuit they were built to power; these supplies are sure to discharge their filter capacitors promptly and need no bleeder.

1. Transformer Voltage:

This is just the peak value of V_{out} plus the two *diode drops* imposed by the bridge rectifier, as usual. If V_{out} is to be 20V *after* ripple, then $V_{out(peak)}$ should be two volts more: around 22V. The transformer voltage then ought to be about 23V.

When we specify the transformer we need to follow the convention that uses V_{rms}, not V_{peak} (V_{rms} defines the *DC* voltage that would deliver the same power as the particular waveform). For a sine wave, V_{rms} is $V_{peak}/\sqrt{2}$, as you know.

In this case,
$$V_{peak}/\sqrt{2} \approx 23V/1.4 = 16V_{rms}$$
This happens to be a standard transformer voltage.

Text sec. 6.12, p. 329

If it had not been standard, we would have needed to take the next higher standard value, or use a transformer with a 'tapped primary' that allows fine tuning of the step-down ratio.

2. Capacitor:

Here is the "ripple" waveform, again. We have labeled the drawing with reminders that Δt depends on circumstances: so, Δt varies under the changed conditions suggesed in the *questions* that conclude this exercise.

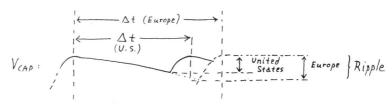

Figure X3.1: Ripple

Using
$$I = C \, dv/dt$$
we plug in what we know, and solve for C. We know—

- "dV" or ΔV, the ripple, is 2V;
- "dt" or ΔT is the time between peaks of the input waveform: $1/(2 \cdot 60Hz) \approx 8$ ms;
- "I" is the peak output current: 1A. (This specification of load *current* rather than load *resistance* may seem odd, at the moment. In fact, it is typical. The typical load for an unregulated supply is a *regulator*—a circuit that holds its output voltage constant; if the regulated supply drives a constant resistance, then it puts out a constant current despite the input ripple, and it thus draws a constant current from the filter capacitor at the same time).

Putting these numbers together, we get:
$$C = I \cdot \Delta t / \Delta v = 1A \cdot 8 \cdot 10^{-3}s / 2V = 4000\mu F$$
Big, but not unreasonably big.

3. Fuse Rating

This supply steps the voltage down from 110V to 16V; the current steps up proportionately: about 7•. So, the output (secondary) current of 1A implies an input (primary) current of about 140mA.

But this calculation of *average* input current understates the heating effect of the primary current, and of the primary and secondary currents in the transformer. Because these currents come in surges, recharging the filter capacitor during only a part of the full cycle, the currents during the surges are large. These surges heat a fuse more than a steady current delivering the same power, so we need to boost the fuse rating by perhaps a factor of 2, and then another factor of about 2 to prevent fuse from blowing at full load. (It's designed for emergencies.)

This set of rules of thumb carry us to something like—

> fuse rating (current) ≈ 140mA • 2(for current surges) • 2 (not to blow under normal full load)
> ≈ 1.1A.

A 1A slow-blow would do. Why slow-blow? Because on power-up (when the supply first is turned on) the filter capacitor is charged rapidly in a few cycles; large currents then flow. A fuse designed to blow during a brief overload would blow every time the supply was turned on. The slow-blow has larger thermal mass: needs overcurrent for a longer time than the normal fuse, before it will blow.

4. Bleeder Resistor

Polite power supplies include such a resistor, or some other fixed load, so as not to surprise their users. Again the value is not critical. Use an R that discharges the filter cap in no more than a few seconds; don't use a tiny R that substantially loads the supply.

Here, let's let RC = a few seconds. ==> R = a few/C ≈ 1k

Before we go on to consider a couple of new issues, let's just draw the circuit with the values we have chosen, so far:

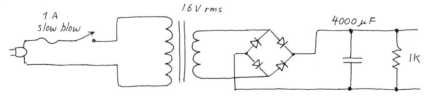

Figure X3.2: Power supply: the usual circuit, with part values inserted

5. Transformer Current Rating

This is harder. The transformer provides brief surges of current into the cap. These heat the transformer more than a continuous flow of smaller current. Here is a sketch of current waveforms in relation to two possible ripple levels:

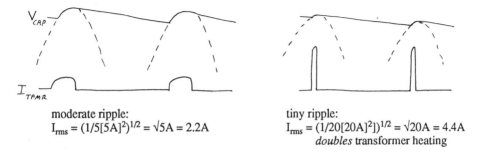

moderate ripple:
$I_{rms} = (1/5[5A]^2)^{1/2} = \sqrt{5}A = 2.2A$

tiny ripple:
$I_{rms} = (1/20[20A]^2])^{1/2} = \sqrt{20}A = 4.4A$
doubles transformer heating

Figure X3.3: Transformer Current versus Ripple: Small Ripple ==> Brief High-Current Pulses, and excessive heating

The left hand figure show current flowing for about 1/5 period, in pulses of 5A, to replace the charge drained at the steady 1 A output rate. The right hand figure shows current flowing for 1/20 period, in pulses of 20 A.

Moral: a little ripple is a good thing. You will see that this is so when you meet voltage regulators, which can reduce ripple by a factor between 1000 and 10,000. To say this another way, a volt of ripple out of the unregulated supply may look like less than a *millivolt* at the point where it goes to work (where the output of the regulated supply drives some load)!

6. Questions: What difference would you see in the circuit output—

1. If you took it to Europe, where the line voltage is 220V, 50Hz?
2. If one diode in the bridge rectifier burned out (open, not shorted)?

Solutions:

1. In Europe, the obvious effect would be a doubling of output voltage. That's likely to cook something driven by this supply (that's why American travelers often carry small 2:1 step-down transformers). It might also cook transformer and filter capacitor, unless you had been very conservative in your design (it would be foolish, in fact, to specify a filter cap that could take double the anticipated voltage; caps grow substantially with the voltage they can tolerate).

So you probably would not have a chance to get interested in less obvious changes in the power supply. But let's look at them anyway: the output *ripple* would change: ΔT would be 1/50Hz = 10ms, not 8ms.

Ripple should grow proportionately. If load current remained constant, ripple should grow to about 2.5V. If the load were resistive, then load current would double with the output voltage, and ripple would double relative to the value just estimated: to around 5V.

2. The burned out diode would make the bridge behave like a *half-wave* rectifier. Δt would double, so ripple amplitude would double, roughly. Ripple frequency would fall from 120Hz to 60Hz. (This information might someday tell you what's wrong with an old radio: if it begins to buzz at you at 60Hz, perhaps half of the bridge has failed; if it buzzes at 120Hz, probably the filter cap has failed. If you like such electronic detective work, many pleasures lie ahead of you.)

Lab 3: Diode Circuits

Reading:	Finish Chapter 1, including "Power in reactive circuits" (pp 28-29)
	Appendix E
Problems:	Problems in text.
	Additional Exercises 7,8.

3-1. LC Resonant Circuit.

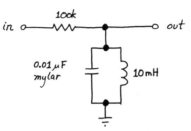

Figure L3.1: LC parallel resonant circuit

Construct the parallel resonant circuit shown above. Drive it with a sine wave, varying the frequency through a range that includes what you calculate to be the circuit's *resonant frequency*. Compare the resonant frequency that you observe with the one you calculated. (The circuit attenuates the signal considerably, even at its resonant frequency; the L is not perfectly efficient, but instead includes some series resistance.)

Use the function-generator's *sweep* feature to show you a scope display of amplitude-out versus frequency. (See Lab 2 notes, if you need some advice on how to do this trick.)

When you succeed in getting such a display of frequency response, try to explain why the display grows funny wiggles on one side of resonance as you increase the sweep rate. *Clue:* the funny wiggles appear on the side after the circuit has already been driven into resonant oscillation; the function generator there is driving an oscillating circuit.

Finding Fourier Components of a Square Wave

This resonant circuit can serve as a "Fourier Analyzer:" the circuit's response measures the amount of 16 kHz (approx.) present in an input waveform.

Try driving the circuit with a square wave at the resonant frequency; note the amplitude of the (sine wave) response. Now gradually lower the driving frequency until you get another peak response (it should occur at 1/3 the resonant frequency) and check the amplitude (it should be 1/3 the amplitude of the fundamental response). With some care you can verify the amplitude and frequency of the first five or six terms of the Fourier series. Can you think of a way to calculate *pi* with this circuit?

Here is a reminder of the Fourier series for a square wave:

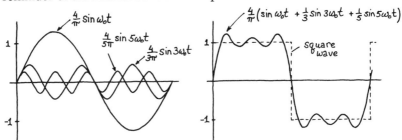

Figure L3.2: Fourier series for square wave

Classier: Frequency Spectrum Display

If you *sweep* the *square wave* input to your 16kHz-detector, you get a sort of inverse frequency spectrum: you should see a big bump at $f_{resonance}$, a smaller bump at $1/3\, f_{resonance}$, and so on.

"Ringing"

Now try driving the circuit with a low-frequency square wave: try 20 Hz. You should see a brief output in response to each edge of the input square wave. If you look closely at this output, you can see that it is a decaying sine wave. (If you find the display dim, increase the square wave frequency to around 100 Hz.)

What is the frequency of this sine wave? (No surprise, here.)

Why does it decay? Does it appear to decay exponentially?

You will see such a response of an LC circuit to a step input whenever you happen to look at a square wave with an improperly grounded scope probe: when you fail to ground the probe close to the point you are probing, you force a ground current to flow through a long (inductive) path. Stray inductance and capacitance form a resonant circuit that produces ugly ringing. You might look for this effect now, if you are curious; or you might just wait for the day (almost sure to come) when you run into this effect inadvertently.

3-2. Half-wave Rectifier.

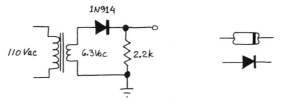

Figure L3.3: Half-wave rectifier

Construct a half-wave rectifier circuit with a 6.3Vac (rms) transformer and a 1N914 diode, as in the figure above. Connect a 2.2k load, and look at the output on the scope. Is it what you expect? Polarity? Why is $V_{peak} > 6.3V$? (Don't be troubled if V_{peak} is a bit more than $6.3V \cdot \sqrt{2}$: the transformer designers want to make sure your power supply gets at least what's advertised, even under heavy load; you're loading it very lightly.)

3-3. Full-wave Bridge Rectifier.

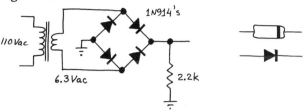

Figure L3.4: Full-wave bridge

Now construct a full-wave bridge circuit, as above. Be careful about polarities—the band on the diode indicates cathode, as in the figure. Look at the output waveform (but *don't* attempt to look at the input—the signal across the transformer's secondary—with the scope's other channel at the same time; this would require connecting the second "ground" lead of the scope to one side of the secondary. What disaster would that cause?). Does it make sense? Why is the peak amplitude less than in the last circuit? How much should it be? What would happen if you were to reverse any one of the four diodes? (*Don't try it!*).

Don't be too gravely alarmed if you find yourself burning out diodes in this experiment. When a diode fails, does it usually fail *open* or *closed*? Do you see why diodes in this circuit usually fail in pairs—in a touching sort of suicide pact?

Look at the region of the output waveform that is near zero volts. Why are there flat regions? Measure their duration, and explain.

3-4. Ripple.

Now connect a 15μF filter capacitor across the output (*Important*—observe polarity). Does the output make sense? Calculate what the "ripple" amplitude should be, then measure it. Does it agree? (If not, have you assumed the wrong discharge time, by a factor of 2?)

Now put a 500μF capacitor across the output (again, be careful about polarity), and see if the ripple is reduced to the value you predict. This circuit is now a respectable voltage source, for loads of low current. To make a "power supply" of higher current capability, you'd use heftier diodes (e.g., 1N4002) and a larger capacitor. (In practice you would always follow the power supply with an active *regulator*, a circuit you will meet in Lab 12.)

3-5. Signal Diodes.

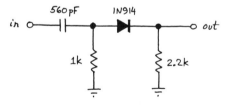

Figure L3.5: Rectified differentiator

Use a diode to make a rectified differentiator, as in the figure above. Drive it with a square wave at 10kHz or so, at the function generator's maximum output amplitude. Look at input and output, using both scope channels. Does it make sense? What does the 2.2k load resistor do? Try removing it.

Hint: You should see what appear to be RC discharge curves in both cases—with and without the 2.2k to ground. The challenge here is to figure out what determines the *R* and *C* that you are watching.

3-6. Diode Clamp.

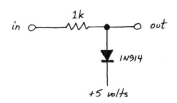

Figure L3.6: Diode clamp

Construct the simple diode clamp circuit shown just above. Drive it with a sine wave from your function generator, at maximum output amplitude, and observe the output. If you can see that the clamped voltage is not quite flat, then you can see the effect of the diode's non-zero impedance. Perhaps you can estimate a value for this *dynamic resistance* (see Text sec. 1.); try a triangle waveform, if you attempt this estimate.

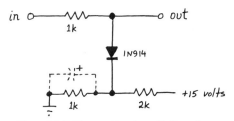

Figure L3.7: Clamp with voltage divider reference

Now try using a voltage divider as the clamping voltage, as shown just above. Drive the circuit with a large sine wave, and examine the peak of the output waveform. Why is it rounded so much? (*Hint*: What is the impedance of the "voltage source" provided by the voltage divider? If you are puzzled, try drawing a Thevenin model for the whole circuit. Incidentally, this circuit is probably best analyzed in the *time* domain.) To check your explanation, drive the circuit with a triangle wave; compare with figure 1.83 in the text.

As a remedy, try adding a 15μF capacitor, as shown with dotted lines (note polarity). Try it out. Explain to your satisfaction why it works. (Here, you might use either a time- or frequency-domain argument.) This case illustrates well the concept of a bypass capacitor. What is it bypassing, and why?

3-7. Diode Limiter.

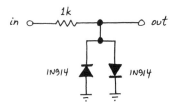

Figure L3.8: Diode limiter

Build the simple diode limiter shown above. Drive it with sines, triangles, and square waves of various amplitudes. Describe what it does, and why. Can you think of a use for it?

3-8. Impedances of Test Instruments.

We mentioned in the first lab that measuring instruments (voltmeters, ammeters) ideally should leave the measured circuit unaffected. For instance, this implies an infinite impedance for voltmeters, and zero impedance for ammeters. Likewise, an oscilloscope should present an infinite input impedance, while power supplies and function generators should be zero-impedance sources.

Begin by measuring the internal resistance of the VOM on its 10V dc range. You won't need anything more than a dc voltage and a resistor, if you're clever.

Next try the same measurement on the 50V dc range. Make sense? (Most needle-type VOM's are marked with a phrase such as "20,000 ohms per volt" on their dc voltage ranges; remember this, from the first class?). For further enlightenment, see the Box on Multimeters (text pp. 9-10).

Now use a similar trick to measure the input *resistance* of the scope. Remember that it should be pretty large, if the scope is a good voltage measuring-instrument. As a voltage source use a 100Hz sine wave, rather than a dc voltage as above.

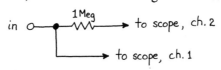

Figure L3.9: Circuit for measuring oscilloscope input impedance

To measure the scope's input *impedance* drive it with a signal in series with 1 megohm (figure 3.9, above). What is the low frequency (f < 1kHz) attenuation? Now raise the frequency. What happens? Explain, in terms of a model of the scope input as an R in parallel with a C. What are the approximate values of R and C? What remedy will make this circuit work as a divide-by-two signal attenuator at *all* frequencies? Try it!

Now go back and read the section entitled "Probes" in the Text's Appendix A. Then get a 10X probe, and use it to look at the calibrator signal (usually a 1V, 1kHz square wave) available on the scope's front panel somewhere. Adjust the probe "compensation" screw to obtain a good square wave. Use 10X probes on your scope in all remaining lab exercises, like a professional!

Finally, measure the internal resistance of the function generator. ***Don't try to do it with an ohmmeter!*** Instead, load the generator with a known resistor and watch the output drop. One value of R_{load} is enough to determine $R_{internal}$, but try several to see if you get a consistent value. Use a small signal, say 1 volt pp at 1kHz.

Ch. 1 Review: Important Topics

Important Topics

1. Resistive Circuits:

    a. voltage dividers
    b. R_{out}, Z_{out}, R_{in}, Z_{out}; Thevenin models

2. RC Circuits

    a. Generally

        i. time-domain vs frequency domain
        ii. step-response vs sine response

    b. Important RC Circuits

        i. integrator, differentiator: (either can be described as a filter that is murdering signal)
        ii. filters

            1. f_{3dB}
            2. phase-shift
            3. phasors (only an *optional* aid to visualizing what's going on)

3. Diode Circuits

    a. rectifiers
    b. clamp
    c. zener voltage reference
    d. power-supply

        i. ripple
        ii. transformer rating: rms
        iii. current ratings: fuse, transformer

4. LC Circuits
Rare in this course: one LC resonant circuit (Lab 3)
But they haunt us as unwanted effect of stray L and C: ringing in circuits; oscillation in a follower, e.g. (see 'unwanted oscillations' in Lab 10)

Ch.1: Jargon and terms

choke	(noun): inductor
droop	fall of voltage as effect of loading (loading implies drawing of current)
primary	input winding of transformer
ripple	variation of voltage resulting from partial discharge of power-supply filter capacitor between re-chargings by transformer
risetime	time for waveform to rise from 10% of final value to 90%
rms	"root mean [of} square[s]". Used to describe power delivered by time-varying waveform. For sine, $V_{rms} \doteq V_{peak}/\sqrt{2}$
secondary	output winding of transformer
stiff	of a voltage source: means it "droops" little under load
V_{peak}	= "amplitude." E.g., in v(t) = Asinωt, "A" is peak voltage (see fig. 1.17, p. 16)
$V_{peak\text{-}to\text{-}peak}$	$V_{p\text{-}p}$: another way to characterize the size of a waveform; much less common than V_{peak}.

CHAPTERS 2, 3

Overview

A novel and powerful new sort of circuit performance appears in this chapter: a circuit that can *amplify*. Sometimes the circuit will amplify voltage; that's what most people think of as an amplifier's job. Sometimes the circuit will amplify only current; in that event one can describe its amplification as a transformation of *impedances*. As you know from your work in Chapter 1, that is a valuable trick.

The transistors introduced in this chapter are called **bipolar** (because the charge carrying mechanism uses carrier of both polarities—but that is a story for another course). In the chapter that follows you will meet the other sort of transistor, which is called *field effect* rather than 'unipolar' (though they were called 'unipolar' at first). The FET type was developed later, but has turned out to be more important than bipolar in *digital* devices. You will see much of these FET logic circuits later in this course. In *analog* circuits, bipolar transistors still dominate, but even there FETs are gaining.

An understanding of transistor circuits is important in this course not so much because you are likely to design with *discrete* transistors as because you will benefit from an understanding of the innards of the *integrated* circuits that you are certain to rely on. After toiling through this chapter, you will find that you can recognize in the schematic of an otherwise-mysterious IC a collection of familiar transistor circuits. This will be true of the *operational amplifier* that will become your standard analog building block. Recognizing familiar circuits, you will consequently recognize the op amp's shortcomings as shortcomings of those familiar transistor circuit elements.

Chapter 2 is difficult. It requires that you get used to a new device and at the same time apply techniques you learned in Chapter 1: you will find yourself worrying about impedances once again: arranging things so that circuit-fragment A can drive circuit-fragment B without undue loading; you will design and build lots of RC circuits; often you will need a Thevenin model to help you determine an effective *R*. We hope, of course, that you will find this chance to apply new skills gratifying; but you are likely to find it taxing as well.

Chapter 3, on FETs is difficult, too, and for similar reasons: you must apply skills recently acquired as you design with a new class of devices. Fortunately, some of what you learn in Chapter 2 applies by analogy to the FETs of Chapter 3: gain as transconductance, for example (g_m: current out per voltage in) is a notion equally useful in the two chapters. Nevertheless, both chapters will make your work hard, and the varieties of FETs will annoy you for a while.

We say this not to discourage you, but, on the contrary, to let you know that if you have trouble digesting Chapter 2, that's not a sign that there's something wrong with you. This is a *rich* chapter.

In Chapter 4 suddenly your life as a circuit designer will become radically easier. Operational amplifiers, used with *feedback*, will make the design of very good circuits very easy. At that point you may wonder why you ever labored with those difficult discrete-transistor problems. But we won't reveal this to you now, because we don't want to sap your present enthusiasm for transistors. In return for your close attention to their demands, transistors will perform some pretty impressive work for you.

Instead of yearning for op amps, let's try to put ourselves into a state of mind approaching that of the transistor's three inventors, who found to their delight, two days before Christmas 1947, that they had constructed a tiny amplifier on a chunk of germanium. They could envision a time when there would be no more vacuum tubes; no more high-voltage supplies; no more power wasted in heating filaments. The world could look forward to the microcomputer, the Walkman—and then the boom-box: the amazingly tiny microcircuit housed in the amazingly huge suitcase. Ah, technology!

Figure IN2,3.1: The first transistor: point-contact type (1947): Nobel prize-winning device, looking wonderfully home-made (it's not *really* made from paper clips, scotch tape and chewing gum). Photo used with permission of AT&T Bell Labs

Class 4: Transistors I: First Model

Topics:

- Two simple views of transistor operation:
  - Simple: $I_C = I_B \cdot \beta$
  - Simpler: $I_C \approx I_E$; $V_{BE} = 0.6\text{V}$

- Applying the models: standard circuits:
  - Follower
  - Current source
  - common-emitter amp
  - push-pull

- Recapitulation: what the standard circuits look like

Preliminary: **Introductory Sketch**

An Intuitive Model:

A transistor is a *valve*:

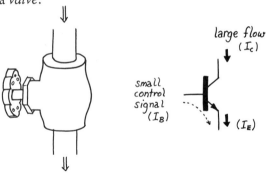

Figure N4.1: A transistor is a valve (not a pump!)

Notice, particularly, that the transistor is not a *pump*: it does not force current to flow; it permits it to flow, to a controllable degree, when the remainder of the circuit tries to force current through the device.

Ground Rules:

Text sec. 2.01

> *Ground Rules:*
> For NPN type:
>
> 1. $V_C > V_E$ (by at least a couple of tenths of a volt)
> 2. "things are arranged" so that $V_B - V_E =$ about 0.6 v (V_{BE} is a diode junction, and must be forward biased)

We begin with **two views of the transistor**: one simple, the other *very* simple. (Next time we will complicate things.)

Text sec. 2.01

> Pretty simple: current amplifier: $I_C = \text{Beta} \cdot I_B$
>
> $\downarrow I_C = \beta I_B$
>
> $I_B \searrow$
>
> $\downarrow I_E = I_C + I_B = (1+\beta) I_B$ 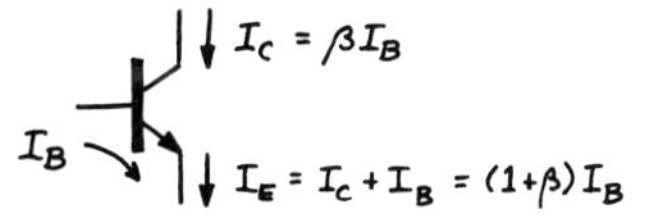
>
> **Figure N4.2**: Transistor as *current*-controlled valve or amplifier

> Very simple: say nothing of Beta (though assume it's at work);
>
> - Call V_{BE} constant (at about 0.6 v);
> - call $I_C = I_E$.

A. The simple view: using Beta explicitly

You need the first view to understand how a ***follower*** changes impedances: small (change in-) current in —> large (change in-) current out:

Text sec. 2.03
Lab 4-2

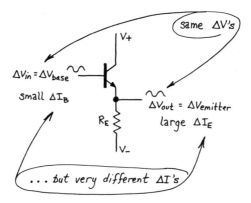

Figure N4.3: How a follower changes impedances

And here is a corny mnemonic device to describe this impedance-changing effect. Imagine an ill-matched couple gazing at each other in a dimly-lit cocktail lounge—and gazing through a rose-colored lens that happens to be a follower. Each sees what he or she wants to see:

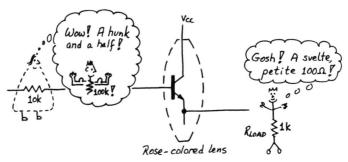

Figure N4.4: Follower as rose-colored lens: it shows what one would *like* to see

Complication: *Biasing*
ext sec. 2.05

We can use a *single* power supply, rather than *two* (both positive and negative) by pulling the transistor's *quiescent* voltages *off-center*—biasing it away from zero volts:

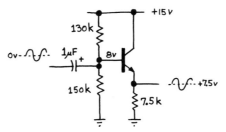

Figure N4.5: Single-supply follower uses *biasing*

The biasing divider must be stiff enough to hold the transistor where we want it (with V_{out} around the midpoint between V_{CC} and ground). It must not be too stiff: the *signal source* must be able to wiggle the transistor's base without much interference from the biasing divider.

The biasing problem is the familiar one: Device A drives B; B drives C. As usual, we want Z_{out} for each element to be low relative to Z_{in} for the next:

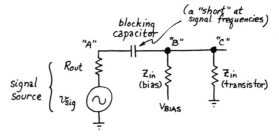

Figure N4.6: Biasing arrangement

You will notice that the biasing divider reduce the circuit's input impedance by a factor of ten. That is regrettable; if you want to peek ahead to complications, see the "bootstrap" circuit (sec. 2.17) for a way around this degradation.

You will have to get used to a funny convention: you will hear us talk about impedances not only *at* points in a circuit, but also *looking* in a particular direction.

Text sec. 2.05

For example: we will talk about the impedance "at the base" in two ways:

- the impedance "looking into the base" (this is a characteristic of the transistor and its emitter load)
- the impedance at the base, looking back toward the input (this characteristic is *not* determined by the transistor; it depends on the *biasing* network, and (at signal frequencies) on the *source impedance*.

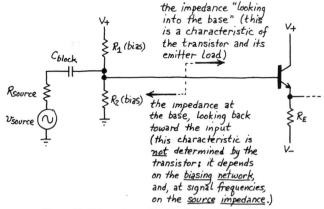

Figure N4.7: Impedances "looking" in specified directions

B. The simplest view: forgetting Beta

We can understand—and even design—many circuits without thinking explicitly about Beta. Try the simplest view:

- Call V_{BE} constant (at about 0.6 v);
- call $I_C = I_E$.

This is enough to let one predict the performance of many important circuits. This view lets one see—

- That a follower follows:

Text Sec. 2.03

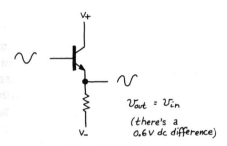

Figure N4.8: Follower

- That a current source provides a constant output current:

Text sec. 2.06

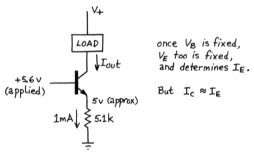

Figure N4.9: Current source

- That a common-emitter amplifier shows voltage gain as advertised:

Text sec. 2.07

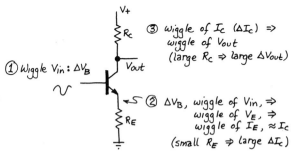

Figure N4.10: Common-emitter amp

- That a push-pull works, and also shows distortion:

Text sec. 2.14,

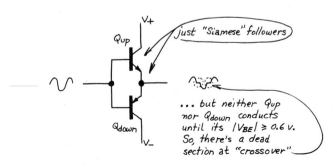

Figure N4.11: Push-pull

Recapitulation: the important transistor circuits at a glance

To get you started on the process of getting used to what bipolar transistor circuits look like, and to the crucial differences that come from what *terminal* you treat as output, here is a family portrait, stripped of all detail:

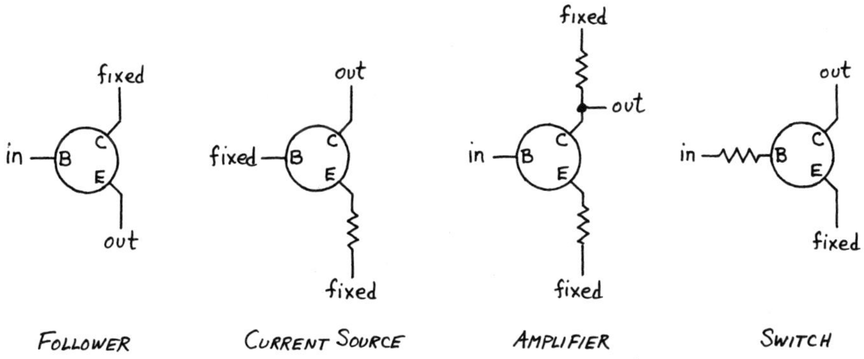

Figure N4.12: The most important bipolar transistor circuits: sketch

Next time, we will begin to use the more complicated *Ebers-Moll* model for the transistor. But the simplest model of the transistor, presented today, will remain important. We will always try to use the simplest view that explains circuit performance, and often the very simplest will suffice.

Ch. 2: Worked Example: Emitter Follower

Text sec. 2.04

The text works a similar problem in detail: sec. 2.04. The example below differs in describing an *AC* follower. That makes a difference, as you will see, but the problems are otherwise very similar.

> ***Problem: AC-coupled follower***
>
> Design a single-supply voltage follower that will allow this source to drive this load, without attenuating the signal more than 10%.
>
> Let V_{CC} = 15 v, let I_C quiescent be 0.5 mA. Put the 3dB point around 100 Hz.
>
>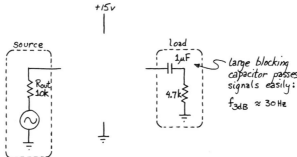
>
> **Figure X4.1**: Emitter Follower (your design) to let given source drive given load

Solution

Before we begin, perhaps we should pause to recall why this circuit is useful. It does *not* amplify the signal voltage; in fact we concede in the design specification that we expect some *attenuation*; we want to limit that effect. But the circuit does something useful: before you met transistors, could you have let a 10k source drive a 4.7k load without having to settle for a good deal of attenuation? How much?

The Text sets out a step-by-step design procedure for a follower, as we have noted already (sec. 2.04). We will follow that procedure, and will try to explain our choices as we go along, in scrupulous—perhaps painful—detail.

1. Draw a skeleton circuit

Perhaps this is obvious; but start by drawing the circuit diagram without part values. Gradually we will fill those in.

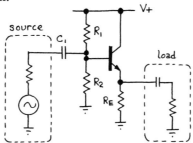

Figure X4.2: Emitter follower skeleton circuit: load is AC coupled

2. Choose R_E to center V_{out}

To say this a little more carefully, we should say,

center $V_{out\text{-}quiescent}$, given $I_{C\text{-}quiescent}$

"Quiescent" means what it sounds like: it means conditions prevailing *with no input signal*. In effect, therefore, *quiescent* conditions mean *DC* conditions, in an *AC* amplifier like the present design.

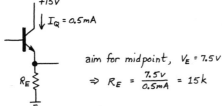

Figure X4.3: Choose R_E to center V_{out}

3. Center V_{base}

Here we'll be a little lazier than the Text suggests: by centering the base voltage we will be sure to *miss* centering V_{out}. But we'll miss by only 0.6 v, and that error won't matter if V_{CC} is big enough: the error is about 4% if we use a 15-volt supply, for example.

Centering the *base* voltage makes the divider resistors equal; that, in turn, makes their $R_{Thevenin}$ very easy to calculate.

4. Choose bias divider R's so as to make bias stiff enough

Stiff enough means, by our rule of thumb, $\leq 1/10\ R_{\text{in-at-base(DC)}}$. If we follow that rule, we will hit the bias voltage we aimed for (to about 10%).

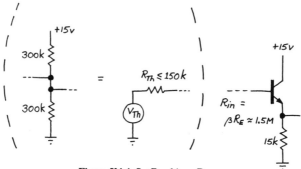

Figure X4.4: Set R_{TH} bias $\ll R_{\text{inbase}}$

$R_{\text{in-at-base}}$ is just $\beta \times R_E$, as usual. That's straightforward. What is not so obvious is that we should *ignore* the AC-coupled *load*. That load is *invisible* to the bias divider, because the divider sets up *DC* conditions (steady state, quiescent conditions), whereas only *AC* signals pass through the blocking capacitor to the load.

That finishes the setting of DC conditions. Now we can finish by choosing the coupling capacitor (also called "blocking capacitor;" evidently both names fit: this cap couples one thing, blocks another).

5. Choose blocking capacitor

We choose C_1 to form a high-pass filter that passes any frequency of interest. Here we have been told to put f_{3dB} around 100 Hz.

The only difficulty appears when we try to decide what the relevant "R" is, in our high-pass filter:

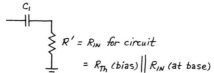

Figure X4.5: What R for blocking cap as high-pass?

We need to look at the input impedance of the follower, seen from this point. The bias divider and transistor appear in *parallel*.

> *Digression on Series versus Parallel*
>
> Stare at the circuit till you can convince yourself of that last proposition. If you have trouble, think of yourself as a little charge carrier—an electron, if you like—and note each place where you have a *choice* of routes: there, the circuit offers *parallel* paths; where the routes are obligatory, they are in *series*. Don't make the mistake of concluding that the bias divider and transistor are in series because they appear to come one after the other as you travel from left to right.

So, $Z_{\text{in, follower}} = R_{\text{TH bias}}$ parallel ($R_{\text{in at base}}$). The slightly subtle point appears as you try to decide what ($R_{\text{in at base}}$) ought to be. Certainly it is $\beta \times$ something. But $\times$ what? Is it just R_E, which has been our usual answer? We did use R_E in choosing R_{TH} for the bias divider.

But this time the answer is, 'No, it's not just R_E,' because the *signal*, unlike the DC bias current, passes *through* the blocking capacitor that links the follower with its load. So we should put R_{load} in parallel with R_E, this time.

The impedance that gets magnified by the factor β, then, is not 15k but (15k parallel 4.7k), about 15k/4 or 3.75k. Even when increased by the factor β, this impedance cannot be neglected for a 10% answer:

$$R_{in} \approx 135\text{k parallel } 375\text{k}$$

That's a little less than 3/4 of 135k (since 375k is a bit short of 3×135k, so we can think of the two resistors as one of value R, the other as 3 of those R's in parallel (using the Text's trick: see Ch. 1 p.6, "shortcut no. 2")). Result: we have 4 parallel resistors of 375k: roughly equivalent to 100k. (By unnatural good luck, we have landed within 1% of the exact answer, this time.)

So, choose C_1 for f_{3dB} of 100Hz. $C_1 = 1/(2\pi\ 100\text{Hz} \cdot 100\text{k})$
$\approx 1/(6 \cdot 10^2 \cdot 100 \times 10^3) = (1/60) \times 10^{-6} \approx 0.016\ \mu\text{F}$.
$C_1 = 0.02\ \mu\text{F}$ would be generous.

Recapitulation

Here, for people who hate to read through explanations in words, is one picture restating what we have just done:

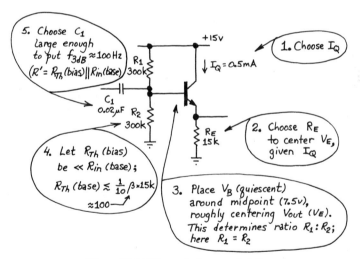

Figure X4.6: Follower design: recapitulation

Lab 4: Transistors I

> **Reading:** Chapter 2.01 – 2.08
> **Problems:** Problems in text
> Additional Exercises 1,3.
> Bad Circuits A, B, D, E

4-1 Transistor Junctions are Diodes

Here is a method for spot-checking a suspected bad transistor: the transistor must look like a pair of diodes when you test each junction separately. But, *caution:* do not take this as a description of the transistor's mechanism when it is operating: it does *not* behave like two back-to-back diodes when operating (the following circuit, made with a pair of diodes, would be a flop, indeed:)

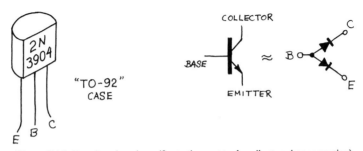

Figure L4.1: Transistor junctions: (for testing, not to describe transistor operation)

Get a 2N3904 NPN transistor, identify its leads, and verify that it looks like the object shown in figure 2.2 of the text (and reproduced just above), by measuring the voltage across the BC and BE junctions, using a DVM's *diode test* function. (Most meters use a diode symbol to indicate this function.) The diode test applies a small current (a few milliamps: current flowing from Red to Black lead), and the meter reads the junction voltage. You can even distinguish BC from BE junction this way: the BC junction is the larger of the two; the lower current density is revealed by a slightly lower voltage drop.

4-2 Emitter Follower

Wire up an NPN transistor as an emitter follower, as shown below.

Drive the follower with a sine wave that is symmetrical about zero volts (be sure the dc "offset" of the function generator is set to zero), and look with a scope at the poor replica that comes out. Explain exactly why this happens.

If you turn up the waveform amplitude you will begin to see bumps *below* ground. How do you explain these? (*Hint*: see V_{BE} breakdown specification in the data sheet for the 2N4400 transistor, Text Appendix K: the '4400 is very similar to the '3904, and in this characteristic is identical.)

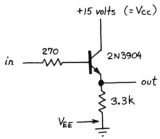

Figure L4.2: Emitter follower. The small base resistor is often necessary to prevent oscillation

Now try connecting the emitter return (the point marked V_{EE}) to –15V instead of ground, and look at the output. Explain the improvement.

4-3 Input and Output Impedance of Follower

Measure Z_{in} and Z_{out} for the follower below:

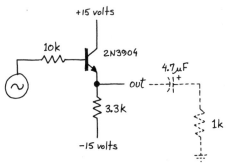

Figure L4.3: Follower: circuit for measuring Z_{in} and Z_{out}

In the last circuit replace the small base resistor with a 10k resistor, in order to simulate a signal source of moderately high impedance, i.e., low current capability (see figure above).

a) Measure Z_{out}, the output impedance of the follower, by connecting a 1k load to the output and observing the drop in output signal amplitude; for this use a small input signal, less than a volt. Use a blocking capacitor — why? (*Hint*: in this case you could get away with omitting the blocking cap, but often you could not).

> *Suggestions for measurement of Z_{out}*
>
> - If you view the emitter follower's output as a signal source in series with $Z_{out_{Thevenin}}$, then the 1k load forms a divider at signal frequencies, where the impedance of the blocking capacitor is negligibly small.
> - The attenuations are likely to be small. To measure them we suggest you take advantage of the *percent* markings on the scope screen.
>
>   Here is one way to do this measurement:
>
>   - center the waveform on the 0% mark;
>   - *AC couple* the signal to the scope, to ensure centering;
>   - adjust amplitude to make the peak just hit 100%;
>   - now load the circuit and read the amplitude in percent.

b) Remove the 1k load. Now measure Z_{in}, which here is the impedance looking into the transistor's base, by looking alternately at both sides of the 10k input resistor. For this measurement the 3.3k emitter resistor is also the "load". Again, use a small signal. Does the result make sense? (See Text sec. 2.03, subsection on impedances.)

When you have measured Z_{in} and Z_{out}, infer your transistor's β.

4-4 Single-Supply Follower

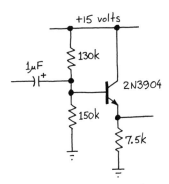

Figure L4.4: Single supply follower. A 270 ohm resistor in series with the base may be necessary if the circuit exhibits oscillations. (This puzzling trick you will find explained in Lab 11, on oscillators.)

The figure above shows a properly-biased emitter follower circuit, operating from a single positive supply voltage. This circuit comes from the example in Text sec. 2.05. Wire it up, and check it for the capability of generating large output swings before the onset of "clipping". For largest dynamic range, amplifier circuits should exhibit symmetrical clipping.

A fine point: in fact, the clipping here may look slightly odd: a bit asymmetric. To see why, watch the *base* with one channel of the scope (*not* the same DC level as at the function generator, n.b.). What happens when the voltage at the base tries to climb above the positive supply?

4-5 Transistor Current Gain

You saw the transistor's current gain, β, at work in 4-3. Now measure β *(or h_{FE})* directly at several values of I_C with the circuit shown below. The 4.7k and 1k resistors limit the currents. Which currents do they limit, and to what values?

Try various values for **R**, using a resistor "substitution box:" e.g., 4.7 M, 1 M, 470K, 100k, 47k. Estimate the base current in each case (don't bother to measure it; assume V_{BE} = 0.6V), and from the measured I_C calculate β *(h_{FE})*.

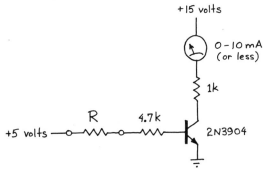

Figure L4.5: Circuit for measurement of β or "h_{FE}"

4-6 Current Source

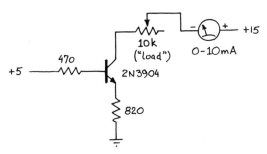

Figure L4.6: Transistor current source

Construct the current source shown above (sometimes called, more exactly, a current "sink").

Slowly vary the 10k variable load, and look for changes in current measured by the VOM. What happens at maximum resistance? Can you explain, in terms of voltage compliance of the current source?

Even within the compliance range, there are detectable variations in output current as the load is varied. What causes these variations? Can you verify your explanation, by making appropriate measurements? (Hint: Two important assumptions were made in the initial explanation of the current source circuit in the Text's fig. 2.21, sec. 2.06.)

4-7 Common-emitter Amplifier

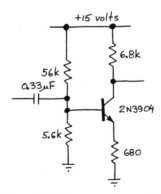

Figure L4.7: Common-emitter amplifier

Wire up the common emitter amplifier shown above. What should its voltage gain be? Check it out. Is the signal's phase inverted?

Is the collector quiescent operating point right (that is, its resting voltage)? How about the amplifier's low frequency 3dB point? What should the output impedance be? Check it by connecting a resistive load, with blocking capacitor. (The blocking cap, again, lets you test impedance at signal frequencies without messing up the biasing scheme.)

4-8 Emitter Follower Buffer (*omit this exercise if you are short of time*)

Hook an NPN emitter follower to the previous amplifier. Think carefully about coupling and bias. Use a 1k emitter resistor.

Figure L4.8: Follower buffering amplifier (details left to you)

Measure output impedance again, using a small signal. Is the overall amplifier gain affected by the addition of the emitter follower?

4-9 Transistor Switch

The circuit below differs from all the circuits you have built so far: the transistor, when on is *saturated*. In this regime you should not expect to see $I_C = \beta \cdot I_B$. Why not? The 2N4400 is a small power transistor, housed in the TO92 package, like the 2N3904, which you have been using in this lab. The 2N4400 can dissipate more power than the '3904, and when *on* shows lower $V_{CE(sat)}$ (see below).

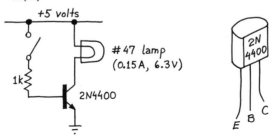

Figure L4.9: Transistor switch

Turn the base current on and off by pulling one end of the resistor out of the breadboard. What is I_B, roughly? What is the minimum required β?

Saturation or "On" voltage: $V_{CE(sat)}$

Measure the saturation voltage, $V_{CE(sat)}$, with DVM or scope. Then parallel the base resistor with 150 ohms, and note the improved $V_{CE(sat)}$. Compare your results with the results promised by the data set out below. (See Appendix G in the text for more on saturation.)

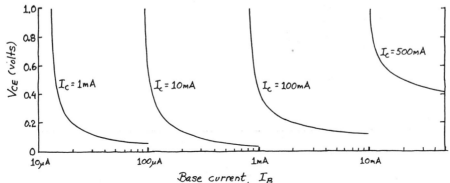

Figure L4.10: 2N4400 Saturation voltage versus I_B and I_C

Why should the designer of a switching circuit be concerned with the small decrease in $V_{CE(sat)}$ that results from generous base drive?

Class 5: Transistors II:
Corrections to the first model:
Ebers-Moll: r_e; applying this new view

Topics:

- *old:*
    - Our first transistor model:
        - Simple: $I_C = \beta \times I_B$
        - Simpler: $I_C \approx I_E$; $V_{BE} = 0.6V$

- *new:*
    - transistor is controlled by V_{BE}: Ebers-Moll view
    - applications: circuits that baffle our earlier view:
        - current mirror
        - common-emitter amp with *no R_E*
        - R_{out} of follower driven by R_{source} of moderate value
    - complications:
        - temperature effects
        - Early Effect

Our first view of transistors held that two truths were pretty much sufficient to describe what was going on—

1. $V_{BE} = 0.6$,
   and
2. $I_E \approx I_C = \beta \times I_B$

This account can take us a long way; $V_{BE} = constant = 0.6V$ is often a good enough approximation to allow understanding a schematic or, say, designing a not-bad current source.

Sometimes, however, we cannot settle for this first view. Some circuits require that we recognize that in fact V_{BE} varies with I_C. In fact, the relation looks just like the diode curve already familiar to you (it differs only in slope):

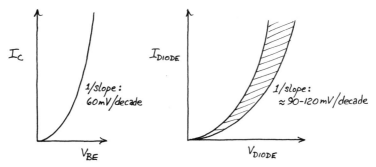

Figure N5.1: V_{BE} *does* vary with I_C, after all. In fact, I_C vs. V_{BE} looks a lot like a diode's curve

You knew, anyway, that the transistor had limited gain, so you would guess that a circuit like the one just below has a gain limited by the properties of the transistor—the I_C vs V_{BE} curve shown above.

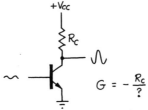

Figure N5.2: Infinite-gain amplifier?

A naive application of the rule $G = -R_C/R_E$ would imply infinite gain; but you know better. Wiggling the input = wiggling V_{BE}, and that produces a limited variation of I_C, which in turn produces a limited variation in V_{out}.

"Intrinsic emitter resistance:" r_e

You can describe this effect handily by drawing it as a little resistor in series with the emitter (for a derivation of r_e, see end of these notes):

Text sec. 2.10
"Rule of Thumb No. 2"

Figure N5.3: "Little r_e"—the intrinsic emitter resistance

This "resistance" we call "little r_e;" please note that it is **not** a resistor planted in the transistor; it only models the limited gain of the device.

Another way to say this is to say that r_e is the slope of the gain curve—but with the curve plotted on its side, with V_{BE} vertical (just so it will have the conventional units of resistance):

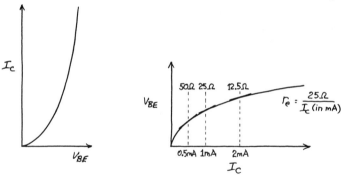

Figure N5.4: r_e is slope of transistor gain curve, if you turn the plot on its side

Evidently the value of r_e varies with I_C. Specifically, here's our rule of thumb:

$$r_e = 25 \text{ ohms}/(I_C \text{ (in mA)})$$

Watch out for the denominator: you must write "1 mA" as 1, not 1×10^{-3}. If you forget this, your answers will be off by even more than what we tolerate in this course!

r_e, "little r e," expresses the Ebers-Moll equation in a convenient form, and you will use this simplifying model more often than you will use the equation.

But we should notice what the Ebers-Moll equation says, before we go on:

Text sec. 2.10

Ebers-Moll

$$I_C = I_S(e^{V_{BE}/(kT/q)} - 1)$$

(grows fast with temperature; more on this later)　　　(negligible)

Ignoring the "-1" term, we can say simply that I_C grows exponentially with V_{BE}.

In addition, we might as well plug in the room-temperature value for that complicated expression "kT/q:" 25 mV. Then Ebers-Moll doesn't look so bad:

> *Ebers-Moll:* (slightly simplified)
> $$I_C \approx I_S \, e^{V_{BE}/25\text{mV}}$$

This equation is most often useful to reveal the relative values of I_C as V_{BE} changes. What happens, for example, if you increase V_{BE} by 18mV? Let's call the old I_C "I_{C_1}," the new one "I_{C_2}:

$$I_{C_2}/I_{C_1} = \{I_S \, e^{V_{BE_2}/25\text{mV}}\} / \{I_S \, e^{V_{BE_1}/25\text{mV}}\}$$

But that is just

$$e^{(18\text{mV}/25\text{mV})} \approx 2$$

This is a number perhaps worth remembering: 18mV ΔV_{BE} for a doubling of I_C; also sometimes handy: 60mV ΔV_{BE} per *decade* (that is, *10X*) change in I_C.

Applying the Ebers-Moll view to circuits

Here, for example, is a circuit that makes no sense without the help of this view of transistors:

* current mirror

Text sec. 2.14
Lab 5-3

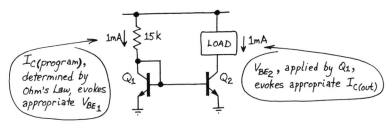

Figure N5.5: Current mirror: Ebers-Moll view *required*

Why is a mirror useful? It makes it easy to link currents in a circuit, matching one to another. It also shows very wide *output compliance*. But for our present purposes it is most useful as a device to demonstrate the power of the Ebers-Moll view.

It's easy to make $I_{C\text{-program}} = I_{C\text{-out}}$, and only a little harder to scale $I_{C\text{-out}}$ relative to $I_{C\text{-program}}$.

You will find in Lab 5 that the mirror departs rather far from this ideal model. *Early effect* and *temperature* effects both disturb it. We will learn later how to fight these problems; for now, let's leave the mirror in its simplest form, as shown above.

Other consequences of this amended view of transistor operation]:

It brings some circuits down to earth:

* a ceiling on gain (a recapitulation): no infinite-gain amps

Text sec. 2.12

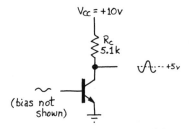

Figure N5.6: What gain? Not infinite

* a floor under Z_{out}

Text sec. 2.11

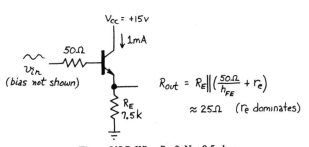

Figure N5.7: What R_{out}? Not 0.5 ohms

Let's look closely at the problem of the *grounded-emitter* amplifier. You knew, anyway, that its gain was not infinite. Now, with the help of r_e, we can evaluate the gain.

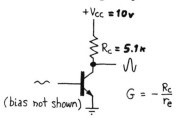

Figure N5.8: Grounded emitter amp again

We can use our familiar rule to evaluate gain here, simply drawing in r_e (at least in our heads). To evaluate r_e we need a value of I_C. There is no single right answer; the best we can do is specify I_C at the quiescent point—where V_{out} is centered.

Roughly, then,
$$G = -5.1\mathrm{k}\Omega/25\Omega \approx -200$$

That's high. But evidently the gain is *not constant*, since I_C must vary as v_{out} moves (indeed, it is variation in I_C that *causes* v_{out} to move.)

Here's the funny "barn-roof" distortion you see (this name is not standard, incidentally) if you feed this circuit a small triangle:

Text sec. 2.12; compare fig. 2.36; Lab 5-2

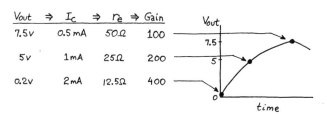

Figure N5.9: Gain of grounded-emitter amp varies during output swing (call it "barn-roof" distortion): Gain evaluated at 3 points in output swing

The plots below show how gain varies (continuously) during the output swing:

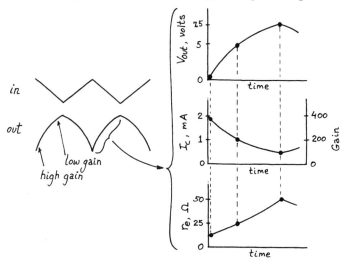

Figure N5.10: During swing of V_{out}, I_C and thus r_e and gain *vary*

This is bad distortion: −50% to +100%! What is to be done?

Remedy: emitter resistor

One cannot eliminate this variation in r_e (—can one?), but one can make its effects negligible. Just add a constant resistance much larger than the varying r_e. That will hold the denominator of the gain equation nearly constant.

text sec. 2.12

With emitter resistor added, gain variation shrinks sharply:

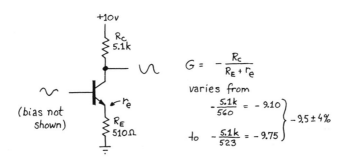

Figure N5.11: Emitter resistor cuts gain, but also cuts gain *variation*

r_e still varies as widely as before; but its variation is buried by the big constant in the denominator.

Circuit gain now varies only from a low of –9.1 to a high of –9.75: a –4%, +3% variation about the midpoint gain of 9.5.

Punchline: emitter resistor greatly reduces error (variation in gain, and consequent distortion). This we get at the price of giving up some gain. (This is one of many instances of Electronic Justice: here, *those greedy for gain will be punished: their output waveforms will be rendered grotesque.*)

We will see shortly that the emitter resistor helps solve other problems as well: the problem of temperature instability, and even distortion caused by *Early effect*. How can a humble resistor do so much? It can because in the latter two cases the resistor is applying *negative feedback*, a design remedy of almost magical power. In the next set of notes, Transistors III, we will look more closely at how the emitter resistor does its job. And in Chapter 4 we will see negative feedback blossom from marginal remedy to central technique. Negative feedback is lovely to watch. Many such treats lie ahead.

If you are in the mood to find negative feedback at work in today's lab, you can find it in the simple-looking circuit fragment: the *program* side of the current mirror:

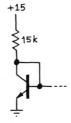

Figure N5.12: Subtle negative feedback: programming side of the current mirror

See if you can explain to yourself how this circuit works. Hint: nearly all of the current flows not in the base path, but from collector to emitter.

Complications: **Temperature effects; Early Effect**

Temperature Effects

Semiconductor junctions respond so vigorously to temperature changes that they often are used as temperature sensors. If you hold V_{BE} fixed, for example, yozu can watch I_C, which varies exponentially with temperature.

But in any circuit not designed to measure temperature, the response of a transistor to temperature is a nuisance. Most of the time, the simple trick of adding an emitter resistor will let you forget about temperature effects. We will see below how this remedy works.

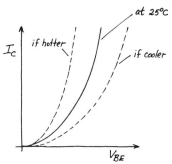

Figure N5.13: Transconductance of bipolar transistor varies rapidly with temperature

Preliminary warning. Do not look for a description of this dependence in the Ebers-Moll equation. That equation (mis-read) will point you in exactly the wrong direction: increasing T should shrink the exponent:

$$I_C = I_S(e^{V_{BE}/(kT/q)} - 1)$$

Don't be fooled: Ebers-Moll equation seems to say I_C falls with temperature. Not so.
But, to the contrary, increasing T increases I_C, and *fast*. Solution to the riddle: I_S grows *very fast* with temperature, overwhelming the effect of the shrinking exponent.

Here are two formulations for the way a transistor responds to temperature:

Text sec. 2.10

Temperature Effects: *two equivalent formulations*

- I_C grows at about 9%/°C, if you hold V_{BE} **constant**.
- V_{BE} falls at 2mV/°C, if you hold I_C **constant** (This is the text's formulation.)

The first formulation is the easier to grasp intuitively: heat the device and it gets more vigorous, passes more current. The second formulation often makes your calculation easier. If you use the second formulation, just make sure that you don't get the feeling that the way to calm your circuits is to build small fires under them!

Example: current mirror

The current mirror misbehaves if the temperatures of the two transistors become unequal. This you will see in the lab. If you heat one of the two transistors, the current out rises; if you heat the other, the current out falls. Thinking through why this happens will help you get used to the two formulations of temperature effects, because this is the rare circuit that invites one to use *both* formulations, since it illustrates both cases: I_C fixed, and V_{BE} fixed. Let's try it.

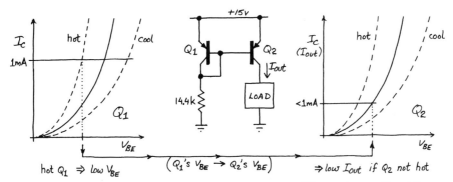

Figure N5.14: Current mirror again: consequence of temperature differences between Q_1 and Q_2

In Q_1, current is held essentially constant by the large voltage across the 15k resistor. So, if we heat Q_1 what happens? The second formulation says V_{BE} falls. In effect, the curve shifts as shown above: at 1 mA, Q_1 now finds a smaller V_{BE} suffices.

Meanwhile, Q_2 feels someone reducing its V_{BE}. We assume Q_2's temperature is unchanged. So what does Q_2 do? It delivers the lower I_C that its curve says is appropriate to the lower V_{BE}.

You can easily talk your way through a similar argument to explain why heating Q_2 while leaving Q_1 at room temperature causes I_{out} to increase.

Remedies: making circuits stable despite temperature changes

1. Compensation

The mirror becomes wonderfully immune to temperature variation if the two transistors stay at the same temperature. One can arrange that by building them on one piece of silicon. The argument that this works is simple: both transistors 'look at the same curve,' if they are heated together. We don't care what those curves look like, only that they match.

Text sec. 2.14
Lab 5-3

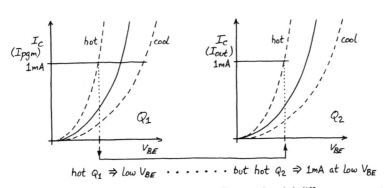

Figure N5.15: Example of temperature *compensation*: Current mirror is indifferent to temperature, if transistor temperatures track

Here is another example, from the text. This circuit *compensates* for changes in one direction by planting a circuit element that tends to change at the same rate in the opposite direction.

Text sec. 2.12

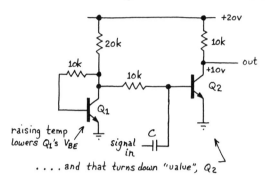

Figure N5.16: A second example of temperature compensation

This circuit (from Text's fig. 2.39) lets the fall of Q_1's V_{BE} squeeze down Q_2 as both get hotter. (The 10k resistor on the base of Q_2 makes the biasing circuit not too stiff: the signal source (presumed to be of impedance « 10k) can have its way, as usual.)

2. Feedback: emitter resistor

The remedy described here is simpler, and more widely used. It is also subtler.

Text sec. 2.12

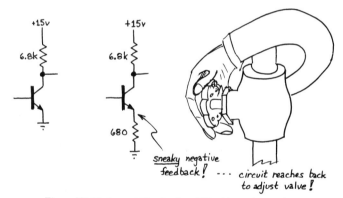

Figure N5.17: An unstable circuit stabilized by emitter resistor

The left-hand circuit is so unstable that it is useless. An 8°C rise in temperature saturates the transistor.

Text sec. 2.12, ex. 2.9

Why does the right-hand circuit work better? How does the emitter resistor help, as I_C grows? Here is feedback at work: the circuit senses trouble as it begins:

- I_C begins to grow in response to increased temperature;
- V_E rises, as a result of increased I_C (this is just Ohm's Law at work);
- But this rise of V_E diminishes V_{BE}, since V_B is fixed. Squeezing V_{BE} tends to close the transistor "valve." Thus the circuit slows itself down.

The remedy is not quite perfect: some growth of I_C with temperature is necessary in order to generate the error signal. But the emitter resistor prevents wide movement of the quiescent point.

2a. Temperature stability *and* high gain

Text sec. 2.13;
Lab 5-2

If you want stability *and* high gain, you can have that combination, by including R_E for DC biasing, but making it disappear at AC. You make it "disappear" by *bypassing* or paralleling R_E with a capacitor, thus:

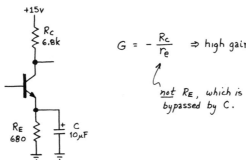

Figure N5.18: Bypassed-emitter resistor: high gain plus temperature stability

This circuit still distorts; note that R_E here remedies only the temperature instability problem.

Here is feedback in a more obvious form, but used to similar effect: this is from Lab 5:

Text sec. 2.12
Lab 5-5

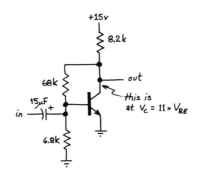

Figure N5.19: DC feedback protects against temperature effects

This looks a lot like operational amplifier circuits that you will see in Chapter 4 (called "Chapter 3" in '81 edition). When you get there you may want to look back at this circuit. Then you will appreciate what now is obscure: the feedback affects DC levels, but *not* circuit gain—not what happens to the signal, in other words. It does not affect the signal because the low output impedance of the function generator overwhelms the relatively feeble feedback signal (that is, *high-impedance* feedback). It is meant to work that way, to keep things simple for us at this stage.

Early Effect

Here is a picture of this effect, which just describes the transistor's departure from the ideal view that it is a current source:

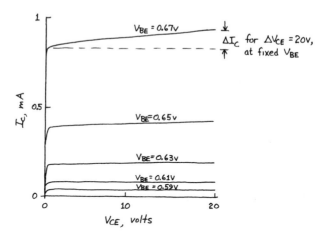

Figure N5.20: Early Effect: graphically, the curve's departure from horizontal

A transistor is a pretty good current source, but it also acts a little like a resistor: I_C grows as the voltage across the device (V_{CE}) grows.

Like the temperature-dependence in the Text's formulation, Early Effect looks like an effect on V_{be}, because I_C is assumed fixed:

Text sec. 2.11

> **Early Effect:**
> $$\Delta V_{BE} = -\alpha \Delta V_{CE}$$
> where $\alpha \approx 0.0001$ (or 10^{-4})
> (I_C is assumed constant.)

Despite the assumption used in this formulation, that I_C is fixed, usually you will see Early Effect causing *variation in I_C*, while *V_{BE} is fixed*. The current mirror provides a good example of the problem, and of ways to beat it.

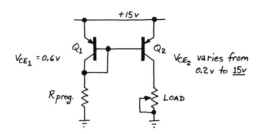

Figure N5.21: Current mirror: Early Effect predicts I_{out} will *not* match $I_{program}$

The flaw in the simple mirror is the likely difference between the voltages across the two transistors: to the extent that the V_{CE}'s differ, Early says the currents will differ.

Quantitatively: try an example: the mirror, powered from +15v. What happens if R_{load} is small, so that most of the 15-volt supply appears across V_{CE} of Q_2? Early Effect predicts I_{out} a good deal larger than $I_{program}$.

How much larger? The quantitative argument is a bit convoluted, as it can be for temperature effects. The difficulty is the assumption that I_C remains constant, while we know that in this case the result is just the contrary: I_{out} is going to *vary* as V_{CE} varies. You need to do a sort of thought-experiment contrary to fact: assume V_{BE} on Q_2 *does* change, then later recognize that it *can't*; see what change in I_C must have occurred to have held V_{BE} constant.

Here's the argument.

- Assume I_C constant;
- The extreme difference between the V_{CE}'s on Q_1 and Q_2 would be about 15 volts (this would occur for R_{load} close to zero Ω). In that case the mismatch of V_{BE}'s predicted by Early Effect would be about 1.5 mV.
- Now we are ready to admit this is not possible: the two V_{BE}'s are equal. One can see that with a glance at the circuit diagram.
- Admitting this is not possible, we ask what difference in I_C's must occur instead. Ebers-Moll gives the answer:
$$I_C = I_S(e^{V_{BE}/(kT/q)} - 1)$$
- The ratio of the higher to lower currents is
$$e^{(V_{BE\text{-}2}/25\text{ mV})} / e^{(V_{BE\text{-}1}/25\text{ mV})}$$
- And this is the same as
$$e^{(V_{BE\text{-}2} - V_{BE\text{-}1})/25\text{ mV}}$$
- In this case that is about
$$e^{1.5\text{ mV}/25\text{mV}}$$
$$\approx e^{0.06}$$
$$\approx 1.06$$

This is smaller than the mismatch we saw when we tried this experiment: we saw a ratio of about 1.12. Probably our estimate of 0.0001 (or 10^{-4}) for the Early Effect α is low (our experiment suggests $\alpha \approx 0.0002$ (or 2×10^{-4})).

But the central point is sound: Early Effect introduces a disappointing error into the otherwise-admirable mirror. We need a remedy.

Here are two:
1. clamp V_{CE} for Q_2, so that the V_{CE} difference does not occur.
    This the clever Wilson Mirror does.

Text sec. 2.13
Lab 5-3

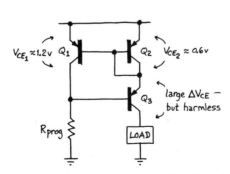

Figure N5.22: Beating early effect: one way: wilson mirror clamps V_{CE}

2. Or Add an emitter resistor that tends to fight the growth of Q_2's I_C. (This works against temperature instability, too, incidentally—in case Q_2 gets hotter than Q_1, as it tends to in the simple mirror above).

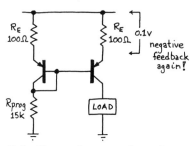

Figure N5.23: Beating Early effect: another way: emitter resistor senses and corrects errors

Using Early Effect to estimate R_{out} for a current source

Usually we settle for calling R_{out} for a transistor current source just "very high." Early Effect lets us calculate it roughly:

Consider a familiar current source: assume the base is driven by a stiff voltage source:

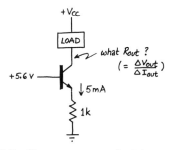

Figure N5.24: Familiar current source: what is its output impedance?

Apply a ΔV, as usual; ask what ΔI results.

- The ΔV appears as a ΔV_{CE}. Let's suppose we apply a ΔV of plus one volt.
- Early effect predicts a consequent ΔV_{BE} of minus 0.1 mV—*if I_C stays constant*. We will assume, for the moment, that I_C does stay constant.
- If V_{BE} shrinks, while V_B is fixed, then V_E must *rise* by 0.1 mV.
- That rise of V_E increases I_C by 0.1 μA.
- So, the output impedance of the current source, defined as $\Delta V / \Delta I$, is 1v / 0.1μA or 10MΩ.

This argument is about right. It does contradict itself in first assuming I_C unchanged, so we should expect it to go wrong when it predicts a *large* change in I_C.

Postscript: Deriving r_e

Deriving the expression for r_e

Text sec. 3.07
(FET chapter)

As the Text explains in Chapter 3 (sec. 3.07) it's not hard to confirm the expression for the value of r_e. The transistor's *gain* is dI_C/dV_{BE}. But if we write Ebers-Moll the way we did above, as

$$I_C \approx I_S\, e^{V_{BE}/25\text{mV}}$$

then

$$d(I_C)/d(V_{BE}) = (1/25\text{ mV})(I_S\, e^{V_{BE}/25\text{mV}}),$$

or, more simply,

$$(1/25\text{mV})(I_C).$$

And r_e, the reciprocal of gain, is

$$25\text{ mV}/I_C\text{ A},$$

which we prefer to write as

$$25\Omega/I_C\text{ (in mA)}.$$

Alternative 'derivation:' Tess o' Bipolarville

Here is an alternative argument to the same result: imagine a lovely milkmaid seated (in the summer twilight) on a stool, tugging dreamily at the emitter of a transistor whose base is fixed. She has pulled gently, until about 1 mA flows. What *delta*-current falls into her milkpail, for an additional tug?

Figure N5.25: Dreamy milkmaid discovers the value of r_e, experimentally

If the base is anchored, tugs on the emitter change V_{BE} a little; in response, I_C changes quite a lot. The squirts of ΔI_E (which we treat as equivalent to ΔI_C) reveal a relation between ΔV_{BE} and ΔI_C. The quotient

$$\Delta V_{BE}/\Delta I_C$$

is r_e, which behaves just like a resistor whose far end is fixed.

"Yes," muses the charming milkmaid, her milkpail now filled (with *charge*, in fact; but she hasn't noticed). "Just as I thought: r_e, though only a model, does behave for all the world like a little resistor. So that's why we draw it that way." And with that she rises, little suspecting what her discovery portends, and carries her milkpail off into the gathering dusk.

Reconciling the two views: $I_C = \beta \times I_B$ and $I_C \approx I_S\, e^{V_{BE}/V_T}$

In case you're troubled by the thought that our two views may not be consistent, here's a picture to reassure you:

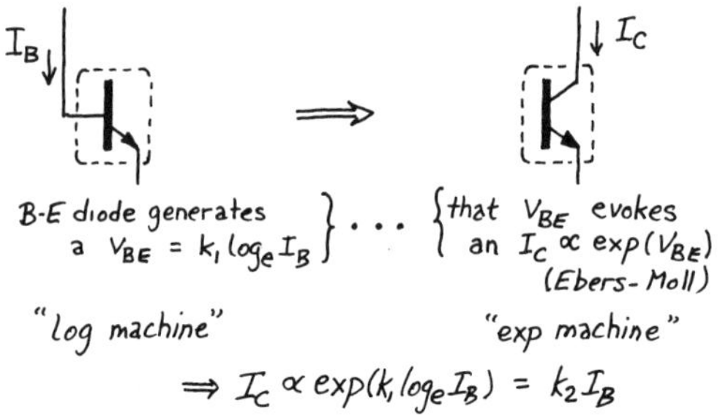

{B-E diode generates a $V_{BE} = k_1 \log_e I_B$} ... {that V_{BE} evokes an $I_C \propto \exp(V_{BE})$ (Ebers-Moll)}

"log machine" "exp machine"

$$\Rightarrow I_C \propto \exp(k_1 \log_e I_B) = k_2 I_B$$

Figure N5.26: Beta... and Ebers-Moll descriptions of transistor gain *reconciled*

This picture is wrong to the extent that β is not constant: the left-hand block—the 'log machine'—is not a very good machine; the right-hand block—the Ebers-Moll exponentiating machine—is good. If this diagram does not make you feel better, forget it; you don't *need* to reconcile the two views if you don't want to.

Ch. 2: Worked Example: Common Emitter Amplifier

Presliminary: A Design Procedure

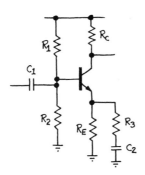

Figure X5.1: Skeleton Circuit

Setting *DC* conditions

1. Choose R_C to center V_{out}, given $I_{C \text{ (quiescent)}}$
2. Choose R_E to put V_E somewhere around 1 volt, for temperature stability
3. Let R_{Th} for the bias divider be about 1/10 (R_{in} transistor, which is $\beta \times R_E$). As for the follower recently designed, the *AC* path to ground is to be ignored: the path through R_3 is closed to DC, so invisible to the bias divider.
4. Choose R_1, R_2 to put V_{base} at (V_E + 0.6V). R_{Th} is roughly R_1, since the divider is so far unbalanced.

Determining AC performance: Gain (what happens to signal)

1. Choose R_3 (if any) for gain at quiescent point
2. Choose C_2 for f_{3dB}: the relevant "R" is $R_3 + r_e$
3. Choose C_1 as usual; relevant "R" is circuit's Z_{in}, as usual: the circuit's *AC* input impedance, as for the follower: we look through capacitor C_2, and see R_3 as a path to ground.

    In choosing C_1 we need to be generous, since two high-pass filters are at work: those using C_1 and C_2. So, if we made the mistake of putting the f_{3dB} for each filter precisely at our target f_{3dB} for the *circuit*, we would be disappointed: we would find the *circuit's* response down *6dB*.

Bypassed-Emitter Amplifier

> ***Problem: common emitter amp***
>
> Design a common-emitter amp to the following specs:
>
> - $V+ = 20V$
> - $f_{3dB} = 100$ Hz (approx.)
> - gain = -100 at quiescent point.
> - I_C quiescent = 0.5 mA
>
> ***Questions:***
>
> - What is the amp's gain—
>
>   - when the output swings to about $+15v$?
>   - when the output swings to about $+5v$?
>
> - What is the amp's Z_{out}?

Solution

This time we will set out the solution simply by drawing a circuit with explanatory 'balloons.'

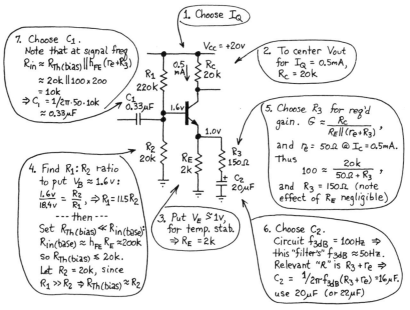

Figure X5.2: Common emitter amplifier: solution to stated problem

Questions:

What is the amp's gain—
  a) —when the output swings to about + 15v?

Here I_C must be at half of its quiescent value (since the drop across R_C is half the quiescent drop). So r_e has doubled, to 100Ω. That takes the denominator up to 250Ω, from 200Ω: up 25%; that should drop gain 20%:

$$\text{gain is down to 80}$$

  b) —when the output swings to about + 5v?

ere I_C is up 50%, so r_e is down to 2/3 of its quiescent value: down from 50Ω to 33Ω, dropping the denominator from 200Ω to 183Ω (or 'about 180'): down about 8%, taking gain up about 10%:

$$\text{gain is up to 110}$$

  c) What is the amp's Z_{out}?

Z_{out} is the impedance of the two paths one sees when 'looking' back at the circuit from the output: the collector resistor parallel whatever the collector itself looks like. But the collector looks like a very large resistor, since the transistor *current* is determined. The transistor is a current source: it allows large ΔV for very small ΔI; in other words, its impedance is very large.

So, Z_{out} (which is not frequency-dependent, and could as well be called R_{out}) is just R_C:

$$R_{out} = R_C = 20\text{k}$$

Lab 5: Transistors II

> **Reading:** Chapter 2.09 – 2.14
> **Problems:** Problems in text.
> Additional Exercise 7.
> Bad Circuits C, F, H, I

5-1. Dynamic Curve Tracer

We will use the setup below to trick the scope into showing a display of *I* versus *V*, first for a diode, then for a transistor.

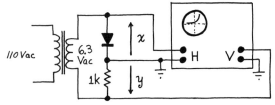

Figure L5.1: Dynamic curve tracer

Wire up the *VI* curve tracer shown above. Explain how it works. Could one use the function generator instead of the 6.3V transformer?

Now let's try it out.

A. *Diode Curve*

Try the curve tracer on a 1N914 diode. Be sure you discover where zero voltage and current are on the screen, by alternately zeroing the input from V and H. Figure out the calibration (mA/div, V/div), and then make a reasonably accurate plot on graph paper. Compare it with the graph you made in lab 1-3. Stare at the diode display, to get a feeling for the Ebers-Moll equation. Then reverse the diode polarity.

Replace the 1N914 (ordinary signal diode) with a 1N749 or 1N751 or equivalent (4.3V or 5.1V zener diode), and plot its characteristic also.

Before leaving this exercise, note the *slope* of the diode curve (forward). $\Delta V_{diode}/\Delta I_{diode}$ should be about 100 mV/decade.

B. Transistor Curve

In place of the diode, install a 2N3904 transistor as shown below. Use the curve tracer as before. You now can watch I_E versus V_{BE}. How close is this to the characteristic of greater interest to us, I_C versus V_{BE}?

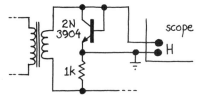

Figure L5.2: Curve tracer applied to *transistor*

Do you see the breakdown voltage of the BE junction?

It is not good for the transistor to have its junction continually broken down, so install the protection diode shown below:

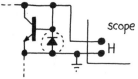

Figure L5.3: Protection diode Added

What is the slope of the transistor's curve? ($\Delta V_{BE}/\Delta I_C$ should be about 60 mV/decade.)

5-2. Grounded Emitter Amplifier

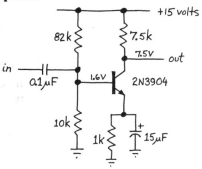

Figure L5.4: Grounded emitter amplifier

Wire up the circuit in the figure above (taken from the text, figure 2.35). First, check the quiescent collector voltage. Then drive it with a small *triangle* wave at 10kHz, at an amplitude that almost produces clipping (you'll need to use plenty of attenuation — 40dB or more — in the function generator). Does the output waveform look like the figure below (text figure 2.34)? Explain to yourself exactly why this "barn-roof" distortion occurs.

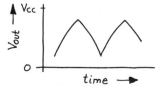

Figure L5.5: Large-swing output of grounded emitter amplifier when driven by a triangle wave

Now remove the 15μF capacitor, increase the drive amplitude (the gain is greatly reduced), and observe a full-swing triangle output without noticeable distortion. Measure the voltage gain — does it agree with your prediction?

Restore the 15µF capacitor, and reduce the function generator output to the minimum possible. Predict the voltage gain at the quiescent point, using r_e. Measure it; does it agree?

5-3. Current Mirror
A. Simple Mirror
1. Discrete Transistors

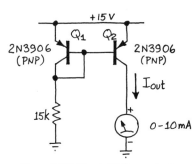

Figure L5.6: Classic PNP current mirror

Build the classic current mirror shown above (same as text figure 2.44). How closely does the output current equal the programming current? (You can calculate the latter without measuring anything.)

The match should be pretty *bad*, because of a combination of temperature and Early effects. Both tend to make I_{out} larger than $I_{program}$. (Do you see why Q_2 predictably runs hotter than Q_1?)

Now try squeezing one of the transistors with your fingers, then the other, to see how much I_{out} is affected by differences in temperature between Q_1 and Q_2. You should find that you can nudge I_{out} up or down by warming one transistor or the other. Convince yourself that the pattern of response that you see makes sense.

2. Matched Transistors

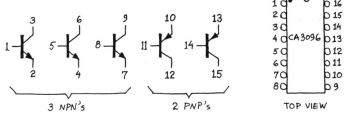

Figure L5.7: CA3096 transistor array: pinout

Now build the same circuit but using a monolithic transistor array, the CA3096. The figure above shows the "pinout" of this integrated circuit. Measure I_{out}, using the matched PNP pair of the CA3096 in the original mirror circuit above. How closely does the output current match the programming current?

This version of the mirror should perform better than the earlier circuit for two reasons. First, the two transistors will run at the same temperature because of their proximity. Second, at a given current the two transistors' V_{BE}'s are pretty well matched: the best version of the CA3096 (the CA3096A) has V_{BE}'s matched to 0.15mV (typical), and 5mV (maximum). (What ratios of collector current do these V_{BE}'s correspond to?)

But you are likely to see I_{out} greater than $I_{program}$ in this improved circuit. Why? Add a 10k potentiometer in series with the current meter, and vary its setting: does I_{out} vary? Why?

B. Wilson Mirror

Early Effect in the preceding circuit predicts considerable variation in I_{out} (we saw about 25% variation when we tried it).

The Wilson Mirror, below, beats that effect. Try it, watching the variation in I_{out} as you vary R_{load}. Watch V_{out} at the same time. How constant is current now? What property of this circuit accounts for the improved performance? (Wilson mirrors are available as 3-terminal integrated circuits, in a T0-92 package.)

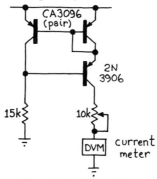

Figure L5.8: Wilson mirror

How good a current source is the current mirror, compared with the current source you built in the last lab? We have made you labor a bit to produce a respectable mirror. *Query*: If mirrors are so ready to misbehave, why are they useful circuits? (In what ways do they perform *better* than the current source you built last time?)

5-4. Ebers-Moll Equation

Wire up the circuit you used in Lab 4 to measure h_{FE} (see the figure below). Again, use the substitution boxes for **R**, to generate collector currents going from a few microamps to a few milliamps. Plot the logarithmic increase of V_{BE} with I_C, and confirm the "60mV/decade" law. (Don't work too hard: perhaps 5 points should reveal the curve.)

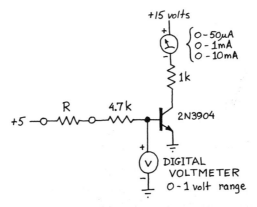

Figure L5.9: Circuit for measuring I_C vs. V_{BE}

5-5. Biasing: good and bad

Generally we use an emitter resistor to stabilize common-emitter amplifiers against temperature effects. The next two circuits show, first, another way to achieve stability, and second, the problem of instability that both schemes try to solve.

a. Biasing With DC Feedback

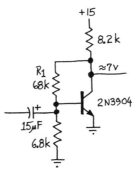

Figure L5.10: Grounded-emitter amplifier stabilized by DC feedback

Wire up the grounded emitter amplifier with dc feedback shown above. This is the clearest application so far of *feedback*—a concept and technique that you will see again and again when you begin to work with *operational amplifiers* (introduced in Chapter 4).

Note that this is *dc* feedback: that is, it stabilizes the quiescent point, but does not affect circuit gain. (In this respect it differs from most of the feedback examples you will meet in Chapter 4.) *Why* is this statement correct? (*Hint*: the feedback from collector to base forms a divider with the output resistance of the signal source.)

This arrangement provides some bias stability. The nominal collector quiescent point is $11 \cdot V_{BE}$, or roughly 7 volts. (Do you see why?) If the quiescent collector voltage were more than that, for instance, the base divider would drive the transistor into heavy conduction, restoring the proper operating point; similarly, the proper operating point would be restored if the quiescent point were to drop.

Check to see if the quiescent collector voltage is approximately correct. Since V_{BE} depends on temperature, you should be able to shift the collector voltage a small amount by warming the transistor between your fingers: which way should it move? In practice this slight temperature sensitivity is not a major drawback; biasing a grounded-emitter stage without such a feedback scheme is considerably more uncertain, as we will now see.

b. An example of <u>bad</u> biasing

In order to appreciate stable biasing it helps to see a crummy circuit in action. Here is one.

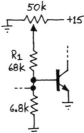

Figure L5.11: Poor biasing scheme for the grounded emitter amplifier

The circuit shown above assumes a particular value of β, violating one of the design rules that the text urges on us.

To build this circuit, you need only disconnect R_1 from the collector (fig. L5.10), and connect R_1 instead to the pot, as shown above. After you have adjust the pot to make the circuit work satisfactorily (symmetrical swing without clipping), replace the 2N3904 (typical β of 100) with a 2N5962 (typical β of 1000), and note the collector saturation ($V_{C\text{-quiescent}}$ too low). Such a bias scheme is very h_{FE}-dependent, and a poor idea.

Leave the 2N5962 in the circuit, and reconnect the 68k resistor (R_1) to the collector (the original circuit). Verify correct biasing, even with this large change in h_{FE}.

To get a preview of additional pleasures in using *feedback*, try modifying the circuit so as to allow *signal* feedback as well as the *DC* feedback demonstrated earlier. We will talk about this in detail in Chapter 4; but to see that something interesting is happening, put a 6.8k resistor in series with the input signal, and note the good linearity at large swing (use a triangle wave, again). Note also that once again we have traded away one good thing (large gain) to get another (linearity, or constant gain).

5-6. Push-Pull

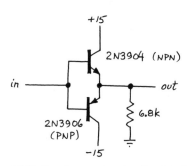

Figure L5.12: Complementary push-pull emitter follower

Explore "crossover distortion" (a notion that may be familiar to audio enthusiasts) by building the push-pull output stage shown above.

Drive the circuit with sine waves of at least a few volts' amplitude, in the neighborhood of 1kHz. Be sure the *offset* control of the function generator is set to zero. Look closely at the output. (If things behave very strangely, you may have a "parasitic oscillation". It can be tamed by putting a 470 ohm resistor in series with the common base lead. This is a trick we used earlier to stabilize the follower: lab exercise 4-4; if necessary, add also a 100pF capacitor from circuit output to ground.)

Try running the amplitude up and down. Play with the dc offset control, if your function generator has one.

Class 6: Transistors III
Differential Amp; Miller Effect

Topics:

- one last important circuit:
    - difference amplifier
- finer points:
    - bootstrap: to boost input impedance—a trick you will see again
    - Miller effect and its remedies: *cascode* amp

Differential Amplifier

Text sec. 2.18
Lab 6-1

The differential amp is the last standard transistor circuit we will ask you to consider. It is especially important to us because it lets us understand the *operational amplifiers* that you soon will meet. These wonderful devices are in fact just very good differential amps, cleverly applied.

Why a differential amp?

A differential amplifier has an internal symmetry that allows it to cancel errors shared by its two sides, whatever the origin of those errors. Sometimes one takes advantage of that symmetry to cancel the effects of errors that arise within the amplifier itself: temperature effects, for example, which become harmless if they affect both sides of the amplifier equally. In other settings the shared error to be canceled is noise picked up by both of the amplifier's two inputs. Used this way, the amplifier picks out a signal that is mixed with noise of this particular sort: so-called "common-mode" noise.

You built a circuit, back in Lab 2, that did something similar: passed a signal and attenuated noise. But that *RC* filter method works only if the noise and signal differ quite widely in *frequency*. The differential amp requires no such difference in frequency. It does require that the noise must be common to the two inputs, and that the *signal*, in contrast, must appear as a difference between the waveforms on the two lines. Such noise turns out to be rather common, and the differential amp faced with such noise can "reject" it (refuse to amplify it), while amplifying the signal: it can throw out the bad, keep the good.

Here is an example of a problem that might call for use of a differential amplifier:

An application: brain wave detector

One can detect brain activity with skin contacts; the activity appears as small (microvolt range) voltage signals. The output impedance of these sources is high.

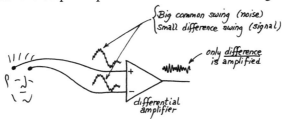

Figure N6.1: An application for a differential amp: brain wave detection

The feebleness of the signals makes their detection difficult. It is not hard to make a high-gain amplifier that can make the signals substantial. But the catch is that not only the signals but also *noise* will be amplified, if we are not careful. We can try to shield the circuit; that helps somewhat. But if the principal source of noise is something that affects both lines equally, we can use a differential amplifier instead—or as well; such a circuit ignores such "common mode" noise.

60 Hz line noise will be coupled into both lines, and is likely to be much larger than the microvolt signal levels. A good differential amp can attenuate this noise by a factor of perhaps 1000 while amplifying the signal by a like amount (that would amount to a "common-mode rejection ratio"—a preference for difference signals—of 10^6: 120 dB.

A Differential Amp Circuit

Here is the lab's differential amp:

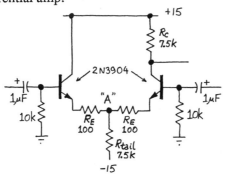

Figure N6.2: Differential amp

The circuit is not hard to analyze, if you use a trick the Text suggests: consider only pure cases—pure *common* signal, then pure *difference* signal.

Let's consider the important properties of this circuit:

Quiescent point:

V_{out}: Before you can predict V_{out} you need to determine currents.

Currents: If the bases are tied to ground, as shown, the base voltages are close to ground, and it follows that point "A" is not far from ground (close to –1 v.). From this observation you can estimate I_{tail} (in the lower 7.5k resistor): it is about 14V/7.5k ≈ 2 mA.

Since the circuit's inputs are at the same voltage, symmetry[1] requires that the 2mA *tail* current split evenly between the two transistors. So I_C in left and right transistors is about 1 mA each. From here $V_{out\ quiescent}$ is easy: centered as usual—but note that it is centered *not* between the supplies (that center would be 0V). Instead, it is centered *in the range through which it can swing*. That is always the deeper goal.

Differential Gain:

Assume a pure *difference* signal: a wiggle up on one input, a wiggle down of the same size on the other input. It follows, you will be able to convince yourself after a few minutes' reflection, that the voltage at "A" does not move.

That observation lets you treat the right-hand side of the amp as a familiar circuit: a common-emitter amp:

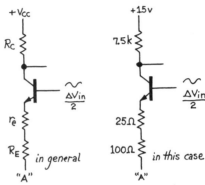

Figure N6.3: Differential gain: just a common emitter amp again

The gain might seem to be

$$G = -R_c / (r_e + R_E).$$

That's almost correct. You need to tack a factor of two into the denominator, just to reflect the way we stated the problem: the *delta v* applied at the input to this "common emitter" amp is only *half* the difference signal we applied at the outset. You also might as well throw out the minus sign, since we have not defined what we might mean by positive or negative difference between the inputs.

So, the expression for differential gain includes that factor of two in the denominator:

$$G_{diff} = R_C / 2(r_e + R_E)$$

Common Mode Gain:

Assume a pure *common* signal: tie the two inputs together and wiggle them. Now A is not fixed. Therefore, *this* common-emitter amp has much lower gain, because R_{tail} appears in the denominator of the gain equation: R_{tail} plus r_e plus R_E.

Again that is *almost* the whole story; another odd factor of two appears, however, to reflect the fact that a twin common-emitter amp—the other side of the differential amp—is squirting a current of the same size into R_{tail}. The result is that the voltage at A jumps twice as far as one might otherwise expect: one can say this another way by calling the effective R_{tail} "2 R_{tail}."

So, the Common Mode Gain is:

$$G_{CM} = -R_C / (r_e + R_E + 2(R_{tail}))$$

---

1. "Symmetry?," you may want to protest. "There's a collector resistor on one side, and not on the other." True. But try to explain to yourself why that does not matter.

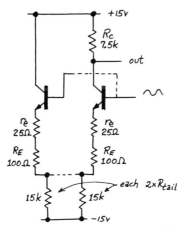

Figure N6.4: Common-mode gain: redrawing the differential amp

The bigger R_{tail}, evidently, the better. A current source in the tail, therefore, provides best common-mode rejection. A respectable differential amplifier normally includes a current source in the tail.

Miller Effect

text sec. 2.19;
lab 6-3

This is a high-frequency effect—a gain roll-off—and you will not often see it in our labs. It presents a real challenge (and headache) to anyone who must design for high frequencies. Manufacturers do what they can, by reducing the capacitance between base and collector; the circuit designer then must use his head so as to eliminate the exaggeration of C_{CB} that Miller describes.

> ### Miller Effect
> In an inverting amplifier, the small capacitance between input and output (base and collector, for a common-emitter amplifier) is effectively enlarged by circuit gain, so that it behaves like a capacitance (1 + Gain) times as large, going to ground.
>
>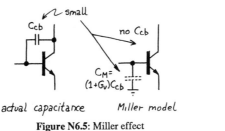
>
> Figure N6.5: Miller effect

Miller noticed that the small C_{CB} (a few picofarads) acts like a capacitance (1 + Gain) times as big as C_{CB}, connected between the base and ground. Since that cap to ground and R_{source} form a low-pass filter, high frequencies lose, and they begin to lose at an f_{3dB} lots lower ("1 + Gain" times lower) than f_{3dB} without Miller Effect.

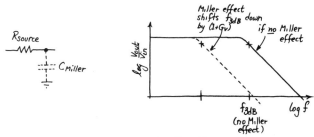

Figure N6.6: Miller effect: "low-pass"

Evidently, R_{source} is important. If you could keep R_{source} zero, you would see *no* Miller Effect. If you can keep R_{source} very small, your circuit may still pass the frequencies it must.

If you don't want to lower R_{source} or cannot do so sufficiently, then you can use cleverness to arrange things so that base and collector of any one transistor do not head in opposite directions at the same time. That arrangement will eliminate Miller Effect.

You will notice that both the *cascode* and *differential* amplifiers use that strategy (p. 84, and below).

Why does the fact that base and collector head in opposite directions increase the apparent size of C_{CB} (compared to what C_{CB} would look like if connected to ground)?

Recall that the size of the current flowing into a cap is proportional to dV/dt across the cap (I = C dV/dt). dV/dt across the cap between base and collector is increased by fact that the collector always perversely races away from the base: when the base wiggles, the collector jumps in the opposite direction. The change in voltage between base and collector is (1 + Gain) times the input change. Since dV/dt across the cap is enlarged, so is I_{cap}. The cap's apparent impedance thus is proportionately decreased. So in the voltage divider that is the low-pass filter, less of the signal survives.

Another way to look at what's happening (if you're not already worn out) is to consider C_{CB} as a path for feedback of a signal that opposes any change at the base. The smaller the impedance of $X_{C_{CB}}$, the more effective the output wiggles will be in killing the input signal. That's negative feedback—useful in other settings (it will do wonderful work for you in Chapter 4)—but harmful when you want lot of gain.

Source Resistance is Crucial

Whichever way you explain Miller Effect to yourself, R_{source} driving the transistor's base is important.

If you look at the effect as a result of *feedback*, then R_{source} matters because it forms half of a divider that determines what fraction of the output wiggle does get fed back: the bigger the fraction, the more effective the output wiggle will be at killing the input signal.

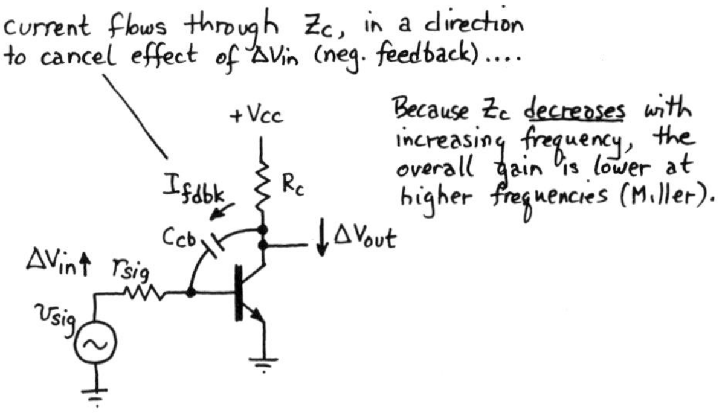

Figure N6.7: Miller effect viewed as feedback

On the other hand, if you see the circuit as simply a low-pass filter, then R_{source} evidently determines f_{3dB}.

Either way, you want R_{source} as *low* as possible. All the clever remedies for Miller Effect do nothing more than put a very low impedance at the base of the transistor whose collector is jumping.

That's true in the cascode, where the additional transistor's base is tied to a DC voltage source.

Text sec. 2.19,
fig. 2.74

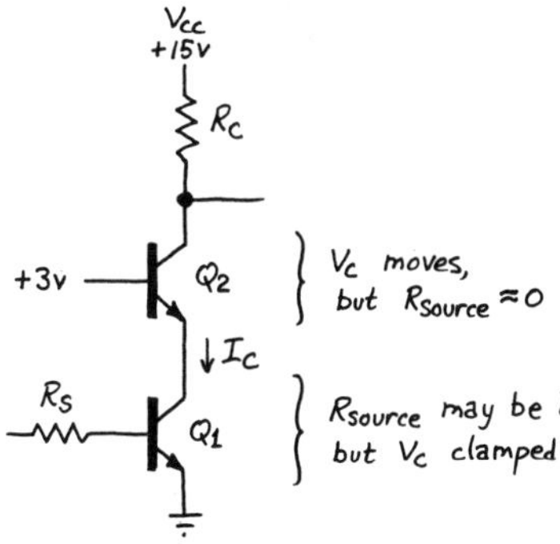

Figure N6.8: Cascode beats miller effect

It's true in the *single-ended-input* differential amp, where the base of the output transistor is tied to ground:

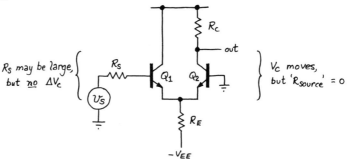

Figure N6.9: Differential amp beats miller effect (single-ended-input only)

It's *only* to the extent that the signal's impedance is larger than that of the cleverly-planted DC source that these Miller-killer circuits get you anywhere.

Ch. 2: Worked Example : Differential Amplifier

Problem: differential amplifier

Design a differential amp to the following specifications:

- Power supplies: ± 15V
- $I_{C\,quiescent} \approx 0.5$ mA
- $\text{Gain}_{Diff} = 25$
- $\text{Gain}_{Common\text{-}mode} \leq 1$

Questions:

1. When you have designed the circuit, estimate the common-mode gain, and the CMRR in dB.
2. How could you improve the CMRR a great deal?
3. What is the input impedance (at signal frequencies), for a *single-ended* input (just ground the other input)?
4. Show how to wire the inputs so as to give best high-frequency response for a *single-ended input* (that means, "not differential; ground one input"). Explain briefly how your arrangement works to beat Miller Effect.
5. If resistors were included on the collectors of *both* transistors that form the differential pair—making the circuit fully symmetrical—would the usual Miller-killer trick work (the trick we hope you just told us about in answering the previous question)? Explain your answer.
6. If the emitter resistors were omitted, what effects on circuit performance would you see? Specifically, what, if anything, would happen to—

    - output quiescent point;
    - gain (you need not calculate the gain; speak qualitatively)

Solution

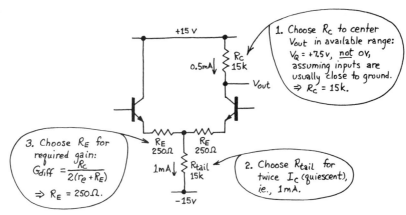

Figure X6.1: Differential amplifier: solution to design problem

Questions:

1. When you have designed the circuit, estimate the common-mode gain, and the CMRR in dB.

$$\text{CMRR} \equiv G_{diff} / (G_{CM})$$
$$G_{CM} = -R_C / (r_e + R_E + 2R_{Tail}) \approx -15k / 2(15k) \approx -0.5$$
$$\text{CMRR} = 25/0.5 = 50$$
$$\text{dB} \equiv 20 \log_{10}(A_2/A_1) \,;\, \text{dB} = 20 \log_{10} 50 = 20(1 + \log_{10} 5) = 20(1 + 0.7) \approx 34\text{dB}$$

2. How could you improve the CMRR a great deal?

> Put a current source in the tail, in place of R_{Tail}. This should push the G_{CM} down to a few parts in a thousand, boosting CMRR proportionately.

3. What is the input impedance (at signal frequencies), for a *single-ended* input (just ground the other input)?

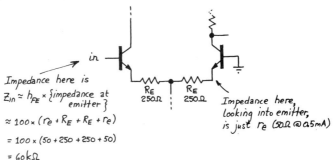

Impedance here is
$Z_{in} \approx h_{FE} \times \{\text{impedance at emitter}\}$
$\approx 100 \times (r_e + R_E + R_E + r_e)$
$= 100 \times (50 + 250 + 250 + 50)$
$= 60k\Omega$

Figure X6.2: Input impedance: single-ended input

4. Show how to wire the inputs so as to give best high-frequency response for a *single-ended input* (that means, "not differential; ground one input"). Explain briefly how your arrangement works to beat Miller Effect.

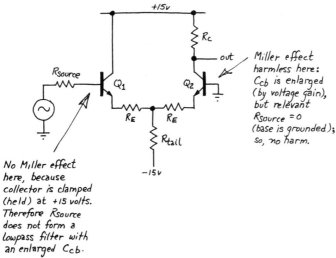

Figure X6.3: Beating Miller Effect

5. If resistors were included on the collectors of *both* transistors that form the differential pair—making the circuit fully symmetrical—would the usual Miller-killer trick work (the trick we hope you just told us about in answering the previous question)? Explain your answer.

> No. Wouldn't beat Miller effect: Q_1's collector then would hop about, degrading frequency response as in the usual common-emitter configuration. This is a good reason *not* to include R_C above Q_1, despite the appeal of symmetry.

6. If the emitter resistors were omitted, what effects on circuit performance would you see? Specifically, what, if anything, would happen to—

- output quiescent point;

   > Nothing. This is set by R_{Tail}. R_E, at 250Ω, is negligible relative to the 15k of R_{Tail}.

- gain (you need not calculate the gain; speak qualitatively)

   > Omitting R_E gives much higher gain, and distortion as well.
   > The circuit is *not* vulnerable to temperature, however, if both Q's stay at the same temperature. So, R_E is not *required* in the differential amp as it is in the common-emitter (except where some other scheme is used to provide temperature stability).
   > (A fine point: if the two transistors are not on the same IC, you might want to put in a collector resistor above Q_1, to make quiescent power dissipations equal. But ordinarily you probably would use a matched pair of transistors, fabricated on one IC.)

Lab 6: Transistors II

> **Reading:** Chapter 2.15 to end
> **Problems:** Problems in text.
> Additional Exercises 2 (this one is hard), 4, 5, 6, 8.
> Bad Circuit G.

6-1 Differential Amplifier

Predict differential and common-mode gains for this amplifier (don't neglect r_e).

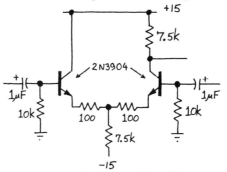

Figure L6.1: Differential Amplifier

Now you will use two function generators to generate a mixture of **common-mode** and *differential* signals.

Preliminaries

One generator will drive the other. This scheme requires that you *"float" the driven generator:* find the switch or metal strap on the lab function generator that lets you *disconnect* the function generator's local ground from absolute or "world" ground. You will find this switch or strap on the back of most generators.

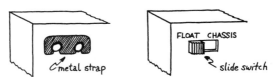

Figure L6.2: "Float" One Function Generator

As you connect the two function generators to your amplifier, you will have to use care to avoid defeating the "floating" of the external function generator: recall that BNC cables and connectors can make *implicit* connections to absolute ground. You must avoid tying the external generator to ground through such inadvertent use of a cable and connector. You may find "BNC-to-mini-grabber" connectors useful: they do not oblige you to connect their *shield* lead to ground.

Composite Signal to Differential Amplifier

Now let the breadboard's function generator (which cannot be "floated") drive the external function generator's *local ground*. Use the output of the external function generator to feed your differential amplifier.

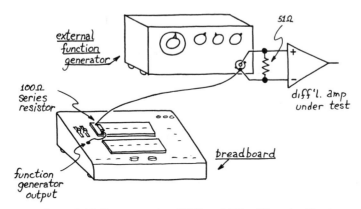

Figure L6.3: Common-mode and Differential Signal Summing Circuit

A. A mediocre differential amp: resistor in 'Tail'
Measure common-mode and differential gain

Suggestion: measuring common-mode and differential gain

Measure *common-mode* gain:
> Shut off the differential signal (external function generator) while driving the amplifier with a signal of a few volts' amplitude. Does the common-mode gain match your prediction?

Measure *differential* gain:
> Turn on the external function generator while cutting common-mode amplitude to a minimum (there is no Off switch on the breadboard function generator).
>
> Apply a small differential signal. (What is the circuit's f_{3dB}?) Does the differential gain match your prediction?

Now turn on *both* generators and compare the amplifier's output with the *composite* input. (To help yourself distinguish the two signals, you may want to use two frequencies rather far apart; but do not let this obscure the point that this differential amp *needs* no such difference: the method you used a few labs ago to pick out a signal while rejecting noise did require such a difference. Do you remember that method?)

This experiment should give you a sense of what "common mode rejection ratio" means: the small amplification of the common signal, and relatively large amplification of the difference signal.

Nevertheless, this circuit still lets a large common-mode signal produce noticeable effects at the output. The improvement in the next step should make common-mode effects much smaller.

B. Improving Common-mode Rejection: Current source in 'Tail'

Replace the 7.5k "tail" resistor with a 2 mA current source:

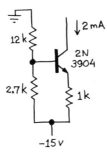

Figure L6.4: 2 mA Current Source for Tail of Differential Amp

This change should make common-mode amplification negligible. (What is common-mode gain if the output impedance of the current source is around 1 M?)

See how this improved circuit treats a signal that combines common-mode and differential signals.

6-2 Bootstrap

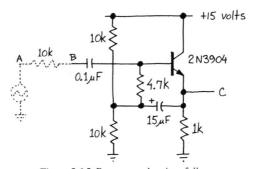

Figure L6.5: Bootstrapped emitter follower

The circuit above illustrates a neat and useful trick, even though (as the text confesses) it is not of great practical importance in this form. You will see the technique of *bootstrapping* used repeatedly (see, e.g., circuits at 4.09 and 7.10).

Begin by connecting up the emitter follower shown above (same as text figure 2.63).

a) First, omit the 15μF capacitor. What should the input impedance be, approximately? Measure it, by connecting 10k in series with the function generator and noting the drop from A to B. Check that the output signal (at C) has the same amplitude as the signal at B (use an input frequency in the range of 10kHz-100kHz).

b) Now add the 15μF bootstrap capacitor. Again measure the input impedance by looking at both sides of the 10k series resistor with the scope. Make sure you understand where the improvement comes from.

Challenge to the zealous: you can do better than say, "The 4.7k looks big." You can estimate *how big* if you note that r_e forms a voltage divider with the emitter resistor, passing a large fraction of delta V_b through as delta V_e. That large fraction lets you estimate delta V across the 4.7k resistor (treat the big cap as a short at signal frequencies). What is the apparent value of the 4.7k, then, seen from the input? (Disappointing footnote: you will not be able to test your calculation experimentally, because you have only a very rough estimate for the transistor's β.)

6-3 Miller Effect

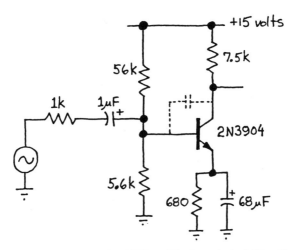

Figure L6.6: Bypassed-emitter NPN amplifier for exploring Miller effect

a) *Miller effect invisible*

Begin by constructing the high-gain (bypassed emitter) single-ended amplifier shown above. Predict the voltage gain, then measure it (short the 1k input resistor). Check that the collector quiescent voltage is reasonable.

b) *Seeing Miller effect*

Now restore the 1k series resistor (it simulates finite generator impedance, like what you might find within a circuit). Measure the *high-frequency* 3dB point.

Now add a 33pF capacitor from collector to base (dotted lines).

This swamps the transistor's junction capacitance of approximately 2pF, exaggerating the Miller effect. Remeasure the high-frequency 3dB point. Explain it quantitatively in terms of the effective capacitance to ground produced by the Miller effect.

c) *Beating Miller effect*

Cascode circuit

Modify the amplifier by adding the second transistor shown below, which clamps the collector voltage of the input transistor. The circuit so modified is called a *cascode*.

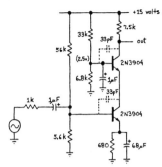

Figure L6.7: Cascode amplifier

Measure the high-frequency f_{3dB} of the cascode amplifier, with the 33pF capacitor still in place between collector and base, and 1k series resistor still in place. Is Miller effect still apparent? How does the cascode work?

6-4 Darlington

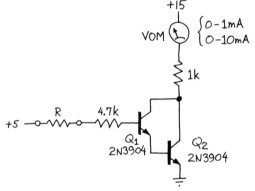

Figure L6.8: Darlington test circuit

Use a substitution box for **R**. Connect two 2N3904's in the Darlington configuration shown above, and measure the circuit's characteristics as follows:

a) Begin by reducing **R** to a few thousand ohms, in order to bring the transistor into good saturation; at this point I_C = 15mA. Measure V_C (the "Darlington saturation voltage") and V_B of Q_1 ("Darlington V_{BE}"). How do they compare with typical single-transistor values? (Measure Q_1 alone, if you don't know, by grounding its emitter.) Explain.

b) Increase **R** into the range of 500k-10Meg, in order to get I_C down to a few milliamps. Measure h_{FE} at collector currents around 1mA and 10mA (as earlier, *measure* I_C, but *calculate* I_B from the value of **R**).

6-5 Superbeta

Substitute a single superbeta 2N5962 for the Darlington pair, and make the same set of measurements, namely V_{BE}, $V_{CE(sat)}$, and h_{FE} at collector currents around 1mA and 10mA. Does it meet the "typical" values graphed in figure 2.78 of the text?

Chapter Review: Ch. 2: Important Topics & Circuits

- **Generally**

    - ground rules: preconditions assumed
    - two models: current amp (β…); voltage amp (Ebers-Moll)
    - biasing: design rule: set *voltage* at base, not *current*

- **Important Circuits**

    - switch
    Here, with transistor saturated, usual β rule does not apply: I_C = typically about $10 \times B$
    - follower

        ♦ impedance-changing: here is one of the few cases where you need to use β in your calculation (but a worst-case β)
        ♦ push-pull: a variation

        common-emitter amp

        ♦ "degenerated" (emitter resistor) (an early view of *feedback*!)
        ♦ distortion <— greed for gain

    - current source

        ♦ current mirror

    - differential amp
    Important for several reasons:

        ♦ it's the guts of any op amp
        ♦ inherently temperature compensated (though not in "tail" current source)
        ♦ can beat Miller effect

- **General Problems (revisited)**

    - temperature effects: remedies:

        ♦ compensation (e.g., mirror; differential amp)
        ♦ R_E as feedback (e.g., degenerated emitter common-emitter amp)

    - Early Effect: remedies: Wilson mirror; r_E
    - Miller Effect: remedies: cascode, differential

Ch. 2: Jargon and Terms

biasing — (Sec. 2.05): setting *quiescent* conditions (see below) so that circuit elements work properly. To *bias* means, literally, to push off-center. We do that in transistor circuits to allow building with a single supply. The term is more general, as you know. (Compare Ch. 1, sec. 1.30, where a diode is *biased* into conduction.)

bootstrap — (Sec. 2,16): In general, any of several seemingly-impossible circuit tricks (source of term: "pull oneself up by the bootstraps:" impossible in life, possible in electronics!). In this chapter refers to the trick of making the impedance of a bias divider appear very large, so as to improve the circuit's input impedance. Also *collector bootstrap*. See sec. 2.17.

bypassed (-emitter resistor)
(Sec. 2.12): In common-emitter amp, a capacitor put in parallel with R_E is said to *bypass* the resistor because it allows *AC* current an easy path, bypassing the larger impedance of the resistor. Used to achieve high gain while keeping R_E large enough for good stability.

cascode — circuit that uses one transistor to buffer or isolate another from voltage variation, so as to improve performance of the protected transistor. Used in cascode amplifier to beat Miller effect, in current source to beat Early effect.

clipping — (E.g, Sec. 2.05 follower design procedure, Step 1): Flattening of output waveform caused by hitting a limit on output swing. Examples: single-supply follower will *clip* at ground and at the positive supply; follower described in sec. 2.03 at p. 55 clips when loaded with 1k.

compliance — (Sec. 2.06): Well defined in text: "The output voltage range over which a current source behaves well...."

Early effect — variation of I_C with V_{CE}. Thus it describes transistors's departure from true current-source behavior.

emitter degeneration (Sec. 2.11): Placing of resistor between emitter and ground (or other negative supply) in common-emitter amp. It is done so as to stabilize the circuit with variation in temperature. (Source of term: gain is reduced or "degenerated." General circuit *performance* is much *improved,* however!)

impedance "looking" in a direction
(Sec. 2.05): impedance at a point considering only the circuit elements lying in one direction or another. Example: at transistor's base impedance *looking back* one "sees" bias divider and R_{source}; *looking into base* one "sees" only $\beta \times R_E$.

Miller effect — exaggeration of actual capacitance between output and input of an inverting amplifier, tending to make a small capacitance behave like a much larger capacitance to ground: *1 + Gain* times as large as actual C.

quiescent (-current, –voltage)
(Sec. 2.07, etc.): condition prevailing when *no* input signal is applied. So, describes DC conditions in an amplifier designed to amplify AC signals. Example: $V_{\text{out quiescent}}$ should be midway between V_{CC} and ground in a single-supply follower, to allow maximum output amplitude (or "swing") without clipping.

split supplies
(Sec. 2.05): Power supplies of both polarities, negative as well as positive. Used in contrast to "single supply."

transconductance
(Sec. 2.09): Well defined in Text. Briefly, $\equiv \Delta I_{\text{out}}/\Delta V_{\text{in}}$.

Wilson mirror
improved form of current mirror in which a third transistor protects the sensitive output transistor against effects of variation in voltage across the load (third transistor in cascode connection, incidentally). (Sec. 2.14, fig. 2.48.)

CHAPTER 3

Class 7: FETs I

Topics:

- FET types
    - we concentrate on N-channel
    - two distinctions, within this category:
        - JFET (junction FET) versus MOSFET (insulated-gate FET)
        - depletion mode (all JFETs, occasional MOSFETs) versus enhancement mode (most MOSFETs, no JFETs)

- Applications
    - current source
    - follower
    - voltage-controlled resistance (less important)

Why FETs?
Text Ch. 3 introduction;
Text sec. 3.05, pp. 124-25

We use them, because they can do some jobs better than bipolar transistors. Consider FETs when you want—

- very high input impedance;
- a bidirectional "analog switch";
- a simple current source (2-terminal);
- a voltage-controlled resistance.

The first of these FET virtues is by far the most important: enormous input impedance.

A Physical Picture
Text sec. 3.01

The operation of a FET is much easier to describe than the operation of a bipolar transistor. You will recall we did not even try to describe *why* a bipolar transistor behaves as it does.

A FET, in contrast, just cries out for diagramming:

Compare Text fig. 3.5

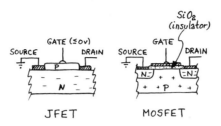

Figure N7.1: JFET (junction FET) versus MOSFET (insulated gate)

A glance at these diagrams will remind you of the FET's greatest virtue, its very high input impedance: The input terminal looks like either an *insulator* (so-called "MOSFET" type) or a *back-biased junction* (so-called "JFET"). So, no *current* flows at the control terminal: you just apply a *voltage*; the channel feels the *field*. Hence the name, of course.

We Will Concentrate on the Most Familiar Types

Text sec. 3.01, p. 114;
3.02, p. 118;
see figs. 3.6, 3.7

As the Text says, FETs come in bewildering varieties of types. We will talk almost exclusively about the types that behave pretty much like the bipolar transistor you are most familiar with, the NPN:

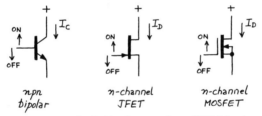

Figure N7.2: Similar control polarities & current flows: NPN & "n-channel" FETs

How a FET Works

Text sec. 3.03; fig. 3.11

Here is a terrifying set of curves, dense with information. We'll look at just two of these curves and the device each describes, and we'll look at these one at a time:

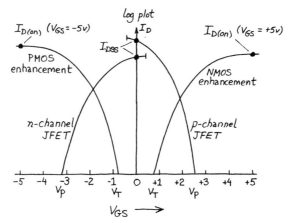

Figure N7.3: I_D vs V_{GS}: a heap of information at a glance

Two "modes:" On until turned Off vs. Off until turned On

Text sec. 3.02, p. 118-19

There are two basic schemes for making a FET. One sort of FET conducts until you stop it; the other blocks current until you make it conduct.

"Depletion Mode"

Text sec. 3.02

One kind of FET conducts until you do something to diminish its conductance ("depleting" the conducting channel). This kind is really easy to picture: it's just a slab of doped silicon; as the name suggests, if you do nothing to it, it *semi-* conducts:

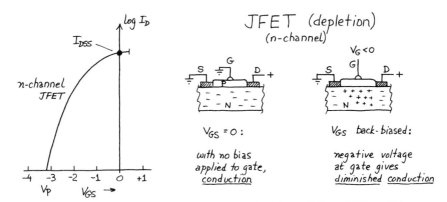

Figure N7.4: Slab of semiconductor: FET (depletion mode) before and after a field is applied at gate

Now apply an electric *field*—through a back-biased diode junction or across an insulator—and thus define a region hostile to conduction, a region of the wrong kind of semiconductor material, *narrowing* the conducting *"channel"*. That's the slab in the second image, above.

All JFETs, and some MOSFETs work this way: they're on until you turn them off.

"Enhancement Mode"

Another kind of FET is designed so that it will *not* conduct unless you apply a field that in effect digs a conducting channel ("enhancing" its conductance).

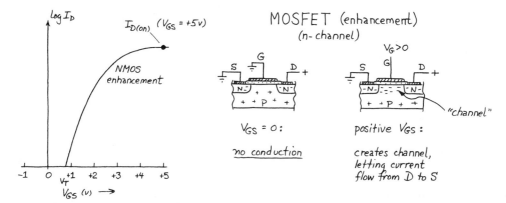

Figure N7.5: Slab of semiconductor, 3 regions: FET (enhancement mode) before and after a field is applied

This arrangement blocks conduction: the drain region is back-biased with respect to the big region (called "body").

Now apply an electric *field* (or "generate" a field, by applying a voltage) so as to create a region *like* the end regions: this *channel* permits conduction between drain and source. We achieve again a continuous slab of semiconductor material.

Since we need to apply a positive voltage, this time (for an N-channel device), we can enhance conductance only for the kind of transistor that uses an *insulator* to isolate the gate from channel: we cannot do the same trick when we rely on a semiconductor junction. A junction so *forward*-biased would *conduct*, and that would ruin the FET's performance.

The stronger the field, the deeper the conducting *channel*.

A picture *cannot*, unfortunately, explain the curious fact that the FET behaves like a *current source* rather than *resistor*, when a substantial voltage is applied across it. Here, as for the bipolar transistor generally, we just call the mechanism magic[1].

Applications

---

[1]. Well, we can't resist a *try* at explaining. As V_{DS} grows, the conducting channel gets warped—pinched narrower toward the drain end than at the source end (since the back-bias there is greater: V_{GD} is larger than V_{GS}).

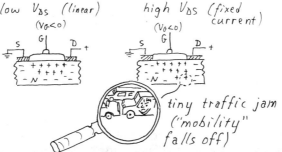

Figure N7.6: Conducting channel gets pinched at one end, when V_{DS} > a few volts; then the bottleneck makes flow-rate (I_D) level off

In that narrow region a kind of traffic jam occurs. The V_{DS} drop occurs over the short length of that pinched section, pushing up the current density in this bottleneck; then, if V_{DS} grows further, increasing the field (drivers at the drain end lean on their car horns) the bottleneck—or traffic-jam—region grows longer, and traffic flow or I_D levels off at its *saturation* value. In short, further increases in V_{DS} cause two opposing effects that nearly cancel: stronger field, but reduced carrier mobility. So, current stays roughly constant.

For a fuller explanation and handsome diagrams see Burns & Bond, <u>Principles of Electronic Circuits</u>, sec. 5.2.

Current Source

Text sec. 3.06

This is the easiest circuit to analyze. If you tie gate to source the transistor has to run at I_{DSS}.

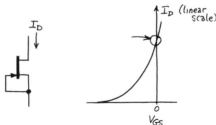

Figure N7.7: Simplest FET current source: runs at I_{DSS}

If you try to build such a current source with a discrete transistor, you'll be stuck with the I_{DSS} of the particular part (not even particular *type*!). That's annoying. (In the lab you will use devices wired this way and housed in a little glass package; but these are easy to use, because the manufacturer has done the pesky job of sorting sorted them by I_{DSS}.)

The FET current source works only so long as you keep V_{DS} greater than a few volts: out of the so-called "linear region," where the FET behaves not like a current source but like a resistor.

Back-biased current source

Text sec. 3.01, fig. 3.2; Lab 7-2

A FET with a bit of back-bias makes a more practical current source: you can adjust its current output by choosing R_S. In addition, it works a little better: its current varies less, for changes in V_{DS}. You can see this on the family of curves below, which show I_D at several values of V_{GS}. (We apologize for the topmost curve—which says that a little *forward* bias increases I_D; our simple model of FET performance does not predict this; we can pretend we didn't notice that top curve, or we can complicate the simple model; let's pretend we didn't notice!)

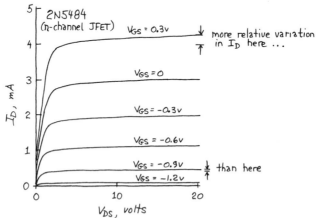

Figure N7.8: A family of curves: I_D vs. V_{DS}: back-bias improves a current source

The circuit below adds a resistor, R_S, to apply some back bias to V_{GS}. The circuit is easy to understand, but choosing R_S so as to achieve a particular current is not easy. If you had a good curve describing the transistor's behavior, you could plot the resistor's behavior on the same diagram, but with the V-axis reversed; the intersection would reveal the output current (Text Appendix F, "Load Lines," sketches this technique.)

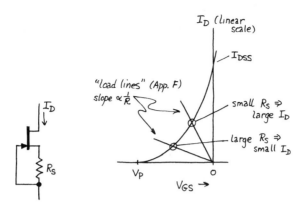

Figure N7.9: Using 'load line' to find current source operating point

In fact, such a transistor curve would be very approximate, so you would have to try the transistor in the circuit, and tinker with R_S values.

Evolving the follower from a current source

Let's set up a current source, but add a terminal from which its gate voltage can be wiggled:

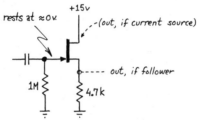

Figure N7.10: Old current source—with an input terminal: "self-biased" follower

Now it's a wiggle-able current source. But if you take the output at the source, you find you have a follower: wiggle at source approximately equals wiggle at input (we'll see shortly how to evaluate this vague "approximately" in this statement). The circuit's input impedance is very good.

FET Limitations: Gain: FET versus Bipolar

Text sec. 3.01, 3.04, 3.05; pp. 130-32

The FET's gain (Δ current out for Δ voltage in: g_m, called "transconductance") is lower than a bipolar transistor's.

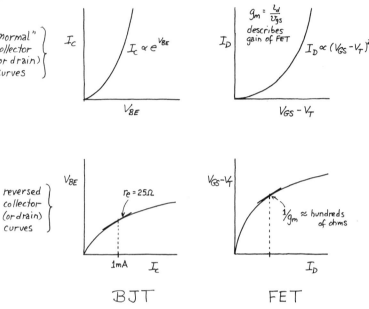

Figure N7.11: Gain: FET vs bipolar transistor ("BJT")

The gain, like the gain of the bipolar transistor, varies as transistor current varies. For the *bipolar*, we said

$$r_e = 25 \text{ ohms}/I_c \text{ (in milliamps)}$$

That is a roundabout way to say that gain varies linearly with current (I_C).

For the **FET**, the gain varies linearly with $V_{GS}-V_T$: (the slope of the parabola $y=x^2$ increases linearly with x).

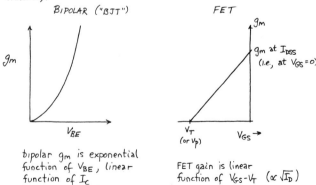

Figure N7.12: Variation of gain: bipolar versus fet

To find the transconductance at any given V_{GS}, just find where you are (what fraction of the way) between the two V_{GS} values that turn the transistor fully *off* and fully *on*. So, for example, if g_m is 2000μmhos at $V_{GS}=0$, and $V_{pinchoff}$ is −2V, if the FET is running at $V_{GS}=-1$V, what's the gain? You're at the midpoint of the gain curve (which is just a straight line); the gain is 1000μmhos (or "1 mA/V," a prettier formulation, but less standard).

The Text says this, but also says

"transconductance increases…as the square root of I_D…." (p. 132).

This is just an alternative formulation of the same rule. Use that formulation if you know the operating current; use the earlier formulation if you know V_{GS}—and always be warned that the data sheet for the transistor gives you only ball-park figures to work with, anyway: the spread of both g_m and $V_{pinchoff}$ is wide (see sec. 3.05, p. 123).

From the FET's mediocre gain—inferior to that of the BJT—several consequences follow:

* FET Amp shows limited Voltage Gain

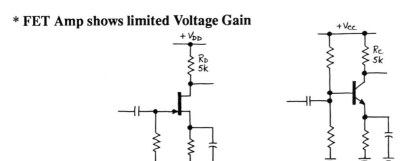

Figure N7.13: FET common-source- vs. BJT common-emitter amplifiers: gain in both cases = $g_m \cdot 5k$; but the g_ms are very different

* Follower Output is Attenuated

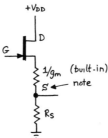

Figure N7.14: FET follower attenuation is easy to predict, using $1/g_m$ exactly analogous to r_e

Remedies

Current source replaces R_S.

Text sec. 3.08

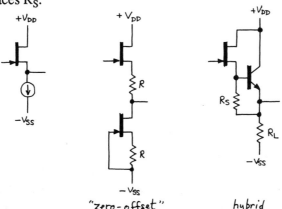

Figure N7.15: Current source in place of R_S solves follower's attenuation problem

* R_{out} is Mediocre

Text sec. 3.08, p. 134

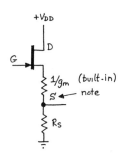

Figure N7.16: R_{out} for FET follower is not very good: = $1/g_m$ parallel R_S

Remedy? Get the help of a bipolar follower. The rightmost follower, in the figure above (this is the Text's fig. 3.27), does that: uses the bipolar both as current source and as a means to drop R_{out}.

Recapitulation: strength and weaknesses of FETs:—and how to have it both ways

FETs are great for high input impedance, but not so hot as amplifiers, and the FET follower's output impedance is much higher than the bipolar circuit's. The manufacturers of integrated amplifiers know this, of course. So you'll not be surprised to learn that as soon as they learned to fabricate both sorts of transistor on one IC, they started building circuits that let us have it 'both ways.'

The LF411 operational amplifier that you will meet in Lab 8, for example, exploits the JFET's high input impedance, and then uses a bipolar common-emitter amp in the high-gain stage, and finally a bipolar push-pull at the output:

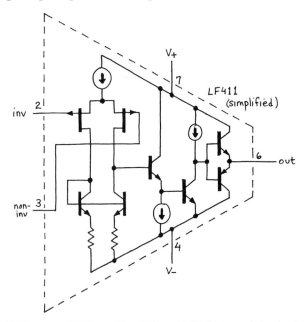

Figure N7.17: IC op amp, LF411: combines FETs and BJTs, letting each do what it does best

The '411's manufacturer was so pleased with itself when it figured out how to put FET and BJT on one chip that it gave the process a name, "BIFET," and got a trademark. But everyone knows this trick, nowadays.

Linear Region: Voltage-controlled resistance, and an application

Text sec. 3.10
Lab 7-4

Earlier we noticed that FET current sources fail if V_{DS} falls too low—and 'too low' is not so low as for a BJT. So a ground rule for use of a FET current source is 'don't let V_{DS} get too low. If you violate this rule, you'll find that your 'current source' begins to behave like a *resistor*.

But sometimes, of course, that's just what you want: a resistor—because it's a nifty sort of resistor: one whose *value* you can regulate with a voltage applied at the gate terminal.

Here are curves to illustrate how you do this trick. The bipolar and FET curves below show roughly similar slopes where the voltage across the transistor is very small:

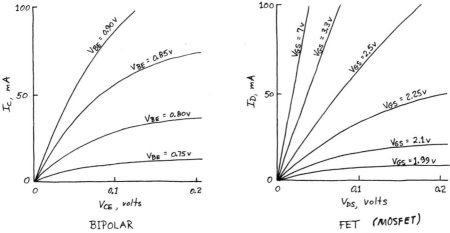

Figure N7.18: More-or-Less linear (resistive) behavior: bipolar & fet

This is the saturation region of the bipolar, and one cannot vary the slope there widely. The FET does better: the curves are straighter, and one can better control the slope (resistance, or 1/resistance).

Here's the lab circuit that applies this behavior:

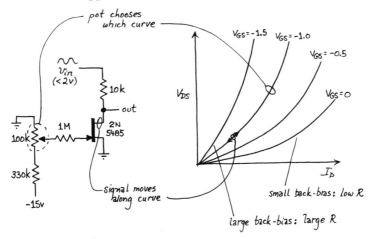

Figure N7.19: FET as variable resistance: V_{DS} must be small: *Note:* resistance curves show the earlier I_D vs V_{DS} curves rotated to make the slope show *resistance* rather than 1/resistance

The pot applies a DC *bias* to the gate; this selects one of the possible curves, with a its nearly-constant slope (thus R value). The signal—which must be small: < (V_{GS}-V_T is the rule)—moves along that curve. Incidentally, the signal voltage, applied as V_{DS}, moves positive and *negative*. This swaps roles of drain and source, which is pretty confusing to think about, but which introduces no large errors for small signal swings. The *resistance* curves remain pretty-nearly straight, when fixed with the *linearizing trick* described in the Text at p. 139, sec. 310, and demonstrated in Lab exercise *7-4 b*. That trick is a wrinkle we will leave to the lab to clarify.

Ch. 7: Worked Examples: Current Source; Source Follower

Two worked examples (related):

1. JFET current source
2. source follower

Problem: current source and source follower

1. Design a JFET current source to deliver 2 mA, given the following typical I_D vs. V_{gs} curve. (Your design will give only approximate results, because of the wide spread of FET characteristics; just do what you can with the *typical* values you are shown.)

   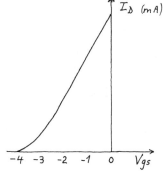

   Figure X7.1: Typical JFET I_D vs V_{gs} curve (this is drawn from 2N5485 data)

2. Now take advantage of the calculation you just made in order to design a self-biased *source follower* to run at $I_{D(quiescent)} = 2$ mA. Let $f_{3dB} \approx 10$ Hz.
   What signal amplitude can this follower reproduce?

3. Given $g_m = 5000$ μmhos at I_{DSS}, what is the follower's—

   a. attenuation?
   b. R_{out}?

4. Design and install a current source to replace R_S. What circuit characteristic(s) does this replacement improve?

Solution:

1. *Current Source*

The curve suggests we need V_{gs} of about –2.8V to get 2 mA to flow. That required value of V_{GS} implies an R value:

$$R_S = 2.8V / 2\,mA = 1.4k\Omega$$

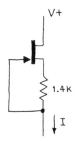

Figure X7.2: Intersection of 2 mA and curve ==> need back-bias of about 2.8V; here is the 2 mA JFET current source

2. *Source Follower*

This is just the current source, but output from source rather than drain, and with an input. The input is AC-coupled:

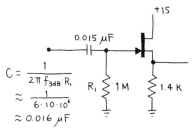

Figure X7.3: Self-biased source follower

Query: what signal amplitude?

Not very large: since V_{out} is quiescent at about 2.8V, that sets the amplitude limit: ≤ about 2.8V.

3. *What attenuation, what R_{out}?*

We must find g_m at the operating point. g_m is *specified* at I_{DSS}, but the circuit does not run there (and no follower can!). Luckily, the relation of gain to operating point is simple. Here are the curves, once again (they are approximate; the 2N5485 data sheet shows an I_D vs V_{gs} function less perfectly quadratic):

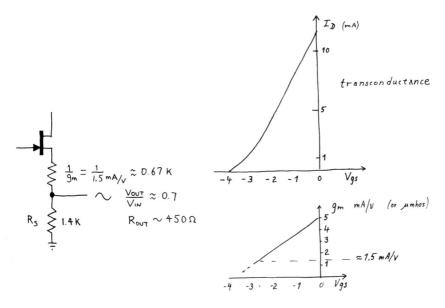

Figure X7.4: Given g_m and V_{gs}, and the curve of I_D vs V_{gs}, we can calculate g_m at the operating point

Here, g_m appears to be about 1.5 mA/V (a clearer way to say "1,500 μumhos"), so $1/g_m = 2/3$ kΩ: ≈ 0.7K.

This gives V_{out}/V_{in} of about 1.4k/2k: about 2/3.

$R_{out} = 1/g_m$ parallel R_S: 0.7k parallel 1.4k: 0.45K: 450Ω.

4. *Design a current source to replace R_S*

We have already done the work: we know what value of R_S will give 2 mA. We can minimize voltage offset ($V_{out} - V_{in}$) by inserting a resistor of value equal to the lower R_S—this is a trick you meet in Lab 7-3.

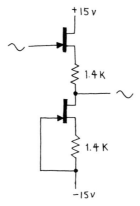

Figure X7.5: Current source replaces R_S: now attenuation and offset are small

Lab 7: FETs I: First View

Reading:	sec. 3.01 – 3.10, re FETs generally, followers and current sources;
	We will return in a later lab (Lab 11) to the use of FETs as *switches*. For today, concentrate on the circuits you will see in this lab: follower, current source, and variable resistor. Of these, the follower probably is the most important.
Problems:	embedded problems

7-1 FET Characteristics

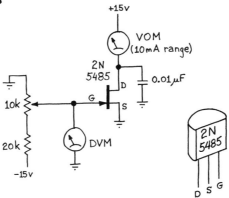

Figure L7.1: FET test circuit. Plot I_D vs. V_{GS}

Measure I_{DSS} and V_P ("pinch-off" voltage = V_T for a JFET) for a couple of samples of 2N5485.

Verify the relation between I_D and V_{GS} shown in the Text's figure 3.15 (semi-log plot). Notice the spread of values even in specimens from the same manufacturer's batch. Check that your values fall within the quoted maximum range:

$$4\text{mA} < I_{DSS} < 10 \text{ mA}$$
$$-4\text{V} < V_P < -0.5\text{V}$$

Note V_P for one of the FETs and use this transistor in the *current source* and *follower* experiments that follow. There it will be useful to know this characteristic for the transistor.

7-2 FET Current Sources

a) *Discrete Transistor Current Source*

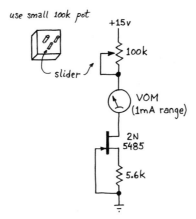

Figure L7.2: FET current source

How good a current source is this? Vary the resistance of the 'load,' and watch V_{DS} with a DVM as you monitor I_{out}.

What is V_{DS} when the constant current behavior starts to break down? This V_{DS} value marks the boundary of the "linear region" and should occur when V_{DS} is near V_{GS}-V_T. Does your FET's "linear region" begin around this value of V_{DS}?

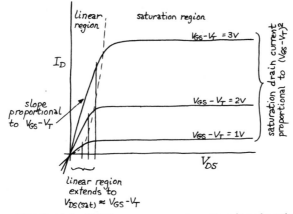

Figure L7.3: Reminder: FET linear versus current-source regions: depends on V_{DS}

Notice that the circuit you have just built is the first *two-terminal* current source you have seen in these labs (that is, a current source that requires no external bias). Such a device is as almost as easy to use as a resistor (why ...*almost*?). FET manufacturers, too, have noticed how handy a device a 2-terminal current source is, and they sell them.

b) *Integrated 2-terminal current source*

The 1N5294 is a JFET with source shorted to gate, but packaged in a small package that looks just like a diode's. These current sources are sorted by I_{DSS}; the '5294 passes about 0.75 mA.

First, try the device in the test circuit you used with the 2N5485 just above. Does this "diode" perform as well as the circuit you built with the '5485? Would you expect it to?

Now, to get a feel for how easy it is to design nifty things with this device, try the circuit below—which at a glance may look foolish.

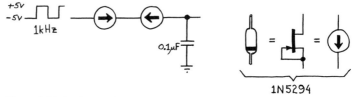

Figure L7.4: Application for two-terminal current source: square-wave to –?- circuit

What waveform out do you expect? Here, incidentally, we are exploiting a fact that ordinarily alarms us: a JFET's gate *conducts* if we forward-bias the gate with respect to either source or drain.

Drive the circuit with a 1kHz square wave of about 5 volts' amplitude. Center the input waveform on zero volts, and note whether the output is centered. If it is not, why is it not? If it is off-center, what stops it from sailing farther off-center?

Now gradually lower the input amplitude until you notice distortion in the output (curvature, near the points). Why does this occur? (*Hint:* At the point where you notice curvature beginning, note the voltage across the FET: V_{DS}.)

7-3 Source Follower

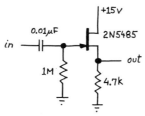

Figure L7.5: Source follower

a) *Simple Follower*

Drive the source follower shown above with a small sine wave at 1kHz. By how much does the gain differ from unity. Why?

From this single observation—the follower's attenuation—you can infer g_m: the FET's transconductance (at this $I_{D_{quiescent}}$). One way to think of g_m's effect is to draw it as an equivalent resistance in the source, exactly analogous to r_e for the bipolar transistor:

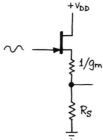

Figure L7.6: Effect of g_m shown as series resistance forming voltage divider in a follower

Compare the g_m that you infer from your follower's attenuation with the g_m shown on the transistor's data sheet, and with the g_m you measured at the start of this lab. Note that g_m varies with I_D, and that the data sheet specifies g_m under the most favorable condition: $V_{GS} = 0V$. Your follower runs with $V_{GS} < 0V$. So, the observed g_m will always be lower than the data sheet's g_m.

You can do better than to say "lower," if you like. Here is the argument (stated in the Text at p. 132):

- Here are the curves for log I_D versus V_{GS}, and for (linear) I_D versus V_{GS}:

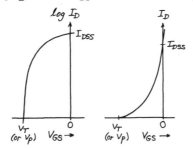

Figure L7.7: log I_D versus V_{GS}; (linear) I_D versus V_{GS}

- The *gain* curve just shows the slope or derivative of those curves. Since I_D varies as the *square* of V_{GS}, the *gain* curve looks like a straight line, reaching its maximum (the specified value of g_m) at I_{DSS} (where $V_{GS} = 0$):

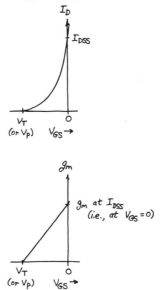

Figure L7.8: Gain varies linearly with $V_{GS} - V_T$

So, if you observe the quiescent value of V_S, you know $V_{GS\text{-quiescent}}$ (V_G rests at ground). Having measured V_T at the start of today's lab, you can see where you must be on the FET's curve.

How close is your estimate of g_m, so derived, to the value of g_m that you observe?

How does the variation of g_m compare with the variation of "g_m" that you saw for a bipolar transistor, back in the lab where you saw distortion in the common-emitter amplifier (exercise 5-2).

b) *Follower with Current Source Load*

Modify the circuit to include a current source load as shown in the figure below. Confirm that this follower performs much better than the simpler circuit.

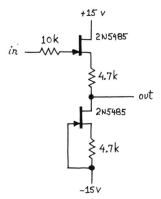

Figure L7.9: Low-Offset source follower with current source load

Measure the gain with a 1V, 1kHz signal (the gain had better be very close to 1.0!).

Attempt to measure the input impedance. If you conclude that R_{in} is around 10M ohms, suspect that you have fallen into a trap!

Measure the DC offset. What accounts for the non-zero offset? Mismatch of FETs or of resistors? What easy circuit changes would let you find out, if you are in doubt?

c) *(Optional:) Matched-FET Follower*

Finally, try the same circuit with a 2N3958 dual FET. The six-lead package is most easily inserted into the breadboard as two rows of three (straddling the central median).

The specified maximum offset for these transistors is 25mV. Is the offset of your circuit as low as that? If not, why not?

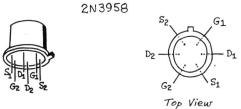

Figure L7.10: 2N3958 matched dual FET

7-4 FET as Variable Resistor

When you tested the FET as *current source*, you found that the circuit failed when V_{DS} shrank so far that the device fell into its "linear region." Here you will build a circuit intended to operate always within that region: it will "fail" if V_{DS} gets large enough to carry the FET into the current source region.

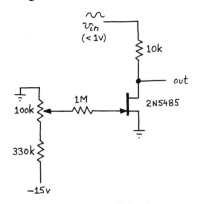

Figure L7.11: FET as voltage-controlled resistor: attenuator circuit

a) *Uncompensated Attenuator*

Drive the circuit above with a *small* sine wave (around 0.2V) at around 1 kHz. Adjust the potentiometer, and notice the results, not only in variable attenuation but also in varying amounts of distortion. (To see distortion clearly, drive the circuit with a *triangle* waveform.) Would you predict such distortion, on the basis of the Text's figure 3.30, reproduced below? How does the circuit treat a larger input waveform?

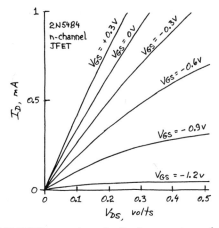

Figure L7.12: FET linear region: a family of curves, at several values of V_{GS}

b) *Compensated Attenuator*

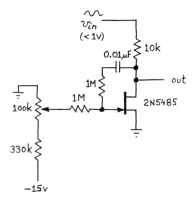

Figure L7.13: Compensated FET-as-resistor

Adding 1/2 v_{DS} to the gate signal turns out to straighten those curves a good deal. (See Text sec. 3.10, p. 139.) The circuit amendment above performs this addition. Take a look at its effect on the shape of V_{out} as you again drive the circuit with a triangle waveform of about 0.2 V amplitude.

c) *Amplitude Modulation*

You may not have been impressed with the preceding circuits, which showed that you could use a potentiometer to vary the attenuation of a signal. That you could do *without* the FET's help! The FET is useful, of course, because its resistance can be controlled by a *voltage*, and not necessarily by the adjustment of a potentiometer.

To put this ability of the FET to work, let a second function generator drive the point that you drove with the pot, so as to let this second signal vary (or "modulate") the attenuation periodically. Let the frequency of modulation be much lower than the signal frequency. Try $f_{modulation}$ of around 50 Hz, and keep the modulation amplitude small: around 0.2 V. For a stable display, you may want to trigger the scope on the modulating signal, not on the composite output.

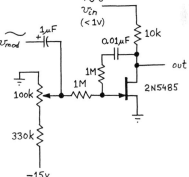

Figure L7.14: Amplitude-modulation, or multiplication of one signal by another

d) *AM Radio (optional)*

You can have some fun with the preceding circuit by turning it into an AM radio transmitter. Drive the point called "V_{in}," in the circuit above, with a sine at around 1 MHz (we'll call this $f_{carrier}$); attach a few inches of wire to the point called "out;" then tune an AM radio located across the room to a quiet place on the dial, and adjust $f_{carrier}$ until you can hear the *modulating* signal (a single tone). (To make the modulating signal more obvious, you may want to sweep it, either by hand or with a function generator's sweep mode.)

CHAPTERS 4, 5, 6

FEEDBACK:
Op amps, oscillators, power supplies:
Overview

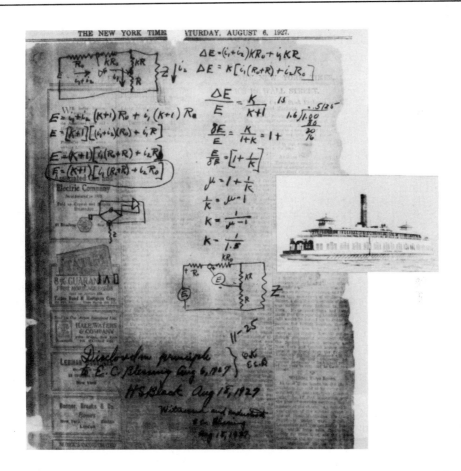

Figure OVR4.1: Harold Black's notes on the feedforward amplifier conceived as he rode the ferry from Staten Island to work, one summer morning in 1927. (Copyright 1977 IEEE. Reprinted, with permission, from Harold S. Black, "Inventing the Negative Feedback Amplifier," IEEE Spectrum, Dec. 1977)

We have been promising you the pleasures of feedback for some time. You probably know about the concept even if you haven't yet used it much in electronics. Now, at last, here it is.

Feedback is going to become more than just an item in your bag of tricks; it will be a central concept that you find yourself applying repeatedly, and in a variety of contexts, some far from operational amplifiers. Already, you have seen feedback in odd corners of transistor circuits; you will see it constantly in the next three chapters; then you will see it again in a digital setting, when you build an analog-to-digital converter in Lab 17, and then a phase-locked-loop in the same lab. It is a powerful idea.

Chs. 4, 5, 6: Feedback: Op amps, oscillators, power supplies

Chapter four begins, as Chapter two did, with a simple, idealized view of the new devices—this time, 'operational amplifiers:' little high-gain differential amplifiers that make it easy to build good feedback circuits. As the chapter continues, we soon feel obliged once again to disillusion you—to tell you about the ways that op amps are imperfect. At the same time, we continue throughout these four labs to look at additional applications for feedback, and we never lose our affection for these circuits. They work magically well. The third op amp lab, Lab 10, introduces the novelty of *positive* feedback: feedback of the sort that makes a circuit unstable. Sometimes that is useful, and sometimes it is a nuisance; we look at cases of both sorts. Several of the circuits that use positive feedback are *oscillators*, a circuit type treated primarily in Chapter 5 (*Active Filters and Oscillators*).

Lab 11 concentrates on FET's, this time used as switches; but this lab includes circuits that use feedback, and among them is the only *active filter* that you will meet in this course.

Lab 12 returns us to circuits that for the most part rely on negative feedback, but these are specialized circuits designed for the narrow but important purpose of providing stable power supplies. The Text devotes a chapter to these circuits (Chapter 6: *Voltage Regulators and Power Circuits*); we give them a lab, and hope that you will feel the continuity between this use of feedback and the more general cases that you met first in Chapter 4. With Lab 12 we conclude the analog half of the course, and with the very next lab you will find the rules of the game radically changed as you begin to build *digital* circuits. But we will save that story till later.

A piece of advice (unsolicited): How to get the greatest satisfaction out of the feedback circuits you are about to meet:

Here are two thoughts that may help you to enjoy these circuits:

- as you work with an op amp circuit, recall the equivalent circuit made without feedback, and the difficulties it presented: for example, the transistor follower, both bipolar and FET, or the transistor current sources. The the op amp versions in general will work better, to an extent that should astonish you.

You have labored through two difficult chapters, 2 and 3, and have learned how to work around annoying characteristics of both sorts of transistor. Now you are entitled to enjoy the ease of working with op amps and feedback.

Chs. 4, 5, 6: Feedback: Op amps, oscillators, power supplies

Here is a picture of you climbing—as you are about to do—out of the dark valleys through which you have toiled, up into that sunny region above the clouds where circuit performance comes close to the ideal:

Figure OVR4.2: Righteous and deserving student, about to be rewarded for his travails with discrete transistors: he climbs into the sunny alpine meadows where feedback blooms

Pat yourself on the back, and have fun.

A second thought:

- Recall that negative feedback in electronics was not always used; was not always obvious—as the Text points out in its opening to Chapter 4, and as Harold Black was able to persuade the patent office (Black comes as close as anyone to being the inventor of electronic feedback).

The faded and scribbled-on newspaper that is shown at the start of these notes is meant to remind you of this second point—meant to help us feel some of the surprise and pleasure that the inventor must have felt as he jotted sketches and a few equations on his morning newspaper while riding the Staten Island Ferry to work one summer morning in 1927. A facsimile of this newspaper, recording the second of Black's basic inventions in the field, appeared in an article Black wrote years later to describe the way he came to conceive his invention. Next time you invent something of comparable value, don't forget to jot notes on a newspaper, preferably in a picturesque setting—and then keep the paper till you get a chance to write your memoirs.

Class 8: Op Amps I: Idealized View

Topics:

- *old:*
  - earlier examples of feedback

- *new:*
  - negative feedback: a notion of wonderful generality
  - feedback without op amps: examples you have seen
  - feedback with op amps
    - ♦ the Golden Rules
    - ♦ Applications: Two amplifiers
    - ♦ Preconditions: when do the Golden Rules apply?
    - ♦ More Applications: improved versions of earlier circuits
      - current source
      - summing circuit
      - follower
      - current-to-voltage converter
    - ♦ A generalization: strange things can sit within the feedback loop:
      - sometimes we want the op amp to hide the strange thing;
      - sometimes we want the op amp to generate (strange-thing)$^{-1}$

Preliminary: Negative Feedback as a general notion

This is the deepest, most powerful notion in this course. It is so useful that the phrase, at least, has passed into ordinary usage—and there it has been blurred. Let's start with some examples of such general use—one genuine cartoon (in the sense that it was not cooked up to illustrate our point), and three cartoons that we did cook up. Ask yourself whether you see feedback at work in the sense relevant to electronics, and if you see feedback, is the sense positive or negative?

Figure N8.1: Feedback: same sense as in electronics? Copyright 1985 Mark Stivers, first published in <u>Suttertown News</u>

The two cases below are meant to raise the question, "To which op amp terminal is the feedback being applied?" (If you have not yet looked at the Text, that question will not yet make sense to you. We assume you know what an op amp is, at this point.)

Figure N8.2: "Negative Feedback": poor usage: which terminal is getting the feedback?

The case below comes closer to fitting the electronic sense of *negative feedback*. In op amp terms (not Hollywood's), who's playing what role?

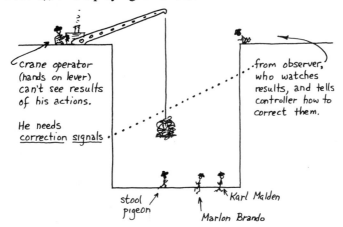

Figure N8.3: "Negative feedback:" a case pretty much like op amp feedback

In conversation, people usually talk as if "positive feedback" is nice, "negative feedback" is nasty. In electronics the truth is usually just the opposite.

Feedback in electronics

Generally speaking, negative feedback in electronics is wonderful stuff; positive feedback is nasty. Nevertheless the phrase means in electronics fundamentally what it should be used to mean in everyday speech.

Harold Black, the first to apply negative feedback to electronic circuits, described his idea this way:

Text sec. 4.26, p. 233

> ...by building an amplifier whose gain is made deliberately, say 40 decibels higher than necessary (10,000-fold excess on energy basis) and then feeding the output back to the input in such a way as to throw away the excess gain, it has been found possible to effect extraordinary improvement in constancy of amplification and freedom from nonlinearity.[1]

Open-loop vs feedback circuits

Nearly *all* our circuits, so far, have operated open-loop—with some exceptions noted below. You may have gotten used to designing amplifiers to run open-loop (we will cure you of that); you would not consider driving a car open loop (we hope), and you probably know that it is almost impossible even to *speak* intelligibly *open-loop*.

Examples of Feedback without Op Amps

We know that feedback is not new to you, not only because you may have a pretty good idea of the notion from ordinary usage, but also because you have seen feedback at work in parts of some transistor circuits:

Lab 5: 5-2, 5-3, 5-5

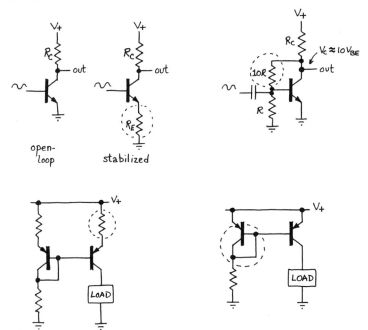

Figure N8.4: Some examples of feedback in circuits we have built without op amps

Feedback with Op Amps

Op amp circuits make the feedback evident, and use a lot of it, so that they perform better than our improvised feedback fragments. Op amps have enormous gain (that is, their *open-loop* gain is enormous: the chip itself, used without feedback, would show huge gain: ≈200,000 at DC, for the LF411, the chip you will use in most of our labs). As Black suggests, op amp circuits throw away most of that gain, in order to improve circuit performance.

---

1. IEEE Spectrum, Dec. 1977

The Golden Rules

Just as we began Chapter 2 with a simple model of transistor behavior, and that model remained sufficient to let us analyze and design many circuits, so in this chapter we start with a a simple, idealized view of the op amp, and usually we will *continue* to use this view even when we meet a more refined model. The *golden rules* (below) are approximations, but good ones :

ext sec. 4.03, p. 177

> **Op amp "Golden Rules"**
>
> 1. The output attempts to do whatever is necessary to make the voltage difference between the two inputs zero.
> 2. The inputs draw no current.

These simple rules will let you analyze a heap of clever circuits.

Applications

Two Amplifiers
ext sec. 4.04, 4.05:
178

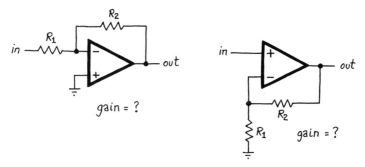

Figure N8.5: Inverting and non-inverting amplifiers

What are the special virtues of each?

- What is R_{in} for inverting amp? (Golden Rule 1 should settle that.)
- Approximately what is R_{in} for the non-inverting amp? (Golden Rule 2 should settle that.)
- The inverting amp's inverting terminal (the one marked "-") often is called "virtual ground." Do you see why? (Why ground? Why "virtual"?) This point, often called by the suggestive name "summing junction," turns out to be useful in several important circuits.

When do the Golden Rules apply?

Text sec. 4.08

Now that we have applied the Golden Rules a couple of times, we are ready to understand that the Rules sometimes do not apply:

- note a *preliminary assumption*: these rules are useful only for circuits that include—

    1. Feedback; and
    2. Feedback of the right flavor: *negative* feedback

- And note the careful wording of the first rule: "the output *attempts*...." This rule is like a guarantee written by a cautious (and prudent) lawyer. It warns a careful reader that the person designing op amp circuits retains an obligation to use his head: apparently there are circuits in which the op amp will be unable to deliver the desired result: it will *attempt* and fail. Let's look at some such cases, to be warned early on.

Try your understanding of the golden rules and their restrictions, by asking yourself whether the golden rules apply to the following circuits:

p. 182, item "2"

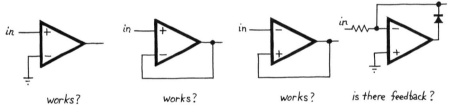

Figure N8.6: Do the Golden Rules apply to these circuits?

And will the output's "attempt..." to hold the voltages at its two inputs equal succeed, in these cases?

p. 182, item "1"

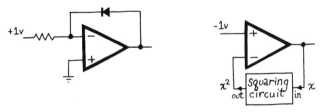

Figure N8.7: Will the output's "attempt..." succeed here?

More Applications: Improved Versions of Earlier Circuits

Nearly all the op amp circuits that you meet will do what some earlier (open-loop) circuit did—but they will do it better. This is true of all the op amp circuits you will see today in the lab. Let's consider a few of these: *current source, summing circuit, follower, and current-to-voltage converter.*

Current Source

xt sec. 4.07, fig. 4.11;
ıb 8-5

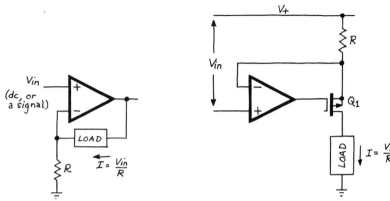

Figure N8.8: Op amp current sources

The right-hand circuit gives you a chance to marvel at the op amp's ability to make a device that's brought within the feedback loop behave as if it were *perfect*. Here, the op amp will hide both the slope of the I_D vs. V_{DS} curve in the "saturation" region (a slope that reveals that the FET is not a perfect current source) and the more radical departure from current-source performance in the "linear" region—a region one must stay out of when using a naked FET. Are you beginning to see *how* the op amp can do this magic? It takes some time to get used to these wonders. At first it seems too good to be true.

Summing Circuit

xt sec. 4.09,
185 fig. 4.19;
ıb 8-7

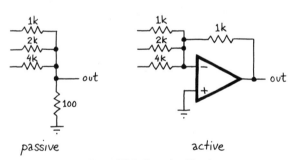

Figure N8.9: Summing Circuits

In the lab you will build a variation on this circuit: a potentiometer lets you vary the *DC offset* of the op amp output.

Followers

Text sec. 4.06;
sec. 4.09, p. 186

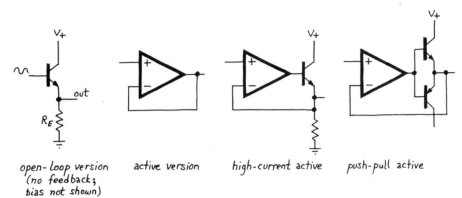

Figure N8.10: Op amp followers

How are the op amp versions better than the bare-transistor version? The obvious difference is that all the op amp circuits hide the annoying 0.6V diode drop. A subtler difference—not obvious, by any means, is the much better output impedance of the op amp circuits. How about input impedance?

Current-to-voltage converter

Text sec. 4.09,
p. 184

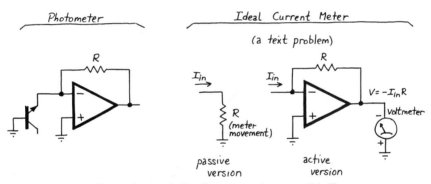

Figure N8.11: Two applications for I-to-V converter: photometer; "ideal" current meter

(*A Puzzle*: if you and I can design an "ideal" current meter so easily, why do our lab multimeters *not* work that way? Are we that much smarter than everyone else?)

Strange Things Can be put Into Feedback Loop

The push-pull follower within the feedback loop begins to illustrate how neatly the op amp can take care of and hide the eccentricities of circuit elements—like bipolar followers, or diodes.

Here's the cheerful scheme:

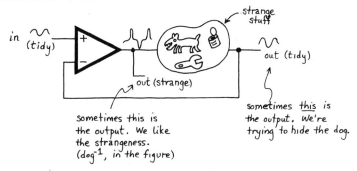

Figure N8.12: Op Amps can tidy up after strange stuff within the loop

In the push-pull follower, we treat the "tidied-up" signal as the output; the strange tricks the op amp output needs to perform to produce a tidy output do not interest us. In other circuits, however, the "strange signal" evoked by the "strange stuff" in the feedback loop may be precisely what *does* interest us. Here are two examples:

compare Text sec. 4.14, 212, fig. 4.35

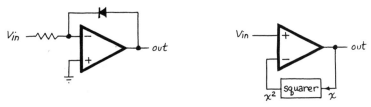

Figure N8.13: Two cases where we plant strange stuff in loop, to get "strange" and interesting op amp output

In both of these cases, far from trying to hide "the dog" (of Fig. N8.12, above), we are proud of him; so proud that we want to gaze at his image, which appears at the op amp output (the image is always dog^{-1}: inverse-dog).

In today's lab you will be so bold as to put the *oscilloscope* itself inside one feedback loop, with entertaining results:

lab 8-6

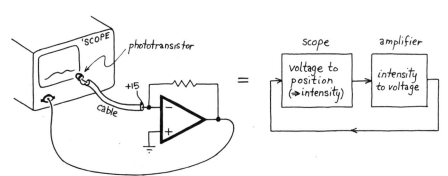

Figure N8.14: Scope brought within feedback loop: adjusts location of CRT beam

You'll see lots of nifty circuits in this chapter. Soon you may find yourself inventing nifty circuits. Op amps give you wonderful powers. (In case you find yourself wanting still more when you have concluded the orgy of cleverness that appears in the *Circuit Ideas* at the end of the Text's Chapter 4, see the books of application notes published by National Semiconductor, among others, or the application notes that follow many op amp data sheets, including the LF411's (see Text appendix K for the '411's data sheet).

Ch. 4: Worked Examples: Op Amps Idealized

Two worked examples:

1. an inverting amplifier
2. a summing circuit

1. • (-100) Amp

> *Problem: Inverting Amplifier*
>
> Design an inverting amplifier with a gain of –100, to be driven by a source whose output impedance is high and uncertain: 100k to 1MΩ.

The inverting amp is easy enough, apart from the impedance issues:

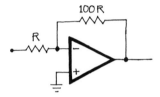

Figure X8.1: Inverting amplifier—but postponing the questions that call for some thought: part values

Will *any* pair of resistor values do, in the ratio 100:1? Does the high and uncertain source impedance matter, here?

A plausible—but wrong—first response might be, 'I don't have to worry about source impedance, because op amps have giant input impedances: that follows from the second golden rule, which says the inputs draw no current.'

You don't fall for that answer, though, because you can see that the golden rule describes how the op amp behaves, whereas what concerns us here is how the op amp *circuit* behaves. In this case, its input impedance is *not* the same as that of the op amp: the circuit input impedance is much lower. It is just R_1:

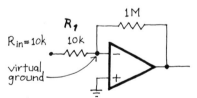

Figure X8.2: R_{in} of inverting amp: as low as R_1

And here's a plausible—but wrong—solution: just make R_1 much bigger than R_{source}: make it, say, 10MΩ. That's a good thought, but it implies that the feedback resistor should be 1000M—1 GΩ, and that is excessive, for reasons we will make sense of only when we admit that op amps are not quite as good as their idealized model.

Easy solution to the impedance problem: a follower

A *follower* solves the problem neatly:

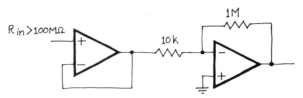

Figure X8.3: Follower buffers the input of the inverting amp

Now we can choose modest resistor values for the inverting amp, like those shown above. At first you may feel uncomfortable tacking in extra op amps to solve circuit problems. We hope you will soon get over this discomfort. A slogan worth remembering will recur in this chapter: *op amps are cheap*. They come 2 and even 4 to a package. One more op amp is no big deal, and often is the best way to refine a circuit.

2. 'Arithmetic'

> **Problem: Summing Circuit**
>
> Design a circuit that forms the following sum of the input voltages A, B:, and C.
>
> $$V_{out} = A + 2B - 3C$$
>
> Again let's make the source impedances high and uncertain: 100k to 1MΩ, just to drum home our earlier point.

And here's a solution. It is so similar to the preceding problem that it does not call for much explanation. Let's do the problem with just comments in "balloons":

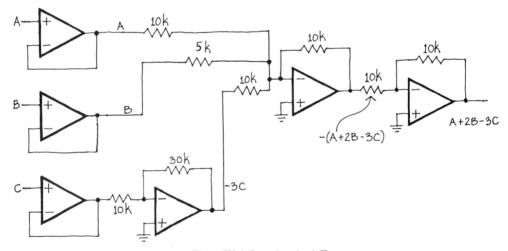

Figure X8.4: Summing circuit II

The function of this circuit—$V_{out} = A + 2B - 3C$—recalls why op amps were given that name: they can do mathematical operations—and other, fancier ones such as multiplication and division as well, with the help of log amps.

Lab 8: Op Amps I

> **Reading:** Chapter 4: 4.01-4.09, pp. 175-187.
> **Problems:** Problems in text.
> Bad Circuits B,D,F,G,I,K,L,M (none of these requires any deep understanding of op amps).

8-1 Open-Loop Test Circuit

Before we ask you to build your first op amp circuit we should remind you of two points:

- first, how the integrated circuit ("IC") package goes into the breadboard. (The package style is called "DIP:" "dual in-line package".)

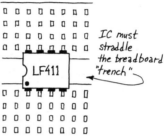

Figure L8.1: How the IC op amp goes into the breadboard: it straddles the trench

- second, a point that may seem to go without saying, but sometimes needs a mention: the op amp *always* needs power, applied at two pins; nearly always that means ±15V, in this course. We remind you of this because circuit diagrams ordinarily *omit* the power connections. On the other hand, many op amp circuits make *no* direct connection between the chip and *ground*. Don't let that rattle you; the *circuit* always includes a ground—in the important sense: common reference called zero volts.

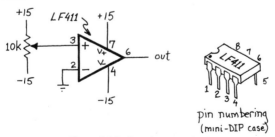

Figure L8.2: Open-loop test circuit

Astound yourself by watching the output voltage as you slowly twiddle the pot, trying to apply 0 volts. Is the behavior consistent with the 411 specification that claims "Gain (typical) = 200V/mV?"

8-2 Inverting Amplifier

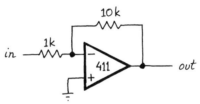

Figure L8.3: Inverting amplifier

Construct the inverting amplifier drawn above. Drive the amplifier with a 1kHz sine wave. What is the gain? What is the maximum output swing? How about linearity (try a triangle wave)? Try sine waves of different frequencies. Note that at some fairly high frequency the amplifier ceases to work well: *sine in* does not produce *sine out*. (We will postpone until next time measuring the *slew rate* that imposes this limit; we are still on our honeymoon with the op amp: it is still *ideal*: "Yes, sweetheart, your slewing is flawless").

Now drive the circuit with a sine wave at 1kHz again. Measure the input impedance of this amplifier circuit by adding 1k in series with the input.

Measure the output impedance (or *try* to measure it, anyway). Note that no blocking capacitor is needed (why?). You should expect to fail, here: you probably can do no more than confirm that Z_{out} is very low. Do not mistake the effect of the op amp's limited *current* output for high Z_{out}. You will have to keep the signal quite small here, to avoid running into this current limit. The following curves say this graphically.

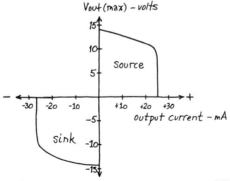

Figure L8.4: Effects of limit on op amp output current (LF411)

These curves say, in compact form, that the current is limited to ±25 mA over an output voltage range of ±10V, and you'll get less current if you push the output to swing close to either rail ("rail" is jargon for "the supply voltages").

Note to zealots: a student who tries very hard to measure R_{out}, by loading the op amp with a very small resistor—say, 1Ω—may see a bizarre result: the output amplitude *grows* under load. If you see that, you are seeing the effect of sneaky *positive* feedback, applied through a voltage divider formed by the 1Ω load and the small resistance of the ground line:

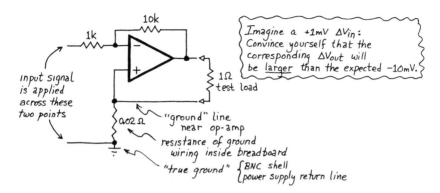

Figure L8.5: Sneaky positive feedback can boost the signal when small R_{load} lets substantial currents flow in the ground lines

Don't *look* for this exotic effect; it's not worth your time. This note is addressed only to the poor student confronted with this strange behavior.

8-3 Non-inverting Amplifier

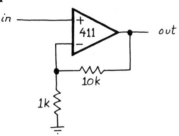

Figure L8.6: Non-inverting amplifier

Wire up the non-inverting amplifier shown above. What is the voltage gain? (It is *not* the same as for the inverting amp you just built.)

Try to measure the circuit's input impedance, at 1kHz, by putting a 1Meg resistor in series with the input. Here, watch out for two difficulties:

- Once again, *beware* the finding, "10 M ohms".
- R_{in} is so huge that C_{in} dominates. You can calculate what C_{in} must be, from the observed value of f_{3dB}. Again make sure that your result is not corrupted by the scope probe's impedance (this time, its capacitance).

Does this configuration maintain the low output impedance you measured for the inverting amplifier? (You can answer this question without doing the experiment, if you redraw the circuit to reveal that, for the purpose of this measurement, it is *the same circuit* as the "inverting amp"!)

8-4 Follower

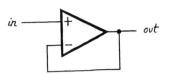

Figure L8.7: Op amp follower

Build the *follower* shown above, using a 411. Check out its performance. In particular, measure (if possible) Z_{in} and Z_{out} (but don't wear yourself out: you already know the answer, if you recognize the *follower* as a special case of one of the circuits you built a few minutes ago).

8-5 Current Source

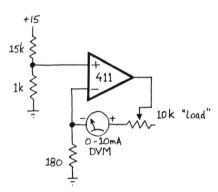

Figure L8.8: Current source

Try the op-amp current source shown above. What should the current be? Vary the *load* pot and watch the current, using a digital multimeter.

Note that this current source, although far more precise and stable than our simple transistor current source, has the disadvantage of requiring a "floating" load (neither side connected to ground); in addition, it has significant speed limitations, leading to problems in a situation where either the output current or load impedance varies at microsecond speeds.

The circuit below begins to solve the first of these two problems: this circuit sources a current into a load connected to ground.

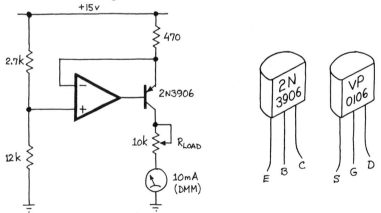

Figure L8.9: Current source for load returned to ground

Watch the variation in I_{out} as you vary R_{load}. To understand why the current source fails when it does, it may help to use a second meter to watch the voltage across the transistor: V_{CE} or V_{DS}.

Try using a bipolar transistor: a 2N3906. Then replace that transistor with a VP01 MOSFET (its pinout is equivalent, so you can plug it in exactly where you removed the 2N3906).

Should the circuit perform better with FET or with a bipolar transistor? Do you find a difference that confirms your prediction? Does the FET's *linear region* restrict the range of circuit performance as it did for the simple FET current source you built in Lab 7? With either kind of transistor the current source is so good that you will have to strain to see a difference between FET and bipolar versions. Note that you have no hope of seeing this difference if you try to use a VOM to measure the current; use a DVM.

8-6 Current to Voltage Converter

a) *Photodiode*

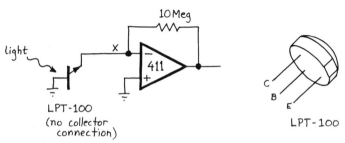

Figure L8.10: Photodiode photometer circuit

Use an LPT100 phototransistor as a photo*diode* in the circuit shown above (you may, instead, use an FPT100 or FPT110: pinouts are the same as for the LPT100; specs are very similar). Look at the output signal (if the dc level is more than 10 volts, reduce the feedback resistor to 1Meg).

If you see fuzz on the output—oscillations—put a small capacitor in parallel with the feedback resistor: 0.001 µF should be big enough for either this or the phototransistor circuit below, even with its smaller $R_{feedback}$. Why does this capacitor douse the oscillation? (*Hint*: what does it do to the circuit's gain at high frequencies?)

What is the average dc output level, and what is the percentage "modulation?" (The latter will be relatively large if the laboratory has fluorescent lights.) What input photocurrent does the output level correspond to? Try covering the phototransistor with your hand. Look at the "summing junction" (point **X**) with the scope, as V_{out} varies. What should you see?

Make sure you understand how this circuit is preferable to a simpler "current-to-voltage converter," a resistor, used thus:

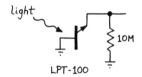

Figure L8.11: A less good photodiode circuit

b) *Phototransistor*

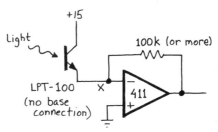

Figure L8.12: Phototransistor photometer circuit

Now connect the LPT100 as a photo*transistor*, as shown just above (the base is to be left *open*, as shown). What is the average input photocurrent now? What about the percentage modulation? Look again at the summing junction.

c) *Applying the Photometer Circuit (optional)*

If you put the phototransistor at the end of a cable connected to your circuit, you can let the transistor look at an image of itself (so to speak) on the scope screen. (A BNC with grabbers on *both* ends is convenient; note that in this circuit *neither* terminal is to be grounded, so do not use one of the breadboard's fixed BNC connectors.) The image appears to be shy: it doesn't like to be looked at by the transistor. Notice that this scheme brings the scope within a feedback loop.

Figure L8.13: Photosensor sees its own image

You can make entertaining use of this curious behavior if you cut out a shadow mask, using heavy paper, and arrange things so that the CRT beam just peeps over the edge of the mask. In this way you can generate arbitrary waveforms.

If you try this, keep the "amplitude" of your cut-out waveform down to an inch or so. Have fun: this will be your last chance, for a while, to generate really silly waveforms: say, Diamond Head, or Volkswagen, or Matterhorn. You will be able to do such a trick again—at least in principle—once you have a working computer, which can store arbitrary patterns in memory in *digital* form. Practical arbitrary-waveform generators use this digital method.

8-7 Summing Amplifier

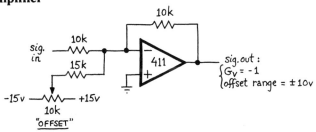

Figure L8.14: Summing circuit: DC offset added to signal

The circuit in the figure above sums a DC level with the input signal. Thus it lets you add a DC offset to a signal. (Could you devise other op amp circuits to do the same task?)

8-8 Push-pull Buffer

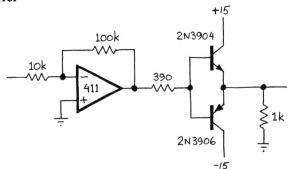

Figure L8.15: Amplifier with push-pull buffer

Build the circuit shown above. Drive it with a sine wave of 100Hz-500Hz. Look at the output of the op-amp, and then at the output of the push-pull stage (make sure you have at least a few volts of output, and that the function generator is set for no dc offset). You should see classic crossover distortion.

Listen to this waveform on the breadboard speaker. But before you drive the speaker you should determine the maximum safe amplitude, given the following power ratings:
> transistors: 350 mW
> speaker: 250 mW.

Now reconnect the right side of the feedback resistor to the push-pull output (as in Text figure 4.22), and once again look at the push-pull output. The crossover distortion should be eliminated now. If that is so, what should the signal at the output of the op-amp look like? Take a look. (Doesn't the op amp seem to be *clever*!)

Listen to this improved waveform: does it sound smoother than the earlier waveform? Why did the crossover distortion sound buzzy—like a higher frequency mixed with the sine?

If you increase signal frequency, you will discover the limitations of this remedy, as of all op amp techniques: you will find a glitch beginning to reappear at the circuit output.

Class 9: Op Amps II: Departures from Ideal

Topics:

- *old:*
  - passive versions of circuits now built with op amps: integrator, differentiator, rectifier

- *new:*
  - three more important circuits (applications):
    - integrator
    - differentiator
    - rectifier
  - op amp departures from ideal
    - offset voltage
    - bias current
    - offset current
    - frequency limitations: open-loop gain; slew rate
    - output current limit

Today we end our honeymoon with the op amp: we admit it is not ideal. But we continue to admire it: we look at more applications, and as we do, we continue to rely on our first, simplest view of op amp circuits, the view summarized in the *Golden Rules*.

After using the Golden Rules to make sense of these circuits, we begin to qualify those rules, recognizing, for example, that op amp inputs draw *a little* current. Let's start with three important new applications; then we'll move to the gloomier topic of op amp imperfections.

1. *Three More Applications:* Integrator, Differentiator, Rectifier

Integrator

To appreciate how very good an op amp integrator can be, we should recall the defects of the simple RC "integrator" you met in Chapter 1.

Passive RC integrator

Figure N9.1: RC "integrator:" integrates, sort-of, if you feed it the right frequency

To make the RC behave like an integrator, we had to make sure that

$$V_{out} \ll V_{in}$$

This kept us on the nearly-straight section of the curving exponential-charging curve, when we put a square wave in. The circuit failed to the extent that V_{out} moved away from ground. But the output *had* to move away from ground, in order to give an output signal.

Op amp version

The op amp integrator solves the problem elegantly, by letting us tie the cap's charging point to 0 volts, while allowing us to get a signal out. "Virtual ground" lets us have it both ways.

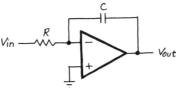

Figure N9.2: Op amp integrator: virtual ground is just what we needed

The op amp integrator is so good that one needs to prevent its output from sailing off to saturation (that is, to one of the supplies) as it integrates error signals: over time, a tiny lack of symmetry in the input waveform will accumulate; so will tiny op amp errors.

So, practical op amp integrators include some scheme to prevent the cap's charging to saturation:

a) One remedy: a large resistor in parallel with the cap (this leaks off a small current, undoing the effect of a small error current in);

Text p. 223

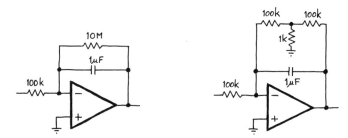

Figure N9.3: Integrator saved from saturation by resistor parallel $C_{feedback}$

Effects of the resistor

Evidently, the resistor compromises performance of the integrator. But we can figure out by *how much*. There are several alternative ways to describe its effects:

The resistor limits DC gain

In the circuit above, where R_{in} = 100k and $R_{feedback}$ = 10M, the DC gain is –100. So a DC input error of ±1 mV —> output error of ±100mV. The integrator still works fine, apart from this error.

The resistor allows a predictable DC leakage

Suppose we apply a DC input of 1V for a while; when the output reaches –1V, the error current is 1/100 the input or "signal" current (because $R_{feedback}$ = 100×R_{in}). This error grows if V_{out} grows relative to V_{in}.

The resistor does no appreciable harm above some low frequency

At some low frequency, X_C becomes less than R, and soon R is utterly insignificant. $X_C = R$, as you know, at f = $1/2\pi RC$. For these components—$R_{feedback}$ = 10M, C = 1µF—that frequency is about 0.01Hz!

A detail: how the resistor T works

Here's a diagram to persuade you that the clever T resistor arrangement does indeed make the 100k resistor look about 100X as large:

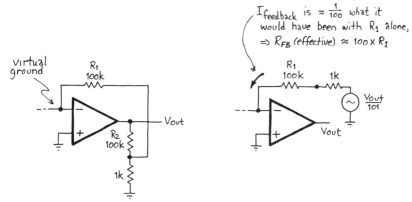

Figure N9.4: How the T arrangement enlarges apparent R values

Neat? The scheme is useful because the lower R_{Thevenin} of the T feedback network has two good effects:

- it generates smaller I_{bias} errors than the use of giant resistors would (we'll discuss this problem, below);
- it drives stray capacitance at the op amp input better (avoiding unintentional low-pass effects in the feedback); this is not important here, but is in a circuit where no capacitor sits parallel to a big feedback resistor.

b) Another remedy: a switch in parallel with the cap (this has to be closed briefly, from time to time).

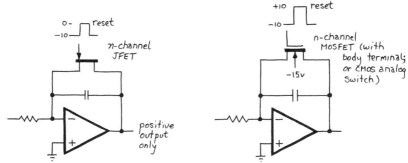

Figure N9.5: Integrator saved from saturation by discharge switch

The switch produces a more perfect integrator, but is more of a nuisance to drive.

Differentiator

Text sec. 4.20;
Lab 9-3

Again the contrast with a passive differentiator helps one appreciate the op amp version:

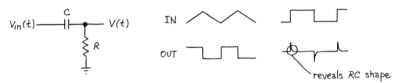

Figure N9.6: RC "differentiator:" differentiates, sort-of, if RC kept very small

To make it work, we must make sure that

Text sec. 1.14, p.25.

$$dV_{out}/dt \ll dV_{in}/dt$$

Again the op amp version exploits *virtual ground* to remove that restriction:

Text sec. 4.20,
pp. 114-25,
figs. 4.51, 4.52

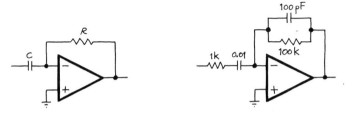

Figure N9.7: Op amp differentiator: simple (idealized); practical

This op amp differentiator is a little disappointing, however: it *must* be compromised in order to work at all. A practical differentiator, shown on the right in the figure above, turns into an integrator (of all things!) at some high frequency.

This scheme is necessary to prevent oscillations (we will look more closely at this topic a class or two hence).

Active Rectifier

Text sec. 4.10,
pp. 187-88, figs. 4.25, 4.26;
Lab 9-4

The simple passive rectifier of Chapter 1 was blind to inputs < about 0.6 v, and put an offset of that amount between input and output. The op amp version hides the diode drop:

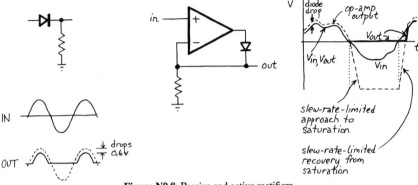

Figure N9.8: Passive and active rectifiers

The circuit shown *saturates* for one input condition. That is poor: produces an output glitch (caused by delay) as it comes up out of saturation. In the lab you will build an improved active rectifier that cleverly stays out of saturation (how does it work?)

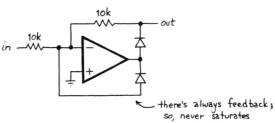

Figure N9.9: Improved active rectifier

2. Op Amp Departures from Ideal

Let's admit it: op amps aren't quite as good as we have been telling you: the Golden Rules exaggerate a bit:

- the inputs *do* draw (or squirt) a little current;
- the inputs are not held at precisely equal voltages.

Here are three circuits that always deliver a saturated output after a short time. They would not if op amps and all components were ideal:

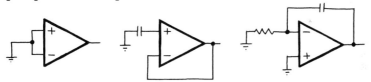

Figure N9.10: Three circuits sure to saturate. why?

Op Amp Errors

We will treat, in turn, the following op amp errors:

- *voltage offset*:
- *bias current*
  offset current
- frequency limitations: *open-loop gain roll-off; slew rate*
- *output current limit.*

Offset Voltage

> **Offset Voltage:** V_{os}
> "The difference in input voltage necessary to bring the output to zero..." (Text, p. 192)

This spec describes the amp's delusion that it is seeing a voltage difference between its inputs when it is not. The amp makes this mistake because of imperfect matching between the two sides of its differential input stage.

Symbol	Parameter	Conditions	LF411A min	LF411A typ	LF411A max	LF411 min	LF411 typ	LF411 max	Units
V_{OS}	Input Offset Voltage	$R_S=10k\Omega, T_A=25°C$		0.3	0.5		0.8	2.0	mV

Figure N9.11: 411 Spec: V_{offset}

You can compensate for this mismatch by deliberately pulling more current out of one side of the input stage than out of the other, to balance things again. This correction is called 'trimming offset,' and you will do it in today's lab. But this trimming is a nuisance, and the balancing does not last: time and temperature-change throw V_{offset} off again.

The better remedies are, instead—

- use a good op amp, with low V_{offset};
- design the circuit to work well with the V_{offset} of the amp you have chosen.

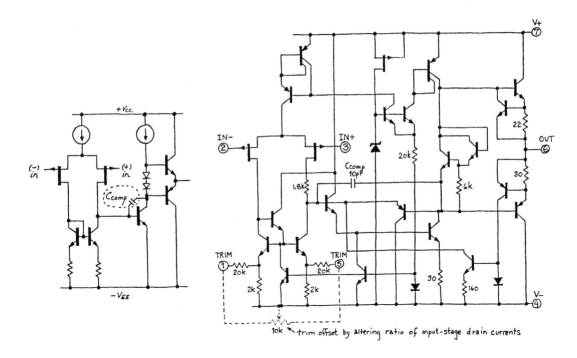

Figure N9.12: Inside the 411: schematics: simplified, and detailed

Bias Current

text sec. 4.11, p. 190;
sec. 4.12, p. 194

Symbol	Parameter	Conditions		LF411A			LF411		Units
			min	typ	max	min	typ	max	
I_{os}	Input Offset Current	$V_s = \pm 15V$ $T_J = 25°C$		25	100		25	100	pA
		$T_J = 70°C$			2			2	nA
		$T_J = 125°C$			25			25	nA
I_B	Input Bias Current	$V_s = \pm 15V$ $T_J = 25°C$		50	200		50	200	pA
		$T_J = 70°C$			4			4	nA
		$T_J = 125°C$			50			50	nA

Figure N9.13: 411 Specs: I_{bias} and I_{offset}

> **Bias Current: I_{bias}**
>
> I_{bias} is a DC current flowing in or out at the input terminals (it is defined as the average of the currents at the two terminals).

For an amplifier with bipolar transistors at the input stage, I_{bias} is base current; for a FET-input op amp like the 411, I_{bias} is a leakage current: it is tiny, therefore, but also grows rapidly with temperature:

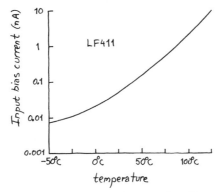

Figure N9.14: 411 bias current: tiny, but grows fast with temperature

The bias current flows through the resistive path feeding each input; it can, therefore, generate an input error voltage, which may be amplified highly to generate an appreciable output error: the Lab exercise uses a high-gain DC amplifier for just that purpose:

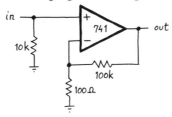

Figure N9.15: Lab circuit: uses high-gain dc amp to make errors measurable

But notice that the lab notes ask you to use a *741* op amp, not a *411*, to make the errors substantial. That requirement suggests that often you will *not* need to worry about the effects of bias current: true, but you should know how to judge whether or not to worry.

To minimize the effects of bias current, match the resistances of the paths that feed the two op amp inputs. Here are examples of circuits that do or do not balance paths:

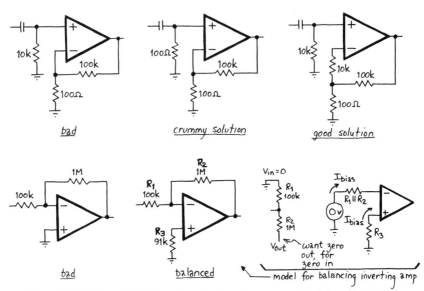

Figure N9.16: Balanced Resistive paths minimize output errors resultinG from bias current

Once you have balanced these resistive paths, I_{bias} no longer causes output errors. But a difference between currents at the inputs still does. That difference is called—

Offset Current

Text sec. 4.11, p. 190;
sec. 4.12, p. 195

> **Offset Current: I_{bias}**
> The difference between the bias currents flowing at the two inputs.

For the 411 the I_{OS} specification is about 1/2 I_{bias}; for the bipolar op amps I_{OS} is smaller relative to I_{bias}. But recall how *tiny* I_{bias} is for the 411 and other FET-input devices.

As noted just above, even when the resistances seen by the two inputs are balanced, an error will occur because of this *difference* in currents. Remedy? Use resistances of moderate value. (< a few 10's of Megohms; recall the argument for the clever *T* resistor trick noted above.)

Note, by the way, that even if bias current were zero, still you would need to provide a DC connection to each op amp input, to define the voltage there; otherwise stray capacitance gradually would charge with leakage currents (in the PC board, if nowhere else). So, these two circuits are bad:

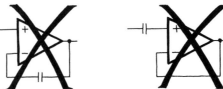

Figure N9.17: One must provide a DC connection to each op amp input

Measuring and correcting effects of V_{offset} and I_{bias}, in lab exercise 9-1

The lab notes suggest that you go through this process in a particular sequence. Make sure you understand why you are asked to proceed as stated there:

You start with a high-gain amp (×1000) that will show large output errors for small input errors. At the outset, the effects of V_{offset} and I_{bias} are *commingled*. You cannot tell, looking at the output, what the effect of either error is, taken by itself. Their effects may even tend to cancel.

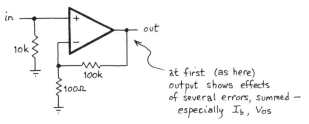

Figure N9.18: Lab 9-1's high-gain DC amp, once more

The procedure suggested goes this way:

- arrange things so that the effects of I_{bias} are negligible. (How?)

    - Measure the output error; infer the input error, and thus V_{OS}.
    - Trim V_{OS} to a minimum.

- arrange things so that I_{bias} causes an input error.

    - measure the output error so caused, and infer the input error, and thus I_{bias}.
    - Alter the circuit so as to minimize the error caused by I_{bias}

- Infer I_{OS} from the remaining output error.

Make sense?

AC Amplifier: An elegant way to minimize effects of I_{bias}, V_{OS} and I_{OS}

Text sec. 4.05,
p. 179, fig. 4.7;
Lab 10-1

If you need to amplify AC signals only, you can make the output errors caused by V_{OS}, I_{bias} and I_{OS} negligible in a clever way: just cut the DC gain to unity:

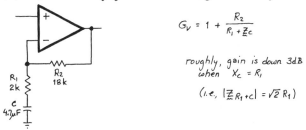

Figure N9.19: AC amplifier: neatly makes effects of small errors at *input* small at *output*

What's its f_{3dB}? If one ignores the "1" in the gain expression, output amplitude is down 3dB when the denominator in the gain expression is $\sqrt{2}(R_1)$. But that happens when $X_C = R_1$, and that happens, as you well know, at $f_{3dB} = 1 / (2\pi R_1 C)$.

This is the same notion you used to choose the emitter-bypassing capacitor, back in Chapter 2:

Text sec. 2-13, p.85

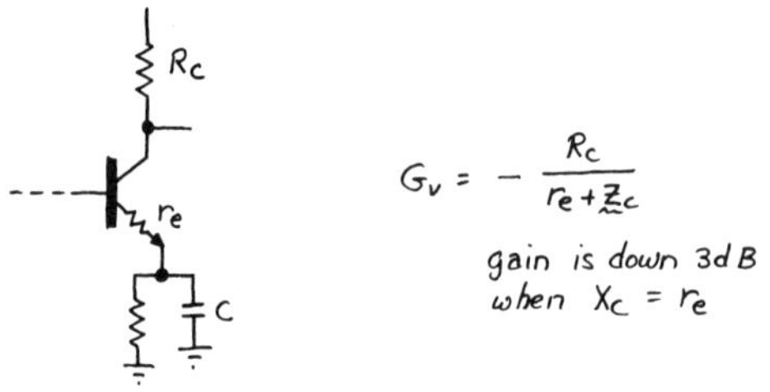

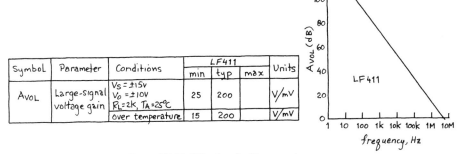

Figure N9.20: Gain is down 3dB when denominator (series impedance of R and C) is up to R√2: true for bypassed-emitter amp, and for op-amp AC amp

Slew Rate & Roll-off of Gain

Text sec. 4.11, pp. 191-92; sec. 4.12, p. 193

These effects turn out to be caused, deliberately, by a gain-killing capacitor planted within the op amp. We will talk about this *compensation* device next time, when we consider op amp stability. For the moment, we will note that the op amp's gain falls off at –6dB/octave (as if the output had been passed through a simple RC low-pass: in effect, it has been!); so the chip's very high gain, necessary to make feedback fruitful, evaporates steadily with increasing frequency—and is *gone* at a few MHz (about 4 MHz for the 411).

Symbol	Parameter	Conditions	LF411 min	LF411 typ	LF411 max	Units
A_{VOL}	Large-signal voltage gain	$V_S = \pm 15V$, $V_O = \pm 10V$, $R_L = 2k$, $T_A = 25°C$	25	200		V/mV
		over temperature	15	200		V/mV

Figure N9.21: 411 gain roll-off: spec and curves

In addition, op amps misbehave in some odder ways: they think, for example, that they see a voltage difference between their inputs even when there is none.

These specifications define an upper limit on the usefulness of *all* op amp circuits; that limit explains why not every circuit should be built with op amps, wonderful though the effects of feedback are.

Output Current Limit

Text sec. 4.11, p. 191; sec. 4.12, p. 194

This is a self-protection trick inserted into the output stage, to protect the small transistors there from the heating that otherwise would result when some clumsy user overloaded the amp. You saw a curve like this one in the first op amp lab:

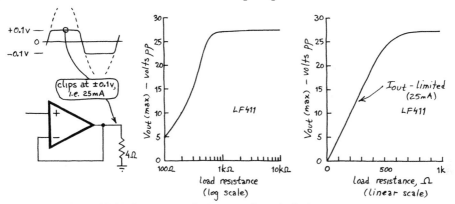

Figure N9.22: Output current limit: output clips under load, despite very low R_{out}

Ch.4: Worked Examples: Integrators; Imperfect op amps

Two worked examples:

1. Integrator design
2. Calculating effect on integrator of op amp errors

---

1. Integrator Design:

> **Problem: Integrator**
> Design an op amp integrator that will ramp at + 1V/ms given a + 1V DC input. Include protection against drift to saturation, and let the input impedance be $\geq 10\text{M}\Omega$.

Solution

Let's start with a sketch, postponing the choice of part values. The *sign* of the output ramp, and the high required input impedance require a couple of extra op amps; but we don't mind: remember?: *op amps are cheap*.

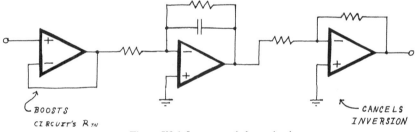

Figure X9.1: Integrator: skeleton circuit

The resistor in the feedback path will limit the DC gain, keeping the op amp output from sailing away to saturation.

A subtle point: why not balance paths for I_{bias}

We seem to be violating our own design rules by failing to provide a resistor to ground on the non-inverting side of the integrator. Here's the argument that says it turns out better *not* to do such balancing in this context:

- we would use a low-bias-current op amp in an integrator; probably we would use an op amp with FET input and the pA currents that are usual for this type. This kind of op amp's I_{OS} will be only about a factor of two better (lower) than its I_{bias}.
- why not take that factor of two, though? Because a side effect of the large R on the non-inverting terminal is increased vulnerability to noise there. In the present case, we would use 100k at that point, whereas now the non-inverting terminal is driven by a good low impedance (ground!).

Now for part values: we want DC gain of around 100, so we don't want R_1 huge: make it, say, 100k; then R_{feedback} can be about 10M.

Given R_1, we can solve for the required C using our usual description of a capacitor's behavior, $I = C\ dV/dt$. I is the current that flows when the 1V input is applied; dV/dt was given us as a design goal. We can solve for C:

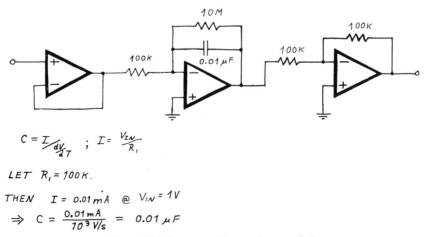

$C = I / \frac{dV}{dT}$; $I = \frac{V_{IN}}{R_1}$

LET $R_1 = 100K$.

THEN $I = 0.01\ mA$ @ $V_{IN} = 1V$

$\Rightarrow C = \frac{0.01\ mA}{10^3\ V/s} = 0.01\ \mu F$

Figure X9.2: Integrator with part values specified

2. Op Amp Errors: effects on an integrator

An integrator will show the effects of even small DC errors, over time. In the next example we will try to calculate the size of those output errors.

> **Problem: Effect on integrator of op amp errors**
>
> What output drift rate would you see in the circuit below, assuming that you use each of the listed op amps. The circuit input is grounded.
>
> $C = I / \frac{dV}{dT}$; $I = \frac{V_{IN}}{R_1}$
>
> Let $R_1 = 10K$
>
> Then $I = 0.1\ mA$
>
> $\Rightarrow C = \frac{0.1 \times 10^{-3} A}{10^3\ V/s}$ $0.1\ \mu F$ 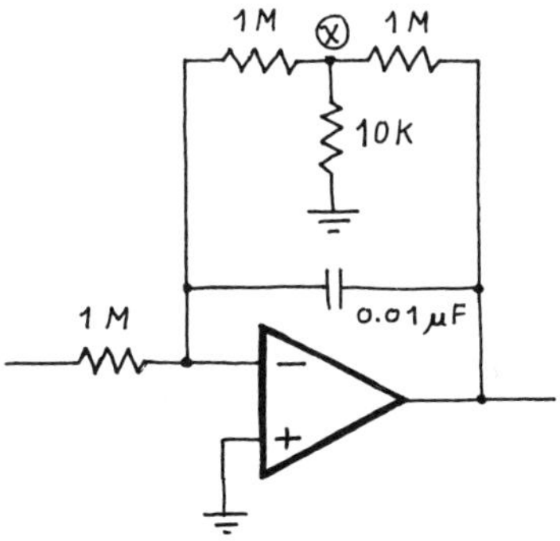
>
> Figure X9.3: Integrator: what drift rates? Proposed remedy: resistor "T" feedback network
>
> | Op amp type | V_{OS} | I_{bias} |
> |---|---|---|
> | 741C | 6mV | 500nA |
> | OP-07A | 25µV | 2nA |
> | LF411 | 2mV | 0.2nA |
>
> What happens if we add the indicated resistor network parallel to the feedback capacitor?

Solution:

The *offset voltage*, V_{OS}, causes the op amp to pull its inverting terminal (by means of the feedback network, of course) not exactly to ground, but to a voltage V_{OS} away from ground (we cannot predict the sign of this error). That error causes current to flow in the resistor; that current can't go into the op amp, so it flows into the capacitor.

The *bias current* flows into (or out of) the op amp's inverting terminal. This produces a drop across the input resistor, which must be canceled by an equal value of current flowing through the integrating capacitor (that is, it produces an output voltage *ramp*).

Worst case, these two currents flowing in the capacitor simply add. So, we get the following results:

Op amp type	V_{OS}	I_{bias}	I ←— V_{OS}	Sum of I's	dv/dt
741C	6mV	500nA	6nA	≈ 500nA	50V/s (50mV/ms)
OP-07A	25μV	2nA	25pA	≈ 2nA	0.2 V/s
LF411	2mV	0.2nA	2nA	2.2nA	0.2 V/s

Figure X9.4: Output errors for a particular integrator made with each of three op amps

Question: "What happens if we add the indicated resistor network parallel to the feedback capacitor?"

Answer:

The Resistor network…

Text sec. 4.19, fig. 4.49, p. 223

The network looks like about 100MΩ: one part in 100 of V_{out} reaches point 'X;' so, the current flowing through the leftmost resistor is about 1/100 what it would be if that resistor alone (1M) were in the feedback path. In other words, if we apply Ohm's Law —$R_{(apparent)}$ = $V_{out\text{-}op\text{-}amp}$ / I— we find that the network behaves like a resistor of about 100M (101M, if you care).

…Its Effect

The '100MΩ' resistor gives the circuit a DC gain of –100. So, instead of sailing off to saturation, the op amp output will begin to drift at about the rate determined in the earlier section of this problem—and then will slow and finally level off at –100×(*input error voltage*).

The input error voltage is the sum of V_{OS} and I_{bias} flowing in the resistance it "sees." What is that? It's the 1M input resistor parallel the other path, which looks like 1M + 10K ≈ 1M. Parallel, the two look like 0.5MΩ:

$$V_{error(in)} = (I_{bias} \times R_{Th\text{-}bias}) + V_{OS}$$

This input error (of undetermined sign) gets amplified by –100. Here are the specific results for the three op amps.

Op amp type	V_{OS}	I_{bias}	V ←— I_{bias} (I_{bias}×0.5MΩ)	Sum	Output Error
741C	6mV	500nA	0.25V	≈ –0.25V	+25V (saturation)
OP-07A	25μV	2nA	1mV	≈ 1mV	±100mV
LF411	2mV	0.2nA	0.1mV	2.1mV	≈ ±200mV

Figure X9.5: Output error (DC) for integrator with feedback resistance added

And here is a sketch of what the output voltage error would look like if we started with no charge on the cap: disaster for the '741, but tolerable results for both of the better op amps:

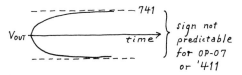

Figure X9.6: Feedback resistor limits integrator's output drift: effect quantified

Lab 9: Op Amps II

> **Reading:** Chapter 4.10-4.22, pp. 187-229.
> **Problems:** Problems in text.
> Additional Exercises 1-4.
> Bad Circuits A,C,H.

This lab introduces you to the sordid truth about op amps: *they're not as good as we said they were last time!* Sorry. But after making you confront op amp imperfections in the first exercise (9-1) we return to the cheerier task of looking at more op amp applications—where, once again, we treat the devices as ideal. On the principle that a person should eat his spinach before the mashed potatoes (or is it the other way round?) let's start by looking at the way that op amps depart from the ideal model.

9-1 Op-amp Limitations

a. Slew Rate

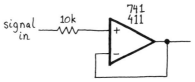

Figure L9.1: Slew rate measuring circuit. (The series resistor prevents damage if the input is driven beyond the supply voltages)

Begin by measuring slew rate and its effects, with the circuit above. We ask you to do this in two stages:

1) *Square wave input*

Drive the input with a square wave, in the neighborhood of 1kHz, and look at the output with a scope. Measure the slew rate by observing the slope of the transitions.

> *Suggestions:*
>
> – Find a straight central section; avoid the regions near "saturation"—near the limits of output swing;
> – Full slew rate is achieved only for strong "overdrive:" a large difference signal seen at the input of the amplifier;
> – The rates for slewing up and down may differ.

See what happens as the input amplitude is varied.

2) *Sine input*

Switch to a sine wave, and measure the frequency at which the output amplitude begins to drop, for an input level of a few volts. Is this result with the slew rate that you measured in part 1), just above?

Now go back and make the same pair of measurements (slew rate, and sine at which its effect appears) with an older op amp: a 741. The 741 claims a "typical" slew rate of 0.5V/μs; the 411 claims 15V/μs. How do these values compare with your measurements?

b. Offset Voltage

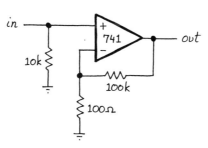

Figure L9.2: Offset measuring circuit

Now construct the ×1000 non-inverting amplifier shown above. Measure the offset voltage, using the amplifier itself to amplify the input offset to measurable levels.

Note: use a *741*, not a *411*, for the remainder of this exercise (9-1). The 411 is too *good* for this exercise: its bias current is so tiny that you would not see appreciable errors attributable to I_{bias}. (You might reasonably infer that you can forget about I_{bias}, simply by choosing a good op amp. Often you can. This exercise means to prepare you for the unusual case in which I_{bias} *does* produce troublesome errors.)

1) *Measure* effects of V_{offset}

The trick here, where you are to look for the effect of *offset voltage*, is to arrange things so that you can measure that effect *alone*, eliminating effects of *bias current*. To do this, you need to think what to do with the "in" terminal so as to make the effects of I_{bias} negligible. The 741's typical I_{bias} is 0.08μA (80 nA).

Compare your measured offset voltage with specs: $V_{OS} = 2mV(typ), 6mV(max)$.

2) *Minimize* the effects of V_{offset}: offset trim

Figure L9.3: 741 offset trimming network

Trim the offset voltage to zero, using the recommended network (figure above).

c. Bias Current

Now remove the connection from "in" to ground that you should have used in part (B) (either a short or a 100Ω resistor). Now the input again is connected to ground only through a 10K resistor. Explain how this input resistor allows you to measure I_{bias}. Then compare your measurement with specs: $I_{bias} = 0.08μA(typ), 0.5μA(max)$.

d. Offset Current

Alter the circuit in such a way that both op-amp input terminals see 10k driving resistance, yet the overall voltage gain of the circuit is unchanged. This requires some thought.

> *Hints*:
>
> – You will need to add one resistor somewhere;
> – That resistor should carry the bias current that is flowing to the inverting terminal.
> – The goal is to let the junction of the two feedback resistors sit at ground while the inverting terminal is allowed to sit below ground, at a voltage equal to that of the inverting terminal.)

Once you have done this, the effects of bias current are canceled, and only the effect of "offset current" (the difference between bias currents at the two op-amp input terminals) remains as an error. Calculate I_{os} from the residual DC level at the output; compare with specs: $I_{os} = 0.02\mu A(typ), 0.2\mu A(max)$.

Note: In the remainder of this lab except 9-4, and in all other op amp exercises, use an LF411 op amp, not the 741. In 9-4, where we ask you to use a *single-supply type*, use the '358. You will not need the 741 again.

9-2 Integrator

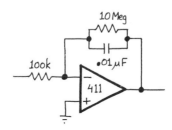

Figure L9.4: Integrator

Construct the active integrator shown above. Try driving it with a 1kHz square wave. This circuit is sensitive to small DC offsets of the input waveform (its gain at DC is 100); if the output appears to go into saturation near the 15 volt supplies, you may have to adjust the function generator's **OFFSET** control. From the component values, predict the peak-to-peak triangle wave amplitude at the output that should result from a 2V(pp), 500Hz square wave input. Then try it.

What is the function of the 10Meg resistor? What would happen if you were to remove it? Try it. Now have some fun playing around with the function generator's DC offset — the circuit will help you gain a real gut feeling for the meaning of an integral!

9-3 Differentiator

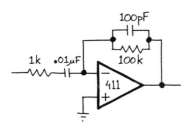

Figure L9.5: Differentiator

The circuit above is an active differentiator. Try driving it with a 1kHz triangle wave.

The differentiator is most impressive when it surprises you. It may surprise you if you apply it to a *sine* from the function generator: you might expect a clean cosine. In fact, some generators (notably the Krohn-Hite generators that we prefer in our lab) will show you a differentiated waveform that reveals the purported *sine* to be a splicing of more-or-less straight-line segments. This strange shape reflects the curious way the sine is generated: it is a triangle wave with its point whittled off by a ladder of four or five diodes. The diodes cut in at successively higher voltages, rounding it more and more as the triangle approaches its peak:

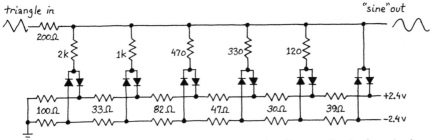

Figure L9.6: Sketch of standard function-generator technique for generating sine from triangle

You may even be able to count the diodes revealed by the output of the differentiator.

A note on stability:

Here we are obliged to mention the difficult topic of stability, a matter treated more fully in a later lab (Lab 10: op amps III). Differentiators are inherently unstable, because a true differentiator would have an overall 6dB/octave rising response; as explained in the text section 4.20, this would violate the stability criterion for feedback amplifiers. To circumvent this problem, it is traditional to include a series resistor at the input, and a parallel capacitor across the feedback resistor, converting the differentiator to an integrator at high frequencies. That is disappointing—and you may notice the effect of this network: it is most evident as a deviation of phase shift, at some frequencies, from the 90° that you would expect. Incidentally, a faster op amp (one with higher f_T) would perform better: the switch-over to integrator must be made, but the faster op amp allows one to set that switchover point at a higher frequency.

9-4 AC amplifier: microphone amplifier

A. Single-supply Op Amp

In this exercise you will meet use a "single-supply" op amp, used here to allow you to run it from the +5V supply that later will power your computer. This op amp, the 358 dual (also available as a "quad" — the 324) can operate like any other op-amp, with $V_+ = +15V$, $V_- = -15V$; however, it can also be operated with $V_- = GND$, since the input operating common-mode range includes V_-, and the output can swing all the way to V_-. Our application here does not take advantage of the single-supply op amp's hallmark: its ability to work right down to its negative supply (ground), incidentally. Often that is the primary reason to use a single-supply device.

> **NOTE**: build this circuit on a private single breadboard strip of your own, so that you can save the circuit for later use: it will feed your computer. This is the first of three such circuits that you will build; you can put them each on a single strip, or you can build them all on a larger board.

Here the 358 is applied to amplify the output of a microphone—a signal of less than 20 mV—so as to generate output swings of a few volts. The "AC amplifier" configuration, you will notice, is convenient here: it passes the input bias voltage to the output, without amplification (gain = 1 at DC).

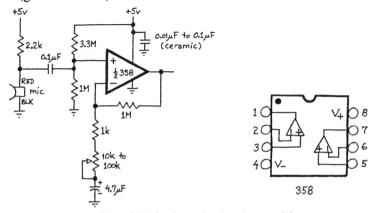

Figure L9.7: Single-supply microphone amplifier

The microphone is an "electret" type (the sound sensor is capacitive: sound pressure varies the spacing between two plates, thus capacitance; charge is held nearly constant, so V changes with sound pressure, according to $Q = CV$); it includes a FET buffer within the package. The FET's varying output current is converted to an output voltage by the 2.2k pullup resistor. So, the output impedance of the microphone is just the value of the pull-up resistor: 2.2k.

If you are troubled by oscillations on the amp output, try isolating the power supply of the microphone, thus:

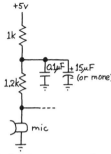

Figure L9.8: Quieting power supply to microphone

You may find, after your best efforts, that your amplifier still picks up pulses of a few tens of millivolts, at 120Hz. The pulses look like this:

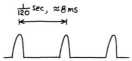

Figure L9.9: Ground noise on PB503 breadboard: caused by current pulses recharging filter capacitor

Probably you will have to live with these, unless you want to go get an external power supply (the adjustable supply you used in Lab 1 will do fine, here). These pulses show the voltage developed in the ground lines when the power supply filter capacitor is recharged by the peaks of the rectifier output. They *shouldn't* be there, but they are hard to get rid of. They appear because of a poor job of defining ground in the PB503 circuit, and you can't remedy that defect without rewiring the innards of the PB503.

9-5 Active Rectifier

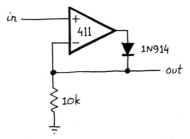

Figure L9.10: Active half-wave rectifier

Construct the active rectifier shown above. Note that the output of the circuit is not taken at the output of the op-amp. Try it with relatively slow sine waves (100Hz, say). Look closely at the output: What causes the "glitch"? Look at the op-amp output — explain. What happens at higher input frequencies?

9-6 Improved Active Rectifier

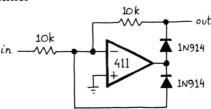

Figure L9.11: Better active half-wave rectifier

Try the clever circuit shown above. The glitch should be much diminished. Explain the improved performance. (You may want to look at the op amp output, to see the contrast with the earlier case.)

9-7 Active Clamp

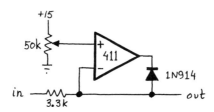

Figure L9.12: Active clamp

Try the op-amp clamp circuit shown above. (Note again that the circuit's output is not taken from the op-amp output; what significance does that have in terms of output impedance?). Drive it with sine waves at 1kHz, and observe the output. What happens at higher frequencies? Why?

Reverse the diode. What should happen?

CHAPTER 4 (continued) & CHAPTER 5

Class 10: Positive Feedback, Good and Bad:
Comparators, oscillators, and unstable circuits; a quantitative view of the effects of negative feedback

Topics:

- effects of feedback: toward a quantitative view
    - feedback: generalized model
    - generalizing the effect of feedback (quantitative account)
- positive feedback:
    - comparator: fast diff amp with versatile output stage
    - hysteresis ("Schmitt trigger")
        - why hysteresis?
        - how much hysteresis?
        - how choose component values for given hysteresis? contrasted with aiming for exact threshold voltages
    - two circuits using positive *and* negative feedback:
        - rc "relaxation oscillator"—easily built from schmitt trigger
        - negative impedance converter (nic)
    - more oscillators, good and bad:
        - good oscillators
            - square wave: the *classic* 555
            - sine wave: wien bridge
        - nasty oscillators: positive feedback that sneaks up on you
            - op amp circuits:
                - why they may oscillate: funny things in the loop
                - remedies: how to prevent oscillation: 'frequency compensation'
            - oscillations without op amps: follower
                - why it oscillates
                - remedies

A. Effects of Feedback: toward a quantitative view

We will leave to a couple of *worked examples* the task of confirming that R_{in}, R_{out}, among other circuit characteristics, are improved by the factor **1 + *AB*** (*A* and *B* are defined below). For the moment, let's settle for an intuitively-appealing sketch of the way feedback does this magic.

Here is the Text's generalized model of feedback:

Text sec. 4.26,
pp. 233, 234,
figs. 4.66, 4.68

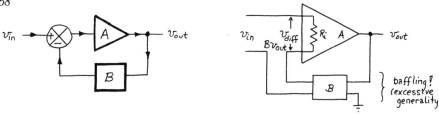

Figure N10.1: General feedback model

It is easier to understand if we redraw the *B* block to show a fraction of V_{out} fed back, as in the usual case.

Text fig. 4.69

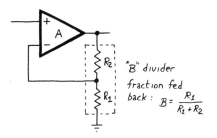

Figure N10.2: Feedback model redrawn

Then

"*B*" = fraction fed back (a characteristic of the circuit, not the chip)
"*A*" = open-loop gain (a characteristic of the chip).

These notions are enough to let us speak quantitatively of circuit characteristics that until now we could only call "big" or "small:" Z_{in}, Z_{out}, constancy of gain. For now, we will stop short of confirming the exact expressions for those characteristics (we will leave that task to a couple of worked examples). Here, we will try to show roughly *how* feedback improves input and output impedances.

Try R_{in} or R_{out} for the non-inverting amp:

Text sec. 4.25, p. 234

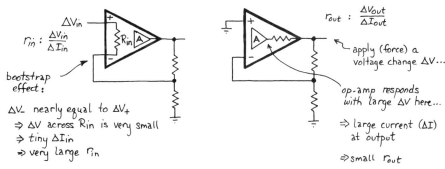

Figure N10.3: Qualitative argument: how feedback improves R_{in} and R_{out}: non-inverting amp

$\underline{R}_{in}$: R_{in} is *bootstrapped*: wiggle V_{in} and the inverting terminal wiggles about the same way; so, ΔV across the input is very small, and so, therefore, is ΔI. R_{in} is improved by the factor $(1+AB)$, it turns out. But this argument becomes silly when R_{in} is 10^{12}, as it is for the 411: then C_{in} dominates the input impedance; most of C_{in} is not bootstrapped, unfortunately.

$\underline{R}_{out}$:   The argument for R_{out} is similar: wiggle the output (apply a little ΔV); the amp responds by moving point X a lot, putting a large voltage across its $R_{out\text{-}chip}$. But this sources or sinks a large current, ΔI_{out}. No doubt you can see the nice result that follows—magically-low effective R_{out}.

A quantitative example

Try R_{in} or R_{out} for the non-inverting amp, assuming the following characteristics (here we anticipate one of the worked examples, which treats this problem in greater detail):

Assumed op amp specifications:

- Op amp's *hardware* R_{in} (differential, between its two inputs) = 1 MΩ
- Op amp's *hardware* R_{out} = 100Ω
- the amplifier has gain of 10
- the chip's open-loop gain is 1000 at 1 kHz

R_{in}, R_{out} each is *improved* by the factor $(1 + AB)$. (This argument is strongly reminiscent of the argument that a bipolar transistor improves what's on the far side of it by the factor $(1 + \beta)$ 675 this is what we called the 'rose-colored lens' effect.)

How large is the factor $(1 + AB)$? A is given as 1000; B is the fraction of the output swing that is fed back: here, it is divided by 10 (that is, $B = 0.1$; this is, of course, how the amp achieves its gain of 10). So, the product $\underline{AB}$ is 100. Therefore the hardware characteristics of the chip are improved by that factor:

R_{in} is boosted from 1MΩ to 100MΩ.
R_{out} is *reduced* from 100Ω to 1Ω.

(Notice that this effect is not quite the same as the effect of the transistor's β: β operated on whatever impedance was on the far side of the transistor; the op amp's $(1 + AB)$ operates on what's inside the feedback loop: one does not look *through* the op amp.)

Important Points:

The virtues of feedback circuit depend on both **A** and **B**: a circuit works best where *AB*, the "loop gain," is large. "A" falls with frequency; "B" is greatest in a follower, least when you look for very high gain (electronic justice, once again!).

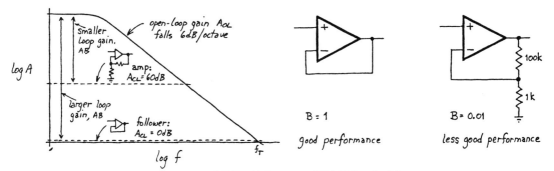

Figure N10.4: "A" Falls with frequency; "B" is highest for follower

B. Positive Feedback

1. *Benign* Positive Feedback
Text sec. 4.23

Positive feedback helps make a circuit decisive: forces the output quickly to extremes (whereas negative feedback moderates the circuit's performance; tends to prevent saturation, tends to stabilize a circuit, keeping the output away from extremes). In a *comparator*, such decisiveness is just what's needed.

What's a comparator? Just a high-gain differential amplifier: such an amplifier "compares" its two inputs, though we have not described its performance that way till now. What's distinctive about the *comparator* is not so much the device as the use to which it is put; an op amp can serve as a comparator, as suggested in the figure below. But—as you will confirm in the lab—it makes a second-rate comparator: it is slow. A chip built to do nothing else—a special-purpose comparator like the '311—works about 100 • faster. Here are the two of these devices (not yet practical circuits, as we will argue below):

Lab 10-3

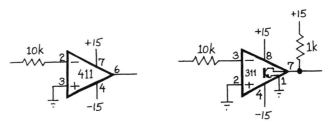

Figure N10.5: Two second-rate comparators: op amp (slow); 311 without feedback (indecisive)

The defects of these comparators appear if you imagine feeding them a noisy input like the one drawn below. Assume that you want the comparator to switch on the big, slow waveform's "zero-crossing." You do not want to switch on the little wiggles. (To make this hypothesis plausible, you might imagine that the slow waveform is 60 Hz, and our goal is to let a computer tell time by counting the 60 Hz zero-crossings.) Try drawing the comparators' response:

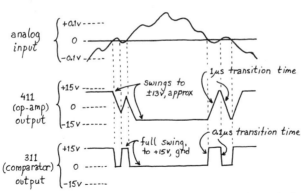

Figure N10.6: A comparator without feedback will misbehave; the op amp is *slow*, as well

Evidently, our clock is going to run fast!

In fact, the problem is even worse than it appears: even if the input waveform is smooth, its gentle slope alone can make the comparator output chatter indecisively as the input inches across threshold (zero volts, in the cases above). You can solve this problem, however.

A Good Comparator Circuit *always* Uses Positive Feedback

It turns out to be easy to make a comparator circuit ignore such small wiggles. Just feed back a small fraction of the output swing—but in a "positive" sense, so that the output swing tends to confirm the comparator's tentative decision. Such *positive feedback* makes the comparator decisive: it pats itself on the back, saying (in the manner of many of us humans), "Whatever you've decided to do must be right." The act of switching tends to reinforce the decision to switch.

Here you can confirm that the comparator circuit with hysteresis does not respond to the wiggles on the input.

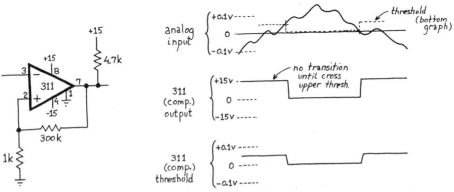

Figure N10.7: A good comparator in action

How Much Hysteresis?

If hysteresis makes the comparator decisive, how much should you design in? As much as possible? What do you lose as you enlarge hysteresis? How should you choose an value for
hysteresis, then?

Figure N10.8: Relation between noise and hysteresis

Two Circuits that use <u>Both</u> Positive and Negative Feedback

RC Relaxation Oscillator

Here's the lab circuit, which you recognize as just the Schmitt trigger (which was given unusually large hysteresis), feeding *itself*:

Lab 10-4;
Text sec. 5.13;
compare p. 285 fig. 5.29

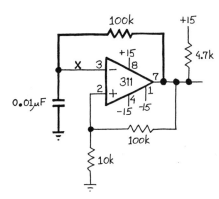

Figure N10.9: RC relaxation oscillator

What are the thresholds? How long to move V_{cap} between the thresholds?

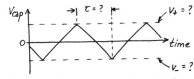

Figure N10.10: V_{cap} travels between the two thresholds: Linear Approximation

A *straight-line* approximation here is pretty good: the endpoint currents are larger and smaller, respectively, than the current when V_{cap} is at ground, so let's use that current: about 15V, then, across the feedback resistor (again, let's neglect the pullup resistor; we can afford a 5% error on one swing).

Negative Impedance Converter ("NIC")

ext sec. 5.03.
2.66 fig. 5.4

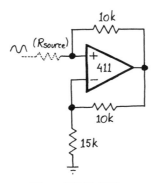

Figure N10.11: NIC

Two possible configurations; use one or other depending on R_{source}. (See below.)

Combines negative and positive feedback. Is it, then a *Golden Rule* circuit? Yes, if we arrange things right (we want it to be such a circuit, because we need to avoid saturated (clipped) outputs).

The circuit is stable *only if negative feedback predominates*. What constraint on R_{source} does that imply?

Wrinkles: effects of *"watching"* the circuit:

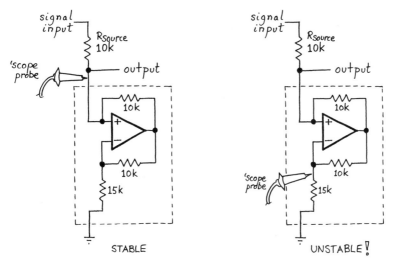

Figure N10.12: NIC: A *Marginally-stable* circuit

- "watch" only circuit output, and circuit is stable;
- "watch" only non-inverting input, and circuit oscillates;
- "watch" both op amp inputs, and circuit is stable again.

Why?

This marginal stability foreshadows troubles you will examine more closely in the lab, and which we discuss below under the heading 'nasty oscillators.'

Oscillators, good and bad

Let's look at two more *nice* oscillators, then switch to the intriguing topic of *nasties*. You will build oscillators of both sorts in lab, today.

A. Good Oscillators

Sometimes, of course, we *want* to build oscillators (in fact, you have built one, in Lab 10). Let's conclude with this cheerier topic. Here are some oscillators you will build in the lab:

1. 555 RC oscillator/timer

Text sec. 5.14

The 555 is nearly as common as the 741 once was: there are small books of 555 application notes. In the lab you will use an improved version: a 7555 (made of CMOS—remember these? MOSFET's—for low power; and capable of running faster than the original, bipolar device).

Because the 555[1] contains a *flip-flop*, a device you have not met, it's a little hard to explain. We'll try, though: here is a diagram of its insides, and then an informal explanation.

Text sec. 5.14,
fig. 5.32, p. 287;
Lab 10-3

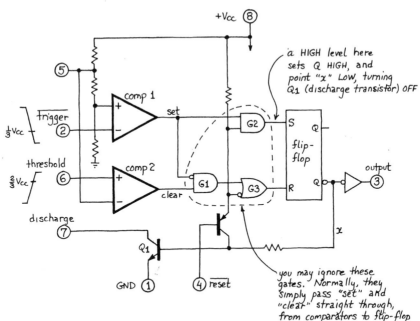

Figure N10.13: 555 oscillator

As in the relaxation oscillator you built last time, the capacitor voltage here moves between two thresholds, always frustrated. Lab 10's RC oscillator used a Schmitt trigger to provide these two thresholds. The 555 uses a simpler scheme: two comparators.

---

[1]. We'll call the chip 555 even though the version you will use is CMOS and likes to be called 7555; for the purpose of this explanation, the version of the 555 does not matter at all.

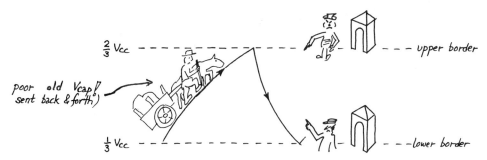

Figure N10.14: A very informal explanation of the 555's operation

The two comparators act like unfriendly guards at the 1/3 V_{CC}, 2/3 V_{CC} borders: when the capacitor voltage crosses either frontier, it finds itself sent back toward the other—like some sad, stateless refugee. The border guards send the capacitor voltage up and down by turning *on* or *off* a *discharge* transistor: this transistor pulls charge out of the cap, or—when off—allows the capacitor to charge up toward the positive supply.

You can vary the 555's waveform *at the capacitor* by replacing the resistors R_A and R_B with other elements: a current source or two, or—for R_A, but not R_B—even a piece of wire. Can you picture the waveforms that result, for the several possible configurations?

2. Sine Oscillator: Wien Bridge

text sec. 5.17;
lab 10-4

This oscillator—the only sine generator you will build in this course—is clever, and fun to analyze.

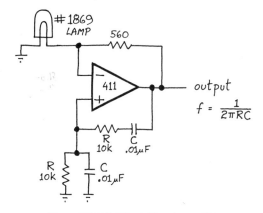

Figure N10.15: Wien bridge sine oscillator

Why does it put out a sine, rather than the usual square wave or triangle (those are the waveforms that are easy to generate)? Somehow it avoids *clipping* (that prevents the usual square output); somehow it selects a particular frequency rather than a large set of frequencies (that delivers a pretty good sine, rather than a complex waveform).

Here, for the purpose of analysis, we have broken the circuit into two fragments, each showing the feedback in one of the two senses: positive or negative:

Positive Feedback

The positive feedback network is frequency-selective, and at the most favored frequency passes a maximum of 1/3 of the output swing back to the + input; it treats all other frequencies—above and below its favored frequency—less kindly. Here's a sketch:

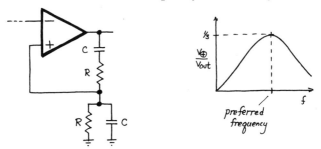

Figure N10.16: Wien bridge positive feedback: fraction of output fed back

At that favored frequency, the phase shift also goes to zero. The preferred frequency—the one at which the oscillator will run—turns out to be $1/(2\pi RC)$.

Negative Feedback

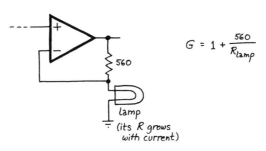

Figure N10.17: Self-adjusting gain: negative feedback of wien bridge, redrawn

The negative feedback—here redrawn to look more familiar—adjusts the gain, exploiting the lamp's current-dependent resistance (the lamp is rated at 14mA @ 10v). Convince yourself that the sense in which the lamp's resistance varies tends to stabilize gain at the necessary value. (What value of gain *is* necessary, to sustain oscillations without clipping?)

The detail of performance that is hardest to explain is the so-called "rubbery behavior" that results if one touches the non-inverting terminal while the circuit is running:

Figure N10.18: "Rubbery behavior" of Wien bridge oscillator, when one touches the non-inverting terminal: at two sweep rates

To explain this, recall that the negative feedback network is stable, but only marginally so because the lamp (with its long time-constant of response) looks like an extreme low-pass filter. So, this circuit resembles the circuits discussed below that raise stability questions: op amps driving heavy capacitive loads. It would not be satisfactory as a general purpose amplifier, because it is so jumpy: it overshoots and then takes a long time to stabilize again.

It is satisfactory in this application, because normally we do not disturb it: we do not poke it with a finger.

B. Nasty Oscillators

ext sec. 4.33, generally

Sometimes negative feedback turns positive. That's bad: makes a stable circuit unstable.

The circuit below looks, at a glance, as if it is getting *negative* feedback, and therefore should be stable:

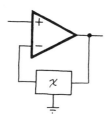

But whether this feedback is *negative*—making the circuit stable; or *positive*—making the output oscillate or saturate—depends, evidently enough, on what is in the box "X".

Figure N10.19: Stable? circuit

Here are two examples of seemingly-negative feedback that is positive in fact: odd things in the box "X" that could cause trouble. One of the examples is silly, the other realistic:

ext sec. 7.07;
ab 10-8

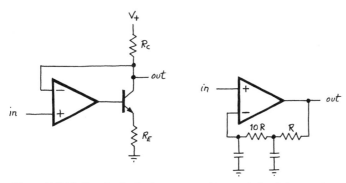

Figure N10.20: Negative feedback can turn positive: crude & subtle examples

Because of this possibility, all feedback circuits oblige one to take precautions to preserve stability. You may have seen some of your circuits oscillate—sometimes in cases where no positive feedback was apparent, and even (as in the case of the follower) where *none* of the conditions for oscillation seemed to be fulfilled.

What are those conditions, by the way? What is required for sustained oscillation?

gain: otherwise the disturbance has to die away (as it does, say, in the LC resonant circuit hit with a square wave); more specifically, as the disturbance appears at the output and is fed back to the input, then appears again at the output, it must not get smaller: net gain around the loop must be at least one.

positive feedback: the circuit must talk to itself, patting itself on the back, saying, 'Good. Do more of what you're doing.'

Let's try these notions on some unstable op amp circuits.

An Unstable Amplifier Circuit
Why does this circuit oscillate?

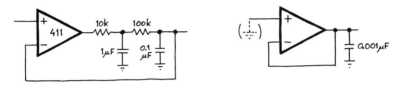

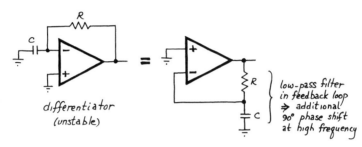

Figure N10.21: Unstable amplifier: more and less obvious cases

Answer: Phase Shifts Add:
Compare Text sec. 4.34, p. 244

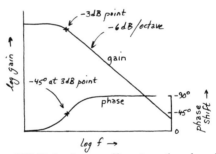

Figure N10.22: Any low-pass filter: attenuation; phase shift

Compare p. 244,
figs. 4.81, 4.83

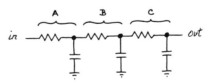

Figure N10.23: Cascaded low-pass filters

If you put several low-passes in cascade, you will find *180°* phase shift at some frequency.

At that frequency, *negative* feedback is transformed into *positive*. Trouble!

But—you may protest—*you* don't plan to put cascaded low-pass filters into a feedback loop, so this isn't a problem you need worry about.

You would be wrong if you said that. Look inside an op amp, and you discover that the several stages constitute low-pass filters in cascade, so that the op amp itself produces just such dangerous phase-shifting.

Here's a plot of phase shifts in an op amp:

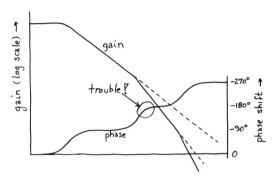

Figure N10.24: Phase shifts in an op amp ("uncompensated")

Remedy: Limit "Loop Gain"

We are stuck with these low-pass filters. At some frequency negative feedback will turn positive. What can we do to keep the circuit stable?

We can rig things so that *gain* is not sufficient, at the deadly 180°-shift frequency, to sustain an oscillation. That means, make sure gain is less than **one** at that frequency, for a disturbance trying to travel around the loop.

This limiting of gain is called *"frequency compensation."*

Here's a diagram meant to summarize the problem, and requirements for stability.

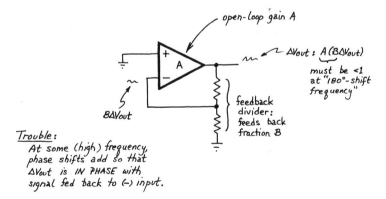

Figure N10.25: Conditions for stability versus oscillation

Assume there is noise around at all frequencies, disturbing the circuit input. That disturbance—call it ΔV_{in}—is amplified to produce a ΔV_{out} A times as big; then some or all of that ΔV_{out} is fed back to the input (a fraction "B" is fed back, so $AB(\Delta V_{in})$ is fed back. If that is as big as the original ΔV_{in}, or bigger, the circuit oscillates.

This view suggests remedies:

- limit the size of A;
- limit the size of B.
- Or limit both. In any case, we need to limit the product, $\underline{AB}$. Let's look separately at the ways to limit $\underline{A}$, then B.

Shrinking A is called "frequency compensation":

Text sec. 4.34,
pp. 245-46,
figs. 4.84, 4.86

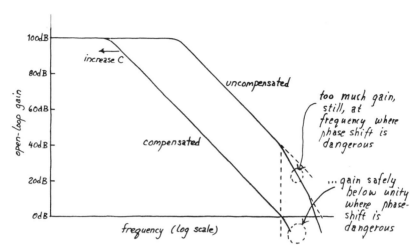

Figure N10.26: Remedy: circuit can be kept stable by killing open loop gain at dangerous (high-) frequencies

Shrinking B you achieve simply by designing a circuit with large voltage gain: one of these circuits is much more likely to oscillate than the other:

Figure N10.27: Remedy: circuit can be made less ready to oscillate by attenuating an output disturbance in feeding it back

*In summary: limit **loop gain, AB***

For stability, you need to hold the ***loop gain*** below unity at frequencies where dangerous phase shifts appear. This was a point we tried to illustrate with figure N10.25. Figure N10.4 showed the way loop gain varies with both A (A falls off with frequency) and with B (B depends on the circuit configuration). Where we first looked at the way loop gain depends on both circuit configuration and chip properties, we were concerned with the quantity AB because it describes the ability of the op amp to improve circuit performance. Here we are again concerned with the quantity AB, but for a reason quite different: AB also describes the ability of the circuit to cause trouble: to oscillate. A high loop gain, AB, gives good circuit performance, and it can cause stability problems. As usual, if you reach far for one thing you want, you may give up something else that you need. Of the two concerns, stability is the more urgent. We must hold AB safely low. We will settle for the circuit performance that results.

Oscillations without Op Amps

Feedback need not be explicit, as it is in op amp circuits, in order to cause trouble. Below is a circuit that oscillates because of very sneaky effects. Wouldn't you nominate this circuit as *least-likely to oscillate*? Isn't it a sure-fire dud?

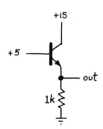

Figure N10.28: Follower: a circuit that "can't oscillate"—but does

Here's the argument that 'proves' that this circuit cannot oscillate:

- Gain is less than one;
- There is no feedback at all, so no positive feedback.

QED: it cannot oscillate. But it *does*. Why?

Redrawing the circuit to show stray capacitance and inductance reveals what must be going on:

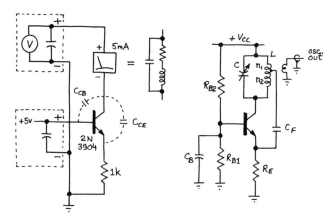

Figure N10.29: The follower *exposed!*: Showing stray C and L; and showing two similar oscillators, one accidental, one purposeful

Redrawn, this circuit is revealed to be nearly identical to a current source whose oscillations are explained in the Text. There, the circuit is likened to a purposeful oscillator called a 'Hartley LC oscillator.' Both of these circuits are shown above.

Class 10 Appendix: Op Amp Frequency Compensation

Appendix (for enthusiasts): *Op Amp Frequency Compensation*

These notes take a look at the way frequency compensation is implemented in an op amp. This is information you can do without, if you like: you can build stable circuits without knowing how the IC designers do their job. But if you're curious, here goes:

Frequency Compensation: a particular example: LF411 op amp

The huge gain of the 411 at DC (200,000) begins to fall at –6dB/octave around 20 Hz! At 4 MHz, gain is down to *1*. This seems strange, doesn't it? Even stranger—if you hadn't heard this before, anyway—the op amp's designers deliberately rolled off the gain. At first glance this looks silly: why build a high-gain amp, then ruin its gain? (Compare a similar riddle, challenging feedback itself: 'Why build a high-gain amp, and then throw away most of its gain with feedback?') You now know the answer: the open-loop gain, A, *must* be rolled off so as to prevent oscillations at the deadly frequency where before phase shifts add to 180°. Here's how the rolling-off is done.

The frequency-compensation is achieved with the help of Miller effect: a capacitor is planted between collector and base in the gain stage. Usually we think of Miller effect as a troublesome enemy; here the effect is harnessed to do some useful work: it magnifies a small capacitance, and cleverly imposes a uniform f_{3dB} on a set of op amps that may show varying *gain*:

Text sec. 4.34, p. 246, fig. 4.85

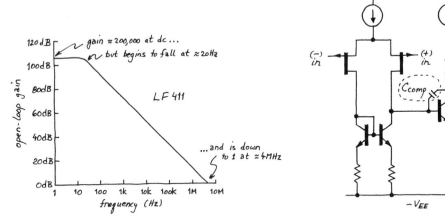

Figure N10A.1: Gain roll-off, and Miller capacitor that achieves this result: generic op amp, and simplified schematic of 411

One can redraw the op amp to show that Miller capacitor as the feedback cap in an *integrator* (of all things!).

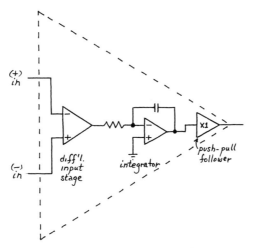

Figure N10A.2: Op amp redrawn to include *integrator*

Suddenly one discovers what sets the op amp's *slew rate*, as well as f_{3dB} for its gain. (You will find the 411 circuit explicated in a separate section called 'Op amp innards'.) The view that the op amp behaves like an integrator also warns us that the device imposes a 90° phase shift; that information helps us to see how even one additional low-pass in the feedback loop could get us to the deadly 180° shift.

Ch. 4: Worked Examples: Effects of Feedback

Two examples:

1. R_{out} for non-inverting amplifier and for inverting amplifier;
2. R_{in} for inverting amplifier used as current-to-voltage converter, or "transresistance" amplifier.

1. Non-inverting amplifier

text sec. 4.34;
gain plot: fig. 4.84, p. 245

> **Problem:** $\underline{R_{in}}$, $\underline{R_{out}}$: *non-inverting amplifier*
>
> Find R_{in} and R_{out} for a non-inverting amplifier made with the op amp described below. Find the values at DC and at 10kHz.
>
> Do not start by *assuming* improvement by the factor *(1 + AB)*; justify this result, for the case of R_{out}.
>
> Here are relevant op amp specs:
>
> - Op amp's *hardware* R_{in} (differential, between its two inputs) = 1 MΩ
> - Op amp's *hardware* R_{out} = 100Ω
> - the amplifier has gain of 10
> - the chip's open-loop gain curve (typical) looks like this:
>
>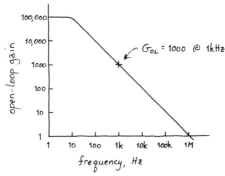
>
> **Figure X10A.1:** Open-loop gain, *A*, for this op amp

Solution:

Class Notes 10 make out the rough qualitative argument. Here, we will hurry on to the quantitative problem.

R_{out} for the non-inverting amplifier

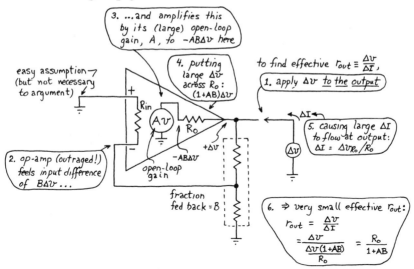

Figure X10A.2: Calculating R_{out} for the non-inverting amplifier

2. Inverting amplifier

compare Text exercise 4.11, 235

> **Problem: $\underline{R_{in}}$, $\underline{R_{out}}$: transresistance amplifier**
>
> 1. Find R_{in} for a *transresistance* amplifier—a current-to-voltage converter made by treating the – input to the op amp as input (as in the photosensor exercise in Lab 7). $R_{feedback}$ is 100k.
>
>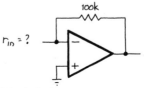
>
> Figure X10A.3: R_{in} for current-to-voltage converter
>
> 2. Calculate R_{out} for a ×10 **inverting** amplifier made with the op amp described by specs shown in question 1, above. Find the values at DC and at 10kHz (consult gain curve, question 1).

Solution:

A qualitative view

R_{in}: This is harder to state than it is for the non-inverting case. The general notion is simple: apply a ΔV to the inverting input, and the op amp output leaps, forcing a large current back at you through the feedback resistor. That large Δcurrent for small ΔV amounts to a very low input impedance.

We will leave the details to this diagram:

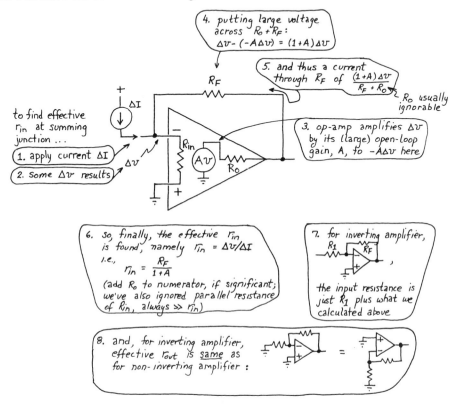

Figure X10A.4: Calculating R_{in} for inverting amplifier, shown as *transresistance amp* (no input resistor), and for inverting amplifier along the way

Inverting amplifier

The curious fact, noted in the diagram above, is that R_{out} for the inverting amplifier is not a new case at all. We can simply redraw the circuit to reveal it as *the same circuit* as the non-inverting amp (to make this evident, ground the input in both cases).

So, R_{out} again is the hardware value of R_{out}, reduced by the factor $1/(1 + AB)$. B is 0.1; A is 100,000 at DC, 100 at 10kHz; the values of AB at those two frequencies are 10,000 and 10, respectively. So R_{out} is

> At DC: $100\Omega / (1 + 10\text{K}) \approx 0.01\Omega$
>
> at 10kHz: $100\Omega / (1 + 10) \approx 10\Omega$

Now you can see why you failed to get any intelligible reading when, back in Lab 7, you tried to measure R_{out} for these amplifiers.

Ch. 4: Worked Examples: Schmitt Trigger

- Design Tips
- Two problems:

    - zero-crossing detector, using single-supply comparator
    - split-supply comparator (as usual), thresholds specified

Schmitt Trigger Design Tips

Setting Thresholds Setting particular thresholds can be a pain in the neck. The process may put you through tedious algebra (see, e.g., Text's problem calling for thresholds at 1V. and 1.5V: p. 232, exercise 4.10). But there are two easy cases:

Two easy cases

Thresholds symmetrical about zero:

For example, if you want thresholds at ± 1V, and output swing is ±15V. The feedback divider pulls threshold equally far above and below ground: the foot of the feedback divider is tied to ground.

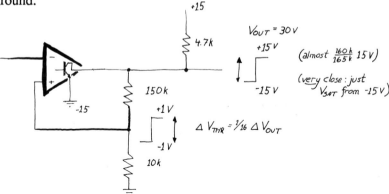

Figure X10B.1: One easy case: thresholds symmetrical about zero

The divider is to deliver 1 V out of 15 => $R_2 \approx 15 \times (R_1)$ ("14X," if you want to be more precise).

The key notion is that the foot of the divider is put at the midpoint of the output swing. You'd get the same tidy result if the thresholds were to be symmetric about 2.5V while the output swung between 0 and +5: say, thresholds at 2.4V and 2.6 V.

Thresholds very-far from symmetrical:

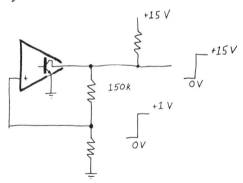

Figure X10B.2: A second easy case: thresholds pulled in only one direction ('very far from symmetrical')

For example, if you want thresholds at 0V and +0.1V, and output swings 0 to +5. Here the feedback divider pulls *up* but not *down*. So just design a divider that pulls the threshold up to 0.1V: $R_1 / (R_1 + R_2) \times 5V = 0.1V$. That means V_{Thresh} is about one part in 50, and R_2 is about 50 X R_1.

An easier task: Determining Hysteresis

Suppose the task given is not *"put thresholds at 1V and 1.1V,"* but instead *"set hysteresis at 0.1 V; put thresholds close to 1V."*

The two formulations look just about equivalent, but the second task turns out to be much easier than the first.

Let's try this task. Suppose output swing is 0 to +5V.:

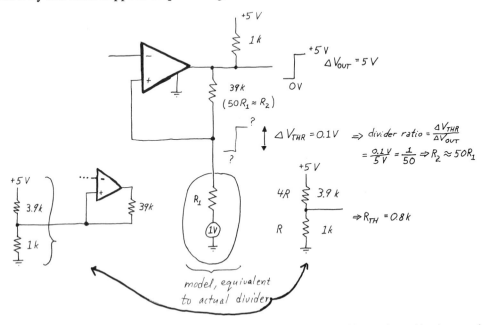

Figure X10B.3: Aiming for particular hysteresis, near a target voltage: Threshold and hysteresis considered *separately*

Here's the process:

> **Hysteresis:** This is determined entirely by the divider ratio. Again we want 0.1V/5V: one part in 50; again $R_2 \approx 50\, R_1$.
>
> **Thresholds "close to 1V.:"**
> This is determined by the voltage to which the foot of the feedback divider is tied (at the moment, let's assume we have a handy source of this 1V, an ideal voltage source; in fact, we are going to use a voltage divider, and then we will treat V_{Th} as that 'voltage at the foot…' If this issue isn't yet worrying you, forget this comment until later).
>
> You'll provide this 1V with a voltage divider—but it has some R_{Th}. Does that mess things up? No: let R_{Th} = the value you want for R_1.

This settling for approximate thresholds may seem like a cheap trick. Often it is not: often what you want is not named threshold voltages, but an approximate threshold, and appropriate hysteresis. Then this shortcut fits the formulated goals nicely.

A Wrinkle: Adjustable Threshold

Can you see a way to make threshold adjustable while holding hysteresis constant? (Possible help: recall that op amps are cheap.)

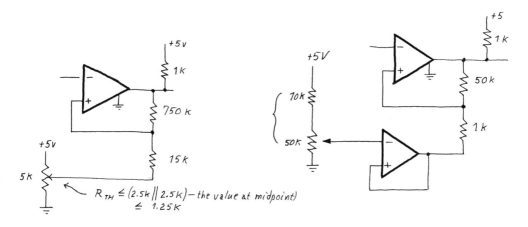

Figure X10B.4: Two ways to hold hysteresis constant despite changes of threshold

Two Problems

1. Zero-crossing detector

> ***Problem: Design a zero-crossing detector***
>
> We want to know when the AC line voltage crosses zero (within a few 100 μs). We are to use a single-supply comparator, powered from +5V, and we have a 12.6V transformer output available; the transformer is powered from the 'line' (60Hz).

Solution:

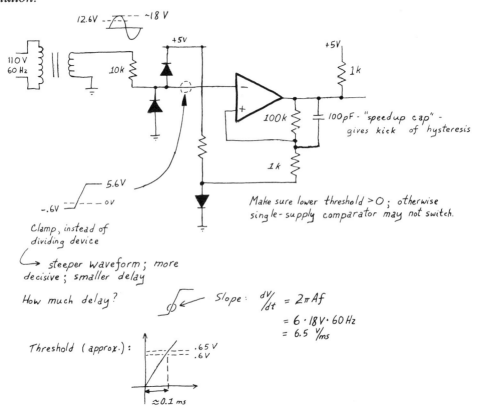

Figure X10B.5: Zero-crossing detector

2. Schmitt trigger, thresholds specified

> ***Problem: Schmitt trigger, thresholds specified, (outside range of output swing)***
>
> Design a Schmitt trigger, using a '311 powered from ±5V, to the following specifications:
>
> - output swing: 0 to +5V
> - thresholds: ±0.1V (approximate)

Solution:

This problem could be painful if we were not willing to approximate; so let's feel free to use approximations. It helps to consider the two cases separately: what happens when the '311 output is low (0V) vs. what happens when the output is high (5V).

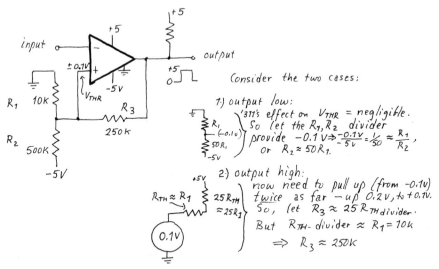

Figure X10B.6: Solution: aiming for particular thresholds, but willing to miss by a bit

Ch. 4: Op Amp Innards

Annotated Schematic of the LF411 Op Amp

You very seldom *need* to know in detail what's going on within an op amp, but it's fun and satisfying to look at a scary schematic and realize that you can recognize familiar circuit elements.

If the sheer quantity of circuitry here didn't scare you off, you might recognize at least the following elements without our help: a differential amp at the input stage; a common emitter amp, next; a push-pull output—and a hall of mirrors! The current limit at the output mimics a trick you will see in the '723 voltage regulator (Chapter 6; Lab and Class 12).

Some of the details are subtle, though, so we've done what we can to explain through our annotations. We offer this schematic both as a reviewing device and as a reward for the hard work you put into Chapters 2 and 3. We hope that this exercise makes you feel knowledgeable—lets you feel that you are beginning to be able to read schematics that would have meant nothing to you just a few weeks ago.

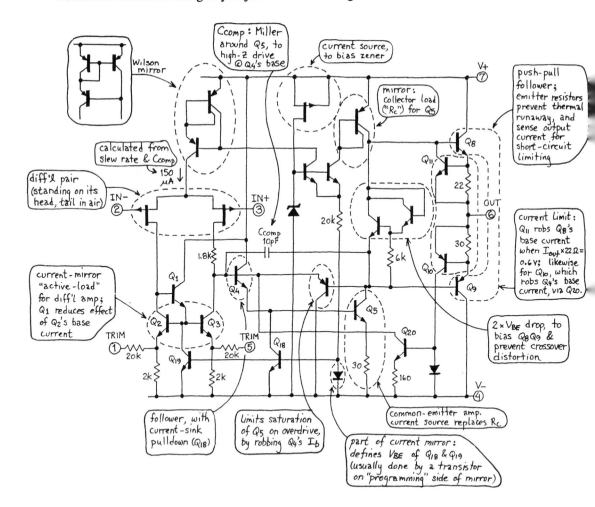

Figure OPIN.1: LF411 schematic, annotated

Lab 10: Op Amps III: Positive Feedback, Good & Bad

> **Reading:** *Chapter 4*: 4.23 to end, pp. 229-261; see especially secs. 4.33-4.35, re op amp frequency-compensation.
>
> *Chapter 5*: For this lab the important part is 5.12 –5.15 (pp. 284-291), secs. 5.17-5.18 (pp. 296-300). The '555 is the most important IC oscillator.
> Read the first part of the chapter for general cultural enlightenment only.
>
> **Problems:** *Chapter 4*: Embedded problems. Additional Exercises 5, 6, 7 (the last two introduce an intriguing circuit called a "negative impedance converter")..
> Bad Circuits I, J, M
>
> *Chapter 5*: Problems 5.8 and 5.9 in text. Additional Exercises (Ch. 5): 3, 4 (tricks with 555's).

Positive Feedback: Good and Bad

Until now we have treated *positive* feedback as evil—or as a mistake: it's what you get when you get confused about which op amp terminal you're feeding. Today you will qualify this view: you will find that positive feedback can be useful: it can improve the performance of a comparator; it can be combined with negative feedback to make an oscillator ("relaxation oscillator": there positive feedback dominates); or to make a Negative Impedance Converter (see exercise AE 4-6, 4-7: there, negative feedback dominates). And another clever circuit combines positive with negative feedback to produce a *sine* wave out (this is the "Wien bridge oscillator").

The more novel part of today's lab lies in the circuits that demonstrate what a pain in the neck positive feedback can be when it sneaks up on you. These circuits oscillate when they should not. Today, of course, they "should," in the sense that we want you to see and believe in the problem of *unwanted* oscillations. On an ordinary day, the oscillations that these circuits can produce would be undesirable, and would call for a remedy. Today you will try first to bring on oscillations, then to stop them. Some of these ways to stop oscillations should go into your growing *bag of tricks*.

A. Positive Feedback: *Benign*

10-1 Two Comparators

Comparators work best with positive feedback. But before we show you these good circuits, let's look at two poor comparator circuits: one using an op amp, the other using a special-purpose comparator chip. These circuits will perform poorly; they will help you to see what's good about the improved comparator that *does* use positive feedback.

Defective comparators: open-loop

Op Amp as Comparator

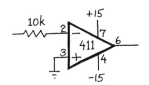

Figure L10.1: Op amp as simple comparator

You will recognize this "comparator" circuit as the very first op amp circuit you wired, where the point was just to show you the "astounding" high gain of the device. In that first glimpse of the op amp, that excessive gain probably looked useless. Here, when we view the circuit as a comparator, the very high gain and the "pinned" output are what we want.

Drive the circuit with a sine wave at around 100KHz, and notice that the output "square wave" output is not as square as one would hope. Why not?

Special-Purpose *Comparator* IC

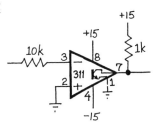

Figure L10.2: 311 comparator: no feedback

Now substitute a 311 comparator for the 411. (The pinouts are *not* the same.) You will notice that the output stage looks funny: it is not like an op amp's, which is always a push-pull; instead, two pins are brought out, and these are connected to the *collector* and *emitter* of the output transistor, respectively. These pins let the user determine both the top and bottom of the output swing. Often one uses +5 and ground here, to make the output compatible with standard digital devices. In the circuits below, you will keep the top of the swing at +15 V, but you will take advantage of your control over the *bottom* of the swing: at first you will set it to ground, later to –15V. Does the 311 perform better than the 411?[1]

---

[1] This question reminds us of a question posed to us by a student a few years ago: 'I can't find a 411. Is a 311 close enough?' This poor person no doubt was recalling the many times we had said, 'Put away that calculator! We'll settle for 10% answers; 2π is 6,' and so on.

Oscillations

A side-effect of the 311's fast response is its readiness to oscillate when given a "close question"—a small voltage difference between its inputs. Try to tease your 311 into oscillating, by feeding a sine wave with a *gentle slope*. With some tinkering you can evoke strange and lovely waveforms that remind some of the Taj Mahal in moonlight—but remind others of the gas storage tanks on the Boston's Southeast Expressway. Judge for yourself.

A good comparator: Schmitt Trigger: using positive feedback

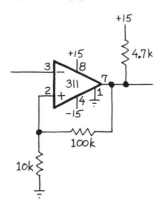

Figure L10.3: Schmitt trigger: comparator with positive feedback (& *hysteresis*)

The *positive feedback* used in the circuit above will eliminate those pretty but harmful oscillations. Predict the thresholds of the circuit above; then try it out.

Notice that triggering stops for sine waves smaller than some critical amplitude. Explain. Measure the hysteresis. Observe the rapid transitions at the output, independent of the input waveform or frequency. Look at both comparator terminals.

Reconnect the so-called "Ground" pin of the 311 (poorly named!) to –15v. (This pin is not necessarily ground, evidently; instead, it is the emitter of the output transistor.) Perhaps you can now see why the chip's designers brought out this pin, as well as why they provided an *open-collector* output.

10-2 RC Oscillator

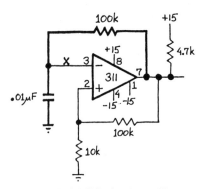

Figure L10.4: RC relaxation oscillator

Now connect an RC network from output to the comparator's inverting input, as shown above. *This feedback signal replaces any external signal source; the circuit has no input.* Here, incidentally, you are for the first time providing *both* negative and positive feedback.

Predict the frequency of oscillation, and then compare your prediction with what you observe.

IC Relaxation Oscillator: 7555

The 555 and its derivatives have made the design of moderate-frequency oscillators easy. There is seldom any reason to design an oscillator from scratch, using an op amp as we did in the proceeding exercise. The 7555 is an improved 555, made with CMOS. It runs up to 500 kHz (versus 100 kHz for the 555), and its very high input impedances and rail-to-rail output swings can simplify designs.

10-3 7555 IC Oscillator (square wave)

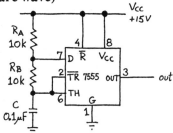

Figure L10.5: 7555 relaxation oscillator: traditional 555 astable

Connect a 7555 in the 555's classic relaxation oscillator configuration, as shown above. Look at the output. Is the frequency correctly predicted by

$$f_{osc} = 1 / (0.7 [R_A + 2R_B] C),$$

as derived in Exercise 5.8?

Now look at the waveform on the capacitor. What voltage levels does it run between? Does this make sense?

Now replace R_B with a short circuit. What do you expect to see at the capacitor? At the output?

An Alternative Astable Circuit

The 7555 can produce a true 50% duty-cycle square wave, if you invent a scheme that lets it charge and discharge the capacitor through a *single* resistor. See if you can draw such a design, and then try it. *Hint:* the old 555 could not do this trick; the 7555 can because of its clean rail-to-rail output swing.

Figure L10.6: 7555 relaxation oscillator: an alternative configuration (your design)

When you get your design working, consider the following issues:

- In what way does the output waveform of this circuit differ from the output of the traditional 555 astable?
- Is the oscillator's period sensitive to loading? See what a 10k resistive load does, for example.

If your design is the same as ours, then the frequency of oscillation should be
$$f_{osc} = 1/(1.4\,RC)$$
—which is the same as for the 'classic' configuration, except that it eliminates the complication of the differing charge and discharge paths.

Does the value of f_{osc} that you *measure* for your design match what you would predict?

Finally, try $V_{CC} = +5V$ with either of your circuits, to see to what extent the output frequency depends on the supply voltage.

10-4 Sawtooth Oscillator

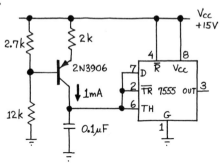

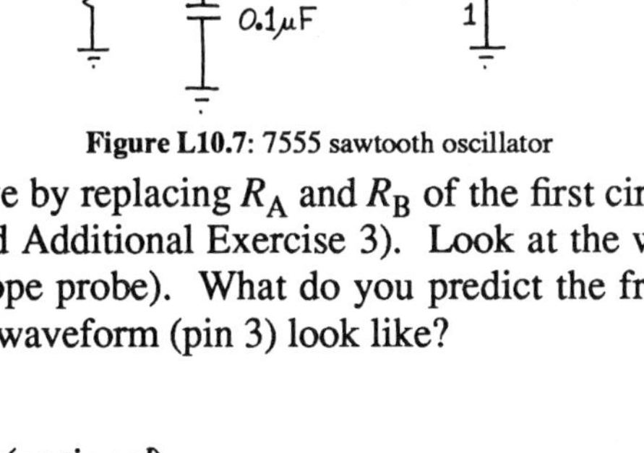

Figure L10.7: 7555 sawtooth oscillator

Generate a *sawtooth* wave by replacing R_A and R_B of the first circuit with a current source, as in the figure above (and Additional Exercise 3). Look at the waveform on the capacitor (be sure to use a X 10 scope probe). What do you predict the frequency to be? Check it. What should the "output" waveform (pin 3) look like?

10-5 Triangle Oscillator (*optional*)

You could generate the sawtooth (at somewhat lower frequency) by replacing the bipolar current source with the two-terminal JFET current source you met back in Lab 7: the 1N5294 (0.75 mA). Would that produce a better or worse sawtooth?

With one more of those JFET current sources—or with one JFET source and a clever use of a *bridge*, as suggested in the Text—you could generate a *triangle* output.

Figure L10.8: Triangle generator, using JFET current source or sources (*your* design)

Would this triangle be better than the one you produced back in Lab 7? If you're in the mood, try out your design for the *triangle* generator.

10-6 Sine Wave Oscillator: Wien Bridge

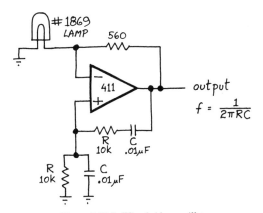

Figure L10.9: Wien bridge oscillator

Curiously enough, the sine wave is one of the most difficult waveforms to synthesize. (Your function generators make a sine by chipping the corners off a triangle!) The Wien bridge oscillator makes a sine by cleverly adjusting its gain so as to prevent clipping (which would occur if gain were too *high*) while keeping the oscillation from dying away (which would occur if gain were too *low*).

The frequency favored by the positive feedback network should be
$$1/(2\pi RC)$$
See whether your oscillator runs at this predicted frequency.

At this frequency, the signal fed back should be *in phase* with the output, and 1/3 the amplitude of V_{out}. The negative feedback adjusts the gain, exploiting the lamp's current-dependent resistance (the lamp is rated at 14mA @ 10v). Convince yourself that the sense in which the lamp's resistance varies tends to stabilize gain at the necessary value. What gain is necessary to sustain oscillations without clipping?

You can reduce the amplitude out by substituting a smaller resistor for the 560 ohms in the negative feedback path.

Try poking the non-inverting input with your finger and note the funny rubbery behavior at the output. Try sweeping the scope slowly as you poke: now you can watch the slow dying away of this oscillation of the sine's *envelope*. Why does the envelope do this?

B. Unwanted Oscillations

10-7 Follower

How can this circuit oscillate?

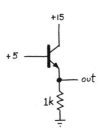

Figure L10.10: Follower as *oscillator*

To answer that riddle one must answer two subsidiary riddles:

- How does the circuit achieve voltage gain (so as to sustain oscillations)?
- How does the circuit provide positive feedback?

The answers depend on the truism that power supplies are not perfect:

- the gain appears if one draws explicitly the inductance always implicit in the power supply, along with stray capacitance:

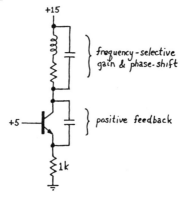

Figure L10.11: Follower redrawn to include feedback, and inductance that can provide gain

- The feedback appears if one notices that a disturbance on the power supply, which here is the collector, can disturb the emitter (through CE capacitance) in a sense that increases the collector disturbance: in other words, here is positive feedback.

This circuit, as redrawn, is nearly identical to the current source whose oscillations are explained in the Text at p. 167 (sec. 5.18, p.298, p.300, fig. 5.46). There the circuit is likened to a purposeful oscillator called a Hartley LC oscillator.

Most of us settle for learning some rules of thumb that stop oscillations when they appear; it is not easy to model a circuit in detail, including stray inductance and capacitance. A sort of electronic Murphy's law holds, however: if the circuit can find a frequency at which it could sustain an oscillation, it will find that frequency and oscillate—irritating its designer.

Build the circuit, and watch the *emitter* (if you try to watch the collector instead, you are likely to kill the oscillation with the scope probe's capacitance). See what "ground" is doing at the foot of the emitter resistor, as well. (To measure this voltage you must ground the scope lead some distance away.)

If the circuit fails to show oscillations, try worsening your supply by making the path for + 15V to your circuit more circuitous: pass it through a wire a couple of feet long. When the path is crummy enough (that is, inductive enough), the oscillation should begin.

Remedies

You saw, we hope, a high-frequency oscillation (in our lab, we saw a sine wave of a couple of volts at about 100 MHz). We can stop it as we stop or prevent op amp oscillations:

- by shrinking the disturbance that is fed back; or
- by diminishing the circuit's (high-frequency-) gain.

Supply Bypass (Shrinking the disturbance fed back)

You can stop the oscillation by quieting the collector: this "shrinks the disturbance," in the terms we used just above.

You can quiet the collector with a capacitor to ground. This cap "isolates" the power lines from the transistor circuit; or, to say the same thing in other terms, provides a local source of charge when the transistor begins to conduct more heavily: the power supply and ground lines need not provide this surge of current, and therefore will not jump in response; instead, the local bucket of charge will provide the needed current. Try a *ceramic* capacitor, about 0.01 to 0.1 µF.

Killing High-Frequency Gain

The gain of this **common base** circuit is

$$R_C \text{ (or } Z_C) / \{ r_e + [(\text{resistance driving base})/ (1 + \beta)] \}$$

Here, with the base driven by a stiff voltage source, the denominator of the gain equation is just "little r_e;" but we can diminish the gain in a way equivalent to the way we cut the gain of the common-emitter amp: by tacking in a resistor that enlarges the denominator of the gain equation. Here, we insert a resistor in series with the *base*; in the common-emitter we achieved an equivalent reduction of gain by inserting a resistor in series with the *emitter*.

Try this remedy: add a resistor of few hundred ohms between the +5 V source and the base. Does this stop the oscillation. Make sense?

10-8 Op Amp Instability: Phase Shift Can Make an Op Amp Oscillate

Low-Pass Filter in the Feedback Loop

The circuit below includes a capacitor you are not likely to install (though you might, in a sample-and-hold: see sec. 4.18, p. 221); more often the capacitance would be an unavoidable part of the load; it might arise because the op amp needed to drive a long cable, for example.

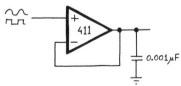

Figure L10.12: Load capacitance can make a feedback circuit unstable

This circuit may not oscillate, at first. If it does not, drive it with a square wave. That should start it, and once started it is likely to continue even when you turn off the function generator, or change to a sine wave.

Remedy: Shrink the Disturbance Fed Back

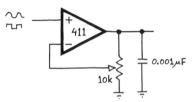

Figure L10.13: Circuit with gain is more stable than similar circuit at unity gain

Modify the follower circuit slightly by inserting a 100k pot in the feedback path. Begin with the pot turned so as to feed the entire output signal back (gain = 1; this is just the follower, again). Again set the circuit oscillating, with a square wave input if necessary.

Now gradually turn the pot so as to shrink the fraction of V_{out} that is fed back. (What are you doing to *gain*, incidentally?) When the oscillation stops, see how the circuit responds to a small square wave (make sure the output does not saturate). If you see what looks like ringing (or like the decaying oscillation of a resonant circuit), continue to shrink the fraction of V_{out} that is fed back until that ringing stops.

We hope that you now have confirmed to your satisfaction that stability can depend on the gain of a feedback circuit. This fact explains why so-called "uncompensated" op amps are available: if you know you will use your op amp for substantial gain, you are wasting bandwidth when you use an amp compensated for stability at unity gain.

Ch. 4 Review: Important Topics

Operational Amplifiers; Feedback

1. **Generally:**
    a. feedback as a general technique vs use with op amps
    b. idealized view of op amps: the "Golden Rules"

2. **Important Circuits**
    a. **Negative Feedback**
        i. amplifiers
            1. non-inverting
               DC, AC
            2. inverting
                a. incidental virtue: "virtual ground" or "summing junction"
        ii. integrator: ideal—but usually compromised for stability
        iii. differentiator: always compromised for stability
        iv. active versions of other familiar circuits:
            1. current source
            2. push-pull
            3. rectifier
            4. clamp
    b. **Positive Feedback**
        i. Schmitt trigger
        ii. relaxation oscillator
    c. Circuits using both positive and negative feedback
        i. NIC
        ii. oscillators
            1. relaxation oscillator
            2. Wien bridge (sine; gain control)

3. **Generally, again: a second view: Departures from Ideal**
    a. important op amp characteristics/ imperfections:
       V_{OS}, I_{bias}, I_{OS}, frequency limits: f_T, slew rate
    b. stability: "compensation"

Ch.4: Jargon and Terms

bias current (I_{bias}): average of input currents flowing at op amp's two inputs (inverting, non-inverting)

frequency compensation
deliberate rolling-off of op amp gain as frequency rises: used to assure stability of feedback circuits despite dangerously-large phase shifts that occur at high frequencies

hysteresis as applied to Schmitt trigger comparator circuits: the voltage difference between upper and lower thresholds

offset current difference between input currents flowing at op amp's two inputs (inverting, non-inverting)

offset voltage op amp's input stage mismatch voltage: the voltage that one must apply in order to bring the op amp output to zero.

slew rate maximum rate (dV/dt) at which op amp output voltage can change

saturation condition in which op amp output voltage has reached one of the two (±) output voltage limits, usually within about 1.5V of the two supplies

Schmitt trigger comparator circuit that includes positive feedback

summing junction inverting terminal of op amp, when op amp is wired in inverting configuration: inverting terminal then sums currents. That is, $I_{feedback}$ is algebraic sum of all currents input to the summing junction

transresistance amplifier
current to voltage converter:

Figure JR4.1: Transresistance amplifer, or current-to-voltage converter

virtual ground inverting terminal (-) of op amp, when non-inverting terminal is grounded. Feedback tries to hold (-) at 0V (hence "virtual *ground*"), and current sent to that terminal does not disappear into "ground," but instead flows through the feedback path (hence the ground is only "*virtual*")

CHAPTER 3 (revisited)

Class 11: FETs II: Switches

Topics:

- *old:*
    - power switching: using bipolar transistor (compared against FET)
- *new:*
    - power switching (and logic): MOSFETs versus bipolars
    - analog switches (small signals, passed by series transistor(s))
        - ♦ CMOS generally beat other FET configurations (and bipolars)
            - imperfections of CMOS analog switches
            - applications:
                - sample & hold
                    - ♦ choosing components for S&H

Power switching, and 'logic' switches

Power switching: turning something On or Off

You built a bipolar switch, long ago (Lab 4); today you'll build the equivalent circuit with a "power MOSFET" (= just "big, brawny" MOSFET).

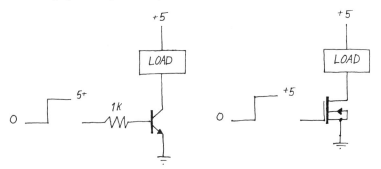

Figure N11.1: Two power switching circuits: bipolar and MOSFET, doing the same job

What's to choose, between the two?

Besides price, the most important issues are:

1. how much power the switch wastes when ON (indicated by R_{on}, for the MOSFET, $V_{saturation}$ for the bipolar); and
2. how easy or hard the thing is to drive.

Text sec. 3.14,
p. 166, table 3.6

How On is On?

Small bipolars win in the first characteristic, but big FETs at high currents do better (remember to include the power wasted as I_B through the 0.6V V_{BE}, to be fair to the FET!) The contrasting specifications are the bipolar's *saturation voltage* (watch out for the nasty fact that "saturation" means something quite different when applied to FETs!) versus the FET's R_{on}.

How hard is the thing to drive?
See Text sec. 3.14,
esp. pp. 161-62

The symbol for the MOSFET should be sufficient to remind you of this transistor's giant input impedance *at DC*. You'll see that in the lab, when you find that your finger can turn the thing on or off (if you touch your other hand to +5, then ground). At higher frequencies, the FET's higher *capacitances* begin to tip things against it.

Another argument for FETs: no 'secondary breakdown'
Text sec. 3.14,
fig. 3.66, p. 160

As the Text figure shows (fig. 3.66), local "current hogging" on the bipolar transistor makes the device behave like a transistor with a lower power rating, once the voltage across the device is appreciable (the figure shows the effect beginning at only 10V). FETs don't behave this way. They are inclined to slough off work instead of hogging it. This doesn't mean a bipolar cannot a particular job; it just means that you have to use a bigger transistor than you might expect to because of this effect.

Effects of FET capacitances on switching
Text sec. 3.17, esp.
fig. 3.70, p. 163;
and see sec. 13.23

The values of input capacitance—C_{gs} and C_{gd}—are big (100's of pF for a MOSFET that can handle a few amps at low R_{on}. But feedback makes things look even worse, exactly as Miller effect gave us trouble at high frequencies: the C_{gd} gets exaggerated by the quick swing of the drain as the device switches: that big dV/dt causes a large flow of current in C_{gd}. You'll see that effect in the lab—causing a strange kink or hesitation in the movement of gate voltage, as the MOSFET switches:

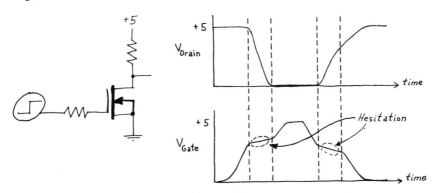

Figure N11.2: Slewing of drain slows movement of gate voltage, as transistor switches

Analog Switches

You can appreciate the CMOS analog switch by trying to achieve what it does, but using some other scheme:

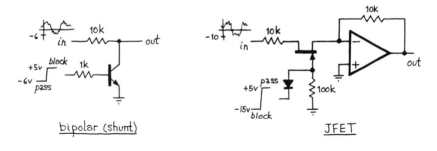

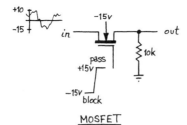

Figure N11.3: Not-so-good analog switches: bipolar; JFET; n-MOSFET

These work (in some respects the alternative FET switches work better than CMOS), but only with restrictions. The switch most widely used is made of CMOS—a complementary" pair of MOSFETs, n- and p-channel:

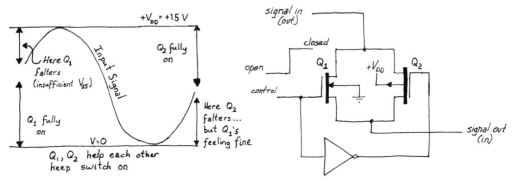

Figure N11.4: CMOS Analog Switch: either n-channel or p-channel alone would have trouble with part of input signal

It is a nifty Jack Sprat circuit, as you can confirm if you imagine applying an input signal that swings all the way from negative to positive supply. Good CMOS switches—like the one you will meet in the lab—handle such wide signal swings happily.

Imperfections

Text sec. 3.11,
pp. 141-42

A perfect transistor switch would be fully ON or fully OFF: it would behave like the mechanical switch. A real transistor switch comes close, but achieves neither ideal, of course. When ON, it looks like a small resistor (the DG403, which you'll use in the lab looks like 30Ω or less); when OFF, it looks like a small capacitance linking input to output (DG403: about 0.5 pF); it also leaks a little: a steady current may flow (DG403: < 0.5 nA). Whether or not these imperfections matter to you depends on the circuit. Consider some cases:

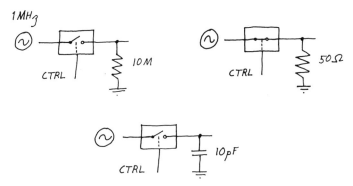

Figure N11.5: Some switch applications: does it matter that the switch does not match the ideal, either ON or OFF?

Applications

Sample & Hold

This is an important circuit, and provides a good test bed for consideration of the effects of switch imperfections.

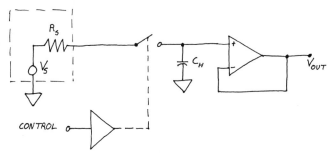

Figure N11.6: Sample & hold: simple scheme

The sample-&-hold is useful whenever an analog value must be held constant for a while. The most frequent need for this function is in feeding an *analog-to-digital* converter, which takes some time to carry out the conversion, and likes to look at a constant voltage while it is making up its mind.

Often a buffer is added on the input side, to drive the cap promptly to the *sampled* voltage. The output buffer prevents leakage of the *held* charge. (In the lab, we omit the left-hand buffer for simplicity, relying instead upon the low source resistance of the function generator.)

Choosing C

What is at stake in the choice of C? What good and bad effects result from use of a large C? —from use of a small C?

Roughly:
A large C makes it easy to *hold*, but hard to *sample*.

Sampling

We want V_{cap} to come close to V_{in}. Two issues here:

Assuming input sits still:
count RC's to get error less than any specified level (5 RC's to get within 1%, for example). What's the relevant R?

Assuming input is moving:
R and C limit the speed of input change that the device can follow during the sampling.

You will find a sample & hold design problem done as a Worked Example. In these notes we will only note the sources of error.

Standard Design Issues

- What size C? (We know now that we lose something at either extreme.)
- How long do we need to *sample*?

<u>What size C?</u>

Big enough to keep droop tolerable: use I = C dv/dt.

- I = leakage currents (sum of all: analog switch, capacitor's self-discharge, op amp buffer's bias current);
- dt or Δt = the hold time—usually just long enough to allow an A/D to complete its conversion;
- dv or Δv = the tolerable change of voltage during conversion: the judgment of what is tolerable musts be somewhat arbitrary; let's suppose we will let the S&H contribute a total error of 1/2 the resolution of the converter. We have already used up about 1/2 this total "error budget" in the sampling stage: V_{CAP} did not quite reach V_{in}, remember? So we can allow droop to soak up the remainder of the budget: 1/4 resolution of the coverter.

This result puts a lower limit on C. The *sampling* concerns put an upper limit, for a given sampling time. As usual, we are caught between two competing concerns.

Sampling

Here the goal is to get V_{cap} close to V_{in}, and do it fast. The non-zero R_{on} of the analog switch limits the speed at which we can charge the storage cap; so does the output impedance of the source—or, if the source is an op amp buffer, then the op amp's output current limit and slew rate best describe the characteristics that slow the cap's charging.

Effects of "charge injection"

Text sec. 3.12, pp. 149-50

When we turn the switch OFF, unfortunately, stray capacitance between the gates of the transistors and the output squirts a dose of charge into the storage capacitor. For the DG403, that does is 60pC.

That doesn't sound big until you look at the ΔV it causes when dumped into a small cap:
$$Q = CV; \Delta V = \Delta Q/C$$
$$= 60pC/10pF = 6 \text{ V!}$$

The hazard of charge injection, therefore, pushes us toward larger cap size; so does the problem of *feedthrough*: the transmission of high-frequency signal when the switch is turned off.

In short, we need a cap a good deal bigger than what we might choose if we considered only the problem of droop.

Summary of S & H Errors

Text sec. 4.15, p. 219, fig. 4.41

Here's the Text's diagram summarizing S&H errors with switching delay added:

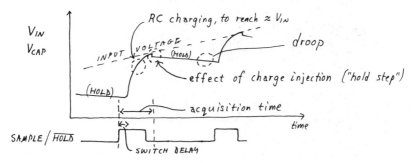

Figure N11.7: Sample & hold errors

The effect of charge injection, called "hold step," here is a negative-going error. The switching *delays* provide a further reason why fairly long sample or acquisition times are necessary. The DG403 takes around 100ns to switch; pretty clearly we would be kidding ourselves if we did a droop calculation and an RC acquisition calculation and concluded that we could acquire the data in 10 ns.

Finally, to give you some sense of what values to expect, here are some typical specifications for good integrated S&H's:

Acquisition Time		# of bits
200 ns	AD7569	8 bits
3µs	AD585	12 bits

Storage Capacitor
100 pF (internal; may be increased by use of external cap)
                                AD585

Figure N11.8: Some typical characteristics: integrated sample & holds

Ch. 3: Worked Example: Sample & Hold

Two worked examples:

1. Design a sample-and-hold to suit a particular A/D converter's demands
2. Figure out what advantages two improved S & H circuits offer

These notes look in greater detail at an issue discussed in class Notes 11.

> *Problem: Design a Sample & Hold to feed a particular A/D*
>
> This sample-and-hold is to feed a 12-bit A/D converter (an analog-to-digital converter that slices a 10V input range into 2^{12} slices: 4k little slices).
>
> Assume:
>
> - the analog switch's R_{on} is 30Ω;
> - we want our sampled voltage error less than 1 bit (one part in 2^{12} of the 10V range);
> - We need to *hold* for 25 μS (during this time, the A/D converter does its job, and wants a nearly-constant input voltage).
>
> What C? How long do we need to *sample*?

Sampling

The voltage corresponding to 1 bit (an error larger than what we allowed ourselves in the Class Note discussion) is 1 part in 2^{12}, or one part in 4K: 0.025%. If we can stand an error of only 1/2 bit (we'll see why *1/2*, in a minute) then we can stand an error of only about 0.0125%. That sounds as if it's going to be a pain to calculate; in fact, it's not: 5 RC's get us to an error under 1%; 0.01% is 1% of 1%—so *10* RC's should do it. Easy?

Now we know R and how many RC's we need. We don't yet know C, and we can't yet choose C intelligently. If we we were worrying about *sampling* only, then surely we would choose a *tiny* C. But we find ourselves pushed the other way as soon as we recall that our circuit also needs to *hold*.

Holding

As we *hold* the sampled voltage, again we can tolerate less than the voltage that corresponds to a 1/2 bit error: 1 part in 8k of the 10V range: 1.25 mV. We can look up leakage currents on data sheets for the op amp and analog switch. That will give us I; we know how much voltage movement we can stand (dV); we know how long we need to hold (dt). We can solve for C.

Leakage currents

The sum of the leakage currents: I_{bias} for the op amp; switch leakage; cap leakage; PC-board or breadboard leakage. We'll ignore the last two.

At room temperature, the maximum values are these:

- I_{bias} = 200 pA;
- switch leakage = 0.5 nA.
- $I = C\, dV/dt \implies C = 0.7 \times 10^{-9}$ (25µs / 1.25 mV)
  = 14 pF

Knowing this value for C, we can look at some other effects:

Effects of "charge injection"

Text sec. 3.12, pp. 149-50

As we noted in the class Notes, this small storage capacitor will get us into trouble, because of charge injection. The DG403's jolt of 60pC would produce a big voltage jump. Here is the calculation, again:

$$Q = CV;\ \Delta V = \Delta Q/C$$
$$= 60\text{pC}/10\text{pF} = 4.3\text{ V}$$

Another concern pushes us in the same direction, toward a larger capacitor. *Feedthrough* would be bad, too, with a 14 pF cap: the DG403 looks like about 0.5 pF between input and output when it's off, so we'd get one part in 30 with the switch off!

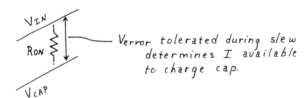

Figure X11.1: Feedthrough problem: we need a substantial cap to protect us from such errors

We'd better back up and crank up C. Much of this effect of charge injection is a constant, which could be subtracted away. Suppose that 20% is a really nasty error, one we cannot subtract away. Then we'd better squeeze this error down under 1/2 bit, too (1/3, if we were being really thorough!): ==> 1.25 mV. To crank even 20% of *6V* down that far requires enlarging C by about a factor of 1000: ==> C = around 0.015 µF. (At this point we might also start looking for a different analog switch!)

How long must we <u>sample</u>?

Earlier, we said 10 *RC*s. Now we know C as well as R, and we can solve for a time. *RC* is 30Ω X 0.015 µF = 0.45 µS, so we need to sample for 4.5 µS.

Limit on slewing

If we tolerate (again) about a millivolt error (toerating this additional error will carry us above a total 1-bit error, incidentally; let's agree to relax our requirements to this extent, to keep things simple), then we can solve for I, knowing R_{on}; that will give us dV/dt, given and C:

$$I = 1\text{ mV} / 30\Omega = 0.03\text{ mA} = 30\text{ }\mu\text{A};$$
$$dV/dt = I/C = 30\text{ }\mu\text{A} / 0.015\text{ }\mu\text{F} = 2000\text{ V/S} = 2\text{ V/ms}$$

That's slow. How fast and big a sine wave you could this S&H follow? This is a calculation you did for op amps, back in Lab 9.

If
$$V(t) = A\sin\omega t,$$
then the slope of the waveform, dV/dt, is
$$A\omega\cos\omega t,$$
whose maximum value is just $A\omega$. Suppose we want to sample a 5V sine wave. Then our limited slew rate, 2V/ms, is equal to $A\omega$. So,

$$f_{maximum} = dV/dt / (2\pi A) \approx 2\times 10^3 / 6 \cdot 5 \approx 60\text{Hz}.$$

Slow, as promised.

The lesson from this exercise seems to be that it is hard to design a good S&H from parts. The commercial units—integrated, or 'hybrid' (small circuits in a small package, but not integrated as a whole)—work better than this S&H we have just designed.

Fancier Sample-and-holds

> *Problem: What do these S&H's offer?*
>
> Here are three sample-and-hold circuits that use feedback to improve performance. Does the feedback help? How? To what extent? 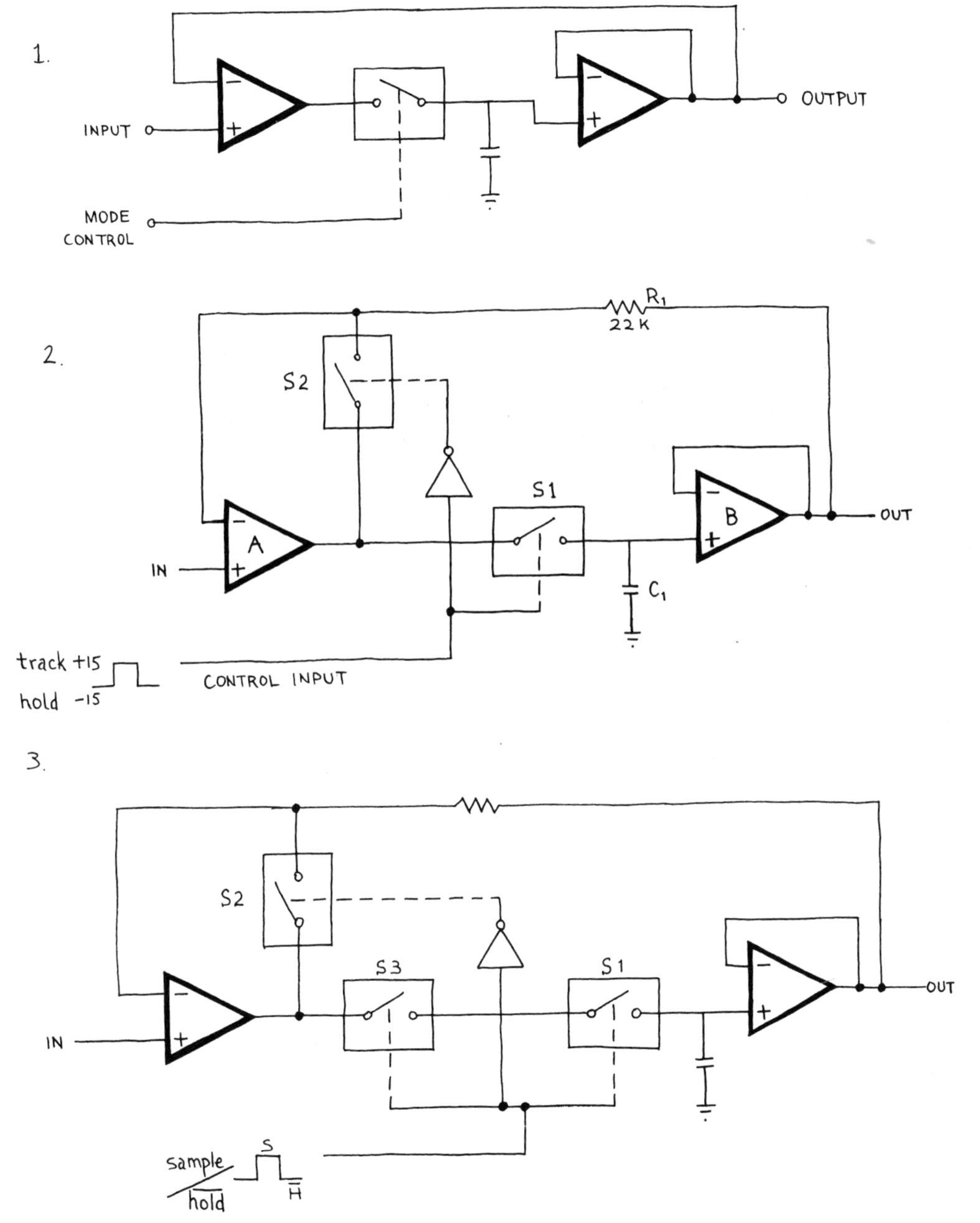
>
> **Figure X11.2**: Three sample & hold circuits improved by feedback

Solution:

1) First Circuit

Overall feedback makes the input op amp try to drive V_{cap} to V_{in}. In order to do that, it will "try to do what's necessary...," as usual. In this case, that means that it will drive the op amp output to saturation, and that will speed the charging of the cap:

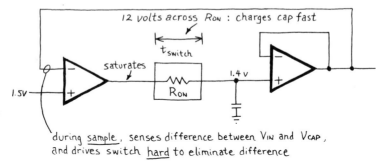

Figure X11.3: How circuit 1 speeds the 'acquisition' of V_{in}. Defects, too.

The trouble with this circuit also appears above: during *hold*, the input op amp loses feedback and must saturate. So sampling is slowed by the op amp's need to get out of saturation (a process that may take a few microseconds) and then to slew to where it ought to be: often it will need to slew from one 'rail' to the other: across nearly 26 volts. A typical '411 would take almost 2µs to make that trip, typically. That's a sad waste of time.

2. Second Circuit

Here a second analog switch is provided, just to prevent saturation of the input op amp during *hold*.

3. Third Circuit

Here another analog switch is inserted in the hope of reducing leakage across the *hold* switch. There's a flaw, though: it's true here that the voltage across the analog switch should be close to zero during hold; but leakage still can occur to the switch *body* and *gates*: these will be at ± 15V. So the better remedy is to get a better analog switch—or even better (and easier) buy someone's integrated sample and hold: it will work better than the one you build up from parts.

Lab 11: FETs II: FET Switches

Reading:	Secs. 3.11 – end of chapter: especially: 3.13: Analog switches & their applications; Sec. 3.14 power switching
Problems:	Embedded problems.

A second FET lab: FET switches

We return to FETs in this lab, now that we have met op amps. The most important of the circuits that uses both switch and op amp—and the principal reason why we chose to postpone this second FET lab—is the *sample-and-hold* (11-5). This circuit is often used when one converts from analog into digital form, to hold the input value steady during the conversion process. It also serves as a good test bed for the analog switch: it reveals the weaknesses of the device.

The lab begins with simpler switching circuits: it introduces you to the power MOSFET as an alternative to a bipolar power switch. But most of the lab is given to trying some applications for the so-called *analog switch* or *transmission gate*: a switch that can pass a signal in either direction, doing a good job of approximating a mechanical switch. This device is not useful as a power switch, but is so easy to use that we hope it will provoke you to invent—or understand—tidy solutions to old problems—like the problem of how to reset an integrator, for example. That's hard if you use discrete transistors; easy if you use an integrated analog switch.

A. FETs as Power Switches

11-1 Power MOSFET

This exercise repeats a task you carried out back in Lab 4 using a bipolar transistor (exercise 4-9). Here you will use a MOSFET to do the job. In some respects it is much better than the bipolar equivalent.

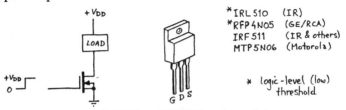

Figure L11.1: Power MOSFET transistor switch

a) Input Impedance

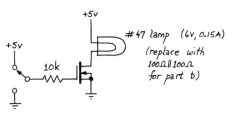

Figure L11.2: MOSFET switching circuit

Build the circuit shown above. Use a #47 lamp as load and confirm that the FET will switch when driven through the 10k resistor at low frequencies (switch the input between 0 and +5v. *by hand*). High input impedance is the MOSFET's great strength, as you know.

To get a more vivid sense of what "high input impedance" means, let the input side of the 10k resistor float, then touch it with one hand, and touch your other hand alternately to ground and the +5 supply. Impressive? Now try letting go of the 10k resistor after switching the FET on or off. Why does the FET seem to remember what you last told it to do?[1] This exercise, frivolous though it seems, foreshadows some of the strange results you will get if you ever forget to tie the input of a MOS logic device either high or low.

b) Power Dissipation

Change the Load, replacing the lamp with a pair of 100-ohm resistors in parallel (why not use a 50 ohm 1/4 watt resistor?).

Measure V_{DS} with the transistor switched on, and infer R_{ON} for this transistor.

How does the FET's power dissipation in this circuit compare with that for a well-saturated bipolar transistor? (See saturation curves for 2N4400, below; see also Text Appendix G on saturation, and notice that base current is much larger than I_C/Beta.) Which type wins at, say 100 mA? (For the bipolar, one should include base current in the power calculation.)

You can, if you like, calculate a characteristic for the bipolar transistor analogous to R_{ON}. For the 2N4400 this would be about 1 ohm at 1 A, 0.5 ohm at 0.5 A, as you can see from the curve below.

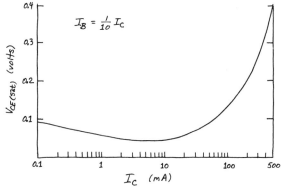

Figure L11.3: Saturation characteristic of a small bipolar power transistor

---

1. Your particular MOSFET may or may not "remember." The maximum leakage rate is rather high. See data sheet.

c) Switching at Higher Frequencies

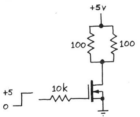

Figure L11.4: FET switch, again

Effect of Input Capacitance

Replace the manual switch that was driving the gate with a signal generator that provides a 0 to +5v square wave. Keep the 10 k resistor in place, for the present. Watch V_{GATE} and V_{OUT} as you increase the driving frequency from about 10 kHz. What goes wrong?

Solve the problem by replacing the 10k resistor with a value that works better at high frequencies. Now see how the switch looks at a few hundred kHz.

You can infer the FET's approximate C_{in} by looking at the shape of the waveform at the gate. The waveform's shape will be a little weird: the gate feels the movement of the drain voltage. Remind you of Miller effect? (See Text at pp. 85-86.) Find a portion that looks like an RC charge or discharge; measure RC. Your C should be large—and probably will be larger by perhaps 4X than the C_{IN} shown on the data sheet, which is specified under conditions different from what you are applying: the specified V_{DS} is 10V; what is V_{DS} where you are measuring RC? The capacitance *changes* during the output swing: see <u>data sheet</u> curve for C_{iss} (= C_{GS} + C_{GD}) and C_{rss} (= C_{DG}) at p. 8-12; see also Text sec. 3.14, especially fig. 3.70 at p. 163, and section 13.23, which includes some useful graphs.

Does the scope probe's capacitance contribute any substantial error? Would this measurement have been much distorted if you had used a BNC cable to feed the scope, in place of a scope probe?

Switching Speed (*optional: if time*)

If you remove the 10k or other series resistor, you will be driving the FET with the function generator's R_{out} of 50 ohms. Now you can measure the transistor's switching speed, with most of the effect of the device's large input capacitance removed. Drive the switch with a square wave at 100 kHz and measure *switching time*. Treat a rise or fall to 2 volts in as the input signal, a fall or rise of output to 2 volts as the switching of the output signal. What delay do you find? Why does $t_{low-high}$ not equal $t_{high-low}$?

d) Temperature Effects (optional: if time)

Wouldn't it be nice if transistors conducted *less* current as they got hot, instead of more? Then the problem of "thermal runaway" would disappear. Circuits would tend to stabilize themselves.

Yes, the FET can behave this way. At low currents, the FET does not do this: it conducts more as it gets warm; but at high currents it conducts less as it heats; and at one value of I_D the transistor shows a magical immunity to temperature effects. (See Text sec. 3.14, p. 162; compare fig. 3.68).

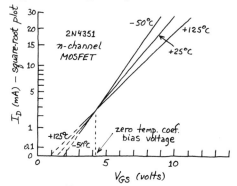

Figure L11.5: FET can show negative or even *zero* tempco

In place of the big MOSFET you have been using, substitute a small power MOSFET, the VN01: it is small enough so that we can put it into the region of the curve (above) where the FET shows a ***negative*** temperature coefficient.

Drive the gate from a pot (we now do not want the transistor fully ON); use a load of two paralleled 100 ohm resistors, again.

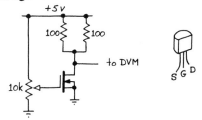

Figure L11.6: Setup to look for negative tempco operating region

Positive Tempco Region

First, adjust V_{gate} until V_{out} is around 4 volts. (Use a DVM; you will need its resolution; these temperature effects are small.) Then warm the transistor by holding it between your fingers. Does I_D grow or shrink? (Watch out for the inversion here between the signs of ΔI_D and ΔV_{out}.) If the heat from your fingers is not sufficient to move V_{out}, borrow some heat from a 100-ohm resistor connected between +5 and ground: press it against the VN01 (the transistor's leads may provide better thermal conductivity than its plastic case).

Negative Tempco Region

Now readjust V_{gate} so as to set V_{out} to about 0.25 V. Again heat the transistor. You should find the voltage moving in a direction that indicates I_D now is *falling* as temperature increases. That's good news.

B. CMOS Analog Switches

The CMOS analog switch is likely to suggest solutions to problems that would be difficult without it. This lab aims to introduce you to this useful device. Schematically, it is extremely simple: it simply passes a signal or does not:

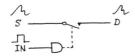

Figure L11.7: Analog switch: *generic*

The switch we are using has especially nice properties: it is switched by a standard logic signal, 0 to +5 (High, +5 = ON). But it can handle an analog signal anywhere in the range between its supplies, which we will put at ±15 volts. It also happens to be a *double-throw* type, nicely suited to selecting between two sources or destinations.

Here is the switch, and its pinout:

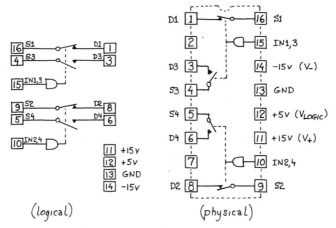

DG403 analog switch; switches shown with IN= 1 (HIGH)

Figure L11.8: DG403 analog switch: block diagram and Pinout

As signal source use an external function generator; as source of the "digital" signal that turns the switch on or off use one of the slide switches on the breadboard. The easiest to use is the 8-position switch: put one of its slides in the ON position; now that point will be high or low, following the position of the slide switch just to the right of the "DIP" ("dual in-line package") switch.

Caution: each package contains two switches. Tie the unused "IN" terminal to ground or to +5 (this makes sure the logic input to the switch does not hang up halfway, a condition that can cause excessive heating and damage.)

11-2 R_{ON}

Ideally, the switch should be a short when it is ON. In fact, it shows a small resistance, called R_{ON}. Measure R_{ON}, using the setup shown below:

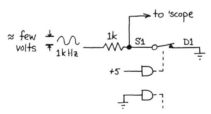

Figure L11.9: R_{ON} measurement

Use a 1 kHz sine of several volts' amplitude. Confirm that the switch does turn On and Off, and measure R_{ON}.

11-3 Feedthrough

The circuit below makes the switch look better: its R_{ON} is negligible relative to the 100k resistor. Confirm this.

When the switch is OFF, does the signal pass through the switch? Try a high-frequency sine (>100 kHz). Try a square wave. If any signal passes through the OFF switch, why does it pass?

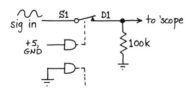

Figure L11.10: More typical application circuit (R_{on} made negligible)

Caution: don't forget that you are looking at the output with a scope probe whose capacitance (to ground) may be more important than its large R_{in}. So long as you don't forget that probe capacitance, you should have no trouble calculating the switch's C_{DS}.

11-4 Chopper Circuit

Here is a cheap way to turn a one-channel scope into two channel (and on up to more channels, if you like). (Query: what are the limitations on this trick?)

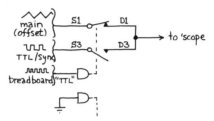

Figure L11.11: Chopper circuit: displays two signals on one scope channel

For a stable display, trigger on one of the input signals, not on the chopper's output, where the transients will confuse your scope.

11-5 Sample & Hold

This application is much more important. It is used to *sample* a changing waveform, *holding* the sampled value while some process occurs (typically, a conversion from analog into digital form).

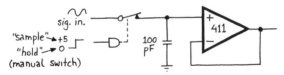

Figure L11.12: Sample & hold

Try the circuit. Can you infer from the droop of the signal when the switch is in *hold* position, what leakage paths dominate? (This will be hard, even after some minutes of squinting at the scope screen; don't give your afternoon to this task!)

Query: how does one choose C value? What good effects, and what bad, would arise from choice of a cap that was a) very large; b) very small?

Can you spot the effect of *charge injection* immediately after a transition on the control input? Compare the specified injection effect and the voltage effect you would predict, given the specification (≤ 60 pC) and the value of your storage capacitor.

Optional: a dynamic view of charge injection

If you inject charge periodically, by turning the switch on and off with a square wave, you can see the voltage error caused by charge injection in vivid form[2] The *held* voltages ride above the input, by a considerable margin; you will notice the margin varies with the waveform voltage. Why?

Good sample & hold circuits evidently must do better, and they do. See, e.g., the AD582 with charge injection of < 2 pC; and see Text sec. 3.12, p. 149.

11-6 Negative Supply from Positive

Here's a little levitation trick, done by shoving "ground" about in a sly way. This kind of circuit is often put onto an integrated circuit that would otherwise require a negative supply.

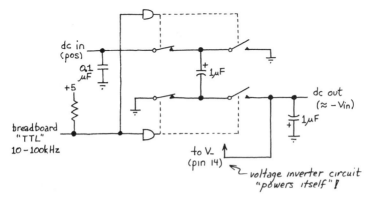

Figure L11.13: Voltage inverter: negative v_{out} from positive V_{IN}

It provides only a small current output, as you can confirm by loading it.

---

2. Thanks to two undergraduates for showing us this technique: Wolf Baum and Tom Killian (1988).

11-7 Switched-capacitor Filter I: built up from parts

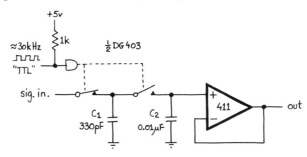

Figure L11.14: Switched-capacitor low-pass filter

This filter's f_{3dB} is regulated by the clock rate. This makes it a type convenient for control by computer. The Text sets out a general formula for f_{3dB} of this filter type; given the values used here, this formula predicts

$$f_{3dB} = (0.3/2\pi) \times f_{clock}$$

Try the circuit, and compare its f_{3dB} with the predicted value. Does the filter behave generally like an RC filter: does it show the same phase shift at f_{3dB}? Does f_{3dB} vary as you would expect with clock frequency? Does the filter fail at the high end of the oscillator's range? Do you see feedthrough of the clock signal?

11-8 Switched-capacitor Filter II: integrated version (*Optional: omit if you are short of time*)

This integrated filter is more complicated than the one you just built up from parts, as you might expect. Integrated switched-capacitor filters are available in several forms. Essentially, they are op amp *active filters* like those described in Chapter 5, except that they use switched capacitors to simulate the performance of a resistor, and thus allow control of the effective *RC*, as you saw in 11-7. Some allow the user to determine the filter type. The one you will meet here is committed as a *low-pass*; that makes it easy to wire. It is a *four-pole Butterworth* filter—like the LC filter you built back in Lab 2. You will use it later in this course, in a lab treating analog-digital interfacing. Here we just want you to discover how easy to use it is, and how sharp its rolloff.

> *Note*: Please build this circuit on a private breadboard: you will use this circuit again in Lab 21.

Like your simple switched-cap filter, this one shows an f_{3dB} proportional to the filter's *clock* rate. That is the feature that later will be especially useful to us (in Lab 21), when we will want to vary f_{3dB} very widely without rebuilding circuits.

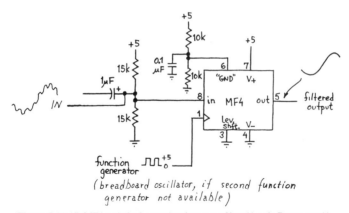

Figure L11.15: MF4 switched-capacitor low-pass filter (4-pole Butterworth)

Build the circuit shown above, using the breadboard TTL oscillator-output to clock the MF4. Let the external function generator provide an analog input to the filter: let the function generator drive pin 8 directly, bypassing the blocking capacitor (which, for the moment, just hangs there, useless; we want to leave it in place so that you can use the MF4 easily when you return to it in Lab 21). Make sure to add a *DC offset* to the input signal, so that the signal stays between about 1V and 4V.

To test the filter, try setting f_{3dB} around 1kHz ($f_{3dB} = f_{\text{clock-filter}}/50$). Feed the filter a sine wave: offset the sine to center on 2.5V; keep the amplitude below about 1V; vary f_{in} from far below f_{3dbB} to far above.

You should be able to see the filter's abrupt roll-off. You will also notice the discrete character of the filter's output: the filter output is shows "steps" rather than a smooth curve.

Two optional filter tests, for fun:

a) Peel away frequency components of a square wave

As you know from your experience in *compensating* scope probes, you can get a portrait of a circuit's frequency response by feeding the circuit a square wave rather than a swept sine. Try this: feed your filter a square wave at around 100Hz, and take f_{clock} gradually down from its maximum. Can you make out the frequency component $3 \times f_{square}$, shortly before you strip the square wave down to its fundamental sine?

b) Fancier demonstration of controllable f_{3dB} (for the energetic)

The prettiest picture of the filter's performance appears if you *sweep* the input frequency, and watch the effect of varying f_{clock} to the filter. f_{3dB} should be variable over a wide range: up to about 2kHz if you use the *breadboard* oscillator as its clock; up to nearly 10kHz if you use a higher clock rate from an external oscillator.

Is *feedthrough* of the chip's clock noticeable at the output? Can you confirm the steep rolloff that is claimed: 24dB/octave?

Ch. 3 Review: Important Topics & Circuits

1. **Generally**

    a. FETs vs Bipolar
    b. FET mechanism: a peek within helps one distinguish—
        i. *depletion-mode* (includes *all* JFETs) <u>versus</u> *enhancement mode* (this is the usual form for MOSFET switch)
        ii. JFET versus MOSFET
            JFET better as input stage of op amps; good for current sources
            MOSFET better as switch; universal in logic, where CMOS (p- and n-channel types combined) is most elegant

2. **Important Circuits**

    a. Follower
        i. Characteristics: versus bipolar follower:
            1. R_{in}; Z_{in}
            2. attenuation
            3. R_{out}
    b. Current Source
       Can be simple (2-terminal); also nice when combined with op amp
    c. Variable Resistor
    d. Switch
        i. power
           Z_{in}: gigantic at DC; but considerable C_{in} (See FET lab II, Lab 11: power switch exercise)
        ii. analog applications
            1. sample & hold
            2. multiplexer
            3. switched-capacitor filter

Ch.3: Jargon and Terms

acquisition time	of sample-and-hold: time to 'acquire' input: that is, to let voltage on storage capacitor come adequately close to V_{in}.
body	in a discrete MOSFET (not an IC) this is the semiconductor material in which the channel is formed. Must be kept back-biased or zero-biased with respect to both drain and source.
charge injection	of sample-and-hold: quantity of charge transferred to FET drain (switch output) when transistor or switch is turned ON or OFF. Expressed in Coulombs.
C_{iss}	common-source input capacitance: FET capacitance seen from gate:

Figure JR3.1: C_{iss}: input capacitance

C_{oss}	common-source output capacitance: FET capacitance seen from drain:

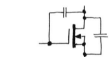

Figure JR3.2: C_{oss}: output capacitance

depletion mode	class of FETs that are fully ON until a voltage at gate (V_{GS}) begins to turn the transistor OFF. Includes all JFETs and occasional MOSFETs.
droop rate	said of *sample and hold*: rate at which voltage on storage capacitor departs from original stored value ($\Delta V/\Delta t$).
enhancement mode	class of FETs that are OFF until V_{GS} begins to turn the transistor ON. Includes some MOSFETs, no JFETs.
feedthrough	of analog switch: extent to which signal passes through an OFF switch; usually expressed in dB of attenuation.
I_{DSS}	drain current that flows when gate is shorted to source: maximum current for a JFET.
JFET	junction FET: gate forms a P-N junction with channel (never to be forward-biased). Always *depletion mode* (ON until turned OFF.)
latchup	pathological condition in which excessive current at input of a FET device (excessive current usually caused by excursion of input beyond a supply rail) causes the device to pass a large current from supply to ground. Destructive.
linear region	region of FET operation, defined by magnitude of V_{DS}, where I_D varies approximately linearly with V_{DS} (that is, the FET behaves like a resistor). Obtains where V_{DS} is small ($<(V_{GS} - V_T)$).

MOSFET	metal-oxide-semiconductor FET: gate is insulated from channel (by "oxide:" SiO_2).
R_{ON}	Drain-source resistance of FET that is fully ON. (Assumes operation in linear region: V_{DS} small.)
sample-and-hold	($\equiv$ 'track-and-hold'): circuit that stores and holds a voltage, 'sampled' from a changing input.
saturation	region of FET operation, defined by magnitude of V_{DS}, where I_D is determined by V_{GS} (FET behaves like a current source). Obtains where V_{DS} not small ($>(V_{GS} - V_T)$). ***Caution:*** this sense is not related to 'saturation' of either a bipolar transistor or an operational amplifier.
substrate	In a FET built on an IC, refers to semiconductor material on which several transistors are built. Circuit designer is obliged to keep this substrate back- or zero-biased with respect to drain and source. (See fig. 3.5, p. 117.) In a discrete FET, $\equiv$ "body."
transconductance (g_m)	characteristic describing FETs gain: ΔI_{out} per ΔV_{in}. Conventional units are µmhos or µS (mho = ohm^{-1}), equivalent to "µA/V," which unfortunately is rarely used. No direct equivalent in bipolar transistor characteristics, but $1/g_m$ is exactly equivalent to the bipolar transistor's r_e.
$V_{GS} - V_T$	extent to which applied gate-source voltage exceeds threshold voltage. This is the quantity—not V_{GS} itself—that determines FET behavior, as V_{BE} determines the behavior of a bipolar transistor.
V_P	pinch-off voltage (also called $V_{GS(off)}$): V_{GS} at which a JFET turns OFF. The definition of OFF is arbitrary: 2N5485, for example, uses I_D = 10nA as 'off' current.
V_T	threshold voltage: V_{GS} at which a MOSFET begins to turn ON. The definition of ON and OFF is arbitrary: some small current defines the boundary. See V_P, above.

Chapter 6

Class 12: Voltage Regulators

Topics:

- *old:*
    - a voltage regulator is "just" {voltage reference + high-current op amp follower}

- *new:*
    - regulator chips
        - 723: an old regulator that reveals what goes on in any regulator
        - 3-terminal regulators: easy to use:
            - fixed: 78xx
            - variable: 317
    - 'crowbar' overvoltage protection
    - switching regulators (very briefly)

Voltage Regulators

One could argue quite plausibly that this is not a topic in its own right; it is only one more application of negative feedback, and could fit quite well into the op amp chapter. Isn't a regulator just a follower driven by a reference?

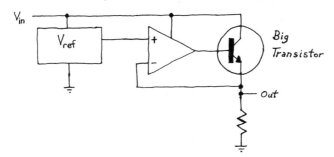

Figure N12.1: Voltage "regulator"

Yes, it is. But this function is needed so often that specialized chips have evolved to do just this job, so that one almost never does use an op amp. And **power** supplies and their regulators are so universal in instruments of all kinds that the Text assigns them a chapter of their own, and we follow this scheme, giving them a day in the lab.

Here is an early regulator chip, still useful as an introduction to the subject, because it includes nearly all the elements of a standard voltage regulator, and makes these elements apparent: it brings out many of their terminals to the IC's pins, so that one is obliged to understand what's going on within the chip to apply it properly. This is the *723*:

Text sec. 6.02,
fig. 6.4, p. 310

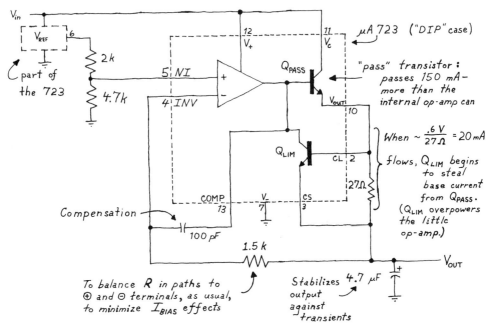

Figure N12.2: 723 regulator

The more fully-integrated regulators, which we will look at in a few minutes, make their innards less evident, since they do not oblige you to take care of so many pins. Once you have done this lab exercise, you are not likely to use a 723 again: it calls for too much thought, and—more important—too many external components.

Note the *current-limiting* scheme: Q_{limit} keeps an eye on the output current, and begins to pull current away from the pass transistor if the output current grows too large. The same trick is used to protect the output stage of op amps. You will recognize the current limit in the output stage of the LF411, for example, below:

*(Full '411 circuit appears
in Notes on Op Am Innards.)*

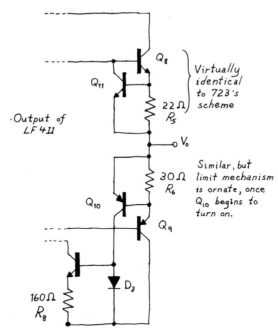

Figure N12.3: Current limit in an op amp: LF411 uses 723's trick

Dropout Voltage

xt sec. 6.02, p. 309,
c. 6.03, p.311;
18 (low dropout), pp. 345-49

Any regulator of this type ("linear," rather than the "switching" type that you will meet below) needs some difference between input and output voltage. This is called the "dropout voltage," because the output drops out of regulation if you don't fulfill this requirement.

Most regulators need two to three volts; specialized low-dropout regulators can get by on a few tenths of a volt. The 723 makes another demand, annoying if you are designing a 5-volt supply: its rather high reference (7.15V, typical) requires a supply of 9.5V; that's unusual. Usually you need worry about only the two or three volts' difference between V_{in} and V_{out}.

Easier:

Now that you have paid your dues by examining the 723, we'll let you consider some regulators that are much easier to use.

Fixed Output: 78xx
Text sec. 6.17

That whole circuit— 723 and discrete components—plus somewhat more, is now available on one chip. The simplest of these regulators, the *three-terminal* fixed-output type, are embarrassingly easy to use:

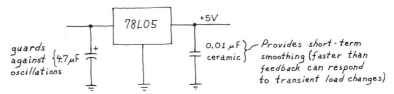

Figure N12.4: 3-terminal regulator

They limit not only their own output current but also their own *temperature*. Such a thermal shutdown protects the regulator when not current alone but *power* is excessive (you'll demonstrate this protection in the lab: *how*?) It also protects you against the effects of inadequate heat sinking.

Variable Output: 317
Text sec. 6.17, 6.18, 6.19,
fig. 6.38, p. 355

The *317* is almost as easy to use as the simpler 78L05, and is more versatile: it allows you to adjust V_{out}. It can also be used to rig up an easy current source.

It likes to hold a *constant* voltage (1.25V) between its two output terminals (it uses no "ground" terminal).

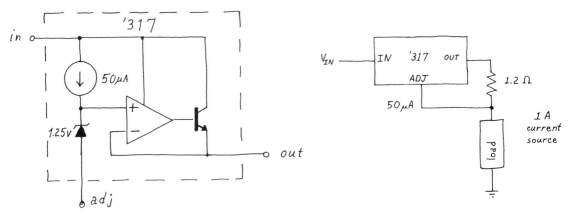

Figure N12.5: 317: adjustable regulator

Wrinkles

Crowbar

ext sec. 6.06

Here's a circuit that shuts down the supply (clamps it to about zero volts) when the voltage climbs too high. (IC's are available to do the voltage sensing, too.)

ext sec. 6.06,
g. 6.8, p. 318

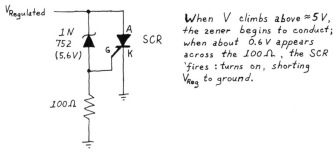

Figure N12.6: Crowbar over-voltage protection

The word "crowbar" apparently refers to the image of someone (courageous? foolhardy?) shutting down a huge power supply by shorting it to ground with a massive piece of steel.

Paralleling Pass Transistors

ext sec. 6.07,
ompare fig. 6.11, p. 321
ith fig. 6.16, p. 324

If you need a high output current—higher than what's available from the power transistor that you want to use—you can put the pass transistors in parallel. But when you do that, remember to prevent one over-eager bipolar transistor from trying to pass all the current:

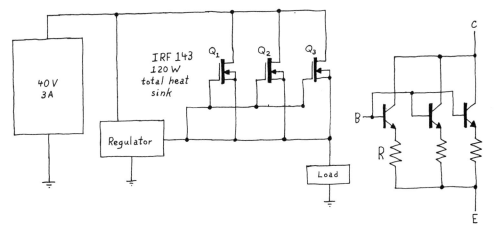

Figure N12.7: Paralleling pass transistors: FET's do it neatly; bipolars less neatly

Remember why the emitter resistors are required? Remember why FET's need no equivalent resistor?

A Different Scheme: Switching Regulators

xt sec. 6.19,
. 6.39, p. 356

These show one great virtue: they do not gulp power as they step voltage down. You can convince yourself with a few seconds' thought that the transistor *switch* that is either fully ON or fully OFF ideally lets the regulator dissipate no power at all. Consider the two cases:

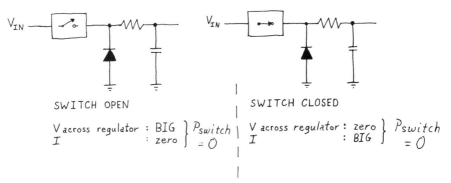

Figure N12.8: Idealized switching regulator: On or Off: zero voltage or zero current

xt sec. 6.19,
357, ex. 6.8

In fact a complete switching regulator typically runs at 75% to 90% efficiency, not 100%, but pretty good. You can answer without picking up a pencil the Text's question, 'What is the best efficiency you can hope for from a linear 5-volt regulator whose input is kept at 12V? (a little high, but only a little, since we need to make sure the regulator doesn't drop out).

A switcher can also step a voltage *up*, or can change the *sign* of a voltage input—or can do both at once. Those tricks are useful for generating a local source of an oddball voltage, in a computer powered with +5V only.

The scheme is to vary the duty cycle of a waveform that squirts charge into a storage "tank"—inductor feeding cap. At high currents, switching makes good sense. Below is a step-down type.

xt sec. 6.19,
. 6.40, p. 357

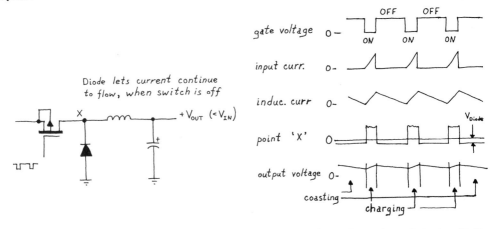

Figure N12.9: Switching regulator: waveforms (the feedback circuitry that carries out the regulation is omitted)

The basic pattern is simple: the switch closes, causing current to begin flowing through the inductor, simultaneously charging the capacitor. When the switch opens, the energy stored in the inductor ($1/2 LI^2$) is transferred to the capacitor, further raising its voltage (or diminishing its droop, if it is heavily loaded, as suggested in the waveforms shown above). Note, by the way, that though the output ripple may seem to resemble that in a linear supply, the ripple frequency is *much* higher: 20 to 100 kHz, versus 120Hz for the linear supply. Some of that high-frequency junk on the line will persist after even the most careful filtering. So you may choose a linear supply when you need to feed a device that dislikes noise.

xt sec. 6.19, 661 ff.

For digital devices, which shrug at low-level noise, switchers are the right choice. See, for example, the computer switching supply that the Text analyzes in detail.

Lab 12: Voltage Regulators

> **Reading:** Chapter 5.01 – 5.06, 5.10 – 5.12, 5.15 – 5.18, pp. 172-83, 187-92, 199-204.
> Problems:
> Problems in text.
> Additional Exercises.
> Bad Circuits (*all* these bad circuits are good and bad, this time!)

12-1. The *723* Regulator

You are not likely to take the trouble to use the 723 in practice, now that more fully-integrated regulators are available. But this 723 circuit shows many of the elements standard to any regulator: a "pass" transistor, controlled by an operational amplifier; a voltage reference; current limiting.

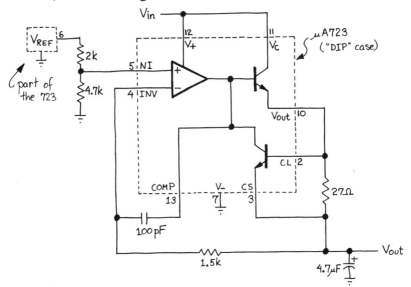

Figure L12.1: 723 voltage regulator IC

Feed the regulator from the variable regulated power supply. Watch how V_{out} varies you adjust V_{in} from 0 to around 20 volts. Can you identify the regulator's ***dropout voltage***? Does it match what the data sheet promises? That question is not a little ambiguous in this case: the 723's dropout voltage is more complicated than usual: its *dropout voltage* in the usual sense—the minimum difference required between input and output voltages—is *not* what will define the minimum input voltage. The 723's dropout voltage of 3V would suggest that the input could fall as low as 8V. Instead, you will see the effect of a slightly different characteristic, the minimum input voltage required to keep the 723's voltage reference running properly. That minimum input is specified at 9.5V. Most regulators do not bother you with this distinction; dropout voltage is just the minimum voltage drop across the regulator, period.

Connect the 1k potentiometer as variable load, and see how V_{out} is affected. What should the maximum output current be? Measure it.

Ripple Rejection

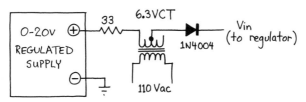

Figure L12.2: Ripple rejection test circuit. Note use of "center-tap" connection to obtain 3.2Vac

To test the circuit's ripple rejection, connect a 6.3Vac transformer in series with the variable DC supply, as in the figure above. (Notice that you are adding *half* that 6.3v as ripple.) The 33 ohm resistor and diode are included to prevent damage in case V_{in} is accidentally shorted to ground; this sort of connection easily blows things out.

First, check out this test circuit by connecting a 2.2k load resistor to ground and looking at the waveform (DC plus 60 Hz ripple) at its output. Note that you must keep the DC level high enough so that the ripple voltage does not fall below the *dropout* voltage; otherwise you will find yourself watching dropout rather than ripple.

Compare the amplitude of the ripple going in and coming out of the regulator. On the output side you may want the extra sensitivity you can get with a BNC cable rather than ×10 scope probe: this is one of the rare cases where the probe is *not* preferable. Even with the "×1 probe" you are likely to find this amplitude hard to judge; the surviving ripple is very small, and gets mixed with other low-level noise, probably including junk on the *ground* line. Is the ratio of ripple$_{out}$ to ripple$_{in}$ close to the manufacturer's promise of –74dB? Does a 4.7 µF cap on the V_{ref} terminal add 12dB as the data sheet promises?

12-2. Three Terminal Fixed Regulator

Figure L12.3: 78L05 3-terminal 5-volt regulator

This device is embarrassingly easy to use. It is so handy, though, that it's worth your while to meet it here.

It protects itself not only with current limiting, but also with a thermal sensor that prevents damage from excessive power dissipation ($I_{out}[V_{in} - V_{out}]$), which could occur even though the current alone remained below the limiting value. You will watch this thermal protection at work, and incidentally will see the effect of **heat sinking** upon the regulator.

Watch this thermal protection by providing a load that draws less than the chip's maximum current of 100 mA. Use two 150-ohm resistors in parallel to ground. (Check, with a quick calculation, that you are not overloading these 1/4-watt resistors.)

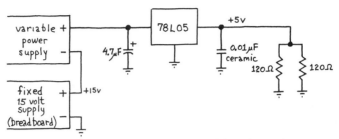

Figure L12.4: 78L05: demonstration of thermal protection

Note: we suggest you *stack* the variable supply on top of the breadboard's fixed +15V supply. This scheme *requires* that you *float* the "negative" (black) terminal of the variable supply; that terminal must *not* be tied to world ground.

To make this demonstration worthwhile, you should try to predict what voltage across the regulator will bring on thermal limiting. That is not at all hard, though it might take you a while to dig the information out of the data sheet (see *Analog Data Sheets*; and see Text sec. 3.04). You will notice that the thermal specifications state a *maximum permissible junction temperature* and give the *thermal resistance* between junction and ambient, which in this case means both the ambient air and the circuit board to which the regulator's leads are attached. You will be using the *TO92* package—plastic—which dissipates substantial heat through its leads (even the length of these leads matters, you'll notice).

To calculate the maximum power the package can dissipate before overheating the junction, plug in the values you find below.

78L05: Thermal Specifications

Package	Typical Θ JC	Max Θ JC	Typical Θ JA	Max Θ JA
TO-39	20°C/W	40C/W	140C/W	190C/W
TO-92			180C/W	190C/W

Definitions:

Θ_{JC} = thermal resistance, junction to case
Θ_{JA} = ...junction to ambient: includes JC and CA (case to ambient) in ser

Assume that the device begins to limit when its junction temperature reaches about 150°C—but don't be shocked if your calculation indicates the device may be tolerating a higher temperature, perhaps as high as 200°. Note units of *thermal resistance*, Θ: °C / Watt.

Experiment: Dropout Voltage and Thermal Self-Protection

Gradually increase V_{in} from close to 0V. Note, as you proceed:

- The dropout voltage; then
- The V_{in} that actually evokes the chip's thermal self-protection.

By the way, *what* does the chip do to limit its power dissipation? How will you know when the chip has begun to take this protective action?

When you notice the chip limiting its own power dissipation, you will be able to call off this self-protection by cooling the chip. Try putting a bit of wet tissue paper to the 78L05's fevered brow. It should recover at once. Then try a push-on heat sink, instead.

Figure L12.5: Push-on heat sink for T092 package

You may have to crank up V_{in} to see the self-protection begin again. Now fan the regulator, or blow on it. Does the output recover, once more?

12-3. Adjustable Three-terminal Regulator: *317*

This regulator allows you to select an output voltage by use of two resistors. You can make a variable output supply by replacing one of the two resistors with a variable resistor. The 317 lets you stock one chip to get all the positive supplies you need (at least, up to 1A output current); it also lets you trim to exactly 5V, for example, if the 78L05's 5% tolerance is too loose. In addition, the 317 is easy to wire as a current source.

In other respects this regulator is much like the 78L05: it includes both current and temperature sensing to protect itself from overloads.

Wire up the circuit in the figure below. Try **R** = 750 ohms; what should V_{out} then be? Measure it.

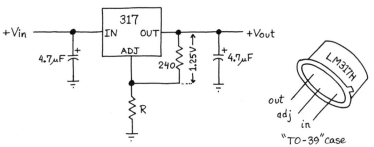

Figure L12.6: 317 voltage regulator circuit

Replace **R** with a 1k pot, and check out the 317's performance as an adjustable regulator. What is the minimum output voltage (**R** = 0)?

12-4 Three Terminal Regulator as Current Source

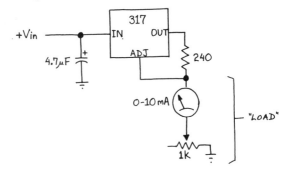

Figure L12.7: Simple current source made with a 3-terminal regulator

As you know, the 317 maintains 1.25V between its *out* and *adj* pins, with very low current at the *adj* pin. Thus the "poor man's current source," above. Try it out. What should the output current be? Check its constancy as the load resistance is varied. How does the circuit work? What limits its performance at extremely low or high currents? What is its voltage compliance?

12-4 Voltage References

Here we will ask you to look again at a *zener* diode, a device that, as you know, can provide a voltage reference. It works, but not very well: as current varies, V_{zener} varies a good deal; in conventional jargon "its knee is quite soft."

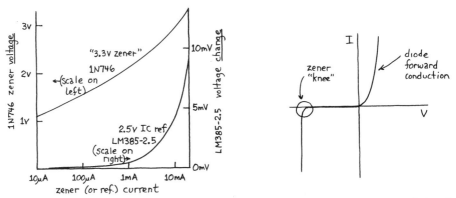

Figure L12.8: Zener's soft knee contrasted with IC voltage reference ('bony knee?'); curves in figure on left show detail of *knee* region; note that vertical scale for the IC reference is expanded by a factor of 250 relative to the ordinary zener!

You can work around that weakness of the zener by biasing it with a current source rather than resistor. Still you would be stuck with its sensitivity to temperature changes.

An integrated voltage reference does much better than a bare zener, as you might expect. Here, we ask you to test a bare zener against an IC 'zener,' an LM385[1]:

---

1. We put *zener* in quotes because the 385 is much more than just a zener; it's quite a complicated circuit, as you can confirm by glancing at the schematic shown in the 385's data sheet (not in these notes; see a *National* Linear data book).

you will vary the current over 3 decades and watch the variation in voltage that results, for each of the two 'zeners.' Here's the setup:

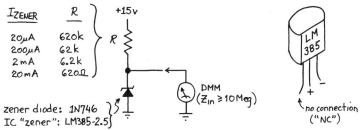

Figure L12.9: Circuit for measurement of variation in V_{out} with variation in current

12-5 "Crowbar" overvoltage protection

Here's a little circuit that can protect against the potentially horrible effects of a power supply failure, by clamping the supply voltage close to ground in case that voltage exceeds some threshold. Here, we have set the threshold around 6 volts: about right as the start of danger in a 5-volt supply (we chose 5 volts because it is the standard computer supply voltage). After your view of the zener's *knee*, you will recognize that this circuit gives us only approximate control of the threshold. You might prefer, in practice, to use an IC overvoltage sensor (some of these include the SCR, as well).

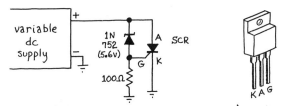

Figure L12.10: Crowbar circuit, and pinout of SCR: "G" = gate; "K" = cathode; "A" = anode

The *silicon controlled rectifier (SCR)*, shown above, is a device you have not seen before in these labs. It behaves more or less like a transistor: to turn it on, you need to provide some current at its gate, which will accept current if you bring its voltage up to around 0.6V above the cathode; the SCR differs from a transistor in *latching itself ON* once it begins to conduct. To turn the device off, you must stop the flow of current by some external means: in this circuit, by shutting off the power supply. Evidently, the SCR is well-suited to this application: trouble turns it *ON*; only someone's intervention then can revive the power supply. Note that the crowbar does *not* shut off if the supply voltage simply attempts to revert to a safe level, such as 5V in the present case.

In a practical circuit, you would add a capacitor to ground at the SCR's gate, to protect against triggering on brief transients. Today, your power supply—which powers nothing *but* this protection circuit!—should be quiet, so we have omitted that protection capacitor.

Try the circuit by gradually cranking up the supply voltage, using either an external variable supply or the breadboard's *15V* supply, which (on the PB503) is adjustable with a built-in potentiometer. (You'll need a skinny screwdriver; the adjustment pot is recessed at the very top of the breadboard's top panel). When you finish this experiment, incidentally, you'll probably want to set the breadboard's adjustable supply back to +15V, since you're likely to expect that voltage next time you use this supply.

Ch.6: Jargon and Terms

crowbar
overvoltage protection circuit: shorts supply output to ground when excessive voltage is detected; remains shorted until reset if done, as usual, with an SCR (see below).

dropout voltage
minimum required voltage difference between input and output of voltage regulator: failure to maintain input voltage high enough to provide this voltage difference will cause the circuit to "drop out of regulation."

SCR
silicon controlled rectifier: a three terminal power switch roughly resembling a bipolar transistor, but with the property that once triggered it remains ON until its current falls to nearly zero.

slow-blow fuse
fuse with long thermal time-constant, designed to ignore transient currents like those that occur on power-up as filter capacitors take initial charge ("inrush current").

CHAPTER 8

Chs. 8 & 9: Digital Electronics: Overview

Digital electronics is so sharply different from analog that it can be studied separately: some courses *begin* with digital, hoping that it can stand on its own. Only later do students learn what is happening inside the little black boxes that they have been stringing together. This view has obvious merits: digital is easier; its fundamental virtue is precisely its indifference to the details of signals. So, layout can be sloppy, one can watch many circuit operations without use of a scope. And anyone can get impressive results from a highly-integrated digital IC.

We have chosen to introduce you first to analog electronics so that, instead, you can begin digital with some sense of what the ICs that you use are doing. Most of the time you do not need any such detailed understanding, and all of us revert to the black box view of digital devices, and then look more closely only when strange things begin to happen—when a CMOS input is left floating, for example. On those occasions, it is useful to know some analog electronics. The digital ICs are of course made up of transistors (mostly MOSFETs), the devices you have just spent considerable effort to understand. You will be rewarded, though perhaps not daily.

We have called digital "easy," and in a sense it is: you will not need a calculator, or calculus. But some students find the subject difficult in another sense: it lacks coherence. In the analog half of the course, some issues and circuit fragments occurred over and over: impedances (can A drive B?); frequency-response; RC circuits (sometimes explicit, sometimes formed of stray elements, as in Miller effect and nasty oscillators). At the end of Chapter 8, in contrast, you may feel that you have been asked to learn a long list of gadgets. If one likened this course to an introduction to a foreign language, then analog would show some structure—rules of grammar, perhaps; Chapter 8 might feel like a vocabulary drill. To some extent it is also *literally* a vocabulary drill: digital electronics, even more than analog, is riddled with jargon: "3-state," "flip-flop," "ripple counter," "one-shot"....

We hope that the microcomputer labs will draw all this information together for you—as we said in the Preface to this book. Then the digital devices may fall into functional categories for you: gates combine signals to make this and that happen; flip-flops store information that otherwise would slip away; and so on.

As you read these two chapters you will need more than your usual skill in picking the important from among the arcane detail. Especially in Chapter 9 you will find much information that is invaluable to a practitioner—someone who wants to design a serial interface, for example—and excessive for a student in a first course. Look to the laboratory reading assignments for guidance here, and—as always—look at the lab exercises themselves to see which circuits have been picked out as most important.

Finally, try for a few minutes—before you get immersed in the details—to appreciate how curious the binary digital scheme is: if someone had proposed this arrangement to you in, say, 1930, what would you have thought? 'Turn voltages into codes; discard all delicate analog electronics—even including the wonders of feedback—and reduce every circuit to a crude *switch* that is simply ON or OFF.' Would you have said, 'Hey, that's a winning idea. I'll bet it takes over electronics'? —or would you have said, 'That's a harebrained mistake'?

When some better scheme is devised, we may look back on binary digital as harebrained; for the moment, it is extremely useful, and you will be pleased to see what powers it gives you: far beyond what you were able to do with a few weeks' worth of analog electronic learning. When microprocessors enter, then you will become powerful indeed. Much fun lies ahead.

Class 13: Digital Gates; Combinational Logic

Topics:

- *old:*
    - a comparator like a '311 resembles a digital gate

- *new:*
    - Why digital?
    - number codes: *two's-complement*
    - Just 3 functions to build any (binary-) digital circuit
      and 'universal gates' make things simpler still: just *one* gate type can build the world
    - 'active' levels: sometimes *low* is defined as *true*.
      'assertion-level symbols' express this circumstance graphically
    - What's in a gate?
        - Two important types: TTL (bipolar), and CMOS
        - logic with CMOS
        - output types: passive pullup, active pullup, 3-state

---

1. Analog versus Digital
What does this distinction mean?

Distinguish "digital" (the more general and more interesting notion) from "binary."
A *binary* logic gate classifies inputs into one of two categories: the gate is a comparator:

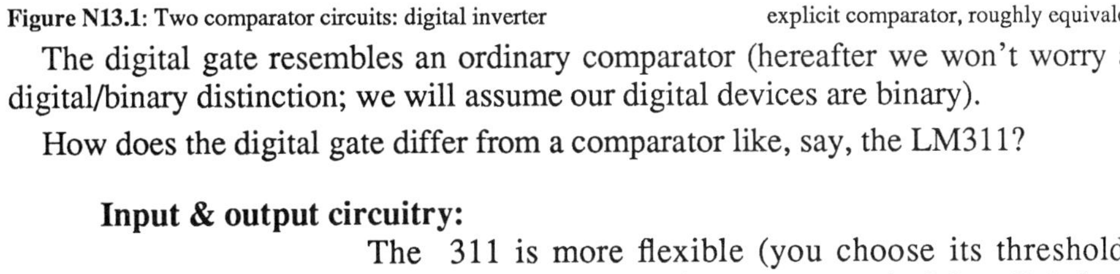

Figure N13.1: Two comparator circuits: digital inverter   explicit comparator, roughly equivalent

The digital gate resembles an ordinary comparator (hereafter we won't worry about the digital/binary distinction; we will assume our digital devices are binary).

How does the digital gate differ from a comparator like, say, the LM311?

Input & output circuitry: The 311 is more flexible (you choose its threshold & hysteresis; you choose its output range). Most digital gates include *no* hysteresis.

Speed: The logic gate makes its decision (and makes its output show that 'decision') about 20 times as fast as the 311 does.

Simplicity: The logic gate requires no external parts, and just power ground and In and Out pins. Typically comes 6 to a 14-pin package.

But why bother with digital?
Sub-questions:
A. (Naive query:) *Is This Transformation Perverse?*

Is it not <u>perverse</u> to force an analog signal—which can carry a rich store of information on a single wire—into a crude binary High/Low form?

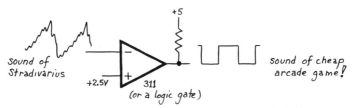

Figure N13.2: Naive version of "digital audio": looks foolish!

Disadvantages of Digital:

Complexity: more lines required to carry same information
Speed: more time to process the numbers that encode the information

Advantages of Digital:
Text 8.01

Noise Immunity: Signal is born again at each gate; from this virtue flow the important applications of digital:

- Allows Stored-Program Computers (may look like a wrinkle at this stage, but turns out, of course, hugely important);

- Allows transmission and also unlimited processing without error—except for round-off/quantization: deciding in which binary bin to put the continuously variable quantity; can be processed out of "real time:" at one's leisure. (This, too, is just a consequence of the noise immunity already noted.)

B. *Can we get the advantages of digital without loss?*

Not without *any* loss, but one can carry sufficient detail of the Stradivarius' sound using the digital form, by using *many* lines:

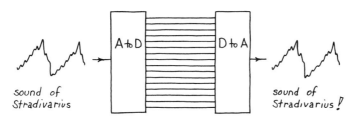

Figure N13.3: Digital audio done reasonably

C. Digital Processing *makes sense*, more obviously, *when the information* to be handled *is digital* from the outset.

Best examples: numbers (e.g., pocket calculator) & words (word processor)

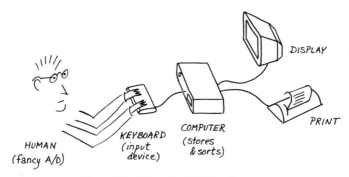

Figure N13.4: Example of all-digital system

Since the information never exists in analog form (except perhaps in the mind of the human), it makes good sense to manipulate the information digitally: as sets of codes.

2. Number codes: *two's-complement*

Binary numbers may be familiar to you, already: each bit (= "*bi*nary digi*it*") carries a weight double its neighbor's:

Text sec. 8.03, p. 474

DECIMAL	BINARY
10^2 10^1 10^0	2^2 2^1 2^0
2 1 3	1 0 1
$200 + 10 + 3 = 213$	$4 + 0 + 1 = 5_{10}$

Figure N13.5: Decimal and binary numbers: each *binary* digit carries twice the 'weight' of its neighbor to the right; each *digital* digit carries 10 times the weight of its neighbor to the right

That's just analogous to decimal numbers, as you know: it's the way we would count if we had one finger. The number represented is just the sum of all the bit values: $1001 = 2^3 + 2^0 = 9_{10}$.

But 1001 *need not* represent 9_{10}. Whether it does or not, in a particular setting, is up to us, the humans. *We* decide what a sequence of binary digits is to mean. Now, that observation may strike you as the emptiest of truisms, but it is not quite Humpty Dumpty's point: we are not saying 1001 means whatever I want it to mean; only that sometimes it is useful to let it mean something else. In a different context it may make more sense to let the bit pattern mean something like "turn on stove, turn off fridge and hot plate, turn on lamp."

And, more immediately to the point, it often turns out to be useful to let 1001 represent not 9_{10}, but a *negative* number.

Text sec. 8.03, pp. 476-77

The scheme most widely used to represent negative numbers is called "two's-complement." The relation of a 2's-comp number to positive or "unsigned" binary is extremely simple:

> the 2's comp number uses the most-significant-bit (MSB)—the leftmost—to represent a *negative* number of the same weight as for the unsigned number.

So, 1000 is +8 in unsigned binary; it is *–8* in 2's comp. And 1001 is –7.

	UNSIGNED	2'S COMP
1 0 1 1	$8+2+1 = 11_{10}$	$-8+2+1 = -5_{10}$
0 1 0 1	$4+1 = 5_{10}$	$\cdots -5_{10}$

Figure N13.6: Examples of 4-bit numbers interpreted as *unsigned* versus *signed*

This formulation is not the standard one. More often, you are told a rule for forming the 2's comp representation of a negative number, and for converting back. This the text does for you:

> To form a negative number, first complement each of the bits of the positive number...then add 1....(p. 477)

It may be easier to form and read 2's comp if you simply read the MSB as a large negative number, and add it to the rest of the number, which represents a (smaller-) positive number interpreted exactly as in ordinary unsigned binary.

This may seem rather odd and abstract to you, just now. When you get a chance to *use* 2's comp it should come down to earth for you. When you program your microcomputer, for example, you will sometimes need to tell the machine exactly how far to "branch" or "jump" forward or back, as it executes a program. The machine doesn't want to be told "forward" or "back" explicitly; instead, it simply adds the number you feed it to its present value; if you feed a negative number (in 2's comp), it hops back. For example:

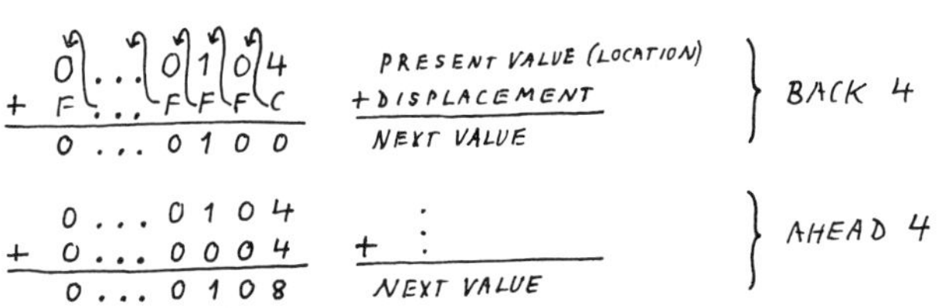

```
    0 . . . 1 0 1 0 4      PRESENT VALUE (LOCATION)    ⎫
  + F . . . F F F C      + DISPLACEMENT              ⎬  BACK 4
    0 . . . 0 1 0 0        NEXT VALUE                 ⎭

    0 . . . 0 1 0 4                                    ⎫
  + 0 . . . 0 0 0 4      + :                          ⎬  AHEAD 4
    0 . . . 0 1 0 8        NEXT VALUE                 ⎭
```

Figure N13.7: Example of 2's comp use: plain *adder* can add or subtract, depending on sign of addend: 68000 *branch* offset

Perhaps this example will begin to persuade you that there is an interesting, substantial difference between "subtracting A from B" and "adding negative-A to B:" the latter operation uses the same *hardware* as an ordinary addition. That's just what makes this a tidy scheme for the microcomputer's *branch* operation.

3. Combinational Logic

Explaining why one might want to put information into digital form is harder than explaining how to manipulate digital signals. In fact, digital logic is pleasantly easy, after the strains of analog design.

Comforting Truth #1

Text sec. 8.04

To build *any* digital device (including the most complex computer) we need only three logic functions:

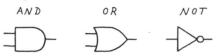

Figure N13.8: Just three fundamental logic functions are necessary

All logic circuits are combinations of these three functions, and only these.

Comforting and remarkable Truth #2

Perhaps more surprising: it turns out that just *one* gate type (not *three*) will suffice to let one build any digital device.

The gate type must be NAND or NOR; these two types are called "universal gates."

Figure N13.9: "Universal" gates: NAND and NOR

DeMorgan showed (in the mid-19th century!) that what looks like an AND function can be transformed into OR (and vise-versa) with the help of some inverters. This is the powerful trick that allows one gate type to build the world.

deMorgan's Theorem

Text sec. 8.07,
p. 483

This is the only important rule of Boolean algebra you are likely to have to memorize (the others that we use are pretty obvious: propositions like $A + A^* = 1$).

> **deMorgan's Theorem** (in graphic rather than algebraic form)
> You can swap *shapes* if at the same time you invert all inputs and outputs.
>
> etc.
>
> Figure N13.10: deMorgan's theorem in graphic form

When you do this you are changing only the **symbol** used to draw the gate; you are not changing the *logic*—the hardware. (That last little observation is easy to say and *hard* to get used to. Don't be embarrassed if it takes you some time to get this idea straight.)

So, any gate that can *invert* can carry out this transformation for you. Therefore some people actually design and build with NAND's alone. (This is less true now than a few years ago, when much was done with discrete gates; these came a few to a package, and that condition made the game of minimizing package-count worthwhile; it was nice to have leftover gates that were "universal". Now one is more likely to do a big design on a logic array—a programmable array of gates; there the rules of the game are different.)

This notion of DeMorgan's, and the "active low" and "assertion-level" notions will drive you crazy for a while, if you have not seen them before. You will be rewarded in the end (the end of this course, in fact), when you meet a lot of signals that are "active low": that is, signals that spend most of their lives close to 5 volts, and go low (close to zero volts) only when they want to say "listen to me!"

Active High versus active Low

Signals come in both flavors. *Example:* Two forms of a signal that say "Ready":

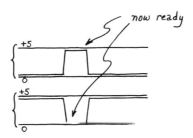

Figure N13.11: Active-High versus Active-Low Signals

Signals are made active low not in order to annoy you, but for good hardware reasons. We will look at those reasons when we have seen how gates are made. But let's now try manipulating gates, watching the effect of our *assumption about which level is True.*

Effect on logic of active level: Active High versus Active Low

As we considered what "1001" *means*, we met the curious fact—perhaps pleasing—that we can establish any convenient convention to define what the bit pattern means; sometimes we want to call 1001 "9_{10};" sometimes we will want to call it "-7_{10}." It's up to us. The same pattern appears as we ask ourselves what a particular *gate* is doing in a circuit. The gate's operation is fixed in the hardware (the little thing doesn't know or care how we're using it); but what its operation means is for us to interpret.

That sounds vague, and perhaps confusing; let's look at an example (just deMorgan revisited, you will recognize):

Most of the time—at least when we first meet gates—we assume that High is true. So, for example, when we describe what an AND gate does, we usually say something like
"The output is true if both inputs are true."

The usual AND truth table of course says the same thing, symbolically. But—as deMorgan promised—if we declare that Zeros interest us rather than Ones, at both input and output, the gate evidently does something different:

Truth table				*Altered* description of its operation, appropriate if we treat		
A	B	A·B		A	B	A·B
0	0	0		T	T	T
0	1	0		T	F	T
1	0	0		F	T	T
1	1	1		F	F	F

Figure N13.12: "AND" gate (so-called!) doing the job of OR'ing Lows

We get a Zero out if A *or* B is Zero. In other words we have an *OR function*, if we are willing to stand signals on their heads. We have a gate that *OR's lows*, and the symbol drawn above says that. It turns out that often we *need* to work with signals 'stood on their heads': active low.

Note, however, that we call this piece of hardware an *AND gate* regardless of what logic it is performing in a particular circuit. To call one piece of hardware by two names would be too hard on everyone. We will try to keep things straight by calling this an AND *gate*, but saying that it performs an OR *function* (*OR*'ing lows, here). Sometimes it's clearest just to refer to the gate by its part number: 'It's an '08.' We all should agree on that point, at least!

Here's an example—a trifle melodramatic—of signals that are active low: you are to finish the design of a circuit that requires two people to go crazy at once, in order to bring on the third world war:

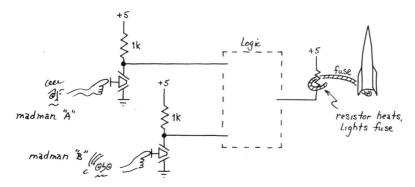

Figure N13.13: Logic that lights fuse if both operators push Fire* at the same time

How should you *draw* the gate that does the job? What is its conventional *name*?

Just now, these ideas probably seem an unnecessary complication. By the end of the course—to reiterate a point we made earlier—when you meet the microcomputer circuit in which *every* control signal is active low, you will be grateful for the notions "active low" and "assertion-level symbol." We will be able to explain in a few minutes *why* control signals typically are active *low*. To provide that explanation we need, first, to look inside some gates.

4. Gate Types: TTL & CMOS

Text sec. 8.07

CMOS versus TTL

TTL—made of bipolar transistors—ruled the world for about 20 years; now its days are numbered, as the ad below is meant to suggest (this ad also reminds us how excited someone can get about a few nanoseconds: an advantage of a few tens of ns over ordinary CMOS is what Zytrex had to offer).

Text sec. 9.01

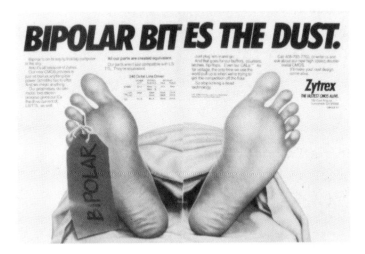

Figure N13.14: Reflections on mortality: RIP Zytrex (1985-87)

Ever heard of Zytrex, forecaster of TTL's doom? Probably not. *Sic transit gloria Zytrecis.* (Is there, perhaps, a moral to this story?)

Gate Innards: TTL vs CMOS

Text sec. 8.09, 8.10

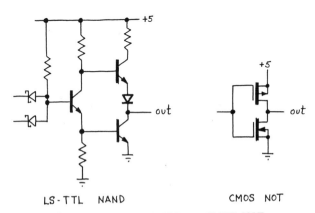

Figure N13.15: TTL & CMOS gates: NAND, NOT

A glance at these diagrams should reveal some characteristics of the gates:

Text sec. 9.06

Inputs: You can see why TTL inputs *float high*, and CMOS do not.

Threshold: You might guess that TTL's threshold is off-center—low, whereas CMOS is approximately centered.

Output: You can see why TTL's high is *not* a clean 5 V, but CMOS' is.

Power consumption:
You can see that CMOS passes *no* current from +5 to ground, when the output sits either high or low; you can see that TTL, in contrast, cannot sit in either output state without passing current in (a) its input base pullup (if an input is pulled low) or (b) in its first transistor (which is ON if the inputs are high).

Thresholds and "Noise Margin"

Text sec. 8.02,
9.01, 9.02

All digital devices show some noise immunity. The guaranteed levels for CMOS and TTL show that CMOS has the better noise immunity:

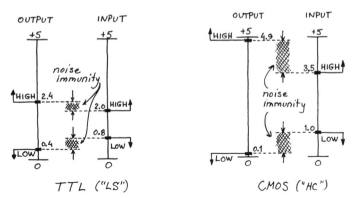

Figure N13.16: Thresholds & noise margin: TTL versus CMOS

Curious footnote: as the text points out, TTL and NMOS devices are so widely used that some new families of CMOS, labeled **74xCTxx**, have been taught TTL's bad habits on purpose: their thresholds are put at TTL's nasty levels ("CT" means "CMOS with TTTL thresholds"). We will use a lot of such gates (**74HCTxx**) in our lab microcomputer, where we are obliged to accommodate the *NMOS* microprocessor, whose output *High* is as wishy-washy as TTL's. When we have a choice, however, we will stick to straight CMOS.

Answer to the question, 'Why is the typical control signal active low?

We promised that a look inside the gate package would settle this question, and it does. TTL's asymmetry explains this preference for active low. If you have several control lines, each of which is inactive most of the time, it's better to let it *rest high*, occasionally be *asserted low*. This explanation works only for TTL, but the conventions were established while TTL was supreme, and they will linger for some time longer, so long as there is some TTL and NMOS around.

Here's the argument:

A TTL *high* input is less vulnerable to noise than a TTL *low* input
The guaranteed noise margins differ by a few tenths of a volt; the *typical* margins differ by more. So, it's safer to leave your control lines safe most of the time; now and then let them dive into danger.

A TTL input is easy to drive High
In fact, since a TTL input *floats* high, you can drive it essentially for 'free:' at a cost of no current at all. So, if you're designing a microprocessor to drive TTL devices, make it easy on your chip by letting most the lines rest at the lazy, low-current level, most of the time.

Both these arguments push in the same direction; hence the result—which you will be able to confirm when you put together your microcomputer, where *every* control line is active low.[1]

Output Types

active pullup

Text sec. 8.11

All respectable gates use *active pullup* on their outputs, to provide firm Highs as well as Lows:

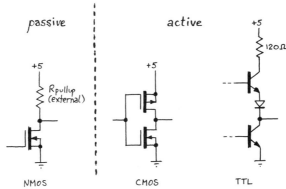

Figure N13.17: Passive versus active pullup output stages

You will confirm in the lab that the *passive-pullup* version not only wastes power but also is *slow*.

---

[1] The phrase 'control line' may puzzle you. Yes, we are saying a little less than that every *signal* is active low. Data and address lines are not. But every line that *has* an active state, to be distinguished from inactive, makes that active state *low*. Some lines have no *active* versus *inactive* states: a data line, for example, is as 'active' when low as when high; same for an address line. So, for those lines designers leave the active-high convention alone. That's lucky for us: we are allowed to read a value like "1001" on the data lines as "9;" we don't need to flip every bit to interpret it! So, instead of letting the *active low* convention get you down, count your blessings: it could be worse.

Open-collector

*Text sec. 8.11,
p. 489, fig. 8.21;
see also sec. 4.23*

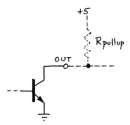

Figure N13.18: Open collector: rarely useful

Once in a great while "open collector" is useful. You have seen this on the 311 comparator.

Three-state

Very often *three-state* outputs are useful: these allow multiple drivers to share a common output line (then called a "bus"). These are very widely used in computers.

Beware the misconception that the "third state" is a third output voltage level. It's not that; it is the *off* or disconnected condition. Here it is, first conceptually, then the way we'll build it in the lab:

*Text sec. 8.11,
p. 488, fig. 8.19*

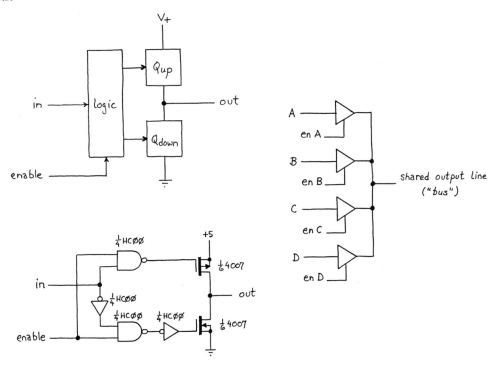

Figure N13.19: Three-State output: conceptual; the way we build it in the lab; driving shared bus

Logic with TTL and CMOS
Text sec. 8.09

The basic TTL gate that we looked at a few pages back was a NAND; it did its logic with diodes. CMOS gates do their logic differently: by putting two or more transistors in series or parallel, as needed. Here is the CMOS NAND gate you'll build in the lab, along with a simplified sketch, showing it to be just such a set of series and parallel transistor switches:

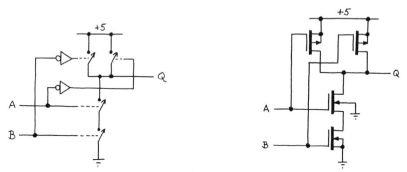

Figure N13.20: CMOS NAND gate

Speed versus Power consumption
Text sec. 9.02,
p. 569 fig. 9.2

The plot below shows the tradeoffs available between speed and power-saving. A few years ago the choice was stark: TTL for speed (ECL for the very impatient), CMOS (4000 series) for low power. Now the choices are more puzzling: some CMOS is very fast, though TTL remains a bit faster. GaAs is fastest. As you can see from this figure, everyone is trying to snuggle down into the lower left corner, the state of Nirvana where you get fast results for almost nothing.

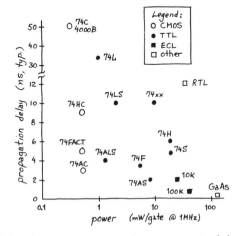

Figure N13.21: Speed versus power consumption: some present and obsolete logic families

Ch. 8: Worked Example: Multiplexers

These notes look at ways to design a small *multiplexer* in 3 ways:

- using ordinary gates;
- using 3-states;
- using analog switches

Multiplexing: generic

Text sec. 8.14, p. 495

The notion of multiplexing, or *time-sharing* is more general and more important than the piece of hardware called a multiplexer (or "mux"). You won't often use a mux, but you use multiplexing continually in any computer, and in many data-acquisition schemes.

Here's multiplexing in its simplest, mechanical form: a mechanical switch might select among several sources. The sources might be, for example, four microphones feeding one remote listening station of the secret police. Once a month an agent comes to the house and flips the switch to listen to another of the four mikes. The motive for multiplexing, here, as always, is to limit the number of lines-+ needed to carry information.

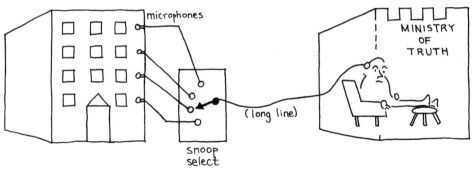

Figure X13A.1: Mux in mechanical form: to let four sources share one output line

This case is a little far-fetched. More typically, the 'scarce resource' would be an A/D converter (see Ch.9, Lab 18). And the most familiar example of all must be the *telephone* system. Without the time-sharing of phone lines phone systems would look like the most monstrous rats' nests: picture a city strung with a pair of wires (or even one wire) dedicated to joining each telephone to every other telephone! Here's what even a network of 8 phones would look like; and here's a computer, sketched with and without multiplexing—which in that context is called sharing of a "bus" (you will hear much more of this notion, soon):

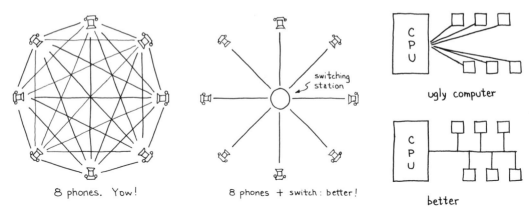

Figure X13A.2: Arguments for multiplexing: limit the number of wires running here to there. Two cases: telephone, computer

Multiplexing: hardware

A mechanical multi-position switch can do the job; so can the transistor equivalent, a set of analog switches. Logic gates, too, can do the job—though these allow flow in only one direction, unlike the other two devices. The digital implementation requires a little more thought, if you haven't seen it before.

We need two elements:

1. *Pass/Block* circuitry, analogous to the closed/open switch
2. *Decoder* circuitry, that will close just one of the pass/block elements at a time

Let's work up each of these elements, first for the implementation that uses ordinary gates.

Pass/Block   An AND gate will do this job, more or less: to *pass* a signal, hold one input high; the output then follows or equals the other input; to *block* a signal, hold an input low. This case is a little strange, because this forces a *low* at the output—not the same as opening a mechanical switch.

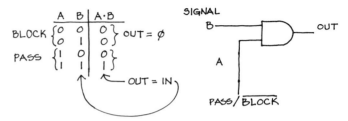

Figure X13A.3: AND gate can do the Pass/Block* operation for us

Joining outputs The outputs of the AND gates may *not* be tied together:

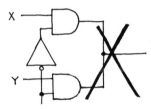

Figure X13A.4: Don't do this! The outputs of ordinary gates may *not* be tied together

Instead, we need a gate that 'ignores' lows, passes any highs (since the 'blocking' AND's will be putting out lows). An OR behaves that way.

So, here's a 2:1 mux made with gates:

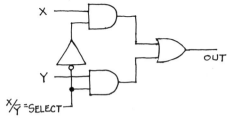

Figure X13A.5: 2:1 mux

Decoder If the mux has more than 2 inputs, we need a fancier scheme to tell *one and only one* gate at a time to *pass* its signal. The circuit that does this job, pointing at one of a set of objects, is called a *decoder*. It takes a binary number (in its *encoded* form: compact: n lines encode 2^n combinations, as you know) and translates it into 1-of-n form (*decoded*: not compact, so not a good way to transmit information, but often the form needed in order to make something happen in hardware).

The decoder's job is to detect each of the possible input combinations. Here, for example, is the beginning of a 2-to-4-line decoder:

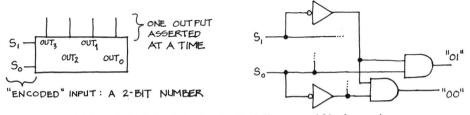

Figure X13A.6: 2-to-4-line decoder: block diagram; partial implementation

Decoders are useful in their own right (you will use such a chip in your microcomputer, for example: a 74HCT138). They are *included* on every multiplexer and in every multiplexing scheme.

Worked example: Mux (Text problem 8.17)

Text sec. 8.14,
ex.. 8.17, p. 496

> **Problem: 4:1 Mux, designed 3 ways**
>
> Show how to make a 4-input multiplexer using (a) ordinary gates, (b) gates with 3-state outputs, and (c) transmission gates. Under what circumstances would (c) be preferable?

Solution

Decoder

All three implementations require a decoder, so let's start with this.

We need to detect all four possibilities; might as well start, as in the earlier example, by generating the complement of each *select* input. (Does it go without saying that we need 2 *select* lines to define four possibilities? If not, let's say it. If this isn't yet obvious to you, it will be soon.)

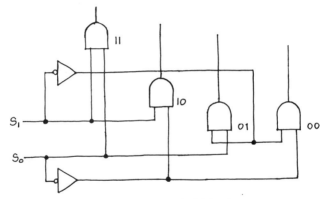

Figure X13A.7: 1-of-4 decoder

Pass/Block gating

Ordinary Gates

You know how to do this job with AND gates. This is the same as before, only we have more than the pair we used last time:

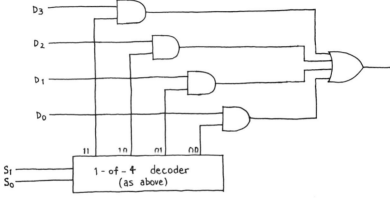

Figure X13A.8: 4-1 mux, gate implementation

Transmission gates

This is easier. (You *do*, of course, need to remember what these "transmission gates" or "analog switches" are. If you don't recall them, from way back in Chapter 3 and Lab 12, go back and take a second look.)

Again we use the decoder. This time we *can* simply join the outputs, since any transmission gate that is *blocking* its signal source *floats* its output, unlike the ordinary logic gate, which drives a *low* at its output when blocking.

Three-states

This is a snap after you have done the preceding case. Again use the same decoder, and again you *can* join the outputs. A 3-state that is off, or blocking, does not fight any other gate. That's the beauty of a 3-state, of course.

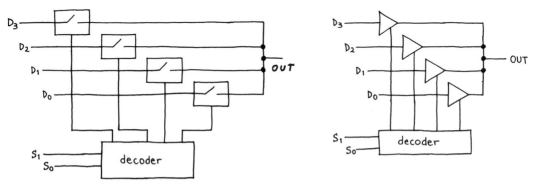

Figure X13A.9: 4-1 mux: transmission-gate and 3-state implementations

Question:

> "Under what circumstances would (c) [the transmission-gate implementation] be preferable?"

It is preferable *only* for handling *analog* signals. (For this purpose it is not just *preferable*; it is *required*!) For digital signals it is inferior, since it lacks the all-important virtue of digital devices: their noise immunity, which can also be described as their ability to clean up a signal: they ignore noise (up to some tolerated amplitude), and they put out a signal stripped of that noise and at a good low impedance. The analog mux, like any analog circuit, cannot do that trick; to the contrary, the analog mux slightly degrades the signal: at very least, it makes the output impedance of the signal source worse, by adding in series the mux's R_{on} (around 100Ω).

The analog version, incidentally, can pass a signal in either direction, so it works as a *demultiplexer* as well as mux. But that is not a good reason to use it for digital signals. If you want to demux digital signals, use a *digital* demux!

Ch. 8: Worked Examples: Binary Arithmetic

These worked examples look at four topics in binary arithmetic:

two's complement: versus *unsigned*, and the problem of *overflow*
addition: a hardware design task, meant to make you think about how the adder uses *carries*
multiplication: an orderly way to do it contrasted with a foolish way
ALU & flags: foreshadowing the microprocessor, this exercise means to give you the sense that the processing guts of the CPU are simple, made of familiar elements

1. Two's Complement

Here's a chance to get used to 2's-comp notation. We want to underline two points, in these examples:

- A given set of bits has no inherent meaning; it means what we choose to let it mean, under our conventions (this is a point made in the class notes, as well);
- A sum (or product or other result), properly arrived at may nevertheless be wrong if we overflow the available range.

Both of these points are rather obvious; nevertheless, most people need to see a number of examples in order to get a feel for either proposition. "Overflow," particularly, can surprise one: it is not the same as 'a result that generates a carry off the end.' Such a result may be valid. Conversely, a result can be wrong (in 2's comp) even though no carry out of the MSB occurs. This is pretty baffling when simply *described*. So let's hurry on to examples.

Let's suppose we feed an *adder* with a pair of four-bit numbers. We'll note the result, and then decide whether this result is correct, under two contrasting assumptions: that we are thinking of the 4-bit values as *unsigned* versus *two's-complement* numbers.

Problem: 2's comp

Suppose that you feed a 4-bit adder the 4-bit values A and B, listed below. As an exercise, please write—

1. the maximum value one can represent in this four-bit result—
   - unsigned
   - in 2's complement

2. the sum;
3. then what the inputs and outputs mean in decimal, under two contrasting assumptions:
   - first, the numbers are (4-bit-) *unsigned*;
   - second, the numbers are (4-bit-) *two's-complement*.

4. Finally, note whether the result is valid under each assumption.

We have done one case for you, to provide a model.

IN:	A	B	A plus B	Valid?
Binary:	0111	1000	1111	
Decimal:				
unsigned:	7	8	15	yes
2's-c:	7	−8	−1	yes
Binary:	0111	0111		
Decimal:				
unsigned:				
2's-c:				
Binary:	0111	1010		
Decimal:				
unsigned:				
2's-c:				
Binary:	0111	0100		
Decimal:				
unsigned:				
2's-c:				
Binary:	1001	1000		
Decimal:				
unsigned:				
2's-c:				

Try these. Then see if you agree with our conclusions, set out below:

Solution:

1. *The maximum values that one can represent with 4 bits*

    - unsigned:
    $$15_{10}$$
    - in 2's complement:
    $$+7_{10}, -8_{10}$$

2. *Sums: overflows*

IN:	A	B	A plus B	Valid?
Binary:	0111	0111	1110	
Decimal:				
unsigned:	7	7	14	yes
2's-c:	7	7	–2	no
Binary:	0111	1010	0001	
Decimal:				
unsigned:	7	10	1	no
2's-c:	7	–6	+1	yes
Binary:	0111	0100	1011	
Decimal:				
unsigned:	7	4	11	yes
2's-c:	7	4	–5	no
Binary:	1001	1000	0001	
Decimal:				
unsigned:	9	8	1	no
2's-c:	–7	–8	+1	no

What's the rule that determines whether the result is valid? For unsigned, isn't it simply whether a carry-out is generated? For *two's-comp* the rule is odder: *if the sign is altered by carries*, the result is bad. A carry into the sign bit indicates trouble, if there is no carry out of that bit, and vice versa. In other words, an XOR between carries in and out of MSB indicates a 2's-comp overflow. The microprocessor you will meet later in this course uses such logic to detect just that overflow (indicated by its "V" flag, in case you care, at this point). You can test this *XOR* rule on the cases above, if you like.

2. Adders

ext sec. 8.14,
p. 497-98

You may find it entertaining to reinvent the *adder*. Here, to remind you, is the way a single-bit *full-adder* should behave:

IN:	Carry	A	B	OUT:	Carry	Sum
	0	0	0		0	0
	0	0	1		0	1
	0	1	0		0	1
	0	1	1		1	0
	1	0	0		0	1
	1	0	1		1	0
	1	1	0		1	0
	1	1	1		1	1

Figure X13B.1: Truth table for a *full adder*

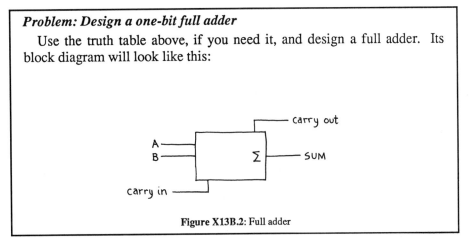

Problem: Design a one-bit full adder

Use the truth table above, if you need it, and design a full adder. Its block diagram will look like this:

Figure X13B.2: Full adder

Solution

The *sum* is A XOR B when Carry-in is low; it's the complement of XOR when Carry-in is high. Does that suggest what the sum function is, as a function of the 3 input variables that include Carry-in? Yes, it's XOR of all 3 variables (output 1 if odd number of inputs 1)k:

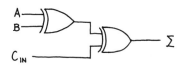

Figure X13B.3: Sum is C_{in} XOR A XOR B

Carry out?

You may be able to see the pattern, after staring a while at the truth table. But you can also enter the values on a Karnaugh map, and let the map reveal patterns: patterns that show chances to ignore a variable or two:

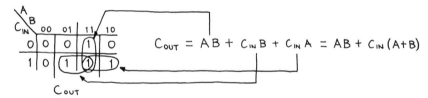

$$C_{OUT} = AB + C_{IN}B + C_{IN}A = AB + C_{IN}(A+B)$$

Figure X13B.4: Carry-Out K map

So, here's what a full adder looks like:

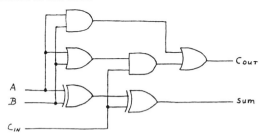

Figure X13B.5: Full Adder

How the pros do it:

You may be interested to see the curious ways that IC adders are designed. They don't look like our circuit, but they do, of course, perform the same logical operations. See, for example, schematics in Texas Instruments data books for 74LS83A and 7482.

It's pretty clear that once we have defined this single-bit function, with a carry-in and carry-out line, we can string these little beads forever, making an adder as big as we like. We will encourage you to look for a similarly-repeatable pattern in designing a multiplier, below.

3. Multipliers

*Text sec. 8.14,
exercise 8.14, p. 493*

Multipliers are much less important than adders; you should not feel obliged to think about how multiplication works. Think about it if the question intrigues you.

One way to discover the logic needed to *multiply* would be to use Karnaugh maps or some other method to find the function needed for each bit of the product. This is the method suggested in Text problem 8.14. For example, to design a 2X2 multiplier, you would need 2 K maps, each with 4 input variables.

But even this small example may be enough to convince you that there must be a better way. A 3 • 3 multiplier (with its 6 input variables) would bog you down in painful K mapping; but larger multipliers are commonplace. Yes, there is a better way.

This turns out to be one of those design tasks one ought *not* to do with K maps, or with any plodding simplification method. Instead, one ought to take advantage of an orderliness in the product function that allows us to use essentially a single scheme, iterated, in order to multiply a large number of bits. The pattern is almost as simple as for the adder.

Binary multiplication can be laid out just like decimal, if we want to do it by hand—and the binary version is much the easier of the two:

```
    DECIMAL           BINARY              BINARY, GENERALIZED

       2 5             1 0 1                      a₂  a₁  a₀
    x  2 4           x 1 0 1                      b₂  b₁  b₀
    -------          -------                   ---------------
      1 0 0            1 0 1                     a₂b₀  a₁b₀  a₀b₀
    + 5 0                0                  a₂b₁  a₁b₁  a₀b₁
    -------         + 1 0 1              a₂b₂  a₁b₁  a₀b₂
      6 0 0          -------            ---------------------------
                     1 1 0 0 1           P₅   P₄   P₃   P₂   P₁   P₀
```

Figure X13B.6: Decimal and binary multiplication examples—and a generalization of the pattern; binary is easier!

Once you have seen the tidy pattern, you will recognize that you need nothing more than 2-input AND gates, and a few *adders*. With this method, you don't have to work very hard to multiply two 4-bit numbers—a task that would be daunting indeed if done with K maps. (As the Text exercise suggests, you need only *half* adders, not *full* adders; but you might choose to use an IC adder like the '83, rather than build everything up from the gate level.) Let's try that problem.

Compare Text exercises E 8-14, AE 8-15

Problem: 3 • 3 Multiplier

Design a 3 • 3 multiplier along the same lines [as in AE 8-14], this time using two 4-bit full adders (74HC83) and as many 2-input gates as you need.

Solution:

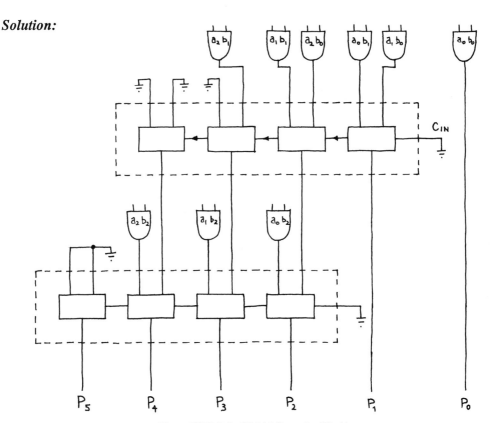

Figure X13B.7: 3 • 3 Multiplier, using IC adders

Whew! There's a circuit you'll want to buy, not build! But you can see that you could easily extend this to make a 4 X 4—or 16 X 16 multiplier.

4. Arithmetic Logic Unit (ALU)

Here, as a wrap-up for this discussion of arithmetic, is a device central to any computer: a circuit that does any of several logical or arithmetic operations. To design this does not require devising anything new; you need only assemble some familiar components.

> *Problem: design a 1-bit ALU*
>
> Here is a block diagram of a simplified ALU. Let's give it a carry-in and a carry-out, and let that output bit be low in any case where a carry is not generated (for example, A OR B).
>
> Two select lines should determine which operation the ALU performs, of the following set (use any select code you like):
> AND OR XOR ADD
>
>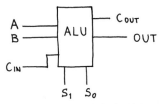
>
> **Figure X13B.8**: 1-bit ALU: block diagram

Solution

All we need is to combine standard gates and an adder with a multiplexer:

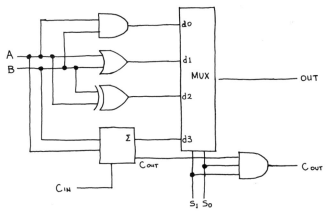

Figure X13B.9: A simple ALU

A refinement: "flags"

A computer's ALU always includes flip-flops that keep a record of important facts about the result of the most recent operation—not the result itself, but summary descriptions of that result:

- was a *carry* generated?
- was the result *zero?*
- is the result positive or negative, in 2's comp notation?
- was a 2's-comp overflow generated?
- —and sometimes a few other items as well (see the 68008's flags, noted in class notes µ3: just one marginally different flag in addition to those we have listed).

Note: You should understand at least the *D* flip-flop before trying this problem.

> **Problem: Add flags to ALU**
>
> Add, to the ALU designed earlier, *flags* to record the following pieces of information concerning the result of the ALU operation:
>
> **Carry**   was a *carry* generated?
> **Zero**   was the result *zero?*
> **Sign**   is the result positive or negative, in 2's comp notation? (assume that the single-bit output of this stage makes up the MSB of a longer word).
> **Overflow**   was a 2's-comp overflow generated? (Make the assumption noted: *this* output is the MSB of a longer word, and thus constitutes the *sign* bit when this word is treated as a 2's-complement value.)
>
> Assume that a timing signal is available to clock the flag flops a short time after the ALU output has settled.

Solution:

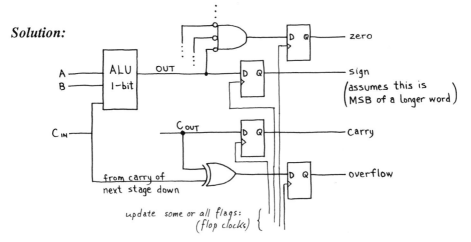

Figure X13B.10: Flags added to ALU

Lab 13: Digital Gates

Reading:	Ch. 8: 8.01-8.12; Ch. 9: 9.01-9.03 *Specific advice:*
Ch. 8:	8.01-8.02; 8.03; among the number codes, ordinary ("natural") binary and 2's complement are important. 8.04-8.12: important topics here: Assertion-level notation (8.07); 8.11, three-state outputs Logic identities (8.12): only a very few are needed, above all deMorgan's: (AB)*=A* + B* ("*" = "bar" or complement).
Ch. 9:	sec. 9.01-9.03 re TTL & CMOS gates: all relevant to today's lab.
Problems:	Embedded problems; try to explain to yourself how the gating shown in *Circuit Ideas* A, B, and E achieves the results claimed (p. 557).

The first part of this lab invites you to try integrated gates, black boxes that work quite well, to carry out some Boolean logic operations.

The later sections of the lab ask you to look within the black box, in effect, by putting together a logic gate from transistors. Here, the point is to appreciate why the IC gates are designed as they are, and to notice some of the properties of the input and output stages of TTL and CMOS gates. We will concentrate in this lab, as we will throughout the course, on CMOS. To overstate the point slightly, we might say that we will treat ordinary TTL as an important antique.

Preliminary

Some *ground rules* in using logic:

1. Never apply a signal beyond the power supplies of any chip. For the logic gates that we use, that means
   keep signals between 0 and + 5 volts.
   (This rule, in its general form—'stay between the supplies'— applies to analog circuits as well; what may be new to you is the nearly-universal use of single supply in digital circuits.)
2. Power all your circuits from +5v. and ground only. This applies equally to CMOS (in its contemporary forms most widely used) and to TTL.

Logic Probe

The logic probe is a gizmo about the size of a thin hot dog, with a cord on one end and a sharp point on the other. It tells you what logic level it sees at its point; in return, it wants to be given power (+5v and ground) at the end of its cord. (N.b., the logic probe does *NOT* feed a signal to the oscilloscope!)

If you find a BNC connector on the probe cord, connect +5V to the center conductor, using one of the breadboard jacks. If, instead, you find that the cord ends with alligator clips or a pair of strange-looking grabbers, use these to take hold of ground and +5. You may have to peer closely at the leads to see the bit of red peeping through the black casing on the + 5 lead (in the case of the Hewlett-Packard probe).

How to Use the Probe

Once the probe is powered, the probe lamp (near the tip) glows. Touch the tip to ground, then to +5v. You will be able to distinguish +5 v and ground from "float" (simply "not driven at all; not connected"). This ability of the probe is extremely useful. (Could a voltmeter make this distinction for you?)

Use the probe to look at the output of the breadboard function generator when it is set to *TTL*. Crank the frequency up to a few kilohertz. Does the probe wink at the frequency of the signal it is watching? Why not?

LED Indicators

The eight LED's on the breadboard are buffered by logic gates. You can turn on an LED with a logic high, and the gate presents a conveniently high input impedance (100k to ground).

In order to appreciate what the logic probe did for you earlier, try looking at a quick pulse train, using an LED rather than logic probe: use the breadboard oscillator (TTL) at a kilohertz or so. Does what you see make sense? You may now recognize that the logic probe stretches short pulses to make them visible to our sluggish eyes: it turns even a 10 ns pulse into a wink of about one-tenth of a second.

Switches

The PB503 includes three sorts of switch on its front panel:

2 debounced pushbuttons
(over on the left, marked PB1, PB2)

These deliver an *open collector* output, and that means that they are capable of pulling to ground *only*. To let this level go to a logic *high*, you will need to add a *pullup* resistor, to 5V:

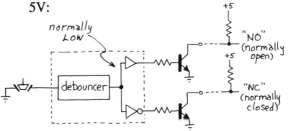

Figure L13.1: Pullup required on open-collector output (debounced pushbuttons on PB503)

**an 8-position DIP switch,
fed from a +5V/0V slide switch
(marked DIP switch S1)**

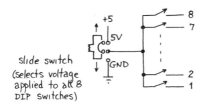

Figure L13.2: DIP switch in-line with output of *one* common slide switch

This looks as if it could give you 8 independent outputs, but it can't; at least, not conveniently. The DIP switch is simply an in-line switch that delivers the level set by the slide switch, if closed—and *nothing* (a *float*: neither high nor low) if the DIP switch is open. To get 8 independent levels you need 8 *pullup* resistors. That's a nuisance, so most of the time you'll probably want to use this to provide just *one* logic level. To get that, *close* the DIP switch; to avoid fooling yourself later, probably it's a good idea to leave all the DIP's switches closed.

**two uncommitted
slide switches (SPDT)**

These are on the lower right, and are *bouncy* (not debounced, anyway). To make them useful, tie one end to ground, the other to +5, and use the *common* terminal as output. You might as well wire these now, use them today, and then leave them so wired for use in later labs.

A. Applying Integrated Gates

13-1 Input & Output Characteristics of Integrated Gates: TTL & CMOS

a) V_{out}

Use the two slide switches to provide 0 or +5 v to the two inputs of a NAND gates, using both TTL and CMOS, simultaneously.

74XX00 pinout:

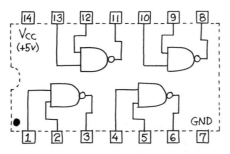

Figure L13.3: NAND gates: TTL & CMOS

The TTL part is

74LS00[1]

The CMOS part is

74HC00[2]

Note: for the CMOS part (but not TTL), tie all the six unused input lines to a common line, and temporarily ground that line.

Now note both *logic* and *voltage* levels out, as you apply the four input combinations. (Only one logic-out column is provided, below, because here TTL and CMOS should agree.)

INPUT		OUTPUT: Logic Levels	Voltages:	TTL	CMOS
0	0				
0	1				
1	0				
1	1				

b) Floating Input

1- TTL

Disconnect both inputs to the NAND, and note the output *logic* level (henceforth we will not worry about output voltages; just logic levels will do). What input does the TTL "think" it sees, therefore, when its input floats?

2- CMOS

Here the story is more complicated, so we will run the experiment in two stages:

1- Floating Input: *effective logic level in*:

Tie HIGH one input to the NAND, tie the other to 6 inches or so of wire; leave the end of that wire floating, and watch the gate's output with a logic probe as you wave your hand near the floating-input wire. (Here you are repeating an experiment you did with the power MOSFET a few labs back.) Try touching your hand, as you do this waving, to + 5 v., ground, the TTL oscillator output. We hope that what you see will convince you that floating CMOS inputs are less predictable than floating TTL inputs, although we urge you to leave *no* logic inputs floating.

---

1. LS stands for "low power Schottky," a process that speeds up switching. At the time when this *LS* prefix was chosen (1976) TTL was thought to go without saying; thus there's no *T* in the designation, in contrast to CMOS, the late-bloomer, which always announces itself with a *C* somewhere in its prefix: HC, HCT, AC, ACT, etc. See the 74HC00, just below.

2. The "74" shows that the part follows the part-numbering and pinout scheme established by the dominant logic family, Texas Instruments' 74xx TTL series; "C" indicates CMOS; "H" stands for "high speed": speed equal to that of the then-dominant TTL family, 74LS.

2- Floating Input: *Effect on CMOS power consumption*:

You may have read that one should not leave unused CMOS inputs floating. Now we would like you to *see* why this rule is sound (though, like most rules, it deserves to be broken now and then).

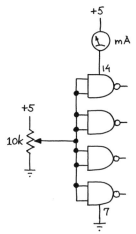

Figure L13.4: Test setup: applying intermediate *input* level raises power consumption

Tie the two NAND inputs to the other six, earlier grounded; disconnect the whole set from ground, and instead connect it to a potentiometer that can deliver a voltage between 0 and +5v. Rotate the pot to one of its limits, applying a good logic level input to all four of the NAND gates in the package.

Now (with power off) insert a current meter (VOM or DVM) between the +5v supply and the V+ pin (14) on the CMOS chip. Restore power and watch the chip's supply current on the meter's most sensitive scale. The chip should show you that it is using very little current: low power consumption is, of course, one of CMOS' great virtues.

Now switch the current meter to its 150 mA scale (or similar range) and gradually turn the pot so that the inputs to all four NAND gates move toward the threshold region where the gate output is not firmly switched high or low. Here, you are frustrating CMOS' neat scheme that assures that one and only one of the transistors in the output stage is on. Both are partially on, and you see the price for this inelegance, on the current meter dial.

Floating inputs thus are likely to cause a CMOS device to waste considerable power. Manufacturers warn that this power use can also overheat and damage the device. In this course we will sometimes allow CMOS inputs to float while breadboarding. But you now know that you should never do this in any circuit that you build to keep.

13-2 Applying NANDs to Generate Particular Logic Functions

Before we look into the gritty details of what lies within a logic IC, let's have some fun with these gates, and try getting used to the remarkable fact that with NANDs you can build *any* logic function.

a) "BOTH"

Use NANDs (CMOS or TTL) to light one of the LEDs when both inputs are high.

Figure L13.5: NANDs to light an LED when both inputs are high (your design)

b) "EITHER"

Use NANDs (CMOS or TTL) to light one of the LEDs when either of the inputs is low (here we mean a plain *OR* operation, not exclusive-or, by the way). (Trick question! Don't work too hard.)

Figure L13.6: NANDs to light an LED when either input is low (your design)

c) XOR *(Optional;* skip this if you feel pressed for time).

Use NANDs (CMOS or TTL) to light one of the LEDs when one and only one of the inputs is high (this is the XOR function). (This task is straightforward with 5 gates, difficult with four. Don't waste much precious lab time on getting down to four; unless the solution happens to strike you at once, this sort of game is better done later, on paper.)

Figure L13.7: XOR built with NANDs (your design)

B. Gate Innards: Looking Within the Black Box

TTL Logic

13-3 TTL ?-Gate

Here is a circuit fragment much like the input stage of an LS00 gate:

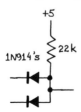

Figure L13.8: Input stage resembling LSTTL gate input

Confirm that the gate performs the function you would expect.

A practical TTL gate follows such a logic stage with buffer stages that provide fast and clean switching, capable of providing a strong and unambiguous logic level out. (We will not build such stages here; if you are curious to see the circuits, see Text at pp.).

CMOS Logic

In the following experiments, we will use two CD4007 (or CA3600) packages; this part is an array of complementary MOS transistors:

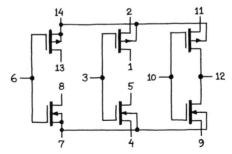

Figure L13.9: '4007 (or "CA3600") MOS transistor array

13-4 Two Inverters

a) Passive Pullup

Build the following circuit, using one of the MOSFETs in a '4007 package. Be sure to tie the two "body" connections appropriately: pin 14 to +5v., pin 7 to ground. (In fact, you will find that this is automatic for the particular FETs in the package that we show below; but you should consider this issue as you use MOSFETs.)

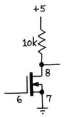

Figure L13.10: Simplest inverter: passive pullup

Confirm that this familiar circuit does invert, driving it with a TTL level from the breadboard oscillator, pulled up to +5v. through a 1k resistor. Watch the output on a *scope* (voltmeter or logic probe will not do, from this point on).

Now crank up the frequency as high as you can. Do you see what goes wrong, and why? Draw what the waveform looks like:

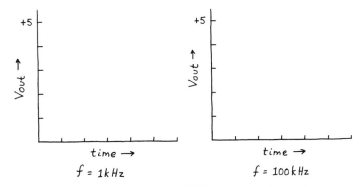

Figure L13.11: Passive-pullup MOS inverter: at two frequencies

b) Active Pullup: CMOS

Now replace the 10k resistor with a P-channel MOSFET, to build the following inverter:

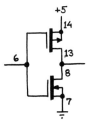

Figure L13.12: Active pullup: CMOS inverter

Look at the output as you try high and low input frequencies.

The high-frequency waveform should reveal to you why all respectable logic gates use *active pullup* circuits in their output stages. (The passive-pullup type, called *open collector* for TTL or *open drain* for MOS, appears now and then in special applications: driving a load returned to a voltage other than the +5v supply, or letting several devices drive a single line, in the rare cases where 3-states (see below) don't do the job better.)

Logic Functions from CMOS

13-5 CMOS NAND

Build the circuit below, and confirm that it performs the NAND function.

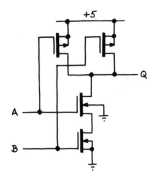

Figure L13.13: CMOS NAND

13-6 CMOS Three-State

The three-state output stage can go into a *third* condition, besides High and Low: *off*. This ability is extremely useful in computers: it allows multiple *drivers* to share a single driven wire, or *bus* line. Here, you will build a buffer (a gate that does nothing except give a fresh start to a signal), and you will be able to switch its output to the OFF state. It is a "three-state buffer."

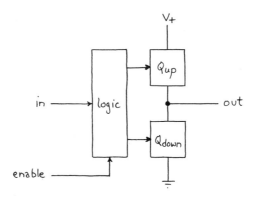

Figure L13.14: Three-state buffer: block diagram

The trick, you will recognize, is just to add some logic that can turn off *both* the *pull-up* and *pull-down* transistors. When that happens, the output is disconnected from both +5 and ground; the output then is off, or "floating." One usually says such a gate has been "three-stated" or put into its "high-impedance" state.

If you're in the mood to design some logic, try to design the gating that will do the job, using NANDs along with the '4007 MOSFETs. Here's the way we want it to behave:

- If a line called Enable is low, turn *off* both the *pull-up* and *pull-down* transistors. That means—

    - drive the gate of the upper transistor *high*;
    - drive the gate of the lower transistor *low*.

- If Enable is high, let the Input signal drive one *or* the other of the upper and lower transistors on: that means—

    - drive the gate of the upper transistor *high* while driving the gate of the lower transistor *low*, and vice versa.

That probably sounds complicated, but the circuit is straightforward. If you're eager to get on with building, peek at our solution, below.

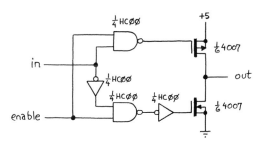

Figure L13.15: 3-state circuit: qualifies gate signals to Q_{up} and Q_{down}, using 74HC00 NAND gates

Test your circuit by driving the input with the breadboard oscillator, while the output is tied to a 100k resistor driven from a slide switch. The slide switch can be set to ground or +5V. Use another slide switch to control the 3-state's *enable*.

Watch the 3-state's output on the scope. Does the 3-state disappear electrically, when you disable the gate?

At the moment, before you have seen applications, this trick may not seem exciting. Later, when you build your computer you will find 3-states at least very useful, if not exciting.

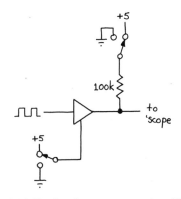

Figure L13.16: Circuit to demonstrate operation of 3-state buffer

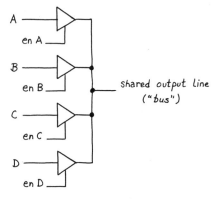

Figure L13.17: Three-states driving a shared "bus" line *do not* bother to build this!)

Class 14: Sequential Circuits: Flip-Flops

Topics:

- *old:*
    - gate output types
- *new:*
    - minimizing combinational logic circuits;
    - first look at sequential circuits: flip-flops.
        - flop types: primitive: *latch*; D; J-K
        - flop applications:
            - counters
            - shift registers

Old topic: (recap) Gate Outputs

Three different types:

ext sec. 9.02, p. 572;
. 566 fig. 9.1

Active-pullup:

Most gates are *active-pullup* types: a transistor pulls the output firmly high; another transistor pulls the output firmly low (not at the same time, of course!).

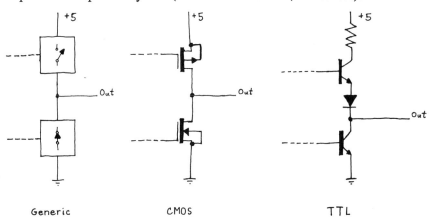

Figure N14.1: Active pullup gates: generic; CMOS; TTL ("totem pole")

three-state:

ext sec. 8.11,
p. 487-88.

Some gates can turn output *off*—that means connect out to *neither* high nor low (this is *not* the same as zero out: King Lear might say, "Nothing will come from nothing...;" but a slogan easier to remember might be that in digital electronics *Zero ain't nothin'*). This output is widely used in computers: it allows multiple devices to share common wiring (a "bus"):

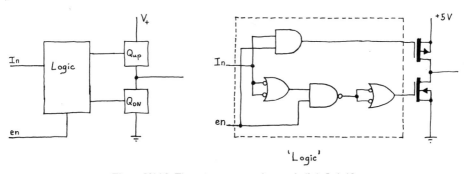

Figure N14.2: Three-state output: what you built in Lab 13

open-collector:

xt sec. 8.11, pp. 488-89

Seldom used (once were used where 3-states now are used; called "open-drain" for MOSFET devices). Useful if output must go to voltage different from V_{CC} or V_{DD}. Also useful to allow the simplest sort of "OR'ing" of driving devices.

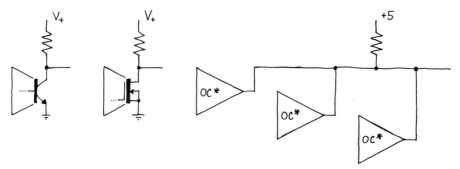

Figure N14.3: Open collector/drain: "wired-OR" connection

Note that ordinary gates must *not* have their outputs tied together; this is a trick reserved for 3-states (and occasionally for carefully-used open-collector).

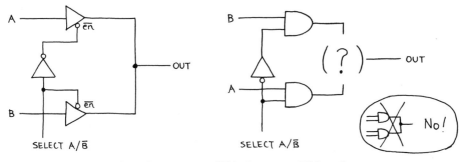

Figure N14.4: Joining outputs: OK for 3-states, not OK for ordinary gates

You can test your understanding of this last notion by designing a *4:1 multiplexer,* using ordinary gates, then 3-states. This is asked in Text exercise 8.17, and is done as a Worked Example.)

A. Combinational Logic: Minimizing

Often you need no such technique:

Most of the time, the combinational logic you need to do is so simple that you need nothing more than a little skill in *drawing* to work out the best implementation. For such simple gating, the main challenge lies in getting used to the widespread use of *active-low* signals.

Here's an example:

"Clock flop if Rd* or Wr* is asserted, and Now* is asserted:

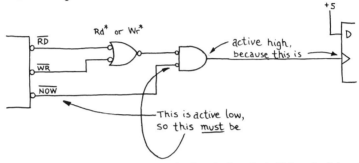

Figure N14.5: Easy combinational logic; some signals active low: don't think too hard; just *draw* it

You can make this problem hard for yourself if you think too hard: if you say something like, 'Let's see, I have a low in and if the other input is low I want the output high—that's a NOR gate...'"—then you're in trouble at the outset. Do it the easy way:

The easy process for drawing clear, correct circuits that include active-low signals:

1. Draw the *shapes* that fit the description: AND shape for the word AND, regardless of active levels;
2. Add inversion bubbles to gate inputs and outputs wherever needed to match active levels;
3. Figure out what gate types you need (notice that this step comes *last*).

In general, bubbles meet bubbles in a circuit diagram properly drawn: a gate output with a bubble feeds a gate input with a bubble, and your eye sees the cancellation that this double-negative implies: bubbles pop bubbles.

Some methods for minimizing

If you do need to implement a more complicated combinational function, however, it's nice to know how.

Here is a function. How would you implement it?

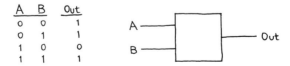

Figure N14.6: A small function to implement: example of need for some *minimization* method

You want to find the simplest implementation. You might proceed in any of several ways:

1. You might simply stare at the table for a while, and discover the pattern. Some people like to work that way.

2. You might write the Boolean expression for each input combination that gives a *1* out:
$$f = X*Y* + X*Y + XY$$
Then you cold use Boolean algebra to simplify this expression: factor:
$$X*(Y* + Y) + XY = X* + XY$$
Thus you discover that you can toss out one variable, X, from two of the terms. After that, if you are on your toes, you recognize a chance to apply another Boolean postulate, and you end up with an agreeably simple equation:
$$f = X* + Y$$
Not at all bad to build—but can we trust the process that got us here? Wouldn't it be easy to miss that last step in the simplification process, for example?

3. You might enter the function on a *Karnaugh map*, and see what simplest form the map delivered:

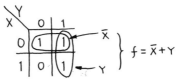

Figure N14.7: Karnaugh map makes chances to simplify *visually* evident

4. The map method is nice just because it does not require cleverness, and makes it hard to overlook a chance to simplify. Let's look briefly at the way these maps work.

Karnaugh maps: rules

Karnaugh maps set out in a *two*-dimensional form exactly the information carried in a truth table: just a description of the way a circuit should behave. The K-map adds nothing.

Here are the rules for the game of K-mapping:

- group *1*'s in blocks of 1, 2, 4, 8, etc. (powers of 2); these groupings must be rectangular.
- make these groupings as large as you can
- read the map to find the variables that describe the region you have grouped; those are the variables (and their levels) that you need to use; other variables have been shown unnecessary.

Here are some examples: bad and good "covers," as these groupings are called:

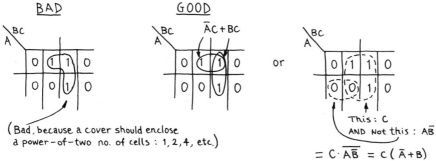

Figure N14.8: Examples of bad and good K-map covers

Why look for big 'covers'? What's at stake?

Karnaugh-mapping may look like a strange, abstract game unless we consider *why* we're trying to draw big blobs on the map. Here is another example, showing a map poorly covered (with timid little covers), then properly covered (with nice big covers). The reward is simpler gating, as you know; but let's look at this particular case. The left-hand map below is covered poorly; the right-hand map is done right.

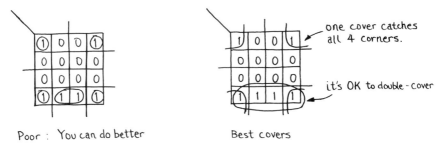

Figure N14.9: One map covered poorly and well

To see *why* big covers are good, contrast the gates you would need to implement the 'poor' covers versus the 'good' covers, above. Incidentally, we are rejecting chances to *factor*, so as to make the contrast simple and stark between the good and the bad.

The gating on the left, below, shows what you would need to implement the clumsy covers above. Each 4-input AND implements one of the isolated *1*s on the map.

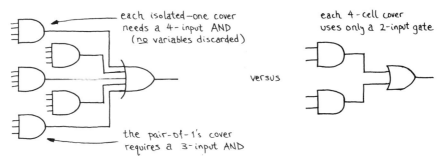

Figure N14.10: Gating required to implement poor versus good set of 'covers'

Summary

Your rewards for finding the best covers are two—related, of course:

big covers   allow *few inputs* per gate;
few covers   allow *few gates*.

A last example that makes Karnaugh Maps look good

xt sec. 8.15,
8.23, p. 500

Before we admit that practically no one uses K maps, let's give a K-map a chance to look good. Here it will discover a little joke planted by Caesar Augustus (or was it Pope Gregory?): the Text's **31-day machine**:

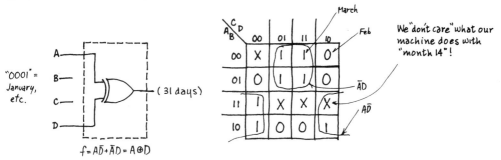

Figure N14.11: 31-day machine: Karnaugh map

This one was a *fluke*! K-maps seldom look so good. In fact, *now* we will admit that you seldom need these maps. Most problems are too easy or too hard for K-maps. So, don't spend your time getting really good at using K maps.

---

2. "Sequential Circuits:" Flip-Flops

All our digital circuits so far have been **combinational**: their outputs were functions of *present* input values (except while waiting for the brief propagation delay).

Now we begin to meet **sequential** circuits—circuits that care about their past (analogous to capacitors, which unlike resistors, cared about *their* pasts.)

Flops are easy to understand. A harder question arises, however: why are *clocked* circuits useful? We will work our way from primitive circuits toward a good clocked flip-flop, and we will try to see why such a device is preferable to the simpler flop.

A. *A primitive flip-flop: the "latch"*
xt sec. 8.16

In the beginning was the cross-coupled latch. It is at the heart of all fancier flip-flops. It looks simple; it also may look fishy, since you can see at a glance that it includes *feedback*. That's what makes it interesting: that's what makes it care about its past.

xt p. 506, fig. 8.47

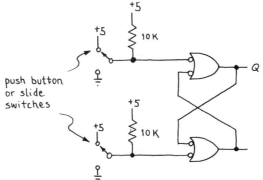

Figure N14.12: Cross-coupled NAND latch

Operation Table:

S	R	Q
0	0	
0	1	
1	0	
1	1	

How does one use this thing? In which state should it *rest*? How flip it, flop it? (That is, how Set it, Reset it: "set" ==> send output high; "reset" or "clear" ==> send output low).

This device works, but is hard to design with. It is useful to **debounce** a switch; but nearly useless for other purposes, in this simplest, barebones form.

How does it *debounce* a switch?

*xt sec. 8.16,
507, fig. 8.49*

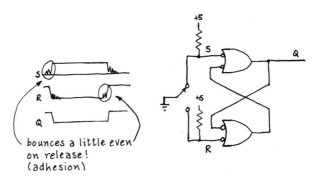

Figure N14.13: Switch bounce: first hit sets or resets flop: bounce does not appear at output

But in other settings this flop would be a pain in the designer's neck. To appreciate the difficulty, imagine a circuit made of such primitive latches, and including some feedback. Imagine trying to predict the circuit's behavior, for all possible input combinations and histories.

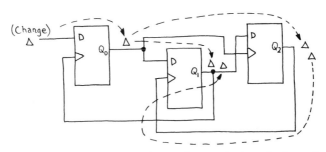

Figure N14.14: Example meant to suggest that asynchronous circuits are hard to analyze or design

Designers wanted a device that would let them worry less about what went on at all times in their circuits. They wanted to be able to get away with letting several signals change in uncertain sequence, for part of the time.

This was the basic scheme they had in mind:

Figure N14.15: Relaxing circuit requirements somewhat: a designer's goal

Toward a Good D Flip-Flop

*xt sec. 8.17,
507, fig. 8.31*

The *gated S-R latch*, or **transparent latch** takes a step in the right direction. This circuit is just the NAND latch plus an input stage that can make the latch indifferent to signals at S and R: an analog to a camera's *shutter*.

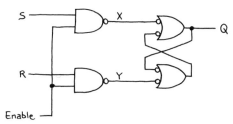

Figure N14.16: Clocked or gated S-R latch ("transparent" latch)

This achieves more or less what we wanted; but not quite:

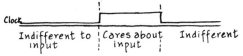

Figure N14.17: Transparent latch moves in the right direction: now we can stand screwy input levels part of the time

The trouble is that this circuit still is hard to design with. Consider what sort of clock signal you would want, to avoid problems with feedback. Ticklish.

A Good D Flip-Flop: Edge-Triggered

*xt sec. 8.17,
 . 508-09*

Because the simple NAND latch is so hard to work with, the practical flip-flops that people actually use nearly always are more complex devices that are called *edge-triggered*.

Edge-triggered flip flops

These flops care about the level of their inputs only during a short time just *before* (and in some rare cases after) the clock edge.

An older design, called by the nasty name, *master-slave*, behaved nastily and was rendered obsolete by the edge-trigger circuit. The master-slave survives only in textbooks, where it has the single virtue that it is easy to understand.

The behavior called *edge-triggering* may sound simple, but it usually takes people a longish time to take it seriously. Apparently the idea violates intuition: the flop acts on what happened *before* it was clocked, not after. No, this behavior does not violate causality. How is this behavior possible? (*Hint:* you have seen something a lot like it on your scopes, which can show you the waveform as it was a short time *before* the *trigger* event. How is *that* magic done?)

*Text sec. 8.17,
p. 509, fig. 8.53*

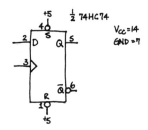

Figure N14.18: 74HC74 edge-triggered D flip-flop

Here are its crucial timing characteristics:

*Text sec. 8.17,
p. 5.11, fig. 8.55*

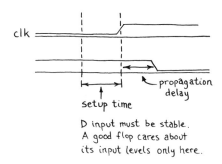

Figure N14.19: D-flop timing: setup time and propagation delay; hold time should be *zero*

A D flop only *saves* information. It does not transform it; just saves it. But that simple function is enormously useful.

Triggering on Rising- versus Falling- Edge

The D flop shown responds to a *rising* edge. Some flops respond to a *falling* edge, instead. As with gates, the default assumption is that the clock is *active high*. An inversion bubble indicates *falling edge* clock:

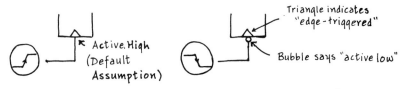

Figure N14.20: Clock edges: rising or falling edge (*never both*)[1]

---

1. As usual, we have to qualify the "never" slightly: there are some *one-shots* that trigger on both edges: e.g., the '423 shown in Text sec. 8.20, and the 8T20. The "never" does seem to hold for flip-flops, though.

Edge- versus *Level-sensitive* flops & inputs

Some flip-flops or flop functions respond to a clock-like signal, but not quite to the *edge*: that is, the device does not lock out further changes that occur after the level begins. The *transparent* latch is such a device. We met this circuit a few pages back.

Examples of *level-sensitive inputs*:

1) A "transparent latch":

The character displays we use in this lab work this way (called "HP Displays" in lab notes):

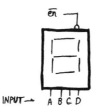

Figure N14.21: Level-sensitive: transparent latch:

Output follows or displays input while E* is low.

2) Reset on a flip-flop

text sec. 8.17, .509

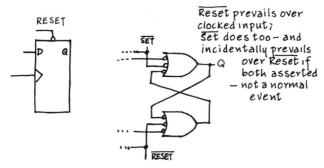

Figure N14.22: Jam Reset*: it reaches into the *output latch*, and prevails over the fancy edge-trigger circuit that precedes the out latch

So-called "jam type" takes effect at once. This scheme is universal on flip-flops, not quite universal on counters. But a reset is never treated as a clock itself—even in the cases where it 'waits for the clock'—the synchronous scheme. The edge-triggering scheme is reserved for *clocks*, with only the rarest exceptions (like the IBM PC's interrupt request lines: sec. 10.11, p. 701).

Edge triggering usually works better than level-action (that is, makes a designer's tasks easier); edge-triggering therefore is much the more common scheme.

*Another type, less important: **J-K Flop***

Text sec. 8.17, p. 509

The D flop is by far the most important, but you should understand the J-K, which is a D plus a little feedback and gating. It can do 4 tricks, instead of two:

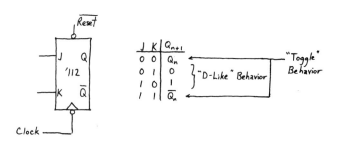

Figure N14.23: J-K Flop, and its operation table

You won't often design with this kind of flop, but the J-K's Toggle/Hold* behavior is important to us in one setting: it makes the design of *synchronous natural-binary counters* straightforward. You'll meet these devices in a minute.

Applications

Counters

Ripple counters

Text sec. 8.17, p. 511

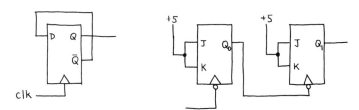

Figure N14.24: Simple "counters": "divide-by-two"; "divide-by-four" (*ripple* type)

Synchronous counters

Text sec. 8.25, p. 524

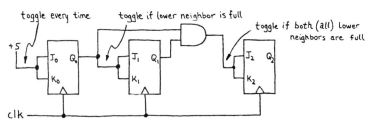

Figure N14.25: Synchronous counter

The signature of the synchronous counter, on a circuit diagram, is just the connection of a common clock line to all flops (<==> syn-chron: same-time). The synchronous counter—like synchronous circuits generally—is preferable to the ripple or *asynchronous* type. The latter exist only because their internal simplicity lets one string a lot of flops on once chip (e.g., the "divide by 16k" counter we have in the lab—the '4020, or the more spectacular '4536: divide-by-16M!).

"State machine" : counters generalized
ext sec. 8.18

One can use D flops plus some gating to design a circuit that will step through any sequence of *states* that you choose ("state" means simply the set of levels on the flop outputs). We will do more of this a couple of labs from now.

p. 514, fig. 8.60

present state   ==>   next state

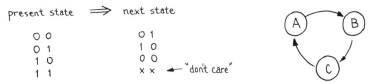

Figure N14.26: Gates & D flops can make an arbitrary "state machine:" first glimpse

Shift-Register
text sec. 8.26

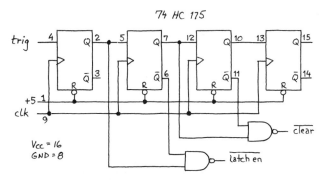

Figure N14.27: Shift-register (lab circuit)

A shift-register generates predictable, orderly delay; it shifts a signal *in time*; it can convert a *serial* stream of data into *parallel* form, or vice-versa.

In the lab, you will use such a circuit to generate a pulse: to act like a so-called "one-shot." A NAND gate added to the shift-register does the job. (We use two, because in Lab 16 we need a double-barreled one-shot). See if you can sketch its timing diagram, and then what happens if you wire a NAND gate, as in the lab, to detect $Q_0 Q_1{}^*$.

Ch. 8: Worked Examples: Combinational Logic

Two Worked Examples:

1. a case so easy that you need no systematic method at all
2. a harder case, which can be solved either systematically (with K maps, for example), or by 'brainstorming.'

1. Glue Logic

Here's another of those common combinational networks often called 'glue,' because these odds and ends of gates stick together the big chips.

> **Problem: turn on a memory under stated conditions**
>
> Enable the memory (at a pin called CS*) under the following conditions: if—
>
> - TimeOk* is asserted and A_{19} is high and Block* is not asserted, or if—
> - Busgive* and Ready* are asserted.
>
> Use *two-input* gates; draw them using *assertion-level symbols*, and label the gates with their part numbers, as follows:
>
> | | |
> |------|----|
> | NAND | 00 |
> | NOR | 02 |
> | NOT | 04 |
> | AND | 08 |
> | OR | 32 |

Solution:

It's easy if you stay cool: just draw the shapes that fit the words ("and," "or"), then take care of the active levels. Don't think about the name of the gate you're drawing till it's all done.

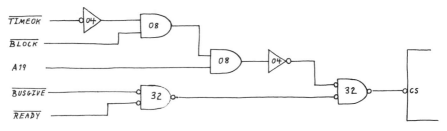

Figure X14.1: Gates to enable memory under the stated conditions

2. Digital Comparator (A > B)

> ### Problem: 2-bit Greater-Than Circuit
>
> Use any gates you like (XORs are handy by not necessary) to make a circuit that sends its output high when the 2-bit number A is greater than the 2-bit number, B. Here is a black box diagram of the circuit you are to design:
>
> **Figure X14.2**: 2-bit greater-than circuit

Note that this problem can be solved either systematically, using K maps, or with some cleverness: consider how you could determine equality of two one-bit numbers, then how to determine that one of the one-bit numbers is greater than the other; then how the more- and less-significant bits relate; and so on). We'll do it both ways.

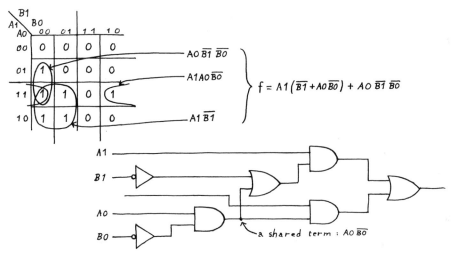

$$f = A1(\overline{B1} + A0\,\overline{B0}) + A0\,\overline{B1}\,\overline{B0}$$

Figure X14.3: 2-bit greater-than circuit: Systematic solution

And here's a more fun way to do it:

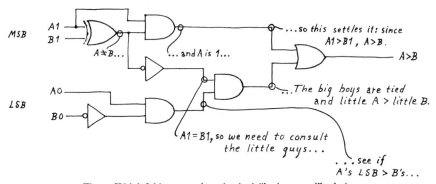

Figure X14.4: 2-bit greater-than circuit: A "brainstormed" solution

The brainstormed circuit is no better than the other. In gate count it's roughly a tie. Use whichever of the two approaches appeals to you. Some people are allergic to K maps; a few like them.

Lab 14: Flip Flops

Reading: Ch. 8.12 – 8-19, plus 9.04 re: switch bounce (yes, it's in Chapter 9).
Specific advice:

> In 8.13, re: K-mapping, see if you can understand why Karnaugh's odd *Gray-code* ordering of the variables is necessary. Don't work too hard at K mapping, though.
- Only scan 8.15, re muxes and PAL's: PALs are important devices but we will not meet them in this course;
- Concentrate on 8.16-8.17 re flip-flops and counters;

Problems: Embedded problems, plus Bad circuits C, D, and AE1, AE11-15.

14-1 A primitive flip-flop: NAND Latch

This circuit, the most fundamental of flip-flop or memory circuits, can be built with either NANDs or NORs. We will build the NAND form:

Operation Table

S	R	Q
0	0	
0	1	
1	0	
1	1	

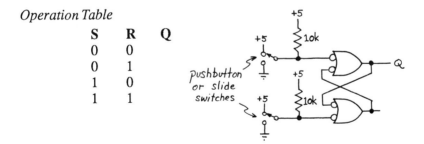

Figure L14.1: A simple flip-flop: cross-coupled NAND latch

Build this latch, and record its operation. Note, particularly, which input combination defines the "memory state;" and make sure you understand why the state is so called.

Leave this circuit set up. We will use it shortly.

Practical Flip-Flops

It turns out that the simple *latch* is very rarely used in circuit design. A more complicated version, the **clocked** flip-flop is much easier to work with.

14-2 D Type

The simplest of the clocked flip-flop types, the **D**, simply saves at its output (Q) what it saw at its input (D) just before the last clocking edge. The particular D flop used below, the 74HC74, responds to a *rising* edge.

The D flop is the workhorse of the flop stable. You will use it 50 times for each time you use the fancier J-K (a device you will meet soon, too).

a) *Basic operations: Saving a Level; Reset*

The D's performance is not flashy, and at first will be hard to admire. But try.

Feed the D input from a breadboard slide switch. Clock the flop with a "debounced" pushbutton (the buttons on the left side of the breadboard will do; note that these switch terminals need *pull-up* resistors, since they have *open-collector* outputs. (This you saw last time; but perhaps you've forgotten!) Dis-assert Reset* and Set* (sometimes called Clear* and Preset*), by tying them high.

Note that the '74 package includes *two* D flops. Tie the inputs of the unused flop high or low (this is just to keep the chip cool, as you recall from last time).

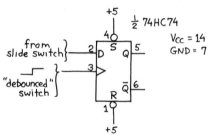

Figure L14.2: D-flop checkout

- Confirm that the D flop ignores information presented to its input (D, for "data") until the flop is clocked.
- Try asserting Reset*. You can do this with a wire; bounce is harmless here. (Why?) What happens if you try to clock in a High at D while asserting Reset*?
- Try asserting Set* and Reset* at the same time (something you would never purposely do in a useful circuit). What happens? (Look at <u>both</u> outputs.) What determines what state the flop rests in after you release both? (Does the answer to that question provide a clue to why you would not want to assert both Set* and Reset* in a circuit?)

b) *Toggle Connection: Feedback*

The feedback in the circuit shown below may trouble you at first glance. (Will the circuit oscillate?) The *clock*, however, makes this circuit easy to analyze.

In effect, the clock breaks the feedback path. (See Text sec. 8.17, pp. 510-11 and class Notes 14, re difficulties that feedback introduces into non-clocked sequential circuits.)

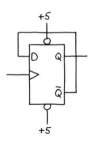

Figure L14.3: D Flop biting its own tail

build this circuit and try it.

- First, clock the circuit manually.
- Then clock it with a square wave from the function generator (the breadboard generator is less good than an external generator, with its higher f_{max}). Watch Clock and Q on the scope. What is the relation between f_{clock} and f_Q? (Now you know why this humble circuit is sometimes called by the fancy name "divide-by-two.")
- Crank up the clock rate to the function generator's maximum, and measure the flop's *propagation delay*. (To do this, you will have to consider what voltages In and Out to use, as you measure the time elapsed. You can settle that by asking yourself just what it is that is "propagating.")

14-3 J-K Type

You might get away with never using this flop type, but you need to understand its behavior in order to understand standard binary *counters*.

The J-K's strength is its versatility. It can mimic the other important flop types: D and T or "toggle". (A "toggle" type does not simply toggle willy-nilly like the D-flop circuit you just wired. It is smarter than that: it can toggle or hold its last state, depending on the level you feed it. You will see such behavior in a few minutes.)

a) *Checkout*

Verify the J-K's behavior (described in Text at p. 509).

Dis-assert Preset* and Clear*; drive the J and K inputs from slide switches; clock the flop with a *debounced* signal.

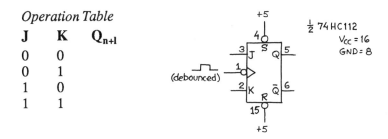

Operation Table

J	K	Q_{n+1}
0	0	
0	1	
1	0	
1	1	

Figure L14.4: J-K checkout

b) *Applications: Mimicking Other Types*

The J-K is—or was (when designers used flops in small packages rather than in big arrays)—especially useful for complex designs, because, as noted above, it can mimic any other kind of flop. Here we will watch it do its chameleon-like tricks:

a- *Type A Flop (= "____ –Type")*

Operation Table

In	Q_{n+1}
0	
1	

Figure L14.5: J-K as ____ –type flop

Build the circuit above and record its behavior on the operation table. What type of flop is the J-K imitating?

b- *Type B Flop (= "____ –Type")*

Operation Table

In	Q_{n+1}
0	
1	

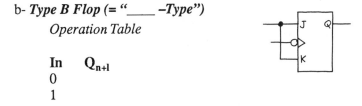

Figure L14.6: J-K as ___ –type flop

14-4 J-K in Counters: Ripple and Synchronous Counters

A. Ripple Counter

The preceding J-K circuit, like the earlier D-with-feedback, can be made to toggle on every clock—or "divide by two."

Cascading two such circuits lets you divide by four; and so on. Build this circuit:

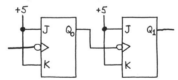

Figure L14.7: Divide-by-four ripple counter

- Watch the counter's outputs on two LEDs while clocking the circuit at a few Hertz. Does it "divide by four?" If not, either your circuit or your understanding of this phrase is faulty. Fix whichever one needs fixing.
- Now clock the counter as fast as you can, and watch Clock and first Q_0 then Q_1 on the scope. Trigger on Q_1.
- Watch the two Q's together and see if you can spot the "rippling" effect that gives the circuit its name: a lag between changes at Q_0 and Q_1.

B. Synchronous Counter

Now alter the circuit to the form shown below. This is a *synchronous* counter.

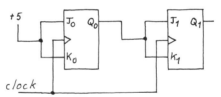

Figure L14.8: Synchronous divide-by-four counter

See if you can use the scope to confirm that the *ripple* delay now is gone.

14-5 Switch Bounce

Here is a storage scope[1] photograph showing a microswitch pushbutton bouncing its way from a high level to low:

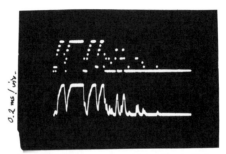

Figure L14.9: Microswitch bouncing from high to low (pulled up through 100k)

---

[1]. Incidentally, this "storage scope" was an ordinary scope fed by a microcomputer of the kind you will build later in this course. The computer took samples during the bouncing process, stored them in memory, then played them back repeatedly to give a stable display. You will have a chance to try this, if you like, during the final lab sessions.

To see the harmful effect of switch bounce, clock your divide-by-four counter with a (bouncy-) ordinary switch such as a microswitch pushbutton. The bouncing of the switch is hard to see; but its effects should be obvious.

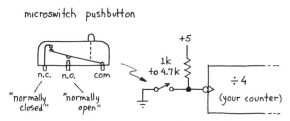

Figure L14.10: Switch bounce demonstration

14-5-a Watching Switch Bounce (*Optional: for scope enthusiasts*)

Switch bounce is hard to see because it does not happen periodically and because the bounces in any event do not occur at exactly repeatable points after the switch is pushed.

You can see the bounce, at least dimly, however, if you trigger the scope in Normal mode with a sweep rate of about 0.1 ms/cm. You will need some patience, and some fine adjustments of trigger level. Some switches bounce only feebly. We suggest a nice snap-action switch like the microswitch type.

14-5-b Eliminating Switch Bounce: Cross-Coupled NAND's as Debouncer

Return to the first and simplest flip-flop (which we hope you saved), the cross-coupled NAND latch, also called an "R-S" flop. Add pullup resistors, and as input use the bouncy pushbutton. Ground its *common* terminal. (This circuit is described in Text sec. 817, p. 507.)

Why does the *latch*—a circuit designed to "remember"—work as a debouncer?

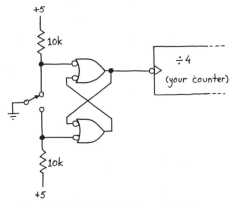

Figure L14.11: NAND latch as switch debouncer

14-6 Shift Register

> *Note*: Please build the circuit below (a *digitally-timed one-shot* that evolves from the shift-register) on a *private breadboard*; you will use this circuit next time.

The *shift-register* below delays the signal called "IN," and *synchronizes* it to the clock. Both effects can be useful. You will use this circuit in a few minutes as a *one-shot*—a circuit that generates a single pulse in response to a "Trigger" input (here, the signal called "IN").

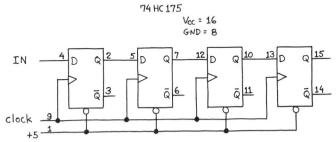

Figure L14.12: Shift register

Clock the circuit with a logic signal from an *external* function generator; use the breadboard's oscillator to provide "IN." Let f_{clock} be at least $10 \cdot f_{``in"}$.

a. *One Flop: Synchronizer*

- Use the scope to watch IN, and Q_0 (Q of the leftmost flop); trigger the scope on IN.
- What accounts for the *jitter* in signal Q_0?
- Now trigger on Q_0, instead. Who's jittery now?
- Which signal is it more reasonable to call jittery or unstable? (Assume that the flops are clocked with a system clock: a signal that times *many* devices, not just these 4 flip-flops.)

b. *Several Flops: Delay*

- Now watch a later output—Q_1, Q_2, or Q_3, along with IN. (We'll leave the triggering to you, this time.)
- Note the effect of altering f_{clock}.

c. *Several Flops Plus NAND Gates: Double-Barreled One-Shot*

Add two NAND gates, as shown below, and watch those gates' outputs along with TRIG. Again note the effect of altering f_{clock}.

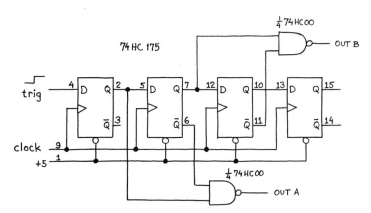

Figure L14.13: Digitally-timed (synchronous) one-shot (double-barreled)

Checkout

- Slow-motion: first use a manual switch to drive *Trig*, and set the clock rate to a few Hertz. Watch the one-shot outputs on two of the breadboard's buffered LEDs. Take *Trig* low for a second or so, then high. You should see first one LED then the other wink low, in response to this low-to-high transition.
- Full-speed: when you are satisfied that the circuit works, drive *Trig* with a square wave from one function generator (the breadboard's) while clocking the device with an external function generator (at a higher rate).

To make sure you understand what this circuit is doing, you may want to draw a timing diagram, showing TRIG, clock, the four flop outputs, and the output of the two NAND gates.

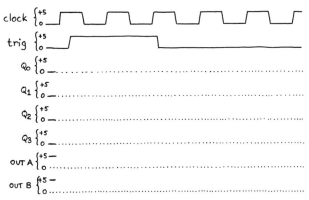

Figure L14.14: Timing diagram for digitally-timed one-shot

What are the strengths and weaknesses of this one-shot relative to the more usual *RC* one-shot?

Class 15: Counters

Topics:

- *old:*
    - Flip-flop timing characteristics
        - why clocks
        - what accounts for *setup time*
    - counters from flip-flops: ripple versus synchronous

    new:
    - counters
        - fancier (IC-) counters
        - "fully synchronous" counters: sync clear, sync load
        - using *load* to make arbitrary ÷ *n* counter
        - Counting as digital design strategy

Old Topics

1. *Flip-flop characteristics: Recapitulation*

Why clocks? Because 'breaking the feedback path' eases design and analysis of sequential circuits.

For example—

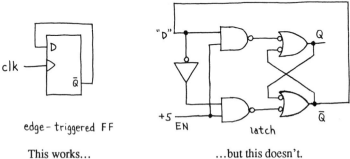

Figure N15.1: Clocked device (edge-triggered) makes feedback harmless: instability is impossible

The circuit using the transparent latch oscillated at around 40MHz, when we built it in the lab.

The beauty of edge-triggered, synchronous circuits

The *clock* edge is a knife cleanly slicing between *causes* of changes—all to the left of the clock edge, in the timing diagram below[1]—and the *results* of changes—all to the right of the clock edge.

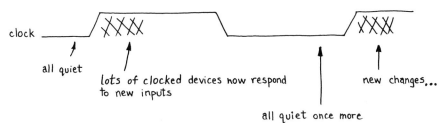

Figure N15.3: Edge-triggered *synchronous* circuit timing

A particular example

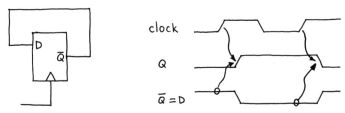

Figure N15.4: D flop biting its tail: feedback is harmless

This works fine, unless you try to push the clock speed very high. In that case, trouble reappears:

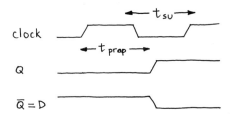

Figure N15.5: Even the wonderful synchronous scheme fails if you try to run it too fast: t_{setup} violation

This fails, because D changes during t_{setup}. This produces an unpredictable output; may even hang up, refusing to make up its mind for a strangely long time ("metastable").

---

1. We assume zero *hold* time; some older devices, like the 7474, have a non-zero hold time; that's nasty, and the newer devices usually show the better characteristic, *zero* hold time. E.g., 74LS74 and 74HC74. There is nothing magical about achieving $t_H = 0$: the IC designers simply adjust the relative internal delay paths on the *data* and *clock* lines so as to shift the "window" of time during which data levels matter. Here's the idea:

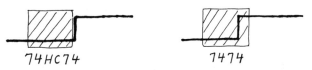

Figure N15.2: An IC designer can trade hold-time against set-up time: two 74xx74 designs

Setup time

Sounds like a technical detail, but it's a concept you need in order to get timing problems right.

Detailed view (in the special case of the 7474 D flop):

setup time is the time required for a change of D level to work its way into the flop—to the points labeled X and Y, below

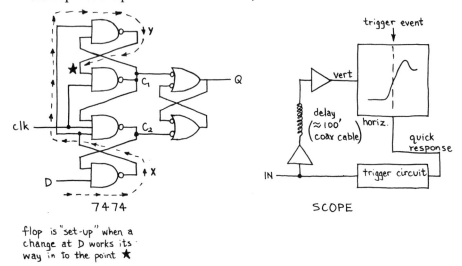

flop is "set-up" when a change at D works its way in to the point ★

Figure N15.6: 7474 D flip-flop: two signal "pipelines": flip-flop setup; oscilloscope trigger's equivalent trick

The new information needs to work its way through a "pipeline" that is two gates deep, in order to set up the flop to act properly on a clock. You might guess that this process could take as long as two gate delays, around 25 to 20 ns, and that is in fact about how long the flop's specified setup time. (In fact, internal gate delays are less than the delay of a packaged gate; but the scale remains about right.)

Counters from flip-flops

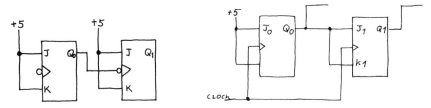

Figure N15.7: Simple counters: Ripple and Synchronous

The ripple counter is the easier to design and build, but the synchronous is in all other respects preferable.

Here is a scope photograph showing two integrated counters in action. The top 4 traces show one counter (Q_2, Q_1, Q_0, clock), the lower 4 traces show another (same pattern). Which is the ripple counter, which synchronous? To which clock edge does each counter respond?

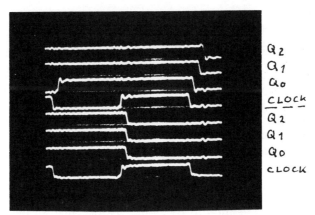

Figure N15.8: Ripple versus synchronous counter: 74LS93 vs 74LS163. Sweep rate = 50 ns/div.

Virtues of synchronous counter:

- settles to valid state faster, after clock (wait for just *one* flop delay)
- shows no false intermediate states

Virtues of ripple counter:

- simple. Therefore, can fabricate more stages on one chip than for synchronous.
- in some applications, its weaknesses are harmless. E.g.,
    - frequency dividers (where we don't care about relative timing of in and out, and don't want to look at the several Q outputs in parallel, but care only about the relative frequencies, out versus in)
    - slow counters (driving a display for human eyes, for example: we can't see the false states)

New topics:

1. Integrated Counters

There is no excuse for *building* a counter from flip-flops, outside a teaching lab. Nifty integrated counters are available. These devices include features that make them easy to use:

Cascading

Any respectable counter allows "cascading" several of the devices so as to form a larger counter. To permit cascading, *synchronous* counters include **carry** pins: a carry IN and a carry OUT.

> **Carry IN**   This is an enable: when asserted, it tells the counter to pay attention to its clock. Notice that all chips are tied to a common clock line. Do *not* drive one counter's *clock* with a carry out. If you do, you're making a ripple scheme, not a fully-synchronous counter.
>
> **Carry OUT**   This signal warns that the counter is *about to roll over* or overflow. In the case of a natural-binary up counter, it detects the condition *all ones* on the flop outputs. Notice that this signal must come *before* the roll-over, not after, because it has to tell the next (more-significant) counter what to do on its next clock.

Here's how easy it is to cascade three of the big counters you are using in today's lab, for example:

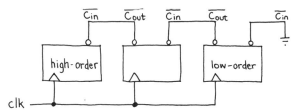

Figure N15.9: Cascading three integrated counters: *easy!*

If you're really on your toes, you will recognize that the Carry OUT* signal shows not only that this counter is full, but also that all less significant counters also are full. What very-simple logic on the chip manages to determine all that?

Loading

Many counters allow you to load a value "broadside" into its flops:

> *Load*: When you assert LD*, the counter is transformed into a simple register of D flops: on the next clock edge, those flops simply take in the values presented on the data inputs. (This description fits so-called "synchronous" load; "asynchronous" or "jam" load also is available; it works like the jam clear described below.)

When you release LD*, the counter becomes a counter once more. This may be hard to grasp, when you simply hear it stated. We'll look at an example of the use of load in a few minutes: Lab 15's '÷ n' circuit.

Clearing

Clearing could be called a special case of *loading*, but it is so often useful that nearly all counters offer this function (more than offer *load*), and they offer it in two styles:

asynchronous or "jam" clear
The clearing happens a short time after Clear* or Reset* is asserted (say, 5 to 10 ns); the clearing does *not* wait for the clock.

synchronous clear The clearing is *timed* by the clock: on assertion of Reset* *nothing happens* until the clock edge.

Query: which sort of clear would you like in a programmable divide-by-n counter, like the one you will build in the lab? (You will find this question explored in a Worked Example.)

Here are some standard counters, with their clear functions (some new counters offer *both* sorts of clear and load—synchronous and asynchronous—on four pins: e.g., 74ALS560). You will find a table, 8.10, at p. 563 showing a great variety of counters and noting the features each offers.

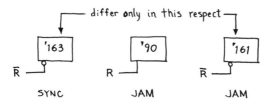

Figure N15.10: Three integrated counters: some offer jam clear, others synchronous

In a separate set of notes you will find some counter-application problems detailed. We won't look at all of those now. Instead, we will concentrate on the counter applications you will meet in the lab.

Using the counter Load* function; two lab circuits

The simple operation of loading the counter with a pushbutton turns out to raise some fussy timing problems, because the lab counters offer a *synchronous* load and clear[2]. Usually, that is good; here is it a bit of a nuisance:. Do you see why the lower "do-nothing" gate is necessary?

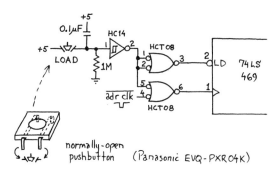

Figure N15.11: Lab debounce and load logic

Lab Displays

While we're showing you oddities that you'll meet in the lab, we should mention the fancy displays you will meet today. They are more than the usual '7-segment' display; they include the *decoder* that translates a 4-bit code into the lighting of the proper pattern of 20 small LEDs; and they include a *transparent latch*.

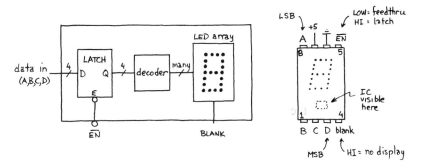

Figure N15.12: Hexadecimal display: includes latch and decoder

---

2. These counters are oddballs: 74LS469's: 8-bit up/down counters, with 3-state outputs. Here's how they're odd:
  - *TTL*—nasty old power-hungry TTL (you'll notice they run warm to the touch). Why? Simply because such counters are not available in CMOS;
  - They're not even standard TTL, despite their standard-sounding part number: they're made by the manufacturer of a programmable gate array (a PAL), and are simply an array, programmed at the factory. The manufacturer saw a gap in the list of standard TTL functions, and filled it.

Lab's Divide-by-N Counter

The circuit below is more interesting: it takes advantage of the Load* function to make a counter that will count through the number of cycles set at the keypad: 4 bits from the keypad feed the counter's data inputs.

Notice that the counter is rigged to count *down*. If you load a value *n*, how many states does the machine run through? (That is, what does it "divide by"?) The answer to this question may answer the riddle you meet when you use the Carry-out* to drive a speaker: '*Why does this silly instrument play out of tune?*'

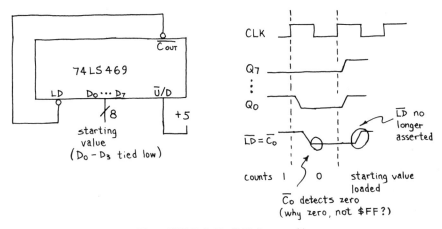

Figure N15.13: Lab's divide-by-n machine

2. Counting as a Digital Design Strategy

Because it is easy to make a digital circuit that counts, it often turns out that a good way to make a digital device designed to measure some quantity is to build a *stopwatch* to measure the duration of a cleverly-generated pulse.

More specifically, here's the idea:

1. build a counter that counts clock edges (this is a sort of 'watch');
2. add gating that lets you start and stop the watch (making it a 'stopwatch');
3. build some circuitry that provides a waveform whose period is proportional to a quantity that interests you (call this 'Input');
4. use the stopwatch to measure period, and thus to measure 'Input.'

Here are two examples to illustrate the technique:

a. Digital Voltmeter (or 'A/D converter; voltage input')

Text sec. 9.20, fig. 9.54, p. 625

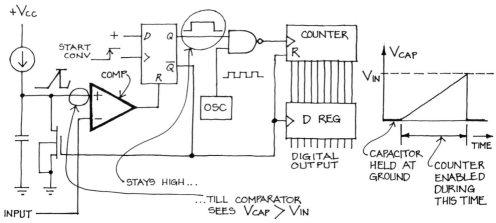

Figure N15.14: Example 1: measure period to measure *voltage*

b.
Digital Capacitance Meter (Lab 15)

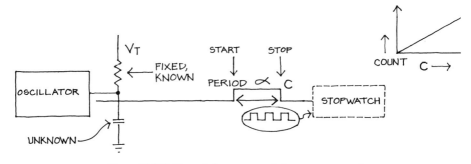

Figure N15.15: Example 2: measure period to measure *capacitance*

The idea is very simple. Questions of *timing* raise the only interesting issues. For a look at such questions see the Worked Examples on counter use.

Ch. 8: Worked Examples: Counter Use

Worked examples, with notes:

1. Modifying count length: divide-by-*13* counter
2. Counting as design strategy:

   a. sonar range detector (period counter I)
   b. reaction timer (period counter II)

1. MODIFYING COUNT LENGTH: strange-modulus counters

Once upon a time—about 10 years ago—designing with counters was a chore that sometimes entailed designing the counter itself, if the counter was not entirely standard. That's no longer true. IC counters make your work easy. In these notes we'll look at two sorts of problem: first, the easier of the two: making a counter divide by some funny number; second, the more interesting of the two problems: using the counter to make an instrument that measures something.

Divide-by-13 counter

The *fully-synchronous loadable* counter makes it now almost as easy to rig up a *divide-by-13* counter, say, as to pull a *divide-by-16* from the drawer. Not quite so easy; but almost.

Synchronous versus Asynchronous Load or Clear

It's not hard to *state* the difference: a *synchronous* input "waits for the clock," before it is recognized; *asynchronous* or *jam* inputs take effect at once (after a propagation delay, of course); they do not wait for the clock. Either one overrrides the normal counting action of the counter.

But it is hard to see why the difference matters without looking at examples. Here are some.

Modifying count length

> **Problem: Divide-by-13 Counter**
>
> Given a ÷ 16 counter (one that counts in natural binary, from 0 through 15), make a ÷ 13 counter (one that counts from 0 through 12). Decide whether you want to use Clear or Load, and whether you want these functions to be synchronous or asynchronous.

a) A poor design: use an asynchronous clear

Here's a plausible but bad solution: detect the unwanted state *13*; clear the counter on that event:

Text sec. 8.29,
compare problem 8.36,
p. 545

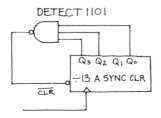

Figure X15.1: A poor way to convert natural binary counter to ÷13

Why is this poor?

The short answer is simply that the design obliges the counter to go into an *unwanted* or false state. There is a glitch: a brief invalid output. You don't need a timing diagram to tell you there is such a glitch; but such a diagram will show how long the false state lasts:

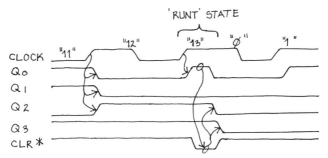

Figure X15.2: Poor "÷13" design: false *14th* state between 12 and 0

In some applications you might get away with such a glitch (ripple counters, after all, go through similar false transient states, and ripple counters still are on the market). You could get into still worse trouble, though: the CLR* signal goes away as soon as state "13" is gone; the quickest flop to clear will terminate the CLR* signal; this may occur before the slower flops have had time to respond to the CLR* signal; the counter may then go not to the zero state, but instead to some unwanted state (12, 9, 8, 5, 4, or 1). That error would be serious; not just a transient.

b) A proper design: synchronous

text sec. 8.29;
compare fig. 8.88, p. 545

A counter with a synchronous Clear or Load function—like the 74HC163—allows one to modify count length cleanly, without putting the counter into false transient states, and with no risk of landing in a wrong state. It is also extremely easy to use.

To cut short the natural binary count, restricting the machine to 13 states requires a little logic to detect the *12* state—not the 13, as before. On detecting 12, the logic tells the counter *before* the clock to clear on the next clock.

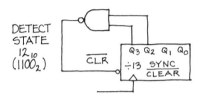

Figure X15.3: Synchronous ÷ 13 from ÷ 16 counter

Here, in case you need convincing, is a timing diagram showing the clean behavior of this circuit.

text sec. 8.29;
compare the very similar timing diagram for the parallel load '163:
fig. 8.89, p. 546

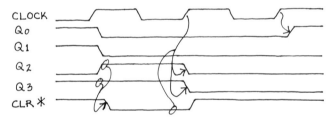

Figure X15.4: Proper count modification: using synchronous Clear*

The synchronous-clear ÷ 13 circuit works nicely. It's too bad, though, that it requires a NAND.

The LOAD function

Use of the Load function instead of the Clear can achieve nearly the same result with a single inverter instead of the NAND. The Text spells out this solution in section 8.29, and looks closely at the timing of this circuit. The use of Load rather than Clear to define the number of states saves gating, but has some funny side effects. Either—

- it obliges one to use a strange set of states (starting from three, say, and counting up to 15, then loading four again in order to define 13 states); this would be all right if the frequency alone interested you, but it would not be all right if you wanted to see the counts 0 through 12.
- Or it requires use of a *down* counter (load the initial value; count to zero, use the Borrow signal to load once more). (This is the technique we use in Lab 16 to make a counter of variable modulus; there, where only frequency concerns us, the technique works fine.)

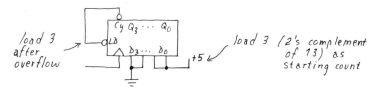

Figure X15.5: Divide-by-13 counter, using synchronous *Load* function

We should not make too much of these odd effects: the number of states a counter steps through always bears a slightly funny relation to the value loaded or detected: if you load and count down, *states = (count + 1)*; if you load and count up, *states = (2's comp of count loaded(!))*; if you detect and clear, *states = (count-detected + 1)*. So, things are tough all over, and it doesn't matter much which scheme you choose.

See Text Table 8.10, p. 563.

Synchronous load and clear functions are nice: they support the ideal of fully-synchronous design. Such *clears* are available on a very few recent **registers**, as well as on most respectable counters (some new counters offer *both* sorts of clear and load, on four pins: e.g., 74ALS560; see Text table 8.10, p. 563 for other examples of such fancy counters).

Synchronous functions have not simply *replaced* asynchronous, because sometimes the synchronous type is a decided nuisance. See the ornate gating required in order to let one use a pushbutton to load the '469 address counters in Lab 15. That Load function is synchronous, so we need to generate a clock, timed properly with respect to the Load* signal. A *jam* load would have been just right, there. But most of the time, synchronous functions remain preferable.

2. Usg a counter to measure period
—and thus many possible input quantities

Counting as a Digital Design Strategy

We have noted already, in the Notes for class 15, that we can build a variety of instruments using the following generic two-stage form:

- an application-specific 'front end' to generate a pulse whose duration is proportional to some quantity that interests us;
- a 'stopwatch' that measures the duration of that front-end pulse.

Here we will look at a couple of examples of circuits that fail in an attempt to use this arrangement, and then we will go through a longer design exercise where we try to do the job right.

a) Two failed attempts

1- Digital Piano-key Speed-sensor

Here is a flawed scheme: part of a gadget intended to let the force with which one strikes the key of an electronic instrument determine the loudness of the sound that is put out. The relation between digital count and loudness shows a nasty *inverse* relation.

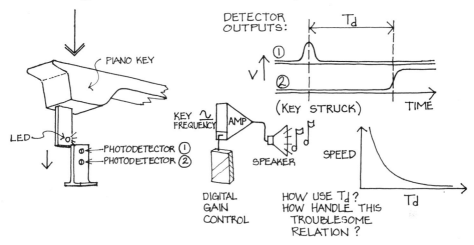

Figure X15.6: Troublesome Example 1: measure period to measure *how hard a key was hit*

We offer this as a cautionary example: probably this piano design is a scheme worth abandoning!

2- Getting the details right can be difficult: a bad circuit from the Text

Here a small error makes this big circuit useless.

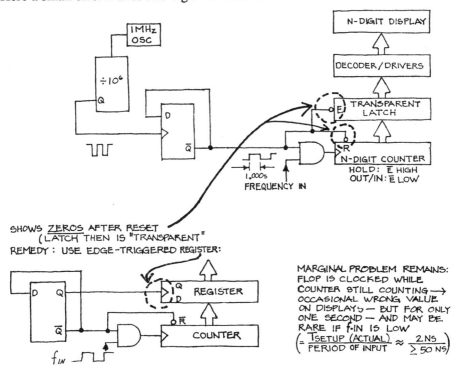

Figure X15.7: A transparent latch causing mischief again

b) Trying to do it right: Sonar ranger, a worked example

Now that you're getting good at designing these circuits, let's do an example more thoroughly, this time working out the details. This is a question from an old exam.

> ### Problem: counter Application: <u>sonar</u> ranger
>
> The Polaroid sonar sensor[1] generates a high pulse between the time when the sensor transmits a burst of high frequency "ultrasound" and the time when the echo or reflection of this waveform hits the sensor. Here's a timing diagram to say this graphically:
>
>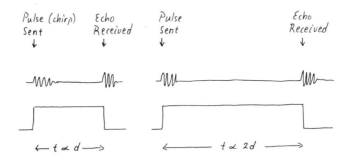
>
> **Figure X15.8**: Timing of sonar ranging device
>
> So, the duration of the high pulse is proportional to the distance between the sensor and the surface from which the sound bounces. Only a couple of bursts are sent per second.
>
> Design hardware that will generate a count that measures the pulse duration, and thus distance. Let your hardware cycle continually, taking a new reading as often as it conveniently can. Assume that the duration of the pulse can vary between 100 μs and 100 ms. You are given a 1 MHz logic-level oscillator.
>
> In particular:
>
> - Make sure the counter output is saved, then counter is cleared to allow a new cycle.
> - Choose an appropriate clock rate, so that the counter will not overflow, and will not waste resolution.

---

[1]. used in many Polaroid cameras and also sold separately; we use one in our lab

A Solution

Step one is just to draw the general scheme; a block diagram. It looks almost exactly like all the earlier examples. The difference is only the device that generates the *period* that our "stopwatch" will measure.

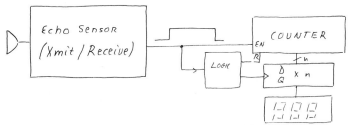

Figure X15.9: Sonar device: digital readout. A block diagram

The rest of our task is only to take care of the timing details. As usual, *timing* is the only challenge in the design task.

We need to save the final count, then clear the counter, at the end of each cycle. You do exactly this in Lab 15, but there your job is complicated by the circumstance that the register you use is the nasty *transparent latch* on the display chips. Here we will choose *edge-triggered* flip-flops instead. Nearly always, edge-triggered devices ease a design task.

Let's suppose we mean to use an 8-bit counter like the one we have been using in the lab, 74LS469. It has no Clear so-called, but we can use its Load function to do the job: just load zeros. It also offers an enable function, called "CB_{in}*"—carry-borrow-in*. That will be handy, letting us start and stop the counter with a logic level.

How's this look?

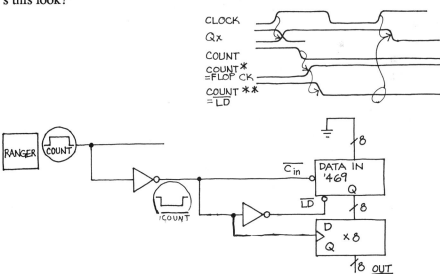

Figure X15.10: Trying to make sure the count is saved before it is cleared

The timing looks scary, doesn't it? Will the count get saved, or could the *cleared* value (just zeros) get saved? The timing diagram above says it's OK: the D flops certainly get their data before the clearing occurs.

To get some practice in thinking through this problem, look at another case, a little different: this hypothetical counter uses an *asynchronous* ("jam") clear.

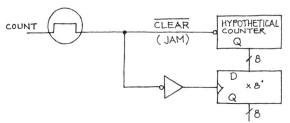

Figure X15.11: ...Another example: failing at the same task

Probably this will work like an elaborate equivalent to the following cheaper circuit:

Figure X15.12: Sad equivalent to the bad save & clear circuit

We should stick with the earlier circuit, then: the one that clocks the D flops slightly before clearing the counter. A ghost of a problem remains, however. Whether we should worry about it depends on whether we can stand an occasional error.

The problem is that we clock the D register while the counter still is counting. Sometimes the counter Q's will be changing during the *setup time* of the D register. That can lead to trouble. The most bothersome trouble would be to have some of the 8 D flops get *old* data (from count n) while others get *new* data (from count n+1). That's worse than it may at first sound: it implies not an error of *one* count, but a possibly huge error: imagine that it happens between a count of 7FH ("H" means "hexadecimal") and 80H: we could (if we were very unlucky) catch a count of 8FH. That's off by nearly a factor of two.

This will happen *very* rarely. (How rarely will the Q's change during setup time? *Typical* time during which flop actually cares about the level at its data input ("aperture," by likeness to a camera shutter, apparently) is 1 or 2 ns (versus 20 ns for *worst case* t_{setup}); if we clock at 1 MHz, that dangerous time makes up a very small part of the clock period: 1 or 2 parts in a thousand. So, we may get a false count every thousand samplings. If we are simply looking at displays that does not matter at all. If, on the other hand, we have made a machine that cannot tolerate a single oddball sample, we need to eliminate these errors.

A very careful solution

Here is one way to solve the problem: *synchronize* the signal that stops the counter (with one flop); delay the *register* clock by one full clock period, to make sure the final count has settled:

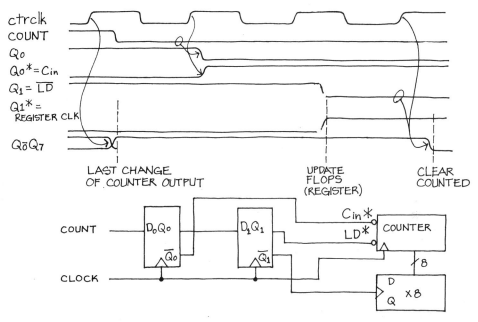

Figure X15.13: Delaying flop clock, to make sure we don't violate setup time

Perhaps you can invent a simpler scheme.

A nice addition: <u>overflow flag</u>

Can you invent a circuit that will record the fact that the sonar ranger has *overflowed*—so as to warn us that its latest reading is not to be trusted?

Hint: flip-flops remember. A good circuit would clear its warning as soon as it ceased to apply: when a valid reading had come in.

Here's one way: (I find this *hard!*):

- Let the *end* of the Carry* pulse clock a flop that is fed a constant high at its D input. Call the Q of this flop *overflow*.
- When the period finally does end, let the *overflow* Q get recorded in a *Warning* flop that holds the warning until the end of the next measurement.
- Meanwhile, to set things up properly for the next try, let the end of the period clear the *overflow* flop, so that it will keep an open mind as it looks at the duration of the next period.

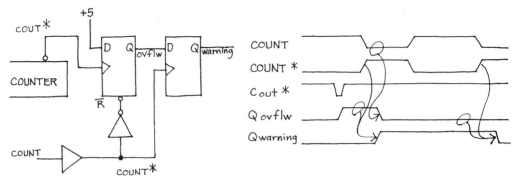

Figure X15.14: One flop records fact of overflow; another tells the world

This scheme looks pretty ornate. See if you can design something tidier.

The details of this design are fussy, but we will see one of the methods used here several more times: feeding a constant to a D input, so as to exploit the nice behavior of an *edge-triggered* input. You will see this again in a "Ready" key, Lab µ3, and then in interrupt hardware, Lab µ5.

Recapitulation: full circuit of sonar ranger

Here, with few words, is a diagram of a no-frills solution to the sonar ranger problem. The diagram omits the overflow warning logic drawn above, and the 'very careful...' logic.

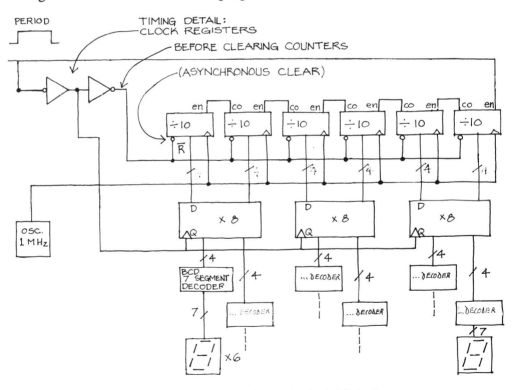

Figure X15.15: Sonar period-measuring circuit: full circuit

f. A Worked Example

ext exercise AE 8-3 (at end of chapter)

Let's conclude with a similar but easier problem, taken from the Text.

> **Problem: Reaction Timer**
>
> Design a reaction timer. "A" pushes his button; an LED goes on, and a counter begins counting. When "B" pushes her button, the light goes out and an LED display reads the time, in milliseconds. Be sure to design the circuit so that it will function properly even if A's button is still held down when B's button is pushed.
>
> Assume that you are given a 1 MHz oscillator. Provide a *Reset* pushbutton.
>
> **Figure X15.16**: Reaction timer: block diagram

REACTION TIMER: Solution

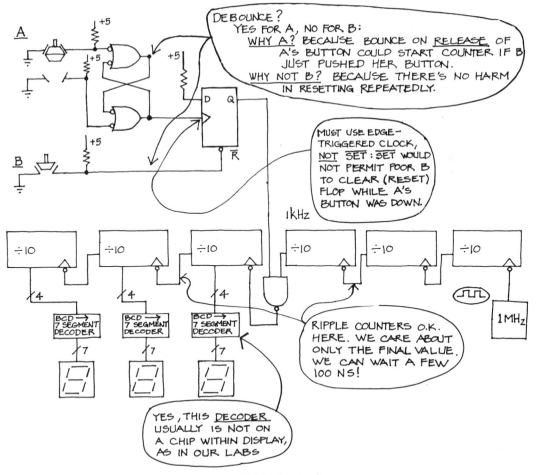

Figure X15.17: Reaction timer

Lab 15: Counters

Reading: Ch. 8, sec. 8.17, re *counters*; 8.18- 8.29.
Specific Advice:

- Important topics:

    - 8.24 latches & registers
    - 8.25 counters
    - 8.26 shift-registers: less important
    - Look closely at 8.29, timing example

- Postpone:

    - 8.27 registered PALs[1]

Problems:
Embedded problems, plus AE2—AE10, AE12, Bad Circuits B, E, F

Today you move up from the modest "divide-by-four" of last time to a 16-bit "fully synchronous" counter. We will let it show off some of its agreeable features, notably its *synchronous load*, and then we will put it to use in two circuits: a programmable *divide-by-n* machine, then a *period-measurer*, which can operate as a capacitance meter, with just a little help.

At first you will use only scope and logic probe to watch the counter's performance; then you will add hexadecimal displays that should make the counter's behavior more visible. A keypad will let you control the counter and load it.

> *Note:* the keypad is not a standard commercial part. It can be made up from the schematic attached to these Lab notes, or it can be ordered in complete form. See *Parts List* for ordering information.

Next time, you will use this counter and display to provide an *address* to a memory. The keypad will let you write 8-bit values into any memory location. In a later lab, counter and memory will serve as foundation of the microcomputer. So, today you are beginning to build your little computer.

---

[1] Paul says a registered pal is just a *spouse*; but Monolithic Memories/AMD, who hold the PAL trademark, disagree.

> *Note:* You must build today's circuits on your own private breadboard: a set of 5 or 6 breadboard strips mounted together. This breadboard will become the foundation of the microcomputer you will soon be putting together.

15-1 Integrated 8-bit Counter

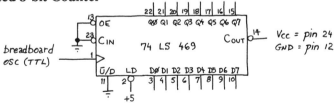

Figure L15.1: Integrated 8-bit up-down counter: 74LS469

This 8-bit counter[2] includes most of the features standard to integrated counters, plus a few that are less standard:

- It counts up or down—and that is handy for our purposes: soon (in Lab 16) you will use the counter to take you to a particular memory location or address;
- it includes three-state outputs: this feature will let us tie the counter outputs directly to the computer bus;
- it can be "loaded" with an initial value. Today, we will use *Load** to let us make a *divide-by-strange-number* counter; later, in the micro labs, we will use *Load** to let us hop to a particular starting point in the memory's address space.
- it includes carry-in* and carry-out* pins that make "cascading" these chips easy, so as to form a bigger counter. Later in this lab you will cascade two '469's to form the 16-bit address counter needed in the micro labs.

Watch clock and Q_0, then clock and Q_7 (triggering in both cases on the Q); then watch Q_0 and Q_7: do you see any delay of the higher-order Q relative to the lower-order, as you did in even the small ripple counter you built in Lab 15? Now take a look at $C_{out}*$.[3]

15-2 Cascaded 16-bit counter

The '469 is as easy to *cascade* as IC counters usually are: all you need do is connect $C_{out}*$ from one stage to $C_{in}*$ of the next; you could keep doing this almost indefinitely (except for the accumulation of carry delays if you made the chain *very* long). Try this out, cascading a second '469:

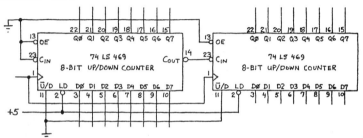

Figure L15.2: Two '469 8-bit counters cascaded to form a synchronous 16-bit counter

---

2. This counter happens to be made up from a logic array (a *PAL*): the manufacturer of the array saw a gap in the set of standard TTL functions, and used one of its arrays to fill that gap. Note the standard-sounding part number: 74LS....
3. The manufacturer calls this carry "Carry-Borrow-Out" ($CB_{out}*$) to reflect the fact that this counter can count *down*. We will use the shorter term "Carry" in these notes.

Note that the carry out of the low-order counter does *not* drive the higher counter's *clock*, as in a ripple counter. (How would the pair of counters behave if you did make that connection?)

Keypad and Load

Now let's add the keypad to the circuit: bring in its DIP connector, letting the keypad feed 8 bits of data to the *middle* 8 bits of the 16-bit counter. Let C_{out}* from the low-order counter drive that (8-bit-) counter's LD* pin ("LD" stands for "load," as you probably know).

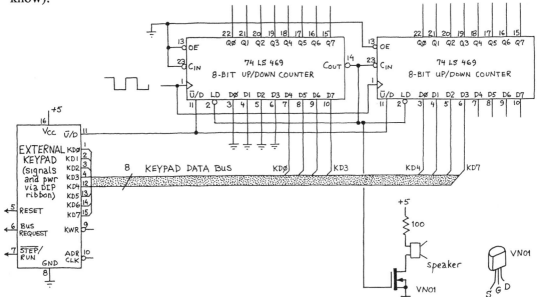

Figure L15.3: 16-bit counter: keypad feeding counter's data inputs, and C_{out}* driving LD*; one C_{out}* drives speaker (see below)

Asserting the LD* pin makes the counter behave like a register of 8 D flops rather than a counter: on the next clock, the '469 loads its 8 flip-flops with the data at its 8 inputs (labeled D0..D7, above). LD* is tied to Cout*, so these signals are asserted only when the counter is about to 'roll over;' since the counter is set to count *down*, it reloads each time it hits *zero*. So, this circuit lets the keypad set the number of counts that occur between *loads*; thus the keypad sets period and frequency of the Cout* pulse waveform. *Note*: Temporarily tie the Up*/Dn pin *high* for this exercise.

Watch C_{out}* and Q_3 of the low-order counter as you vary the keypad input value. Does the response fit what you expect?

15-3 Hearing the effect of Load*: strange modulus counter

Now, to make more vivid the power of this loadable counter to vary its *modulus*—the number of states it steps through—let's listen to the counter's output frequency: let C_{out}* drive a transistor switch (a power MOSFET is easiest), which in turn drives the breadboard's speaker, as shown in the figure above. If you want to annoy your neighbors with a louder tone, let C_{out}* drive a toggling flip-flop: the 50%-duty-cycle signal that comes out of the flop makes more noise than the narrow C_{out}* pulse does.

The keypad (its low nybble) now determines the number of cycles (or clocks) the counter steps through: 16 X key-value. You can see this effect if you clock the counter with the breadboard's TTL oscillator and watch the Load* pin on the scope.

The keypad now should act like the keyboard of a crude musical instrument. Do you hear the pitch fall by an *octave* when you change from key X to key 2X? (At one frequency extreme, your instrument goes out of tune. Why?)

15-4 Adding Displays

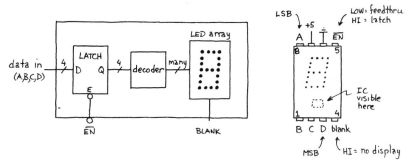

Figure L15.4: Hexadecimal display: pinout & functions

The display chips that you will now install do quite a lot: they are not simply 7-segment LED displays. They include—

- the LEDs (lots more than 7 segments: *20* little dots, instead);
- a decoder that translates the 4-bit input into lightings of the appropriate LEDs;
- a 4-bit latch;
- a blanking function.

Each display shows the *hexadecimal* character (0 through F) that corresponds to the four-bit value fed to it.

The latch is not edge-triggered but "transparent:" so long as you hold En* low, information passes straight through, as if there were no latch present; when En* goes high, the latch holds the last value it saw. Today we will keep the latches in their "transparent" mode (En* tied low).

The *blank* function (at pin 4) does what it sounds as if it should: it blanks the display if held high. The pin has very *strange* input characteristics, however: unlike any other logic input (TTL or CMOS) it floats *low*, and one must source a large current into Blank (2 mA) in order to drive the pin high.

Now *install* the four hexadecimal displays as shown in figure L15.5 (next page). Note, however:

- do not alter the counter clock wiring at this point: the breadboard oscillator should continue to clock the counters.
- don't let the "bus" shown in the diagram confuse you: the bus is a useful notion but note that to connect a particular display line to the 'bus' may mean, in concrete terms, simply that one installs a wire linking a pin on the display to one Q of the counter. It is true that both points now are on the bus; but you will not *see* any evidence of that. The bus is a concept, not a thing—unless you happen to be using a breadboard that carries a set of common lines dedicated to this bus.

Watching C_{out}* in slow motion

Now the displays make it easy to see the present *state* of the counter. Try using the breadboard's oscillator to carry you to the state in which the low-order counter is filled (showing hex characters *FF*). Take control of the Up*/Dn pin to help you steer your way to that count, then walk slowly up a few counts, down a few counts, crossing and recrossing the boundary between $00FF and $0100. Watch C_{out}* as you do this, and see whether that pin behaves as you expect: what *state* of the low-order counter does it detect when you are counting *up*? ...when you are counting *down*?

15-5 Completing keypad-to-counter wiring

The additional connections shown below let you control the counter from the keypad. Make these changes, and then check that the circuit performs as it should.

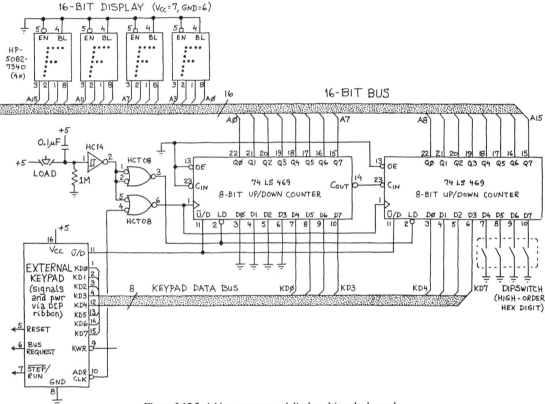

Figure L15.5: Address counter and display; driven by keypad

Most of the changes are self-explanatory. *Note* that you should now *disconnect* the line that tied U*/D high; reconnect U*/D to the keypad (at pin 11).

The logic that drives the LD* pin is *not* self-explanatory, so let's take a closer look at this part:

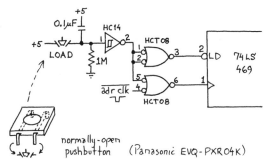

Figure L15.6: Debounce & load logic

The complications result from the fact that the counter's Load* function is *synchronous*: that is, it takes effect only when the counter is clocked. Ordinarily, synchronous operation is good; here, it is a nuisance, requiring us to generate a properly-timed *clock* every time we want to load by hand. The logic shown debounces the pushbutton and delays the release of Load* until the clock edge comes up. The loading occurs upon *release* of the LOAD* pushbutton.

Checkout

When the keypad, counter and displays are wired, check that you can control the address counter from the keypad:

- INC should increment the counter;
- DN should decrement the counter;
- RPT should let you race about in memory (Up by default; Down if you hold DN while pressing RPT).
- The LOAD* pushbutton should load the counter's middle two hex digits with the value entered at the keypad, when the LOAD* button is *released*.

The 4-position DIP switch driving the counter's top 4 bits determines which "segment" of address space (of 16 possibilities) Load* lands you in.

When you are satisfied that your counter works, your circuit is ready for use in the next lab, where it will serve—as you know—to provide an *address* to memory. That is the most important of the jobs we want the counter to perform. But in the time that remains in this lab we hope you will take the opportunity to exploit the counter for a couple of applications that should be both instructive and fun: capacitance meter, and state machine.

A. Counter Application: "Stopwatch" as Capacitance Meter

A very slight alteration of the 16-bit counter you wired last time will let you start and stop the counter: you need only add a manual switch to control the level of the low-order Carry in*. So altered, your counter would be a primitive stopwatch. The addition of a few flip-flops and two NAND gates can make this stopwatch more convenient: first by letting you latch the counter output into the displays (that way you need not watch the counting-up process), then by clearing the counter automatically after the result has been latched.

That circuit can measure the length of time a signal spends low (as we have wired it). This period-measuring circuit then could be put to any of a number of uses. We ask you to use it to measure the period (or half-period, to be a little more accurate) of the waveform coming from a 555 RC oscillator. If you then hold the "R" constant and plug in various

C's, you will find that you have built a *capacitance meter*. We hope you will find this a satisfying payback for about a half-hour's wiring.

When you have had some fun with the circuit, we ask you to restore your counter and RAM to their earlier form: so, flag the changes you make to the circuit as you go along—you may want to plug in odd-looking wires at the points where you remove a wire. (There are only a few of these points.)

15-6 Stage One: Simple Stopwatch

Remove the lines that now drive Clock and LD* of both '469's; remove the line that grounds C_{in}*.

Drive Clock with a TTL signal from the function generator (not the *breadboard* generator; you will soon need a frequency higher than its 100 kHz); temporarily tie LD* high.

Confirm, first, that you can start and stop the counter by taking C_{in}* low, then high, using a manual switch. Then set the keypad to 00 and check that you can clear the counters by taking LD* low (while the clock is running). You now have a clumsy stopwatch.

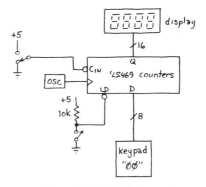

Figure L15.7: Simple stopwatch

15-7 Stage Two: Automatic Period Meter

The stopwatch becomes a period meter if we add the automatic output-latching and counter-clearing mentioned above. For this purpose we need a circuit that will generate pulses timed thus:

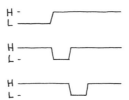

Figure L15.8: Required latch and clear pulses

You may recognize this pair of pulses, evoked by the rise of Count*, as precisely the output of the digitally-timed one-shot that you built in Lab 14—and saved, we hope.

The first pulse, LatchEn*, will update the HP displays: these transparent latches will take in new information when En* is low; they will hold that information after En* goes high. (Notice that we need to generate a *pulse*, not the usual *edge*, to make use of this transparent latch; here's an example of the clumsiness of such pseudo-clocking: edge-triggering is much neater.) The second pulse clears the counter when its result is safely stored in the latches.

Generating the required pulses: double-barreled One-Shot

Pull out the one-shot that you built in Lab 14:

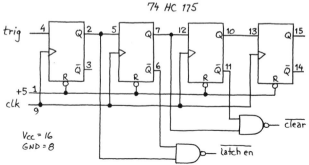

Figure L15.9: Digitally timed one-shot ("double-barreled")

Use the signal called LatchEn* to drive the En* of all four address displays (you must of course remove the prior connection to ground). Let Clear* drive LD* of both counters. Clock the one-shot and counter with the same signal, from the function generator. (The frequency now may be as high as you like. Try 1 MHz).

Drive the one-shot input (*Trig*, above) *and* the counter's C_{in}* with the manual switch. You should find that the circuit measures the duration of the time you hold C_{in}* low. (If you clock at 1 MHz, then the duration is measured in microseconds, of course.) Note that the count you see is in *hexadecimal*—a little unfamiliar to most of us ten-fingered creatures.

15-8 Capacitance Meter

You can transform your period meter into a capacitance meter in a few minutes by wiring a 7555 oscillator and feeding its output to the period meter, in place of your manual switch:

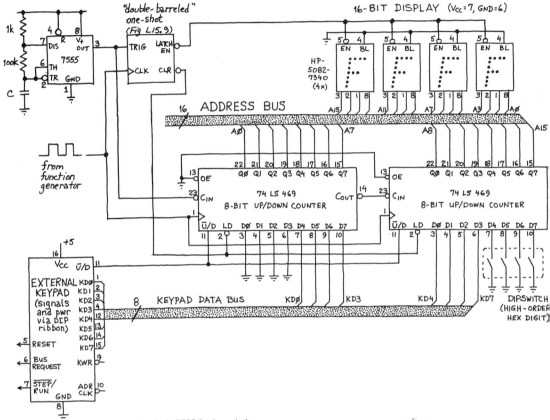

Figure L15.10: 7555 feeds period-meter: counter now can measure capacitance

Try a 0.01 µF capacitor, and rig the clock rate so that the counter reads "0010." (You may see some flicker in the display; the counter's result *must* be uncertain to one bit.)

Try adding another 0.01 in parallel. Does the meter appear to remember the effect of paralleling capacitors? If the value is not precisely what you would predict, you will have to decide whether to trust the meter or the nominal value on the capacitor. Have some fun trying other C values.

Can you be sure the counter is not overflowing? Can you devise a circuit that would alert you to an overflow by lighting an LED and leaving it on until you reset it by hand?

When you have had enough of your C-meter, restore the counter and display connections to their former state, so that we can use the counter next time for its usual purpose: to provide an address for the memory that you will install in the next lab.
Specifically, remember—

- enable the address displays continuously: tie En* low
- remove the switch on the counter LD* pin that allows you to select between Program and Run*.
- remove the one-shot and 7555. You will not need either of those circuits again.

Elements of today's circuit: private breadboard; keypad

Here we will make a few suggestions on wiring this big circuit, and then will describe the keypad that you begin to use today, and that you will rely on throughout the micro labs.

Breadboard, & Circuit layout

Don't worry about finding the best way to lay out your circuit. Aim to be moderately neat (the too-tidy sometimes get less done than the moderately untidy; the slobs do get into trouble). Being 'neat' means primarily keeping wires short: wiring that is short and close to the board is likely to survive a bump of the hand—and a board that you find reasonably pretty will help you enjoy the many hours you will spend handling this little circuit.

Ribbon Cable

Color-coded ribbon cable is available for bussed signals. It makes good sense to follow the resistor color code (though some people avoid black, reserving it for ground; they color Data Line zero *brown*, rather than black). Use the large black cable stripper for this cable, if you have one in the lab. It will strip up to 6 wires at once. It is probably best to separate the last few inches of a ribbon cable of 8 or more lines into groups of no more than *4* lines: that makes it easier to place the wires into the breadboard, and reduces the chance that the lines at either end of a wide cable will pop out of the board.

Quieting the Power Supply

Sprinkle *decoupling capacitors* liberally through your circuit, at least one set per breadboard strip. These should be ceramic capacitors of 0.01 to 0.2 µF, between +5v and ground, to kill high-frequency noise caused by current surges on ground and +5. In addition, you should use a big tantalum capacitor (1 to 15 µF) as well as a small ceramic where power enters the board.

Keypad

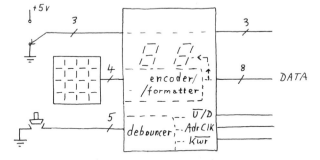

Figure L15.11: Keypad: sketch

The keypad's main function is to provide a byte of data, which you will be able to load into memory. (When the memory becomes part of a computer, the values that you load from the keypad will constitute programs and data.) The keypad also provides some useful control functions: its debounced control buttons (those that run up the left edge of the board) can be used to clock the counter, and can be used to assert the lines that execute a *write* to memory. We will also exploit the keypad to help us load an initial value into the counter.

A pinout and a description of the keypad's several outputs follows.

Keypad: Pinout and Functions

The keypad's main job is to convert two successive key pressings into an 8-bit value.

In addition, however, the keypad provides five function buttons and three slide switches. Most, but not quite all of these are self-explanatory.

Here is the pinout of the connector that links the keypad to your computer. The connector is a 16-pin DIP form, shaped just like an IC. It is also *powered* like an ordinary digital IC: at the corners.

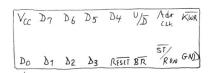

Figure L15.12: Keypad connector pinout

- **Three Slide Switches** These provide logic levels to control three functions important in the microcomputer labs. You can ignore these lines today, but later they will drive Step*/Run, Busrequest* and Reset*. Their circuitry is odd enough to require some explanation:

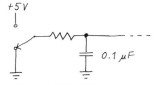

Figure L15.13: Slide switches

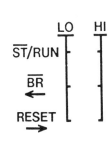

This RC is included because of the special vulnerability to *cross-talk* of signals carried on a ribbon cable. Note that these slide-switch signals travel on the ribbon cable *between* other control signals. We want to make sure that an edge on one line does not generate a false edge on another line nearby in the cable.

The capacitor protects against such cross-talk in two ways:

- The capacitor slows the edge of the slide switch signal, so as not to *cause* cross-talk to an adjacent line;
- the capacitor also provides a low-impedance that tends to kill cross-talk that might be impressed by an edge appearing on an adjacent line.

Figure L15.14: Function keys

- Write*

    The KWR* line stays low so long as the Write* button is down. The signal is debounced. (We call the signal KWR* to distinguish it from a later Write* signal that will come from the CPU.)

- W & I

    Write and Increment operates as Write* does, and in addition gives a rising edge on the address-counter clock line (AdrClk), a few milliseconds after release of the button. Thus the write occurs, and then the address counter is incremented to allow writing to the next location.

    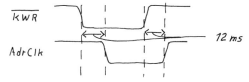

    Figure L15.15: Write* & increment timing

- INC

    Clocks the address counter, by providing a low pulse of about 2 µsecond on the AdrClk line.

    Figure L15.16: AdrClk signal

- RPT

    Repeats AdrClk signal at 20 Hertz (the keypad can be jumpered to make this rate 10Hz).

- DN

    Does two jobs: when pressed once, decrements counter; when held down along with RPT, counts *down* at 20 Hz. (DN drives the address counter's Up*/Dn line *high*.)

Schematic of Keypad circuit follows

Lab 15: Counters

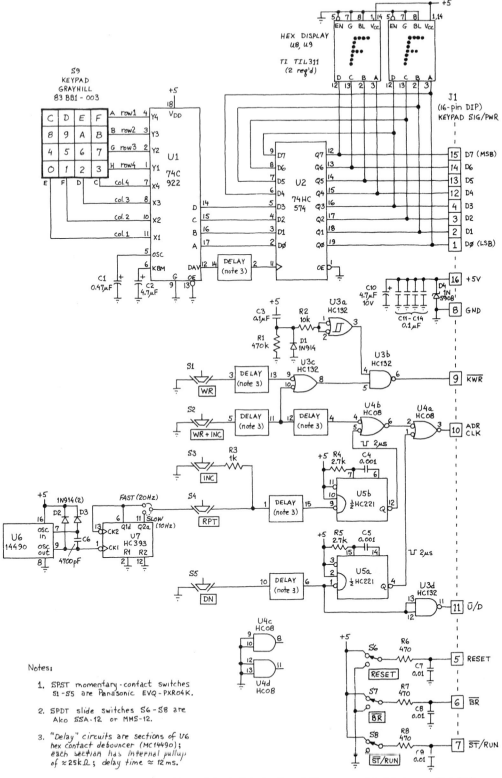

Figure L15.17: Schematic of keypad circuit (this circuit available as a completed device; see Parts list)

Class 16: Memory; Buses; State Machines

Topics:

- buses: where 3-states are required
- memories
    - defined: a way of *organizing* stored bits
    - types: ROM/RAM, dynamic/static, etc.
    - characteristics: most important: *access time*
- state machines
    - generally
      even a humble J-K from D's demonstrates the general technique
    - using gates vs using memory
    - fancy example: vending machine (this is done as a worked example, using memory)

1. Buses

These are just lines that make lots of stops, picking up and letting off anyone who needs a ride. The origin of the word is the same as the origin of the word for the thing that rolls along city streets.

Power buses carry +5 and ground to each chip; data and address buses go to many chips in a computer.

When are three-states needed?

Answer: whenever more than one *driver* (or "talker," to put it informally) is tied to one wire.

The presence of more than one *receiver* (or "listener") does *not* call for 3-states. (Think of a telephone party line, if you find yourself confused on this score.)

Test yourself: in the circuit below, where are 3-states needed?

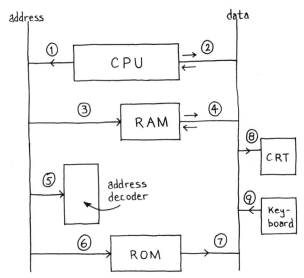

Figure N16.1: Where are 3-states needed?

Example of 3-state use in today's lab: Data Bus

And here's a part of the circuit you will build today. One 3-state is shown; another is implicit.

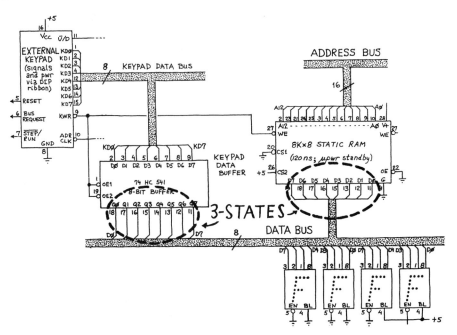

Figure N16.2: Example of 3-state use in Lab 16: data buffer meets memory

The '541 keeps the keypad from monopolizing the data bus. What keeps the RAM from hogging the bus all the time? (Note that the RAM's OE*, its explicit 3-state control, is tied low—asserted—all the time).

Answer: the RAM's WE* pin turns off the RAM's 3-states when asserted (low).

2. Memory

text sec. 11.12

A *memory* is an array of flip-flops (or other devices each of which can hold a bit of information), organized in a particular way. A memory is used most often, as you know, to hold data and program for a computer; it can also be used to generate a combinational logic function—as you will see in Lab 16. So applied, it resembles the *logic arrays* or *PALs* described in the Text.

text sec. 8.15

Organization in "memory" form

8 bits can be stored in a *register*, as you know. The register is just a collection of flip-flops, with all D's and Q's brought out in parallel.

Those 8 bits can also be stored in *memory* form. In that case, the inputs and outputs can be multiplexed onto one line.

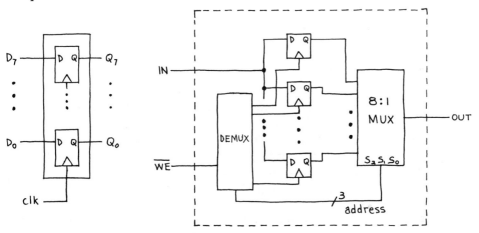

Figure N16.3: Register versus memory organization of stored bits

At first glance the memory looks silly: a lot of extra internal hardware just to save a few pins? Let's count pins:

Function	Register	Memory
data In	8	1
		or 1 for In & Out
data Out	8	1
clock/write	1	1
power, ground	2	2
address	0	3
total	19	7 or 8

Figure N16.4: Pin count: register versus memory

The memory advantage doesn't look exciting—until you crank up the number of stored bits. Then the memory reveals that it is the *only* feasible way to store the information (try counting pins on a 1Mbit memory reorganized as a *register*, for example). In fact, memories nearly always squeeze away one more pin: they combine input and output into one pin. (Can you invent the internal logic that allows this?)

It also mates nicely with present computers, which usually process only one item of information at a time.

That's all there is to the *concept* of a memory. Now on to the *jargon*.

Memory types; memory jargon
Text sec. 11.12, again

One can slice the universe of memory types in several ways:

storage technology flip-flops: called "static" RAM, *versus* capacitors: called "dynamic" RAM.
One can tell that these names must have been invented by the manufacturers of the dynamic memories: the ones that use capacitors. Cap memories aptly could be named "forgettories" (to steal a word from the Text), since they forget after a couple of *milliseconds*! These memories need continual reading or pseudo-reading, to "refresh" their capacitors. Despite their seeming clumsiness, *dynamic* RAMs enjoy the single advantage that their remembering unit is very small and uses little power; so, most large memories are dynamic.

can one write to the memory?
If Yes, easily, it's called a **RAM**—a misnomer, since *all* these semiconductor memories are "random access memories" (excepting only "bubble memories").
If No, it's called a **ROM** ("read only memory:" this name fits).
If No, not unless you work pretty hard, it's called a **PROM** ("programmable read-only memory").
PROMs subdivide further:

- If you cannot erase it—say, if you program it by blowing little fuses—its just a PROM;
- If you can erase it with ultraviolet light, it's called an **EPROM** (pronounced "Ee-Prom"). Here the information is stored on a capacitor—but the discharge time is 10 to 20 *years*!
- If you can erase it electrically (that's handy), it's called an **EEPROM** (pronounced "Ee-Ee-Prom") or **EAROM** ("electrically alterable...").

does the chip forget when the power is turned off?
If yes, as is usual for a RAM, it's called "*volatile*."
If No, it's non-volatile: usually this is a trick reserved for ROM's; some RAMs achieve non-volatility sneakily:

- battery-backup: a low-power memory, made of CMOS, can hold its data for years on microamps. Hence the sneaky "chubby RAM"—not its official name:

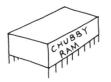

Figure N16.5: Chubby RAM acts like RAM, but remembers for 10 years. You can guess why the package is fat.

A variation on this scheme: a battery *socket* with battery and power-down/up circuitry.

— other backup schemes are used, too, such as 'shadow-RAM': it's a RAM with a non-volatile memory attached, which can be updated from the RAM before power-down.

How memories are specified:

Organization: "#-of-words X word-length." For example, the RAM you will use in the lab is 8K X 8: 8192 words (or locations), each 8 bits wide.

The most important memory characteristic, once you have settled on a type and size, is *access time*: the delay between presentation of a valid address ('your' job) and delivery of valid data (the memory's job).

Static RAM Timing

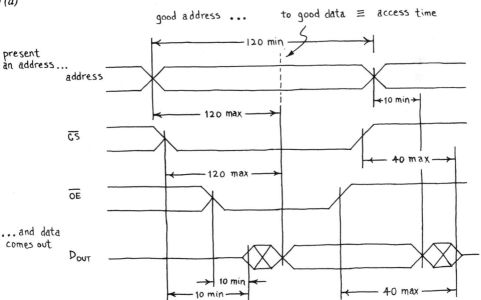

Figure N16.6: Read timing for 120ns static RAM

When you have wired your microcomputer you will be able to *see* the access time: the delay between chip enable, in that case, and good data. If you're feeling energetic at that point, take a look.

Multiple RAM enables

A diagram of what's inside the RAM may help explain its strangely-complicated enabling scheme.

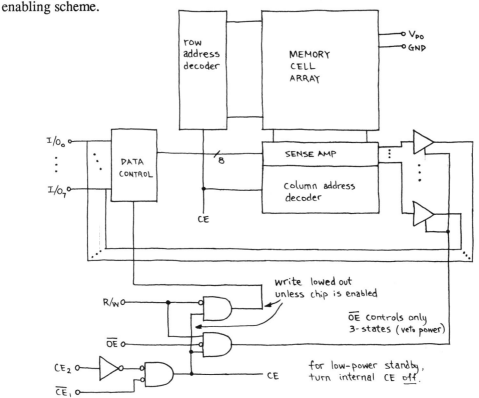

Figure N16.7: RAM block diagram, showing what the 3 enables do

---

3. "State Machine:" New name for old notion.

A *counter* is the most familiar of state machines: a sequential device that walks through predictable sequences of "states." (A state, you will recall, is defined as the set of outputs on the device's flip-flops.) That sounds complicated; but if you apply the notion to a simple counter you will see that the idea is simple and, indeed, familiar. Here, as a reminder, is a 2-bit counter and the "state diagram" that shows, rather abstractly, how the counter behaves:

LABEL	$Q_1 Q_0$
A	0 0
B	0 1
C	1 0
D	1 1

Figure N16.8: 2-bit counter: flow and state diagram

Designing Any Sequential Circuit: Beyond Counters

text sec. 8.18,
g. 8.57, p. 512

Here is a diagram showing any clocked sequential circuit.

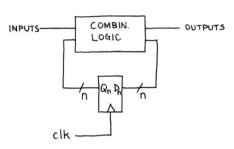

Figure N16.9: Sequential circuit: general model

As you can see, the D flops will take the levels at the D inputs and transfer them to their Q's after the next clock. Thus the circuit Q's will go from their *present state* to the values on the D's on the next clock; the values on the D's thus show the circuit's *next state*.

The Text shows how to use this notion to design a divide-by-three counter. Here we propose to begin with a simpler assignment:

1) Designing a J-K Flop

Our aim is simply to design a circuit that behaves like a J-K flip-flop, given a D flop. The technique is the same: describe what the circuit should do after the next clock, given its present state and what is coming in on its inputs.

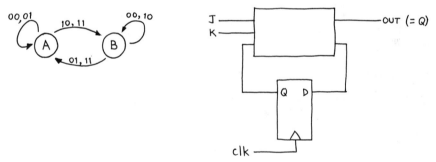

Figure N16.10: A particular sequential circuit: J-K behavior from D flop + logic

"External Inputs:" J K	Present State Q_n	Next State $Q_{n+1} = D$
0 0	0	0
	1	1
0 1	0	0
	1	0
1 0	0	1
	1	1
1 1	0	1
	1	0

Figure N16.11: Present-state — next-state table for J-K flop

One can scan for the logic pattern, or one can draw this Next-State function (the function that is to feed the D input) on a Karnaugh map:

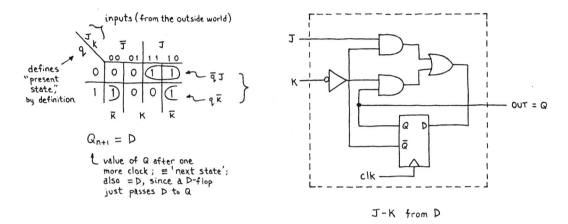

Figure N16.12: K-map of D input as function of Q, J, K; a gate implementation

2) Designing a Divide-by-Three Counter

The Text does this problem, and you may have done something very similar a couple of labs ago, so we will only sketch this problem, here:

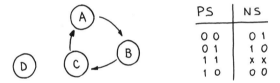

Figure N16.13: A particular sequential circuit: divide-by-three counter, from d flop + logic

This circuit, you will notice, uses *no* external inputs, but requires two flops.

3) Designing a Divide-by-Five Counter

Now you are getting the idea, and you can see that any other counter presents fundamentally the same design problem. Here is a slightly bigger counter:

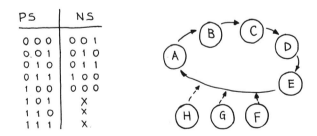

Three unused states :
Where should they go ?
Where should they not go ?

Figure N16.14: A particular sequential circuit: divide-by-five counter

4) Circuits requiring Complicated Combinational Logic: a <u>Memory</u> can replace the combinational gating

A divide-by-256 counter is not hard, as you know, because the behavior of a binary counter is so orderly; you can design one stage and then cascade like stages. But imagine trying to do the job with D flops, the way we have done the ÷3 and ÷5 counters: the Q's present 8 variables: quite unmanageable as a K-map problem!

It would be nice to have another way to generate the logic that feeds the flops. It turns out we do have another way—described in another set of notes: the use of *memory*. With such a device we need hardly use our heads at all; instead, we use the *address* lines of the memory to carry the *input* variables and present-state bits; we use the *data* lines of the memory to deliver the *output functions* while also generating the *next-state* information.

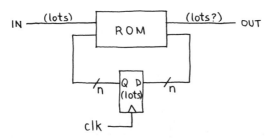

Figure N16.15: Combinational logic, in sequential circuit, implemented with memory

On the general circuit diagram, this change is not substantial; but to the designer the change is large: your design task is now much simplified. You can design arbitrary state machines, and you can toss in extra memory data bits, if you like, in order to implement arbitrary output functions as part of the deal. The memory does the work for you. You will find an example of this technique worked as the 'vending machine' Worked Example.

Recapitulation: Several Possible Sequential Machines

- If there are *no* external inputs, the circuit may be a counter (it could be binary or decade, for example; it could even count in some strange sequence; but it can perform just one trick; its next state will depend upon its present state alone).
- If there is *one* external input: then this could be an up/down counter, for example, or—more generally—a device capable of two tricks.
- If there are *eight* external inputs: then this sequential machine would be capable of running many alternative sequences: 256 of them. Such a machine sounds far out, but you will meet it soon in the form of a microprocessor (and the one you will use is fed *16* inputs). The 8 or 16 input lines select one out of its large repertoire of tricks at a time; these input lines are said to carry "instructions" to the processor. A sequence of instructions, as every toddler nowadays knows, is called a program.

Ch. 8: Worked Examples: State Machines

Three examples:

- a washing machine controller, made with counter and decoder
- a small but conventional state machine, using gates and flops: detect a particular sequence on an input line
- a bigger state machine, but implemented in a wonderfully easy way: Text's vending machine, done with memory rather than with gates (PAL or discrete).

The washing machine controller sidesteps the usual process, and demonstrates a technique useful for very simple sequencers. The second example—the sequence detector—shows a useful machine, but one that turns out to be susceptible to a subversively-easy implementation by another method. The last example shows off nicely how easy ROM can make the design of state machines; this method is an easier variation on the usual scheme, which would use a logic array to the job we do here with a ROM.

Lurking behind all these examples lies the competitor for any relatively slow sequential operations: the microprocessor. It would do the vending machine job easily, and probably would be preferable there. It might well be too slow for the task of detecting a sequence coming in on a serial line. So, there is room for sequential machines like those we describe here.

---

1. Sequencer from Standard Counter

Before we take you through the more traditional design process, let's admit that one sometimes can sidestep such work by combining a counter and decoder.

Text sec. 8.17,
Decoder: see pp. 496-47, fig. 8.35

Let's suppose our task is to control a washing machine:

> **Problem: Washing machine controller**
>
> Suppose we want to replace the usual wheel-and-cam timer that controls a washing machine with a digital sequencer (we want to be able to boast *"digitally controlled!"*—even though our design may work no better than the old mechanical one).
>
> We are given a 1 kHz clock, and the machine is to walk through this sequence:
>
> | Operation | How long? |
> |---|---|
> | fill | 100 seconds |
> | wash | 200 seconds |
> | rinse | 200 seconds |
> | drain | 100 seconds |
>
> Show how to use a ÷ *16* counter plus a *3-to-8 decoder*, and other parts as needed, to generate the signals that could control the switches called "fill, wash, rinse, and drain." At the end of a cycle, let all functions shut off. A manual Reset* pushbutton should start a new cycle.

A solution

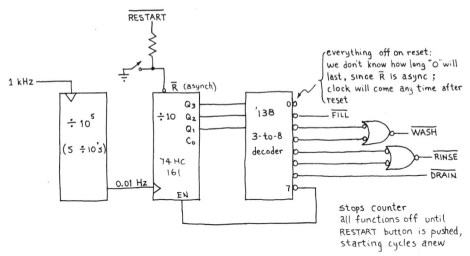

Figure X16.1: Sketch of counter use in washing machine sequencer

This circuit starts up in an annoying way: it may wait as long as 100 seconds after the Reset* button is pushed before beginning to fill. You may want to invent a way to eliminate this annoyance.

Glitches

We don't need to worry about nanosecond timing issues in this application: the sluggish electromechanical controls here (solenoid valves and motors) aren't fussy. In another setting, however, you might need to check the timing of this circuit: do two *decoder* outputs ever overlap? Do two *circuit* outputs? Is there ever a time when *no* output is asserted?

Even without looking at the specifications for the decoder we can see that the answer to some of these questions is *yes*: the delay path for "fill" and "drain" is shorter than the delay path for "wash" and "rinse." That should produce a brief overlap in one case, gap in the other.

2. Sequence Detector

Problem:

> **Sequence Detector**
> Design a *synchronous* circuit that detects the sequence *011* coming in on one line. The levels are to be sampled as usual, on the rising edge of the clock, and 011 on 3 successive clocks should send the circuit's output high for one clock period.

You might think of this gadget as part of a sort of paging system, with each person (or device) alerted when his (or its) code appears. To keep things simple, we have made this device *synchronous*, as usual: that means we'll use clocked flops in the feedback loop, not just the generic 'delay elements' we are forced to consider when we design circuits that lack clocks. A practical sequence-detector would more probably work the other way: it would be *asynchronous*. Such a device is harder to design, and in this course we will dodge that topic entirely.

State diagram

Here is a graphic way to say, 'It detects 011.'

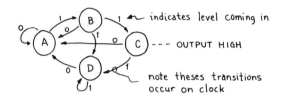

Figure X16.2: State diagram for a machine to detect the sequence *011*

State Table

We can say this once more, this time making a table that relates *Present State* to *Next State*. We do this, first, speaking of the states with their arbitrary and rather abstract labels, A, B..., and then with the Boolean combinations into which we choose (arbitrarily) to encode those states: 00, 01...:

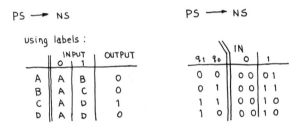

Figure X16.3: Present-state—next-state tables: using labels, and using Boolean combinations

Getting equations from state table

Since we intend to use *D* flops (not J-K), our design task is nearly over: the *Next State* table defines what we should feed to the *D* inputs (since a D flop simply passes to Q what it sees at D, on each clock). Here is the PS-NS table broken into two pieces, one for each of the variables: Q_{n+1} or D_n (these are the same: the subscript "n+1" means, "After one more clock"):

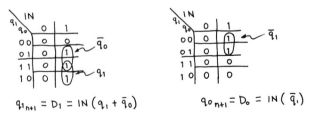

$$q_{1_{n+1}} = D_1 = \text{IN} \cdot (q_1 + \bar{q}_0)$$

$$q_{0_{n+1}} = D_0 = \text{IN} \cdot (\bar{q}_1)$$

Figure X16.4: PS-NS table broken into two pieces — now simply Karnaugh maps, thanks to our choice of binary codes for the several states.

Output

That's all there is to it, except that you need some logic to make an *output* appear in a particular state: here, state C. A two-input AND gate will do, here:

Figure X16.5: Output function: just detects the 'Got it!' state (state C)

Full Circuit

Here is what the whole circuit would look like:

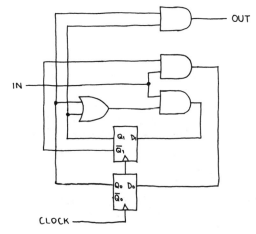

Figure X16.6: *011* detector: full circuit (conventional state machine design)

An easier way, for this application: shift register

The state machine we just designed can be mimicked with a simple shift-register. That's too bad: it seems to undermine our claim that state machines and the design procedure we just went through are useful. It need not do that. It means only that this application is not well-suited to the laborious method just set forth. Here's the shift-register circuit that does about the same job:

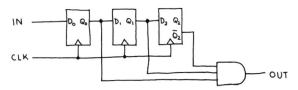

Figure X16.7: Shift-register wired to detect a particular sequence: 011.

A shift register cannot usually stand in for a traditional state machine. It is not so versatile. The shift register happens to be good at this operation—essentially a serial-to-parallel translation. And even in this application, the shift-register calls for a lot of flops: n flops to detect a sequence n-bits long (versus $\log_2 n$ flops for a conventional design).

3. Improved Vending machine, done with ROM rather than gates

The text works through the design of a vending machine, in order to demonstrate both the notion of clocked sequential machine and the way that PALs can ease the programming of sequential devices. That example can be done very easily if a memory replaces the PAL. The PAL is cheaper and faster than the ROM, but requires a Boolean minimizer; the ROM implementation requires only a pencil and paper, and no cleverness at all.

Let's state the problem anew, with refinements:

Text sec. 8.27, p. 533-537

Problem:

> **Vending machine**
>
> Duplicate the design of the Text's vending machine, but this time use a ROM in place of gates, and let the machine also give correct change along with a bottle of pop.
>
> The original problem, you will recall, used a coin sensor that put out a strobe and a 2-bit code defining the input coin. 25 cents is to cause a bottle to drop. (This is an antique machine!)

Now let's do the problem, in stages:

State Diagram

We need to add states to the original state diagram: the machine needs to "know" how much money it has received above the 25-cent price: "25-cents-or-more" must be divided into all the possible amounts "more," so as to deliver the right change.

Compare Text Fig. 8.80, p. 535

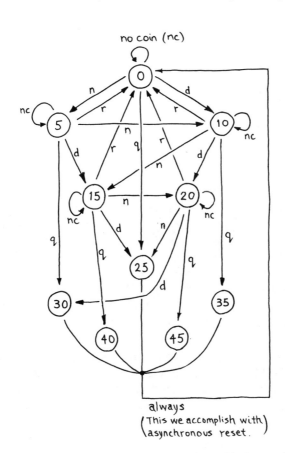

Figure X16.8: A few states added to original state diagram, to let machine keep track of change that it owes

We need four more states. What differences will that make to the design task?

How many flops?

Now there are ten states. 3 flops no longer will suffice (as they did in the Text example); we need *four* flops. They permit 16 states, but we don't mind wasting 6 (we haven't any choice, in fact!).

A preliminary circuit diagram

Now we can sketch the circuit, postponing just a few details. The diagram looks almost exactly like a generic state machine's, but it shows how many inputs and outputs we need, and shows signal sources and destinations:

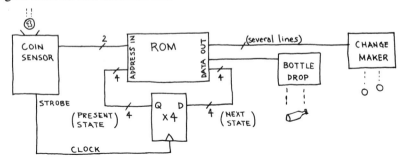

Figure X16.9: Vending machine sketched: shows the general design, and how many inputs and outputs to expect

ROM contents

Now, to make the machine perform we need to build that block at the top of the generic diagram labeled "combinational logic," a block that usually is implemented with gates (whether PAL, as in the Text example, or discrete). Here the task would be formidable if done with gates: one more flip-flop implies one more function coming *out* of the combinational block; that's not bad. But the added flop also implies an additional *input* signal going to the logic block: there now are *seven* input variables: no longer manageable, even with sweat, by means of Karnaugh maps. Perhaps you could subdivide the problem. A computer minimizer like CUPL used in the Text example could handle those seven variables easily; but the required gates no longer would fit within the particular PAL used in the Text's example.

These problems are soluble (use a more complex PLD). But we can sidestep all these difficulties by using a ROM. We'll just store as *data* whatever *next state* we want; that will feed the D flops. Clock them, and we'll land at that next state. Very simple. In addition, when we want an output, we get it almost free just use an additional bit in the *data* word.

Here's the idea:

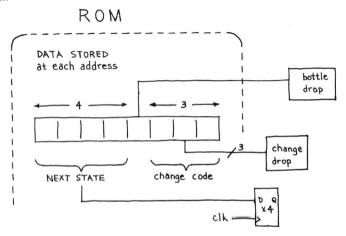

Figure X16.10: ROM providing *next state* and *output* information

Now all that remains is to load the ROM. We know, from the state diagram, the Present-State—>Next-State pattern; we need to translate that into binary codes. Let's start by listing the present states, and assigning them arbitrary binary codings:

State and coin codes

Money so far	Labeled State	Binary code (arbitrary)		Coin codes	
0	A	0000		slug	00
5	B	0001		5	01
10	C	0010		10	10
15	D	0011		25	11
20	E	0100			
25	F	0101			
30	G	0110			
35	H	0111			
40	I	1000			
45	J	1001			

Figure X16.11: Codes assigned to states and coin inputs

Now let's plug these state and coin codes into the Present-State —> Next-State tables (we'll do only some of this: how long *is* the full memory table, by the way?).

Present State —> Next State: encoded

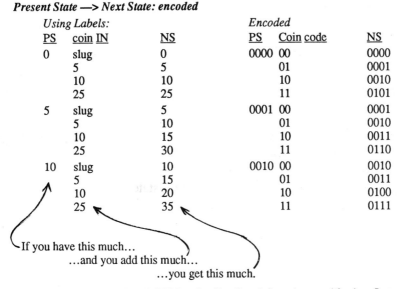

Figure X16.12: Sample ROM entries: *Next State* information stored for three *Present States*

...and so on. Since the table is long, the process of writing out all next-state entries is tedious; but this work calls for no cleverness at all. We *know* the ROM can do the job (no need to count product terms, as in the PAL implementation). The ROM is perfectly indifferent to the patterns we put in.

Outputs

To get a pop bottle and change out of this machine, we need to detect particular machine *states* and then make things happen: *drop bottle*, or *drop (particular amount of-) change*. That may sound as if it implies that we need some AND gates; but no, the memory can do this job for us, too.

To make change for us we can use hardware that delivers one coin (of the chosen denomination) for one pulse (perhaps the pulse drives a solenoid). Such a scheme lets us return appropriate change by putting out a bit pattern that releases this and that coin. (To keep things simple, let's give ourselves *two* stacks of dimes, so that we can put out 20 cents' change in one operation.)

Here's the idea:

STATE	CHANGE NEEDED	DROP POP?	CHANGE CODE $d_1\ d_0\ n$
20	–	0	000
25	–	1	000
30	5	1	001
35	10	1	010
40	15	1	011
45	20	1	110

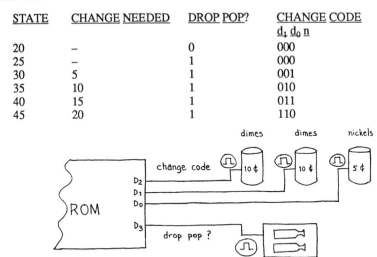

Figure X16.13: Vending machine's output functions: drop pop & change codes

Getting the machine to deliver the bottle and appropriate change now requires no more than adding *4 data bits* to the stored words.

How big is the required memory?

Not very big:

- ***state* information** 4 bits: this calls for 4 *data* bits to define the next state, and 4 *address* lines to select the next state entry.
- ***input* information** 2 bits (this code defines *which coin* has just come in): these are address lines, and do not imply need for additional data bits. Incidentally, the coin sensor also provides a *strobe* signal indicating that a valid coin-code is available. This strobe is used as clock, and is not treated as an "input" to the state machine, though it is a necessary signal.
- ***do-this, do-that*** 4 *data* bits carry this code (1 for bottle drop, 3 for change code). These bits imply no need for additional address lines.

The memory needs 8 data lines, then, and 6 address lines: it would be called a *64 X 8* ROM: very small.

Reset

We have simplified our circuit, relative to the one implemented in the Text example, by providing an asynchronous Reset: a pushbutton that drives Reset* on the flops, and also kicks out all the coins accumulated on this pass.

The *drop pop* signal can do two other jobs for us:

1. it can cause the accumulated coins to be gulped into the machine: the customer has his bottle; he no longer needs a refund.
2. it can Reset the whole machine, setting up the next pass: a fresh start for the next customer.

Whole Circuit

Here, collecting the several subcircuits in one diagram, is the whole machine. This *hardware* description falls short of describing the circuit fully, of course—since the brains of the circuit lie in the ROM code!

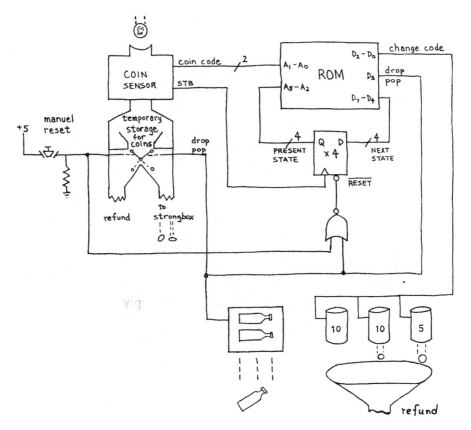

Figure X16.14: Vending machine, including change-maker: full circuit

Postscript on ROM versus gates & PALs, and both versus Microprocessors

Having promoted the ROM implementation as easiest on your brain, we must admit again that this method is unusual. People who manufacture gadgets—like vending machines—don't mind making their digital designer work a few extra hours in order to save a half dollar per unit, and PALs are cheaper than ROMs. In addition, they run faster. That doesn't matter in this application but does matter in many others, of course.

A microprocessor could do this task perfectly well and might well be the best choice, as we said at the outset. The micro could take care of lots of other operations at very little additional cost—it could drive a display, for example, to show the total accumulated change, or even drive a talker chip that might say, "Thirsty?" It is also the implementation easiest to modify, though the ROM comes in a pretty-close second.

Lab 16: Memory; State Machines

Reading:	Ch. 8 to end (applications): Look again at sec. 8.18; read 8.27, omitted last time, concentrating on state machines.
	Specific Advice:
	Concentrate on the outlines of the vending machine example in 8.27 (but don't worry about the details of PALs or CUPL's notation).
	Ch. 11:
	sec. 11.12, <u>re:</u> memory (pp. 812-821)
Problems:	As an exercise, try to implement the *vending machine* state machine, using not a PAL as in the Text, but *memory* and flip-flops. This problem is done as a *worked example*. (The present lab demonstrates such a use of memory.)

This lab is in two parts:

- The first part asks you to add memory—an 8K • 8 RAM—to the counter circuit you wired last time. This memory and its manual addressing and loading hardware will form the foundation of the small computer you soon will build.
- The second part invites you to play with the machine you have constructed, by using it to make a versatile *state machine*. These machines provide an introduction to the microprocessor that you will meet in Lab 18.

The first part is essential: you cannot proceed with the micro labs until you have installed the RAM. The second part is not essential: it demonstrates a concept fundamental to computers, but you can proceed without building these circuits. If you are very short of time, evidently you should do the first part and then squeeze in what you can of the second part.

A: RAM

The memory that your 16-bit counter is to address is a big array of CMOS flip-flops: 64K (= 65,536). These flops are arranged as 8K "words" or address locations, each word holding eight bits. The memory is "static:" it uses flip-flops, not capacitors to store its values; it therefore requires no refreshing. It is "volatile," in the sense that it forgets when it loses power; but it requires so little power that it can easily be kept alive with a pair of AA cells, as you later will show, when you begin to use the RAM to hold programs. Today we will omit this battery backup, because you are not likely to write anything valuable during today's session.

Because its outputs are equipped with three-states, the memory does not need separate In and Out lines. Its common data lines serve as outputs until WE* is asserted; at that time the 3-state drivers are turned off and data lines serve as inputs: the RAM begins to listen. The data bus circuit shown takes advantage of this tidy scheme: we turn on *our* three-state drivers (an 8-bit buffer: the '541) with the same line that asserts WE*. Thus the RAM shuts off its 3-states just as we turn ours on, and vice versa.[1]

16-1 Data Buffer & RAM, Linked to Counter

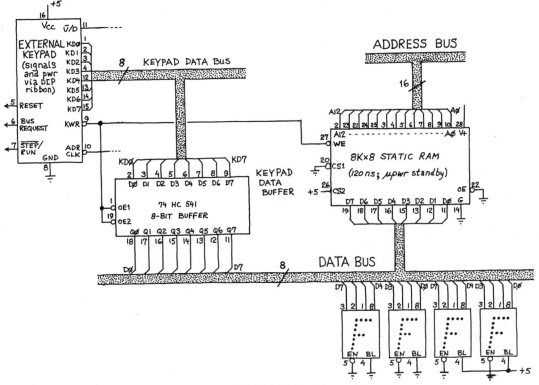

Figure L16.1: Data Buffer & Display; Memory Pinout

Install the RAM and the 3-state buffer that sits between keypad and data bus. This buffer prevents clashes between the keypad and other devices that sometimes drive the data bus: the RAM today, but soon also the CPU and peripherals.

Note: a pinout label for the RAM appears in the pinout section, Appendix D of this Manual. We suggest that you photocopy that label and paste it onto the RAM. The label will save you a good deal of pin-counting.

The data display shown is larger than what is necessary today: it can show 16 bits, but we need only 8 until the processor enters; therefore we *blank* the other 8 bits, for the time being. We suggest that you wire all 16 bits today, however, since you will need them later. Note that the two 8-bit displays (each a pair of two Hex display chips) are driven in parallel with the same 8 lines from the data bus. This probably looks silly, now. It will make sense later—when the CPU enters: for when the CPU drives 16 bits of data to these displays, it does so in two *8-bit* passes. (We will take advantage of the display *latches* included within

---

[1]. This is roughly true, but not true to the nanosecond: the RAM, made of relatively slow CMOS, turns off more slowly than our driver turns on; so, the two drivers may clash for ten to twenty nanoseconds. But this "bus contention" is brief enough to be harmless.

Checkout

Confirming that the memory remembers is as easy as it sounds:

- Set up a value on the keypad by hitting two keys, say *A*, *B*. *AB* appears on the keypad display.
- Push the WR* button: *AB* should appear on your breadboarded *data display*. If it does, you have wired the 3-state buffer properly (the '541): the *AB* is driven from keypad to data bus.
- When you release WR*, *AB* should remain on the display: on release of WR*, the RAM resumes driving the data bus, showing us its stored byte.
- Change the keypad value, without hitting WR*. The breadboard data value should not change.
- Increment the address, then decrement; confirm that the value stored earlier remains.

Works? Congratulations.

B. State Machines

16-2 A simple State Machine: Divide-by-3 Counter

A counter is a special case of the general device, *state machine*. But since it is a familiar case, and relatively easy to design, let's review the notion by asking you to design and build a divide-by-three counter, using D flops and whatever gates you need. (If you are ambitious, feel free to use J-K flops instead of D's: they will require more work on your part, but may save you a gate or two. D's are perfectly adequate; they illustrate the technique just as well as J-K's do.)

To make the problem fun—so that you don't just open the book to page 513 and find the Text's solution—let's change the sequence a bit: let's make the counter count *down*, from 3 to 1:

Present State $q_1 q_0$	Next State $Q_{1(n+1)} Q_{0(n+1)} = D_1 D_0$
00	XX (don't care)
01	11
11	10
10	01

Figure L16.2: Present-state — Next-state table for divide-by-3 down counter

We have written this list in a funny order so that it looks like a Karnaugh map without further manipulation. By grouping *1*'s on this table, perhaps you can make out what logic is required.

Incidentally, we spoiled the game for you, slightly: we decreed that you'll need two flip-flops, and that one state is wasted (the "don't care"). But you knew that, anyway. When you finish your design, you should check—on paper—to see that the unused state did not by chance get assigned as $XX = 00$. Why?

Figure L16.3: Divide-by-three down counter: your design

This example was simple (though perhaps not *easy*: no task is easy the first time you have to try it). The state machines described in the remainder of the lab are less simple—but much easier to design: they require no gating at all, leaving all that work to the RAM, instead.

16-3 RAM-Based State Machine: I: Register Added, to let RAM data define Next State

We offer this exercise for two reasons: to illustrate a way to make complex state machines; and to foreshadow the *microprocessor*. We hope to let you feel the truth that the processor is only a specially-fancy version of devices you have seen before: it is a fancy state machine, cousin to plain counters; it differs in being spectacularly more versatile than any counter.

We assume that when you begin this lab you have wired a circuit that allows you to address any location in RAM, and to load a value from the keypad into that location.

As you have begun to prove to yourself (with lab exercise 16-2), one can build an arbitrary "state machine" by feeding *next state* values to the D's of some flops, then clocking the flops. Here you will get a chance to try out the very easiest way to generate those next state values: with a memory.

The RAM is well adapted to this job: address evokes data; that's in the nature of a memory. Applied in a state machine:

ADDRESS (= present state) ———> (evokes) ———> DATA (= next state).

Here's the idea—and then the corrected version, with a register of edge-triggered D flops "breaking the feedback path."

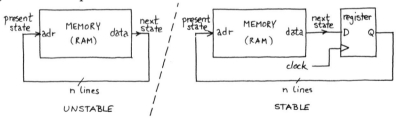

Figure L16.4: RAM is well adapted to the present —> next task; it needs a register of flops to make it stable

The flops are solving a problem you have seen before. The toggling D flop (a circuit you built in Lab 14) is stable if edge-triggering breaks the feedback path; it is not stable when enabled continuously.

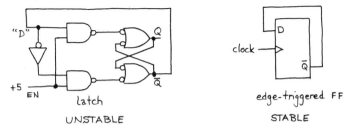

Figure L16.5: A reminder of a similar case: the familiar toggling D-flop using similar hardware to achieve stability: bad and good circuits

Ordinarily, one would use a D register to hold the "present state." (Some chips integrate memory and register onto a chip.) To save ourselves extra wiring, we use our pseudo-D register, the 74LS469 8-bit counter already installed.

So long as the '469's LD* pin is asserted, you will recall, the '469 does not count; instead, it Loads the data presented at its 8 data inputs. When we want the counter to serve as a register, therefore, we will simply hold LD* low.

We cannot *tie* that line permanently low, however, because we need to be able to use the counters as usual while we are putting data into memory ('setting up the program,' we might call this process). Therefore we must add a *temporary* Run*/Program" switch that asserts LD* only we when we want to *Run* the state machine.

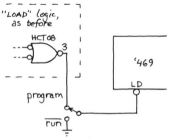

Figure L16.6: Run*/Program Switch (temporary: remove after this lab)

And here is the full counter circuit (the RAM is not shown):

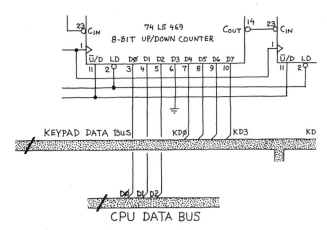

Figure L16.7: Octal D Register ('469 with LD* grounded) Fed by 3 bits of RAM Data: now feedback loop is closed

The three lines driven from the data bus are driven, in this case, by the RAM data lines d0...d2. Refer to figure 16.1 for the RAM pin numbers.

Sample 'Program'

Load the following data into RAM, at the bottom 8 locations (this requires setting the high-order addresses to zeros, using the keypad).

Address (hexadecimal)				Data (arbitrary)	next address
A15-A13	A12-A8	A7-A4	A3-A0	D7-D4	D3-D0
0	0	0	0	0	3
			1	D	5
			2	A	1
			3	B	2
			4	0	7
			5	F	4
			6	D	3
			7	0	6

Figure L16.8: Data and Next-Address Loaded into RAM

Try the circuit and program above, using the following procedure:

1. Load the data into RAM; make sure that the address counter lies somewhere in the range 0-7;
2. Flip the newly-added toggle switch to the Run* position.
3. Clock the '469, using "INC" button on the keypad.

The data bus should show you the next address in the low nybble, and a sequence of letters in the high nybble. See if you can make out a message in the MSD 'arbitrary' data. (The messages in this lab are pretty silly; but then we have only the letters A through F to work with.) You can eliminate the distracting "next address" information from the display by *blanking* the right-hand display: temporarily wire its pin 4 to +5 v. (No resistor needed.)

16-4 State Machine II: External Control Used to Select Sequence

You can select between two alternative 'programs,' while these programs are running, by using an external line to take control of one of the RAM's address bits. Here is a diagram in the generalized "state machine" form, designed to remind you what we are doing:

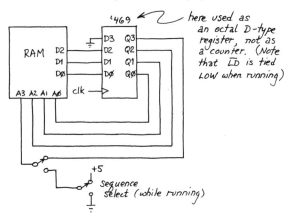

Figure L16.9: State Machine: External Control Added

In order to allow ourselves to drive A3, we insert a toggle switch: in one position, it feeds a constant *low* to the d_3 input; this we use when putting in the values that 'program' the state machine. In the other position, the switch allows us to determine the level of D3, using a second switch, here called "sequence select."

So, we use the switches as follows:

- while putting in values: *setup/go* switch in *setup* position (d_3 constant, at 0V);
- while running the state machine:

    - *setup/go* switch in *go* position
    - *sequence select* switch in either position: can be changed at any time, to select one sequence or the other.

The 'State Machine' circuit, so far

Here, to help you get some context for the small changes you have been making to your circuit, is the way your 'state machine' circuit looks: three bits define the changing 'present state' (out of the '469'); the RAM in turn delivers three bits of data, d0-d2, defining the machine's 'next state.' The other bits are constant, except for four data bits out of the RAM, which we use to display silly messages.

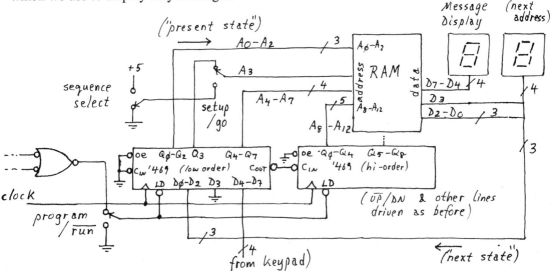

Figure L16.10: State machine circuit: '469s operating as D registers define *present state*; RAM delivers *next state*

Now add the following "message" at memory locations 8 through F (hex) (or write any other message that entertains you). The earlier data is shown below as well; that data, which you entered in the preceding stage, should stay in place. (Again, you can, of course, enter any other message there as well: the point is only to illustrate that you can now select one sequence or another, using the manual switch that now drives A3.)

Address (hexadecimal)				Data (arbitrary):	next address	
A15-A13	A12-A8	A7-A4		A3-A0	D7-D4	D3-D0
(MESSAGE ONE; entered earlier)						
0	0	0		0	0	3
				1	D	5
				2	A	1
				3	B	2
				4	0	7
				5	F	4
				6	D	3
				7	0	6
(MESSAGE TWO; enter now)						
				8	D	F
				9	A	8
				A	E	9
				B	D	A
				C	F	D
				D	E	B
				E	A	C
				F	C	E

Figure L16.11: Two-Sequence RAM Table: External Line Selects Sequence One vs. Two

Trying this circuit, you may find it convenient to use a slow square wave to clock the register (say, 2 Hz), while you confirm that you can select a sequence using the A3 switch. (If you like a story to make sense of these nonsense messages, imagine a power struggle between the proprietor of the Dead Cafe (Gulch, Ariz.) and his embittered former cook. The two of them wrestle over the Message One*/Two switch that controls the sequence of letters to be displayed on the giant single-digit display that stands out in front, blinking in the desert night. Does that make this circuit more exciting?)

We will stop here, hoping that this addition of *one* input that can steer the state machine is enough to let you imagine the way the scheme could be extended: 2 inputs would allow 4 sequences or 'tricks;' and 16 lines (as in the 68008, due to arrive in Lab 18) would allow 64K tricks. So, the 68008 is a big state machine; in principle, given enough time, perhaps you now could design such a microprocessor. But don't worry: that is *not* the next lab exercise.

Ch. 8: Review: Important Topics

1. **Some General Ideas**

    a. Boolean manipulation

        i. deMorgan's theorem the most important non-obvious notion

        ii. "active-low" and "assertion-level symbol" cause confusion, but useful and easy if you relax and don't fight it

    b. Hardware gate properties

        i. Inputs: TTL versus CMOS: impedance; what floated inputs do

        ii. Outputs: TTL versus CMOS; *3-state*; open-collector (rare, except in comparators, where it is standard)

2. **Devices**

    a. Combinational

        i. Gates: just a few standard functions

            1. AND, OR, NOT, XOR and the complemented-output versions

        ii. Larger building blocks

            1. multiplexer
            2. decoder/encoder
            3. adder

    b. Sequential

        i. Flops: D; S-R; J-K

        ii. Flops combined:

            1. shift register
            2. counters

                a. ripple
                b. synchronous

                    i. loadable
                    ii. up/down

            3. State machines
    A generalization of counter; can be made with gating + flops, or with ROM + flops

        iii. one-shots

            1. RC-timed (usual)
            2. digitally-timed (good if a clock is at hand)

Ch. 8.: Jargon and Terms

Active-high /-low   Defines the level (high or low) in which a signal is "True." We avoid the term "True" because many people associate "True" with "High," and that is an association we must break.

Assert   said of a signal. This is a strange word, chosen for its strangeness, to give it a neutrality that the phrase 'Make true' would lack. We can say, with equal propriety, "Assert dtack*," and "assert Ready." In the first case it means 'take it low;' in the second case it means 'take it high' (because the signal names reveal that the first signal is *active low*, the second *active high*)

Assertion-level symbol
   This is a very strange phrase, meant to express a notion rather hard to express: it is a logic symbol, of the *two* that deMorgan teaches us always are equivalent, *chosen to show active levels*. So, a gate that AND's lows and drives a pin called EN*, should be drawn as an AND shape with bubbles at its inputs and output, even though its conventional name may be "OR."

Asynchronous   contrasted, of course, with synchronous: asynchronous devices do not share a common clock, and may have *no* clock at all. Most flip-flops use an *asynchronous* clear function, also called "jam-type."

Clear   force output(s) to zero. Applied only to sequential devices. Same as *Reset*.

Combinational   (≡ "combinatorial") of logic: output is a function of present inputs, not past (except for some brief propagation delay); contrasted with *sequential* (see below).

Decoder   combinational circuit that takes in a binary number and asserts one of its outputs defined by that input number. (Other decoders are possible; this is the one you will see in the Labs.) For example, a '138 is a 3-to-8 decoder that takes in a 3-bit binary number on its 3 select lines; it responds by asserting one of its 8 outputs.

Edge-triggered   describes flip-flop or counter *clock* behavior: the clock responds to a transition, not to a level. By far the most widely used scheme.

Enable   The meaning varies with context, but generally it means 'Bring a chip to life:' let a counter count; let a decoder assert one of its outputs; turn on a 3-state buffer; and so on.

Float   (said of input, or of a 3-stated output): driven neither high nor low.

Flop   lazy baby talk for "flip-flop," which is baby talk for "bistable multivibrator." But babies in this case express themselves better than old-fashioned physicists or engineers, who were inclined to say all those ugly syllables. (Flop is not to be confused, incidentally, with the acronym FLOP, a piece of computer jargon that stands for "Floating Point Operation".)

Hold time   time *after* a clock edge (or other timing signal) during which data inputs must be held stable. On new designs hold time normally is *zero*.

Jam clear, jam load   asynchronous clear, load: the sort that does not wait for a clock.

Latch	strictly, a *transparent latch* (see below), but often loosely used to mean *register* (edge-triggered) as well.
Load	a counter function by which the counter flops are loaded as if they were elements of a simple *D-flop* register.
Multiplexer	("mux"): combinational circuit that passes one out of n inputs to the single output; uses a binary number code input on *select* line to determine which input is so routed. For example, a 4:1 mux uses 2 select lines to route one its four inputs to the output.
One-shot	circuit that delivers one pulse in response to an input signal (usually called *trigger*), which may be a level or an edge. Most *one-shots* are timed by an RC circuit; some are timed by a clock signal, instead.
Preset	Force output(s) of a sequential circuit high. Same as *Set*.
Propagation delay	Time for signal to pass through a device. Timed from crossing of logic threshold at input to crossing of logic threshold at output.
Register	set of D flip-flops, always edge-triggered
Reset	same as *Clear*: means force output(s) to zero
Ripple counter	simple but annoying counter type: asynchronous: the Q of one flop drives the clock of the next. Slow to settle, and obliged to show many false transient states.
Sequential	of logic: circuit whose output depends on past as well as (in some cases) on present inputs. Example: a J-K flip-flop fed a 11 combination: its next state depends on both its history and the JK values. Contrasted with *combinational* (above).
Set	same as *Preset*: means set output high (said only of flops and other sequential devices)
Setup time	time *before* clock during which data inputs must be held stable. Always *non-zero*. Worst-case number for a 74HC: about 20 ns.
Shift register	a set of D flops connected Q-of-one to D-of-the-next. Often used for conversion between parallel and serial data forms.
State	condition of a sequential circuit, defined by the levels on its flip-flop outputs ("Q"'s). A counter's state, for example, is simply the combination of values on its Q's.
State machine	short for "finite state machine"; sometimes, "FSM." A generalized description for any sequential circuit, since any steps through a determined sequence of states. Usually reserved for the machines that run through non-standard sequences. A toggle flip-flop and a ÷10 counter are FSM's, but never are so-called. A microprocessor executing its microcode is a state machine.
Synchronous	sharing a common clock signal (syn-chron means same time). Synchronous functions (LD*, for example, or CLR*) "wait for the clock."
Three-state	describes a gate *output* capable of turning *off* instead of driving a High or Low logic level. Same as *tri-state*.
Tri-state	a trade name for *3-state*: Registered trademark of National Semiconductor.

CHAPTER 9

Class 17: Analog <—> Digital; Phase-Locked Loop

Topics:

old:
- digital vs analog: digital is clean as a whistle;
    - but crossing the boundary raises *analog* issues

new:
- How to convert between Analog and Digital forms
    - getting enough information:
        how fine to make the slices, in amplitude and in time. How many bits for a give resolution: error = *1/2* LSB.
    - conversion methods:
        * D/A: scaled current sources; "R-2R" ladder
        * A/D methods
            - flash
            - integrating
            - feedback: binary search
- Phase-locked loop
    - the scheme is simple
    - ...but the changes of units make the stability problem hard to grasp

An old topic:

analog vs digital

The task of converting from either form to the other can be a challenging task in *analog* electronics: from analog to digital (ADC, A/D) one has to classify an input (usually a voltage), putting it into the correct digital category; often, one must do that fast. Running in the other direction (DAC, D/A) is easier, conceptually; but it is hard to do it precisely (good to many bits), and the digital processing will have added extraneous *non*-information, like 'steppiness' in the sampled and recovered waveform; this junk must be cleaned away.

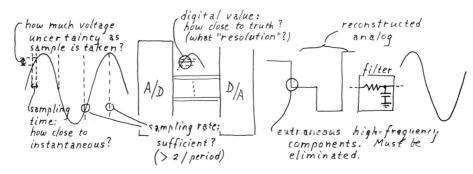

Figure N17.1: An old issue revisited: analog to digital to analog again: includes delicate analog tasks

That cleaning job is a challenging analog task (the low-pass filter that cleans up the output of a digital-audio CD's output must do such a task). So, you can congratulate yourself on having taken the trouble to study some analog electronics, not falling for the notion that electronics is now simply digital.

New topics:

How fine to slice

If you mean to convert a sine wave to digital form, then back again, you run at once into the problem *how much information* do you need to carry into the digital domain. You must be content with a limited number of slices in the vertical direction: amplitude; and in the horizontal direction: sampling rate.

- *Slices across the vertical axis: amplitude resolution*

    This is a pretty straightforward issue. You decide, as usual, how big an error is tolerable. In this course, all the converters you build or use in the lab will be *8-bit* devices; thus we'll settle for 256 voltage slices, each worth between 10 and 16 mV. Commercial digital audio cuts each of our slices into another 256 slices (*16*-bit resolution: 1 part in 64K). Our breadboarded circuits would bury any such pretended resolution in noise, as you know if you recall how big the usual fuzz is on your scope screen when you turn the gain all the way up.

Size of *error* versus size of *slice*

If a voltage range is sliced into n little slices, each labeled with a digital value, the value will be correct to *1/2* the size of the slice: "1/2 LSB," in the jargon. Perhaps this is obvious to you. If not, try this example:

> *Example:*
>
> How many bits are required for 0.01% resolution?

0.01% is one part in 10,000. If a converter spans a 5-volt input range, for example, we mean that when we give a digital answer, like "4.411V," we expect to be wrong by no more than 0.5 mV (1 part in 10,000 of the *full-scale* range: 0.5 mV/5V).

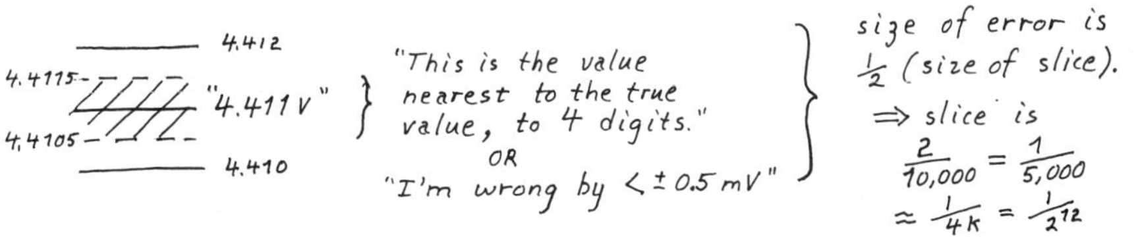

Figure N17.2: What I mean when I claim to be right to 0.01%: I say the answer is 4.411V; I could be wrong by ±0.5 mV

How many bits does the converter need? We can tolerate slices that are *two* parts in 10,000 wide, or 1/5k. 12 bits give 4k slices, and give an *error* of 1/8k: 0.00012, which we can round off to 1 part in 10,000.

- *Slices across the horizontal axis: sampling rate*

    Here the result is surprising, to anyone who has not considered the question before: Nyquist pointed out the curious fact that a shade more than *two* samples per period of a sine wave will carry a *full* description of that sine wave; thus one can reconstruct the original with just that pair of samples. Here's the idea, illustrated with the converter you will see in the micro labs, picking up about *three* samples per period. The filtered output looks like the original (which is not shown):

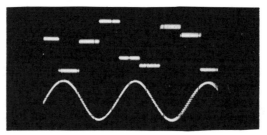

Figure N17.3: Confirming Nyquist's surprising claim: sine wave reconstructed from about 3 samples/period

You will get a chance to reproduce this effect in Lab 21, when you control sampling rate with your microcomputer. Evidently, the sampling rate required depends upon the frequency of the analog waveform; if the waveform includes many frequency components, then it is the *highest* that concerns us; the lower frequencies are easier to catch. Commercial digital audio again provides a useful case in point: that system samples at about 44kHz ("10% oversampling") in order to store and recover signals up to 20kHz.

Effect of an inadequate sampling rate: "aliasing"

If you don't do what Nyquist told you to do, you get into trouble. You don't just fail to get what you meant to get; you get nonsense: sampling at too low a rate, you get a sum and difference frequency; the difference signal shows up as a fake—"aliased"—low frequency. Here's an illustration:

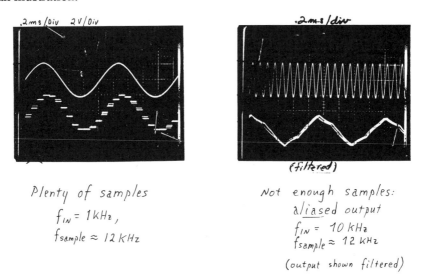

Figure N17.4: Adequate versus inadequate sampling rate; the inadequate rate produces a fake, *aliased* output signal

To protect against aliasing, you need a good low-pass filter ahead of the A/D input, to make sure that the dangerous frequencies never are fed to the converter.

An informal argument for Nyquist

Nyquist's claim becomes less surprising when one recalls that the Fourier series for a square wave would be reduced to a sine at the fundamental frequency, if we stripped away all the the higher-frequency components. But the low-pass that filters a reconstructed sine does just such a stripping job.

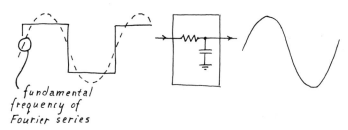

Figure N17.5: A casual confirmation of Nyquist's claim: just a pair of points can define a square wave; a good low-pass could be set to keep only the fundamental, a sine

Enough generalities. Let's get back to hardware: let's look at some ways to carry out the conversion, in both directions.

A. Analog <—> Digital Conversion Methods

1) *D to A (or "DAC")*

A D/A is conceptually simple

All we need is a way to *sum* a binary-scaled set of voltages or currents: an input *1* at the LSB should generate output *V*, the next bit by itself should generate *2V*; both together should generate *3V*; and so on.

You might use an op amp current-summing circuit, and resistors of values R, 2R,...:

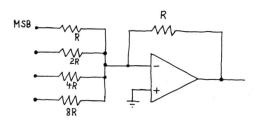

Figure N17.6: A possible D/A—but hard to fabricate

This method rarely is used, because it is difficult to fabricate many values of resistor to the required precision.

R-2R Ladder

Text sec. 9.15,
pp. 614-17,
figs. 9.46, 9.47

Instead, one can get the same result—a binary scaling of currents—with an ingenious circuit called an *R-2R ladder*. The ladder's virtue is that it requires just 2 resistor values, not *n* for an *n*-bit converter. The left-hand circuit, below, uses such a ladder to source current into the summing junction of an op amp, again; the right-hand circuit (the schematic of an IC D/A) omits the op amp, so the output is a *current*. It uses just *two* resistor values, rather than n values for an n-bit converter, as in the op amp scheme.

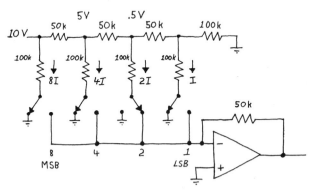

Figure N17.7: D/A using current switching

Here is a way to arrange this scaling of currents in the several sources:

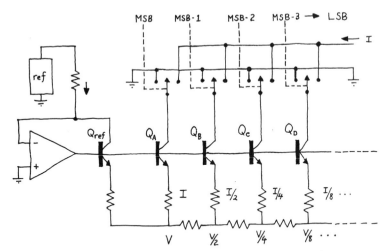

Figure N17.8: Scaled current sources using R-2R network

The ladder is formed out of units that look like the one on the left, below.

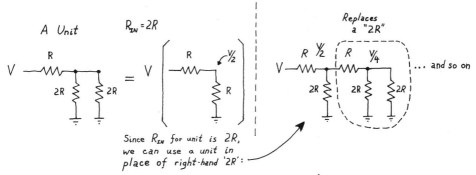

Figure N17.9: How an R-2R ladder divides down a voltage

Since the unit looks like $2R$, seen from its left side, we can plug in another unit in place of its right-hand $2R$—and so on, forever. At the right-hand end of each R, the voltage is down to 1/2 what it was at the left end of the resistor. (In both the D/A circuits diagrammed above, those diminishing voltages are applied across equal valued resistors so as to generate scaled *currents*.)

...But a good D/A is hard to build

Though it is easy to draw a diagram for a D/A, it is difficult to make a D/A that works with good resolution: that is, responds properly to a large number of bits. In this course we will use modest *8-bit* D/A's, which imply resolution down to 40mV, given a 10V-volt range. Our converters use a somewhat smaller range, pushing the analog value of a bit down to 10mV. But that we can manage, even in our breadboarded circuits.

2) A to D (or "ADC")

Converting in this direction allows lots of room for cleverness. Here, first, is a summary of the leading methods:

Method:
parallel	feedback	integrating
(≡ "flash")	binary-search	dual-slope

Characteristics:
Fast,	middling speed	slow
few bits	good resolution	best resolution
		can cancel line noise

Typical applications:
storage scopes	general purpose	DVM
radar processing		

Figure N17.10: A spectrum of A/D methods

1. *Open-loop*

a) *Flash*

Text sec. 9.20, pp. 621-22;
compare fig. 9.49

We will start with the method that is conceptually simplest, though most difficult to fabricate: flash (or parallel) conversion.

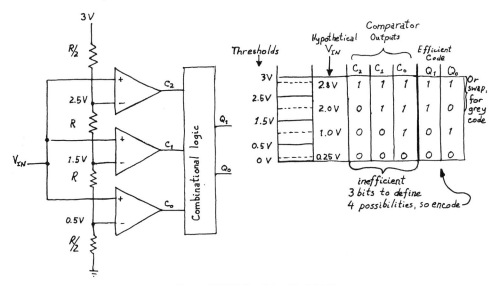

Figure N17.11: Parallel or "flash" A/D

This takes a lot of analog circuitry for a modest number of bits. A 9-bit converter needs 511 comparators. That may not sound like a lot of parts, if you are accustomed to reading about million-transistor digital parts, and 4-megabit memories. But the fact that flash converters now are limited to about 9 or 10 bits underlines the point that *precise, fast analog* circuitry is hard to build, whereas fast digital circuitry is relatively easy: each of the outputs of a 50 ns memory, say, needs to decide only whether the stored data is closer to High or to Low. Each comparator in a 9-bit, 5-volt-range converter, in contrast, needs to make a decision good to about 10 mV; and it needs to do fast—in around 50-odd ns, to run at the same speed as the memory. The comparators' job evidently is much the harder of the two.

b) *Dual-Slope*

Text sec. 9.21,
pp. 626-27

We looked a *single-slope* converter in the class notes on counter applications, and it appears in Text sec. 9.20, Fig. 9.54. The single-slope measured the time a ramp took to reach V_{in}. The dual slope converter is similar: it lets V_{in} determine the size of a current that feeds a capacitor, for a fixed time; when the capacitor voltage has ramped up to a reference, a fixed current discharges the capacitor, and a counter measures the time for the discharge:

ext p. 627, fig. 9.56;
compare p. 626, fig. 9.55

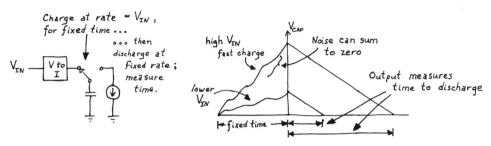

Figure N17.12: Dual slope converter: charge for fixed time, at rate determined by V_{in}; measure time to discharge; integration can cancel periodic noise

The dual slope differs, however, in one important respect from the single-slope converter and from all the other converters we describe in these notes: it is an *integrating* converter, so that variations above and below the average input voltage level tend to get swallowed up, not recorded as the true level. The swallowing-up is perfect if the converter takes in an *integral* number of periods of the interfering signal (the integral of that signal then is zero).

So, you can get the right answer in the presence of periodic interference. In particular, if you expect periodic interference at 60Hz and its harmonics (as usually you can expect) you can let a conversion include an integral number of cycles of this signal; then the bumps above and below the average value will cancel. This is the method used in the lab's DVM's.

2. Closed-loop: *Successive-approximation & tracking*

ext sec. 9.20,
p. 622-24

A closed-loop converter, and in particular, one that uses a binary search, beats either flash or dual-slope as a general purpose converter. The flash costs too much, and provides only mediocre resolution; the dual-slope is slow. The closed-loop converter works like a discrete-step op amp follower: it makes a digital "estimate;" converts that to its *analog* equivalent, and feeds that voltage back; a single comparator decides whether that estimate is too high or too low (relative to the analog input, of course); the comparator output tells the digital estimator which direction it ought to go as it forms its next, improved estimate. Here's a block diagram illustrating the notion.

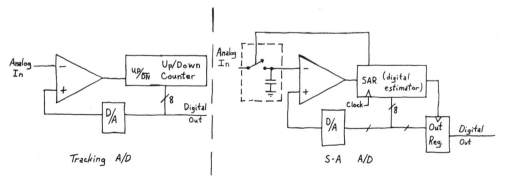

Figure N17.13: Two closed-loop A/D converters: tracking and binary-search

The tracking converter is dopey: it forms and improves its estimate simply by counting up or down. The successive-approximation estimator (SAR) is clever: it does a binary search, starting always at the *midpoint* of the range, and asking the comparator which way to go next. As it proceeds, it goes always to the midpoint of the remaining range. That sounds complicated; in fact it is easy for a machine that is in its nature already binary. The comparator tells the binary-search device *bit-by-bit* whether the most recent estimate was too high or too low, relative to the analog input.

Because of this gradual way of composing its answer, such a converter needs an output register: its digital estimates look funny, until the process is complete. In addition, the analog input should be passed through a *sample-and-hold*, which freezes the input to the converter during the conversion process. We leave this S&H off, in the lab; we lose certainty in sampling time. That translates, however, to considerable *errors*, for any processing of the signal—including the simplest: playback through a D/A and low-pass filter—assumes *periodic* sampling.

Here are a couple of contrasting waveforms showing the outputs of the *feedback* D/A's that show the analog estimates of the two converters. (These are the analog equivalents to the digital estimates, but don't forget that it is the *digital* values that we are after, not these analog equivalents! These are, after all, A/<u>D</u>'s.)

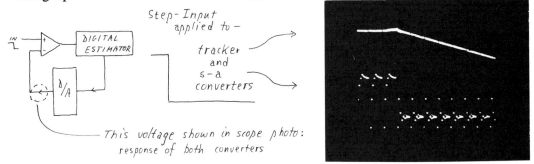

Figure N17.14: Tracker versus binary-search or successive-approximation converter: trying to follow a step input

We present the *tracker* only in order to make the binary-search estimator look good. There are very few applications for which anyone would consider a tracker.

C. A/D Characteristics: speed & resolution

As we suggested earlier in these notes, it is hard to make precise analog parts: hard to make, say, a D/A good to many bits. It's easy to make digital parts, and easy to string them together to handle many bits. It would be easy, for example, to build an up/down counter or SAR of 100 bits. So, it is the *analog* parts—the D/A and comparator—that limit the resolution of the closed-loop converter. They also limit the converter's *speed* :

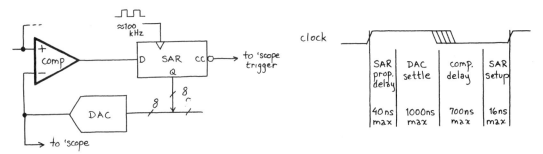

Figure N17.15: Closed-loop A/D: clock period must allow time for all delays in loop

Recapitulation: the important A/D, D/A characteristics

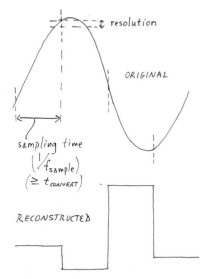

Figure N17.16: Summary of A/D & sampling characteristics

resolution
this is determined by the number of bits: n bits slice into 2^n levels, give error of $1/2(2^n)$

speed
puts a ceiling on sampling rate, and sampling rate, in turn, puts a ceiling on the analog bandwidth (= max. input frequency) that the converter can capture. The converter you build in lab can convert in about 1.6 µs; ==> max sampling rate 600kHz; ==> max analog input frequency of nearly 300kHz

Typically steppy D/A output: reconstructed waveform:
steppiness isn't bad; can be smoothed away by a good filter, so long as we have enough information about the input waveform: a shade over two samples/period at the minimum

A binary search: example

Here, for anyone not already convinced, is a demonstration of the strength of the *binary search* strategy, used by the "successive-approximation register" (SAR) you use in Lab 17. If someone chooses a number in the range 0 through 255 and tells you whether your SAR-like estimates are *too low*, you can get to the answer in 8 guesses. If you write out the binary equivalents of those guesses, you can see how the SAR forms its answer, bit by bit, proceeding from MSB to LSB.

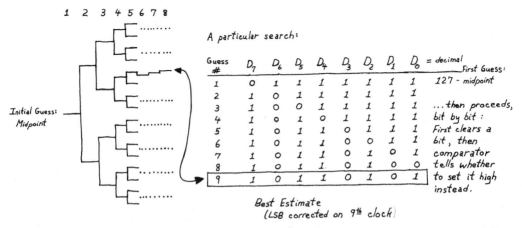

Figure N17.17: Binary search: an example, and how a digital SAR would form the estimate, from MSB through LSB

B. Phase-Locked Loop

text sec. 9.27-9.31

A PLL uses feedback to produce a replica of an input *frequency*. It is a lot like an op amp circuit; the difference is that it amplifies not the *voltage* difference between the inputs, but the frequency or *phase* difference (once the frequencies match, as they do when the circuit is "locked," the remaining error is only a *phase* difference). The phase *error* signal is applied to a VCO in a sense that tends to diminish the phase error toward zero.

This sounds familiar, doesn't it: sounds like a discussion of an op amp circuit.

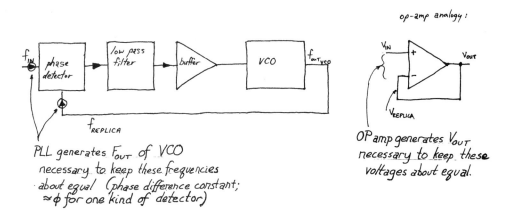

Figure N17.18: Phase-locked loop

The scheme is simple. The only difficulty in making a PLL work lies in designing the *loop filter*. We'll postpone that issue for a few minutes.

Phase Detectors

text sec. 9.27, p. 644

The simplest phase detector, as the Text says, is just an XOR gate—a gate that detects *inequality* between its inputs. (We assume digital inputs; sine wave inputs require phase detectors that are different, though conceptually equivalent.)

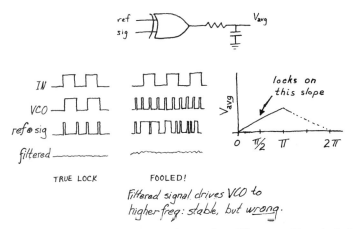

Figure N17.19: XOR phase detector: *needs* some phase difference, and it can be fooled

The CMOS 4046 that you will meet in the lab includes a fancier phase detector—a state machine that generates correction signals during the time when one square wave's rising edge has led the other:

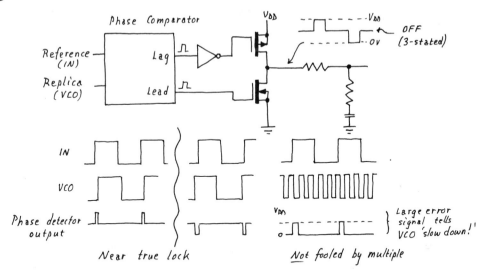

Figure N17.20: 4046 edge-sensitive phase detector; it's smarter than XOR, and cannot be fooled

It uses a 3-state output, and when it's happy—when phase difference goes to zero—it shuts off its 3-state. At that time, the filter capacitor just holds its voltage, acting like a sample-and-hold, really, not like a filter at all.

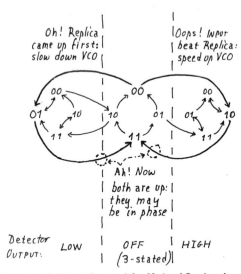

Figure N17.21: 4046 state diagram (after National Semiconductor data)

The detector becomes unhappy when it finds one waveform rising before the other. The state machine notices which rising edge comes first. If A—the input signal—comes up before B, for example, the detector says to itself, "Oh, B is slow. Need to crank up the VCO a bit." So it steps into the right-hand block, where it turns on the upper transistor, squirting some charge into the filter capacitor.

Then B comes up; A and B both are up, so this detector (like the XOR) sees equality, and goes back to the middle block, saying, "Everything looks OK for the moment; A and B are behaving the same way." In the middle block, the output stage is *off* (*3-stated*); the capacitor just holds its last voltage.

This pattern tends to force the VCO up, bringing B high a little earlier. But that reduces the time the detector spends in the right-hand block. The machine spends more and more of its time in the middle block—the "Everything's all right" block. This is where it lives when the loop is locked. You will see this on the 4046, when you watch an LED driven by a 4046 pin that means "The output is 3-stated." That is equivalent to "I don't see much phase difference," or simply, "The loop is locked."

Here is the simple circuit that behaves this way:

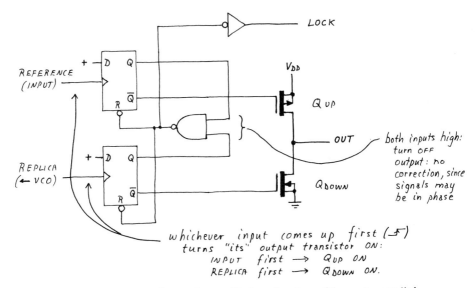

Figure N17.22: 4046 phase detector: just two flip-flops; the winner of the race turns on 'its' output transistor till both inputs go high

Applications

The PLL lets you generate a signal that is a precise multiple of an input frequency. That you will do in the lab:

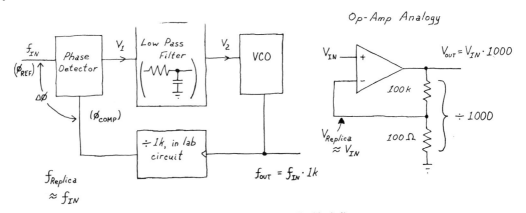

Figure N17.23. Frequency multiplier block diagram

This trick is useful for canceling line noise by averaging (as in integrating A/D/s), or for generating video clocks that eliminate line noise from the display. You will use this circuit in Lab 21, where we need a clock that is a multiple of the *sampling rate* (16 X). We want to be able to vary the sampling rate, so we use the PLL to gives us this multiple. We will use this multiple to set f_{3dB} of a low-pass filter controlled by its clock rate. Here's the scheme:

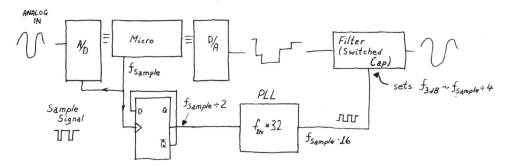

Figure N17.24: How we will use the PLL in Lab 21: low-pass adjusts automatically to knock out frequencies we cannot handle, as sampling rate varies

The PLL also can demodulate an FM signal: just watch the input to the VCO (which is the filtered *error* signal, describing variations in input frequency).

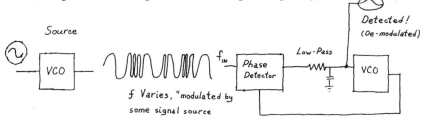

Figure N17.25: PLL demodulating an FM signal

PLL stability: designing the low-pass filter

The PLL is vulnerable to oscillations because of an implicit 90° shift within the loop; an 'integration' performed by the VCO; a simple low-pass filter could impose a total shift of 180° at some frequency, and such a shift could produce oscillations—an endless, restless hunting for *lock*. This is essentially the same as the hazard you recall from our earlier discussions of op amp stability and compensation. The second resistor in the low-pass is designed to eliminate this hazard by pushing the low-pass phase shift toward zero as frequency rises:

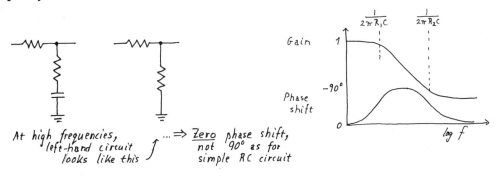

Figure N17.26: "Lead-lag" RC filter: phase shift is limited; goes toward zero at high frequencies

The point that the VCO imposes an integration is hard to get a grip on. Here's a longer paraphrase of the Text's informal argument:

Text sec. 9.29, p. 647

> Imagine that the loop is locked; phase-difference signal out of the phase detector is zero (this works for the kind of phase detector you see in the lab); VCO runs where it should, at a constant frequency equal to f_{in}.
>
> Now apply a ΔV to the VCO input. In response to this step of voltage, the VCO moves to a new frequency. Imagine that it is only a little different from what it was. The *replica* signal—the VCO output—now runs slightly faster; it begins to slide past the *input* signal: the phase difference increases linearly with time. Hence the proposition that the VCO integrates with respect to time (recall the op amp integrator, if you have trouble at this point: remember, its output ramps for a step in?). QED, more or less.

And here's a diagram to say this in summary form. Notice that the variable is treated as *phase*; that's easier to handle than frequency (its time-derivative). It's hard to follow it around the loop because sometimes it rides disguised as a voltage proportional to phase (to phase-*difference*, more exactly); it does that at the input to the VCO.

Text sec. 9.29, p.647, fig. 9.72

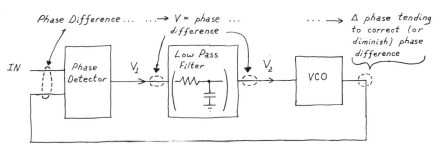

Figure N17.27: Trying to keep track of what goes around the loop: difficult!

Lab 17: Analog <—> Digital; Phase-locked Loop

> **Reading:** 9.03, 9.05 – end.
>
> *Specific Advice*:
>
> - 9.03: interfacing between families: {TTL levels (includes NMOS)} <—> {5-volt CMOS} is the issue important to us.
> - 9.08: driving external loads: you should learn how to drive a) a load 'returned to' a voltage higher than 5 volts, and b) a high-current load.
> - 9.09: NMOS: note similarity of NMOS and TTL output levels.
> - 9.10: opto-electronics: only *scan* this encyclopedic discussion
> - 9.11: noise: this is important; make sure you understand why 'bypass' capacitors are necessary
> - 9.12-9.14: driving buses and cables: same advice as for optoelectronics section.
> - 9.25-9.26: A<—>D conversion: this is the core of the chapter, for our purposes. Some of this material is chatty, aimed at someone shopping for a converter (e.g., 9.22); other parts you should read closely (e.g., 9.15). We think you'll recognize the difference, here. The most important A/D conversion method, for our purposes, is the binary-search or 'successive-approximation' (pp. 622-24), and this is the method you will apply in Lab 17. Read quickly through text pages re other methods for D/A, A/D conversion.
> - 9.27 to end: concentrate on phase-locked loop; make sure you understand the scheme, and some of the applications. Note the issue of stability discussed in 9.29, but we don't expect you to digest all this until the moment comes when you actually need to design the low-pass filter for the loop. The scheme is fundamentally simple, apart from design of the filter.
> - Only *scan* (and enjoy) 9.32 (pseudo-random noise generators) to end.
>
> **Problems:** embedded problems.

This lab presents two devices, both partially digital, that have in common the use of feedback to generate an output related in a useful way to an input signal. The first circuit, an A/D converter, uses feedback to generate the digital equivalent to an analog input *voltage*. The second circuit, a phase-locked loop, uses feedback to generate a signal matched in *frequency* to the input signal—or to some multiple of that frequency.

A. Analog to Digital Converter

The A/D conversion method used in this lab, "successive approximation", or "binary search", is probably the most widely used. It provides a good compromise between speed and low cost. It substitutes some cleverness for the "brute force" (that is, large amounts of analog circuitry) used in the fastest method, "flash" or parallel conversion.

Note that the converter you will build today with four chips normally would be fabricated on a single chip. We build it up, using an essentially obsolete "successive-approximation register" (SAR), so that you will be able to watch the conversion process. In an integrated converter the approximation process is harder to observe because the successive analog "estimates" are not brought out to any pin. We also omit, for simplicity, the *sample-and-hold* that normally is included.

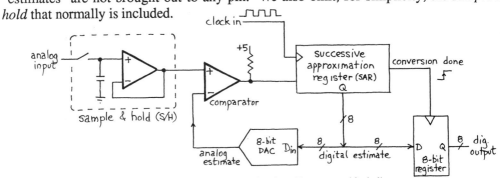

Figure L17.1: Successive-approximation a/d converter: block diagram

17-1 D/A Converter

The process of converting digital to analog is easier and less interesting than converting in the other direction. In the successive-approximation A/D, as in any closed-loop A/D method, a D/A is necessary to complete the feedback loop: it provides the analog translation of the digital "estimate," allowing correction and improvement of that digital estimate. As a first step in construction of the A/D we will wire up a D/A.

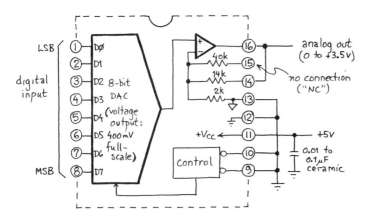

This D/A, the Analog Devices AD558, integrates on one chip not only the D/A, but also an output amplifier and an input latch. The latch is of the "transparent" rather than edge-triggered type, and we will ignore it in this lab, holding the latch in its transparent mode throughout.

Figure L17.2: AD558 D/A

Checkout:

Confirm that the 558 is working, by controlling its two MSBs and its LSB with a DIP switch or simply with wires plugged into ground or +5. Hold the other five input lines low. (You may find a ribbon cable convenient; you will need to feed all 8 lines in a few minutes, in any case.)

Note the relation between switch settings and V_{out}. Full-scale V_{out} range should be about 0 to 3.8 v. What "weight" (in output voltage swing) should D_7 carry? D_6?

Note the weight of the LSB. Here, if you look at the output with a scope you are likely to find noise of an amplitude comparable to that of the signal. If the noise is at a very high frequency, however, it may be quite harmless. How does your DVM appear to treat this noise? How do you expect the comparator to treat it when you insert the D/A into the A/D loop?

Digital In:				V_{out}
D_7	D_6	D_5-D_1	D_0	
0	0	0	0 / 1	(LSB weight)
0	1	0	0	
1	0	0	0	(MSB weight)

Figure L17.3: D/A Checkout: voltage weights for particular input bits

17-2 A/D: Watching the Conversion Process

When you are satisfied that the D/A is working, add the rest of the S-A converter circuit: comparator and SAR:

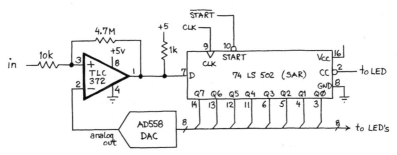

Figure L17.4: Successive-approximation a/d converter: slow-motion checkout: layout

Connect the 8 D/A inputs to the 8 breadboard LEDs, so that you can watch the estimating process.

17-2.1 Slow Motion Checkout

Use a debounced switch for Clock and another switch (which may bounce) for Start* (S*). Watch Conversion Complete* (CC*) on an LED.

- You will have to wire up this 9th LED, with a current-limiting resistor. To choose the resistor value, consider the following points: an LS-TTL output can sink 8 mA, sufficient to light an ordinary LED dimly; 2 mA is enough for a high-efficiency LED; the LED drops about 2 volts when lit.

Note: The behavior of S* may strike you as odd:

- First, the SAR is "fully synchronous": asserting S* by itself has no effect. The SAR ignores S* unless you clock the SAR while asserting S*.
- Second, "Start*" is poorly named. It should be called something like "Initialize*", because conversion will not proceed beyond the initial guess until you *release* S*.

Ground the analog input,[1] and walk the device through a conversion cycle in slow motion. Watch the *digital* estimates on the 8 LEDs, and the analog equivalents, the "analog estimates" on DVM.

As you clock slowly through a conversion cycle, the D/A output (showing the analog equivalent of the SAR's digital estimates) should home in on the correct answer, 0 volts. Do you recognize the *binary search* pattern in these successive estimates? Isn't the *digital* pattern that generates these estimates pleasantly simple? If you get a digital *01* rather than *00*, make sure that you have grounded the input close to the converter; then look closely at ground and +5 lines, watching for noise.

17-2.2 Operation at Normal Speed

Now make three changes:

1. Connect "Conversion Complete*" (CC*) to "Start*" (S*). This lets the converter tell itself to start a new conversion cycle as soon as it finishes carrying out a conversion. (Disconnect the pushbutton that was driving S*, of course.)

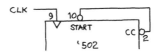

Figure L17.5: CC* wired to drive start*: sar starts self

2. Feed the converter from a pot (2.5k or less: (Why?)), rather than from ground:

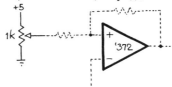

Figure L17.6: Variable DC analog input

3. Clock the converter with a TTL-output oscillator, rather than with the pushbutton. (The oscillator built into your breadboard is convenient.) Let f_{clock} = 100kHz.

---

[1]. Note that the "analog input" is not the non-inverting terminal of the comparator, but one end of the 10k resistor; that resistor is necessary to maintain hysteresis.

Watch the D/A output and A/D input on the scope. (If you want a rock-steady picture, trigger the scope on CC*.) Vary the pot setting (analog input voltage), and confirm that the converter homes in on the input value.

17-2.3 Displaying Full Search "Tree"

You have watched the converter put together its best estimate, homing on the input value. If you feed the converter *all possible* input values, you can get a pleasing display that shows the converter trying out every branch of its estimate "tree."

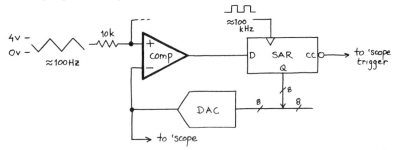

Figure L17.7: Displaying binary search "tree"

Feed the converter an analog input signal from a *second* function generator: a triangle wave spanning the converter's full input range (0 to about 4 v). Let $f_{\text{analog signal}}$ be about 100 Hz. Trigger the scope on CC*.

If necessary, tinker with the frequency and amplitude of the input waveform, until you achieve a display of the entire binary search tree. This can be a lovely display (you may even notice the "quaking aspen" effect); you are privileged to see the binary search in such vivid form—while it remains to most other people only an abstraction of computer science. (The search tree produced by a group of students a few years ago appears as fig. 9.52 on p. 623, sec. 9.20.)

17-2.4 Speed Limit

The A/D completes an 8-bit conversion in nine clock cycles. Evidently, the faster you can clock it, the faster it can convert. The faster it can convert, the higher the frequency of the input waveform that you can capture (you need a shade more than two samples per period, in theory; in practice you may want to take 3 to 5 samples).

How fast can you clock your converter? Here's what must happen between clock edges:

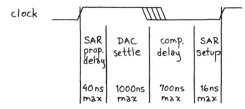

Figure L17.8: A/D speed limit: what must be accomplished within one clock period

These numbers suggest a maximum clock frequency of a little less than 600kHz.

Feed the converter a DC level and gradually crank up the clock rate as you watch the analog estimates on the scope. This time, use the *external* function generator as clock source (the breadboard's top frequency, 100kHz, is too low). At some clock frequency you will recognize breakdown: the final estimate will change, because the clock period will no longer be allowing time for all levels to settle. Chances are, this will happen at a frequency well above the worst-case value of 600kHz (above 1MHz, in most cases we have seen.)

17-3 Completing the A/D: Latching the Digital Output

Up to this point we have been looking at the converter's feedback D/A output. Do not let our attention to this *analog* signal distract you from the perhaps-obvious fact that the feedback-D/A output is not the *converter's* output. We use an A/D in order to get a *digital* output, of course, and on a practical IC A/D, as we observed at the start of these lab notes, the analog estimate is not even brought out to any pin.

We now return our attention to the normal subject of interest: the A/D's *digital* output.

Output Register

An IC A/D normally includes a register to save its output (and incidentally, in the age of microcomputers, such a register routinely includes 3-state outputs, for easy connection to a computer's data bus).

Now we will add an 8-bit register of D flip-flops to complete the A/D. We need to provide a clock pulse, properly timed, that will catch the converter's best estimate and hold it till the next one is ready. Timing concerns make this task more delicate than it looks at first glance.

CC* ("Conversion Complete*") certainly *sounds* like the right signal. It turns out that it is not—not quite. The trailing edge (rise) of CC* comes too late; but the other edge (which, inverted, could provide a rising edge) comes too early.

At the beginning (fall) of CC*, the SAR is putting out its initial estimate for the LSB; it has not yet corrected it (set it high), if such correction is necessary. Thus you would lose the LSB data, getting a constant Low, if you somehow used the start of CC* to latch the output.

At the end (rise) of CC*, the SAR is already presenting the first guess of its next cycle (0111 1111).[2]

Figure L17.9: CC* timing

What we need is a pulse that ends well away from both edges of CC*:

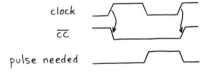

Figure L17.10: Output register clock needed

A single gate can do the job.

Add this gate and feed its output to the clock of the output register (74HC574). Let the register's outputs drive the breadboard's eight LEDs.

---

2. Why? Because the rise of CC* and the initial guess both come in response to the SAR clock, and CC* happens to come up a little *later* (see 74LS502 data sheet: t_{PLH} is slower than t_{PHL}. How's that for fine print?)

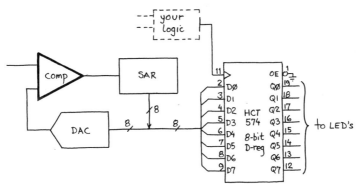

Figure L17.11: Output register added to s-a a/d converter (output-register clocking left open for your addition)

Watch the converter's digital output and confirm that it follows the analog input applied from the potentiometer. It may be indecisive, by 1 bit. Is this indecision avoidable?

17-4 Phase-Locked Loop: Frequency Multiplier

> *Note*: you should build the PLL on your private breadboard—either the micro board or the single strip holding the amplifier you built back in Lab 10.

You will have a chance to apply this PLL, if you like, in Lab 21. There we need to generate a multiple of the frequency at which the computer samples a waveform (*16 ×* the sampling rate) ; we will use that multiple to regulate f_{3dB} of a low-pass filter. This adjustable filter is of the switched-capacitor type, like the one you built in Lab 12.

Thus we will make the filter follow our sampling rate (so as to clear away the spurious high-frequency elements in the D/A's *steppy* output waveform). This future application for the PLL explains some elements of its design: particularly, its very wide capture- and lock-range. We will first apply the loop, however, as if we were using it to generate a multiple of the line frequency, 60 Hz. This is the example discussed in detail in the Text (sec. 9.29), and the circuit below is the one designed in the Text discussion, except that we have altered the VCO component values to permit a wider range of operation.

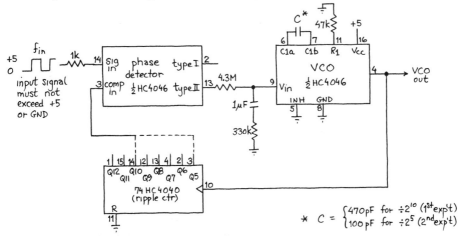

Figure L17.12: Phase locked loop frequency multiplier circuit

Construct the circuit shown in the figure above. The phase detector and VCO are drawn as separate blocks, above; but note that they are *within one 4046 chip*: you do *not* need two 4046s.

Generating a multiple of line frequency: (Type II detector)

Let C_1 = about 470pF, in this stage. Take the *replica* signal from Q_{10} of the 4040 (pin 14). Set the function generator, which drives the input, to around 60Hz. Use the scope to compare this *input* signal with the synthesized *replica*. Confirm that the PLL *locks* onto the input frequency within a few seconds. Are the two waveforms in phase? The *lock* LED should light when the loop is locked (a logic *high* at pin 1 indicates that the phase detector output is *3-stated*: that is, the detector is content, seeing little reason to correct the *replica* frequency).

See if the replica follows the input as you change frequency slowly; then try teasing the PLL by changing frequency abruptly. You should be able to see a brief *hunting* process before the loop locks again. If you replace the filter's 330k resistor with 33k, you should find that the loop hunts longer: you have reduced the *damping*. When you have seen this effect, restore the 330k; it makes the circuit work better, on the whole.

What frequency should be present at the VCO output when the loop is locked? Check your prediction by looking with the scope at that point (pin 4 of the 4046). Why is this waveform jittery?

Now look at the output of the phase detector (pin 13 of the 4046). This is the type II detector described in Ch. 9 (sec. 9.27, pp. 644-45). You'll notice a string of brief positive going pulses, each decaying away exponentially when the detector reverts to its 3-stated condition (you may also see smaller negative-going pulses, to the extent that the circuit is troubled by noise).

Theory predicts that these correction pulses should vanish in the steady state. But the 10 megohm load of the scope probe you are using is discharging the filter capacitor enough to make the pulses. Interpose a 411 op amp follower between the capacitor and the probe and the positive pulses should all but vanish. They never go away entirely, because the time-difference between the rising edges of the *original* and *replica* signals does not go quite to zero, and the correction pulses have a minimum duration of a few microseconds. These pulses look insignificant when you are watching the loop locked to 60 Hz; they make the phase detector look nervous when it is locked to a higher frequency.

Type I Detector (exclusive-OR)

The 74HC4046 includes three phase detectors. The Text describes two of them, in sec. 9.27, pp. 644-45 (types I and II); the third detector on the HC4046, Type III, is simply a variation on the SR latch, with S and R driven by positive-going *edges* on the input and replica lines, respectively. We will *ignore* the Type III detector: *two* detectors seem plenty to consider on a first encounter with a PLL. But feel free to check out the Type III, if you like: its output appears at pin 15.

The Type I detector output is at pin 2, and the inputs are the same as for the type II detector; to use the Type I, simply move the wire from pin 13 to pin 2 (and then to 15 for the Type III, if you *must*!). For both Type I and Type III you should be able to see the fluctuation of the VCO frequency over the period of the input, which you can exaggerate by reducing the size of the 1μF loop filter capacitor.

If you make a sudden, large change in the input frequency, you should be able to fool the circuit into locking onto a *harmonic* of the input frequency (a multiple of the input frequency). For our purpose in Lab 21, such an error would make the circuit useless, so we will use the Type II detector. You are likely to make the same choice in most applications.

Note also the phase difference that persists between input and feedback signal in the locked state. These simple phase detectors *require* such a phase difference; this difference generates the signal that drives the VCO. If the phase difference ever goes to zero (or to π), the loop loses feedback: it can no longer correct frequency in both senses, as required. The Type II detector is altogether classier: it requires no errors to keep the loop locked, for it is able to use the capacitor as a *sample-and-hold* rather than as a conventional filter, when locked.

When you have finished looking at the behavior (and *mis*behavior) of the Type I detector, revert to the earlier circuit, using the Type II (output at pin 13).

Expanded Lock Range: × 32 rather than × 1k

Now let's set up the PLL as you will want to find it the next time you use it, in Lab 21: change the *tap* on the 4040 from Q_{10} to Q_5 (pin 3). Change the VCO capacitor (between pins 6 and 7) as well: from 470pF to 100 pF (this lets the VCO run faster: though we will generate a smaller multiple of the input frequency, we want a higher VCO range). Now you are generating a modest $32 \times f_{in}$ at the VCO output.

Over what range of input frequencies does the PLL now remain locked? The range should be wide; we need this range in order to make the sampling scheme of Lab 21 flexible. The 4046 is capable of *capture* and *lock* over a huge range: better than 1000:1. We will be content with a range between about 200Hz and 20 kHz. Capture and lock range are the *same* for the Type II detector; for the less clever detectors, lock range (the range over which the loop will *hang on* once it has locked) is larger than capture range (the range over which the loop will be able to *achieve lock*). This ability of the Type II follows from its immunity to harmonics: you can't fool the Type II.

Ch. 9: Jargon & Terms; Review

Jargon & Terms

ADC	A/D: analog-to-digital converter
aliasing	production, by a sampled-data circuit or system, of a spurious output caused by sampling at an insufficient rate (see Nyquist Theorem, below)
binary search	a search strategy in which one begins at the midpoint of the range not yet ruled out; as information about the correct answer arises, the next guess is always the midpoint of range in which the answer may lie.
capture range	of a phase-locked loop: the range of input frequencies over which the circuit can 'lock' to an input (see *lock*, below, and see secs. 9.27, 9.30).
DAC	D/A: digital-to-analog converter
dual slope converter	A/D of integrating type, slow but capable of canceling periodic noise
flash A/D	parallel A/D converter: fastest of converters, using n-1 analog comparators in parallel, to convert to $\ln_2 n$ bits.
lock	of phase-locked loop: condition in which the replica signal is held in a fixed phase-relation to the input (see sec. 9.27).
lock range	of a phase-locked loop: the range of input frequencies over which the circuit can retain lock, once locked. For some phase detector types this range is larger than the capture range.
Nyquist theorem	observes that at least two samples (we prefer to say 'more than two samples') must be taken within each period of an input sine wave in order to gather enough information to characterize the waveform fully. If the input waveform is not a sinusoid, then apply the requirement to the highest-frequency component of the waveform.
SAR	successive approximation register: the estimating block of a binary search or successive-approximation A/D converter

Ch. 9: Review: Important Topics

1. Hardware gate properties
    a. Inputs: TTL versus CMOS: impedance; what floated inputs do
    b. Outputs: TTL versus CMOS; *3-state*; open-collector (rare, except in comparators, where it is standard)

2. A <—> D conversion specifications:
    a. resolution: # of bits
    b. analog frequency limit: need > 2 samples/period of highest (Nyquist)

3. Phase-locked loop

CHAPTERS 10 & 11

Chs. 10, 11: Microcomputers and Microprocessors: Overview

As we reach computers we reach a subject that in one sense is familiar to most students; practically everyone, these days, has done some programming, and some of you no doubt are sophisticated in the subject. But in another sense computers as we view them in this course may be new even to an experienced programmer. We take a worm's-level view of the machines, concentrating on hardware, and thinking about hardware even when we do program. So, these chapters on microcomputers should confirm the digital hardware learning that you have been engaged in since Lab 13.

The computer that you build in this series of labs will be capable of running fancy programs—but you will not have time to write anything fancy for it. Our interest will stay with the intimate relation between program and hardware. You will see things happen in response to every few lines of code that you write. You will see, in gritty detail, what happens if you write faulty code; you will see what interesting misbehavior results if your hardware is even a "bit" wrong—if a high-impedance input is left floating, for example, or if a couple of data lines are interchanged.

We hope—as we said back at the start of these notes—that putting together this little machine and then setting it to work will render vivid and concrete notions that otherwise might remain abstract: *bus*, *bus contention*, *processor speed limitations*, *subroutine call*, *interrupt*, for example. You will work hard for these rewards. You may sometimes wish you had a ready-made computer to work with—so that you could simply hang peripherals onto it. Students have told us repeatedly, however, that they did not favor such an approach. They hove told us—looking back—that they found the labor worthwhile; that they would have been sorry to miss the sense of accomplishment and close acquaintance with the machine that comes from struggling up from the chip level, and then teaching the machine every line of code that it knows. We hope you will feel a similar sense of satisfaction, and of ownership. This is the part of the course when students begin to ask for extra time in the lab, and begin to show their circuits to friends—and by the end there is usually someone who feels sentimentally obliged to take a photograph of his little machine before saying goodbye to it. ('Saying goodbye' is a euphemism, actually, for 'taking to pieces;' a melancholy event a little like Hal's gradual lobotomizing in the movie 2001.

Chapter 11 speaks of the processor that you will use in the lab (the 68008), and in that sense is the more relevant of the two chapters. But it is Chapter 10 that introduces concepts (buses, interrupts, and even assembly-language, though the language happens to differ from the one your processor understands). Chapter 11 concentrates on a no-holds-barred, sophisticated application for the 68008 processor. That *signal averager* is not a circuit or program you will want to grapple with in your first days of assembly-language programming. Save it for the time when you have begun to get the notions firmly in place and are ready to appreciate some sneaky designs (like the ROM that appears on power-up, and then disappears) and some programming tricks as well.

To help you get used to assembly-language you will probably prefer the simplest programs available. The first of these appear in some of the early exercises of Chapters 10 and 11 (though some among these early problems require either cleverness or experience); another set of simple programs appear in the lab programs: Lab 19's listings, and in the worked example called 'Ten Tiny Programs.' These lab programs are worth studying not because they are ingenious, but just because they are *not*. They are simple, and you will get a chance to run them, watching in detail how they behave.

The last lab invites you to do something satisfying with the machine that you have worked hard to build. In our course, we treat this *invitation* as just that—not a requirement but an opportunity for enthusiasts. The last lab, Lab 23, may appear to be scheduled for a single day, but it need not be so arranged, and should be allowed extra time if some students are ready to dig into a project.

A good project can serve at least a couple of important purposes: it can let you solidify your understanding of programming (not our primary concern), and of the relation between your programs and the hardware on which they run (that relation *is* one of our central concerns in this part of the course). If the project includes some analog elements, or at least some use of A/D or D/A, even better: in that case it may serve to draw together some of the themes of the entire course, providing a 'review' to put it one way; providing coherence, to put it a little more grandly.

In any event we hope you emerge from your labor with the microcomputer feeling that the microprocessor is neither mysterious nor daunting, but just another device in your bag of tricks—and one that gives very impressive results for a modest expenditure of your effort. It's pretty much true that when you've seen one processor you've seen them all, so you can expect these labs to prepare you for the use of processors in general. We hope that once you have understood your own small machine you will enjoy a new feeling of power.

Class 18: μ1: IBM PC and our lab micro

Topics:

- computers generally versus our little machine: what's missing, what's extra
- turning a microprocessor into a microcomputer or controller; or adding peripherals to a working computer: *hardware*:
  how the processor tells the world what it wants to do: its *control* signals
  two cases:

    - IBM PC bus signals
    - 68008 processor signals

- a *minimal* 68008 computer/controller is very simple:
  simpler than what you will build
- …but details of a particular implementation can be subtle:
  two examples from our lab machine:

    - a timing problem resolved: fixing *busgrant**
    - taking advantage of *dtack** to provide a *single-step* function

---

A. Computers generally versus our lab micro

Our lab microcomputer includes many elements common to all standard computers. It also includes some quirky peculiarities. We hope that you will come to tell which elements fit into each category, because we aim to teach you about computers generally, even as we ask you to wire this sometimes odd machine.

Any old computer

text sec. 10.01, fig. 10.1

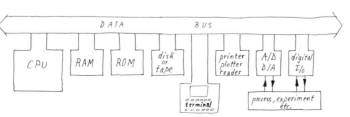

Figure N18.1: General computer block diagram

Our lab computer

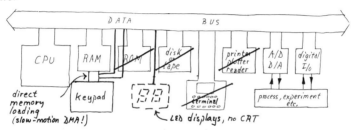

Figure N18.2: Lab microcomputer

You can see what's missing from our machine, and what's extra.

In grittiest detail your lab computer may look dismaying (see the full schematic that appears after Lab 23). It's not hard, though, taken a section at a time, and as you build the machine we will ask you to wire just one small subsection, and then test that. It's also comforting to recall that already you have wired more than half of the full circuit.

B. Control Signals: how a processor says what it's doing

The processor (CPU) determines what happens in the computer, and when. A designer need only combine these signals appropriately to make things happen as the processor expects.

Any processor

Any processor needs to announce, through its output signals, at least the following information:

Timing: one or more signals says when other signals are valid and may be acted upon. If a separate line, often called "strobe."

Direction of transfer: toward the CPU or away. Always called Read/Write. Note CPU-centric convention: "Read," e.g., is a transfer of data into the CPU.

Category: Memory vs I/O:
    Some processors use a signal so-called; others do not, and instead allow one to use address lines to make this distinction (see below).

Which device/location within a category:
    The lines that say this are called "address."

Particular processors/computers

Text sec. 10.13, 11.4

Here are the two examples important to us:

- the IBM PC, which you may well want to hook something to, and which is discussed in some detail in Chapter 10;
- the 68008, a good processor (sibling of the 680xx processors used in the Macintosh computers, for example), and the processor you will use in your lab computer[1].

---

[1]. This comparison—computer versus microprocessor—is a bit of an apples-versus-oranges operation, or at least tangerines versus oranges. For, some signals on the PC bus are not bare processor signals, but instead are formed by logic that combines CPU signals much as our small logic clusters will put together 68008 signals in our little computer. For example, the Intel processors within the IBM PC use an (I/O)/M* signal (other Intel processors use I/O*/M!), but that line—so similar to the simplest scheme for use with a Motorola processor, in which a single address line is assigned exactly this task—does not come out to the IBM PC bus.

	IBM PC	**68008**
Timing:	Included in signals defining category & direction	data strobe* (DS*)
Direction:	Included: e.g., IOR*, IOWR*	(R/W*)
Category:		
I/O	IOR* IOWR*	one or more address lines; designer's choice
Memory	MEMR* MEMW*	
Which location: (within a category)	20 address lines 10 address lines on I/O operations	20 address lines (it is easy but not necessary to dedicate one address line to distinguishing I/O vs memory)

Figure N18.3: Control signals used by IBM PC bus versus 68008 processor: similar

Examples: hooking up peripherals to a microcomputer

We will start by trying a simple decoding task with gates—putting together address and control signals to determine when to turn on memory and when to turn on a single output port. Then we will look at two fuller examples, one for an IBM PC and one for a 68008.

Warmup: decoding control signals for a minimal 68008 controller

Encouraging thought: we don't need to worry about every CPU pin

The 68008 has a lot more pins than any other chip you have met in this course. We can make this thicket of pins less intimidating, however, if we group the pins by category:

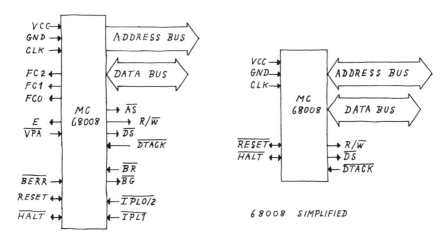

Figure N18.4: 68008 pins, grouped by function

We make things simpler still in our lab computer by ignoring many of these lines; in a minimal computer we could ignore still more, and the shrunken processor sketched on the right, above, is meant to show how simple things begin to look. You will find such a barebones machine described in a Worked Example.

A simple example

Just to make sure you get the idea, try putting together a memory enable and an I/O enable. Assume that address line A_{15} defines the boundary between memory and I/O space. We want to generate two signals: MemEnable* and Out*. Yes, it's as straightforward as it appears:

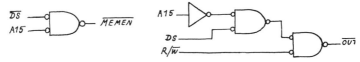

Figure N18.5: Putting together a few 68008 control signals

Two fuller examples: IBM & 68008

Here are two examples from the Text. They show ports defined for an IBM PC and then a 68008.

Text sec. 10.13

Input Port: IBM PC

First, an input port for the IBM PC. This uses *complete* address decoding (that is, it looks at all *10* address lines that the PC uses for I/O.

Text sec. 10.07, p. 689,
figs. 10.9, 10.10

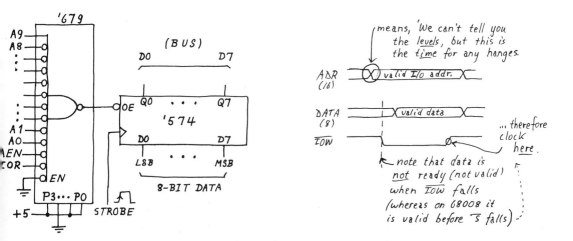

Figure N18.6: IBM PC input port, and I/O timing

AEN is a signal asserted only during the very exceptional case of a DMA (direct memory access) operation; you can think of it as equivalent to a Normal* signal (*low* ==> normal, note). Your lab computer uses a very similar signal to condition the enabling of both memory and I/O devices: it requires that two signals called "function codes," fc_1 and fc_0, shall not both be high; that 11 would mark another exceptional event—as exceptional[2] as DMA for the PC—an *interrupt acknowledge*; we won't discuss interrupts until we reach Lab 22.)

The '679 is a comparator—of an especially clever design: it looks at 12 lines; it compares them to 12 reference lines. That much is standard in a digital comparator. The nice trick here is the saving of pins: instead of using 12 lines to define the reference value, the '679 uses just *4* pins (labeled P3-P0) to *encode* that value: these four lines determine *how many of the 12 reference values are zero*. For example,

P_3- P_0	$A_1\ A_2 \ldots A_{12}$
0 0 0 0	1 1 1 ... 1
0 0 0 1	0 1 1 ... 1
0 0 1 0	0 0 1 ... 1

Neat?

---

2. We're speaking relatively: there may be many interrupts per second, but there are few relative to the number of instructions executed.

Input, Output: 68008

Here is a similar circuit, vaguer concerning address decoding. The logic block takes in all 20 address lines, but in a small computer you would not look at all *20* address lines in order to decide whether or not to turn on this device. What is gained, what lost by looking at few than all address lines?

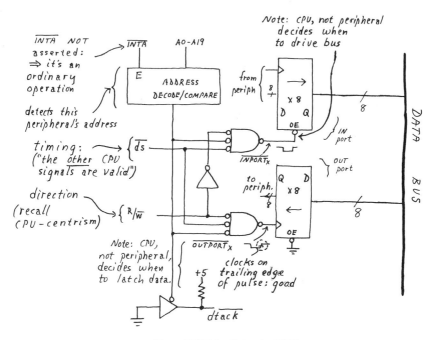

Figure N18.7: In, Out ports: 68008

Driving dtack*
Text sec. 11.4, 11.05, p. 765

In the circuit above, the 3-state driving the signal called "dtack*" requires some explanation. Dtack* can be treated more crudely. In today's lab we simply tie dtack* low, most of the time; we do that to keep things simple. The scheme shown above asserts dtack* when this peripheral detects its address. That assertion of dtack* tells the processor "go ahead; everything is all right; the transaction worked" ("dtack" = "data transfer acknowledge"). A design like this one requires that *every* peripheral or memory attached to the computer also drive dtack* when addressed; it also requires additional hardware that will keep the machine from getting hung up if dtack* does not come back. The Text describes such a "timeout" arrangement in its full controller design (see text reference above).

And here is how the 68008's timing looks on a read or write (either I/O or memory):

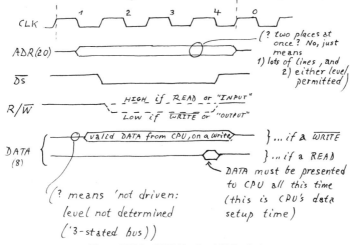

Figure N18.8: 68008 Read and Write timing

These I/O port examples are instructive but not usual in their reliance on *gates*. An address *decoder* is more efficient, and in Lab 19 you will meet such a chip and apply it in precisely this setting: as I/O decoder, defining ports.

d. Some details of the lab computer

Here are a couple of circuit fragments hard to understand at first glance. They illustrate the meeting of *general* and *quirky* features in the computer.

In one sense these features are extremely quirky; you will not see them again. But they can also be likened to two computer operations of greater generality:

- the bg*/busgive* logic is necessary for **DMA**, a technique used widely for fast data transfers, like those between a disk drive and computer memory;
- the dtack* blocker mimics what a slow peripheral would do, to give itself time to respond to the quick CPU.

In both cases what is odd is that we, the humans, are in the role of peripheral, and we operate in *ultra*-slow motion; we are the ultimate in sluggish peripherals.

Fixing Busgrant*

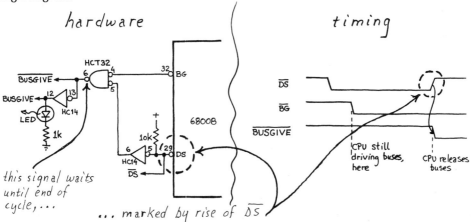

Figure N18.9: Fixing busgrant*: the properly-timed *busgive** just awaits the end of the cycle, marked by the end of ds*

The problem this circuit aims to solve is discussed in the lab notes: in a phrase, it is that bg* comes early. Not until ds* goes high can we rely on the CPU to have turned off its 3-states. Not until then, therefore, should any other 'bus master' turn his on.

Single-step

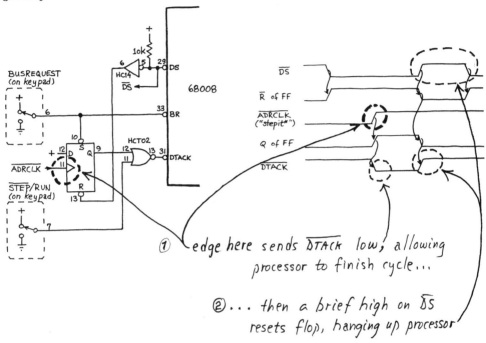

Figure N18.10: The second of two idiosyncratic circuit fragments from the lab computer: single-step logic

A capsule explanation of this circuit might say, simply, that it hangs the computer up in the middle of every cycle, by putting dtack* high; a pushbutton lets the computer finish the cycle, by driving dtack* low again; but dtack* stays low only long enough to finish the cycle. At cycle's end (marked by the rise of ds*), dtack* goes high again, and again the machine is hung up. Even this "capsule" description was a bit of a mouthful.

The value of this odd little circuit is simply that it lets us slow the processor to our sluggish human pace: we let it begin to do some operation, then check whether things are going as we expected. This form of single-step allows us to check both hardware and software, and you will rely on it throughout the micro labs.

Ch 11.: Worked Example: Minimal 68008 Controller

Compare Text sec. 11.05 a much more ornate memory enabling scheme).

> **Problem: Minimal 68008 controller**
>
> Design a microcontroller (that is, an incomplete computer), using a 68008. Give it the following characteristics:
>
> - ROM: 32K
> - RAM: 16K
> - I/O: one input, one output
>
> Ignore the function codes and all other CPU signals not shown below; drive dtack* as simply as possible.
> <u>CPU Signals</u>
>
> - DATA (8)
> - ds*
> - r/w*
> - dtack*
> - ADDRESS (20)

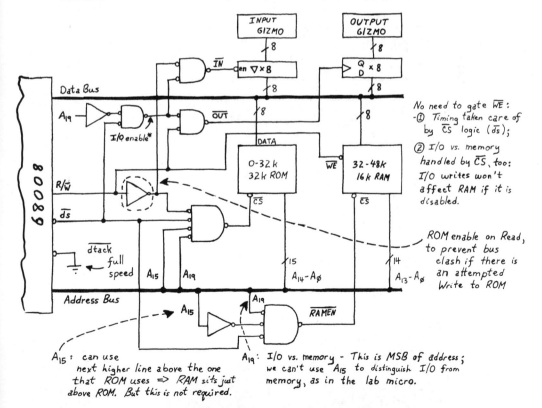

Figure XCTRLR.1: Minimal 68008 controller

The gating shown above is a bit slow. Here's another way to enable ROM and RAM:

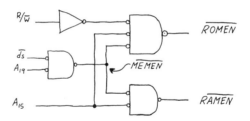

Figure XCTRLR.2: Wider gates allow shorter propagation path, but waste 'shared terms'

A very-wide gate like a PAL (see Chapter 8, sec. 8.15) could do the job still faster. But that's not necessary here. Our processor allows enough time for two and probably three layers of gates within its timing requirements. You will find such timing issues discussed in sec. 11.05.

Lab 18: µ1: Add CPU

Reading:	Chapter 10 re buses and computer organization generally; 10.01, 10.05-10.08; 10.20 (i/o hardware example). Omit the other sections, which treat programming. Chapter 11.4 (bus signals, which recalls chapter 10's similar account of the IBM signals); postpone considering *interrupts*; 11.05 re example of small 68008 computer. Problems: Embedded problems, but omit programming exercises.

Today's is another long lab. Don't get depressed if you need part of the next session to finish this work. But we do hope you can finish most of this work today. The principal reason, of course, is simply to leave yourself as much time as possible near the end of the term in which to enjoy putting your computer to use. A lesser reason is the circumstance that almost the entire reward in today's lab, for the considerable wiring task we ask you to do, appears at the very end of the lab: there you will see your machine run a program for the first time. We get a little melancholy seeing you do too much wiring without reward. Perhaps you do, too.

Today you will again make use of the loadable RAM you built a few sessions ago. This time, you will use it in the way that a memory most often is used: to hold programs and data for a processor (or "CPU"). By the end of this lab your circuit will have evolved into a sort of fetal computer: alive—capable of running a program—but helpless, and not yet useful.

It will be not yet useful, and not truly a computer, because its I/O (input/output) powers will be so severely limited: you cannot speak to the machine while it is running a program (so there is no "I" at all); it can speak to you only by showing you what is on its data bus (so there is nearly not "/O" at all). In the next lab (Micro 2) you will begin to remedy those deficiencies.

Reminders:

1. Try to test each stage as soon as you have built it, so as to keep your de-bugging tasks as simple as possible.
2. We suggest that you build from the lab notes and not from the overall circuit diagram: those notes let you build in stages, and also give you a fair chance of understanding the details of the circuitry. It would be too bad to build this circuit blindly, as if you were an uninformed, alienated technician.
3. Today you will not only add to the earlier circuit; you will also replace some temporary connections made in Lab 15. These changes are indicated on the diagram below (next page). It is easy to forget, say, that the address counter's 3-

states must no longer be enabled all the time, now that the CPU has arrived. It is easy but not permissible: busgive* must replace the ground connection.

The Big Picture:
This Lab's Changes and Additions to the Full Circuit

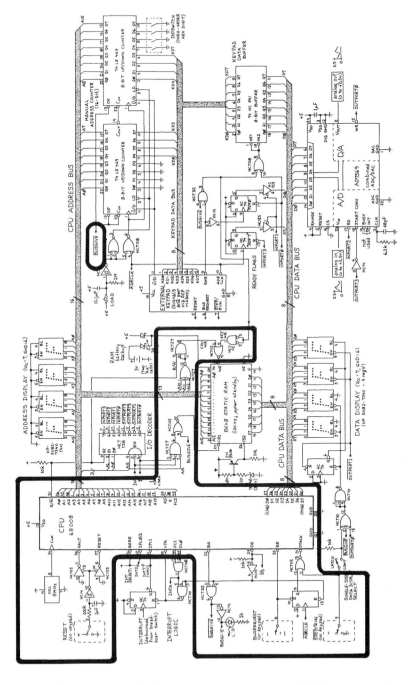

Figure L18.1: Today's changes and additions: the computer evolves from RAM & counter circuit

> *Note:* A full circuit schematic for the lab microcomputer appears in Appendix C.

We urge you *not* to wire from this big document. Students sometimes have made this mistake. Finding the full diagram, they began to wire from the left edge toward the right, more or less in the manner of a McCormick reaper. The results are poor. You find yourself building things that you do not understand. But the Big Picture (as we like to call this big picture!) helps a great deal when you are trying to find your way about in a section of the circuit already wired.

18-1 Clock

Power this integrated oscillator, and confirm that it performs as it should:

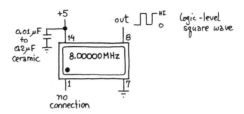

Figure L18.2: 8 MHz Crystal Oscillator

- frequency: 8 MHz.
- output levels: CMOS levels: within about 0.1 v of supplies.
- rise and fall time less than 10 ns (rise & fall defined as time between 0.8 and 2.0 volts).
- duty cycle: 50% is the target (that is, a waveform that is high for half its period); the processor will tolerate a 45% to 55% duty cycle. Treat threshold as 1.4 v.

18-2 Installing Central Processor: Preliminary Test

Here, you will install the processor, leaving many of its lines unconnected, and run a first test to see whether the processor responds properly to your request that it "grant you the buses:" that is, that it 3-state its own internal bus drivers.

Before you begin to wire the CPU—with its formidable *48* pins—we urge you to photocopy a *pinout* label from the *pinout* section of this Manual (Appendix D, p. 612), and paste it onto the CPU. The label will save you much pin-counting.

Installing CPU

Make the connections shown on the diagram below. Pins shown connected to nothing may float. Note particularly that we are not asking you to wire the address and data buses. Leave all those pins unconnected.

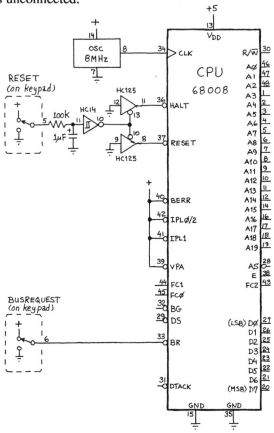

Figure L18.3: CPU Preliminary Wiring: Busgrant* Test (buses not yet wired); pinout label shown on right

Test Procedure

Check to see whether the processor will "give you the buses" in response to your request, using the following procedure:

- Apply the clock signal to the chip and assert br* (pronounced "bus request bar" or "b r bar").
- A few clock cycles after you assert br* (a twinkling, at 8 MHz!) you should get *busgrant** (*bg**). Watch the *bg** pin with logic probe or a breadboard LED.
- In rare cases, your processor may refuse to "give you the bus" until you have first asserted and then released Reset*. So, try Reset* if you ever cannot get the bus.

When the processor passes the *busgrant** test, proceed to the next step, which is to add a few gates that fix the faulty timing of *bg**.

18-3 Fixing *busgrant**

Oddly, the processor's *bg** signal is not properly timed. Therefore we must build some logic to assure proper timing of this signal that tells us we may take control of the buses. (You, the human, need to be able to drive the buses, as you know, in order to load programs into memory, as you have done in earlier labs.)

The 68008 puts out a signal called *bg**, bus-grant*, that *ought* to mean that the CPU has three-stated the buses. It does not quite mean that, however. It means, instead, that the CPU *soon* will release the buses. The logic below delays the effective bus-grant signal, which we call *Busgive**, until the buses surely are free. They come free at the end of the current cycle, marked or timed by the rise of the timing signal *ds** (data strobe bar).

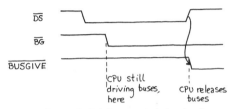

Figure L18.4: *busgrant** timing detail

We suggest you use a green LED (high efficiency) to indicate *busgive**, since the lighting of this LED means "Go ahead, use the buses." But you may, of course, use any LED color that pleases you. Here is a handful of gates to fix the timing of *bg**.

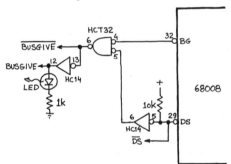

Figure L18.5: *busgrant** to *busgive** Logic

The test that earlier showed you an assertion of *bg** in response to your *br** (bus request) should now show you a lighting of your *busgive** LED.

When *busgive** is working properly, wire *busgive** to the OE* pins (3-state control) of the '469 address counters, and to $C_{in}*$ of the *less significant* '469. The latter connection freezes the counter while the CPU is running programs; we will take advantage of that effect by using *Adr-Clk* while programs are running, without disturbing the address counters. *busgive** replaces the temporary ground connection you should find at both OE* and $C_{in}*$. Soon *busgive** will drive additional points, as well, in logic that you are about to build.

Buses

Now that *busgive** is working, you should join the CPU to the address and data *buses*. The CPU drives the full 8 lines of the data bus, and drives all *16 lines* of our address bus (of which 13 lines go to the RAM; the remainder go only to the display, at present). We do not use the highest four address lines coming out of the CPU: $A_{16} - A_{19}$. Leave them unconnected.

18-4 Memory Enable Logic

Until now, the RAM has been enabled continuously. The gating shown below will enable memory only when the CPU wants it enabled.

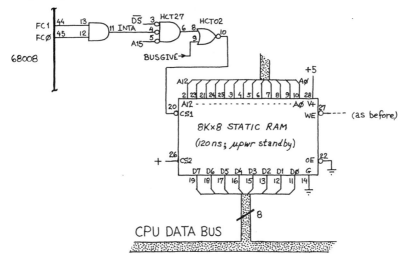

Figure L18.6: Memory enable logic

The RAM is enabled under either of two conditions: we, the humans, want to load or check RAM; or the CPU wants to read or write.

- **busgive:** One of the enabling conditions remains approximately as before: if you have "requested the buses" and the CPU has granted your request by asserting **busgive***, then the RAM will be enabled continuously, as it was in the earlier labs. *Note* that the signal used here is the *complement* of the more-widely used signal called busgive*.
- **CPU write:** But when the CPU controls the buses, as it does whenever it is running a program, the RAM is to be enabled only during a transaction between the CPU and memory (a read or a write operation). The signals that define such a transaction are of two types—

    - *timing*: the time is right;
    - *categoric*: it's this sort of transaction—between CPU and memory.

One can put this more exactly:

- *ds**:
  The RAM enabling is *timed* by this general-purpose timing signal. Ds* says, as we have noted, that the time is right; more specifically, it says that the CPU's output lines can be trusted:

    - the address is valid;
    - R/W* is valid;
    - other odd signals like the "function codes," fc_1 and fc_0, below, are valid;
    - and, during a *write*, ds* also means that the data bus is valid (on a *write*, the CPU drives the data bus).

- A_{15} *low* says that the operation is in *memory* address space, rather than the address space that will be used for *I/O* devices. We'll say a few more words, below, about this distinction between memory and I/O address space.

- *Fc_1 and fc_0*, two of the CPU's so-called "function codes," by which it describes what sort of operation it is carrying out, are included in order to rule out the case "1 1," which indicates that the CPU is responding to an interrupt. In that event, we do not want memory or I/O devices to be enabled by the logic you now are building.

We simply ground the RAM's *OE** ("output enable*") which controls the RAM's three-states, allowing the RAM to drive the bus. This does not mean that the RAM drives the bus constantly, for CS_1* also must be asserted before the RAM will drive the buses.

Here's a diagram of the way the RAM's three enables are used within the chip. OE* does not affect writes, you will notice.

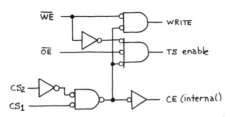

Figure L18.7: RAM enables: internal logic

Memory vs I/O Address Space

The use of A_{15} to define the boundary between memory and I/O causes the first 32K bytes to be treated as memory (and all are occupied by our single 8k RAM, and its 'ghosts'—see Text sec. 1105, p. 766). The next higher 32K bytes, where A_{15} is high, are treated as I/O space. This is wildly extravagant, of course: in fact, we are going to use only *four* of those 32K locations for I/O devices. But we have plenty of address space to play with (1 M addresses). (In fact our design is still more extravagant than this : we assign a full *half-Meg* locations to I/O! Do you see how?)

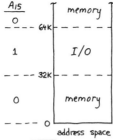

Figure L18.8: Address-space allocation: memory versus I/O

18-5 Memory Write Logic

The simple scheme we used earlier, whereby the RAM's WE* pin was tied to Kwr* (from the keypad) now should be replaced by the more complex scheme shown below.

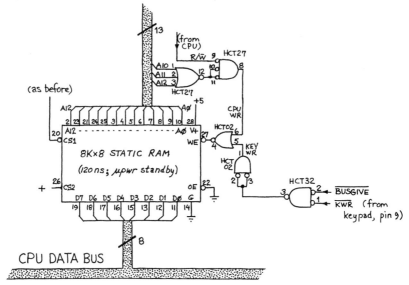

Figure L18.9: Memory write logic

Evidently there are two possible sources of a memory *write** signal, just as there are two possible bases for RAM enabling:

- a manual write (when we humans have the buses);
- a CPU write, when it is running a program.

Most of the scheme is straightforward, but two of the circuit elements call for some explanation:

- *CPU write:*
  Write-protect: the CPU is not allowed to assert the RAM's WE* pin unless A_{12} or A_{11} or A_{10} is high: that means that the CPU can never write into the bottom 1k of memory.

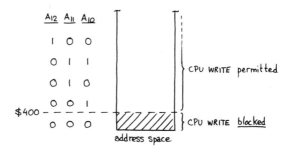

Figure L18.10: Write-protected memory region

This arrangement is designed to protect your programs and also the CPU's "exception vectors:" addresses that tell the CPU where to go in case of reset, interrupt, etc. Even a program that goes crazy will be unable to overwrite your work. (The principal hazard to your programs is a runaway stack—a repeated write operation by the CPU.)

Testing write-protect

You can test write-protect by going to the border between protected and unprotected RAM, and faking a CPU write. Here's the procedure:

- Take the bus, as usual (get "busgive*");
- Apply the highest *protected* address to the RAM: $3FF (an easy way to get there: load address $400, then decrement);
- Manually *ground* the pin connected to the CPU's R/W* line; this temporary grounding mimics or fakes a CPU write, and causes no harm: the CPU shut off its drive of the R/W* line (*3-stated* it) upon giving you the bus;
- Watch the RAM's WE* pin with a logic probe. It should *not* go low.
- Now *increment* the address, so that you now are applying the first *unprotected* address.
- Confirm that the RAM's WE* pin now does what it should.

When you are satisfied that the gating works as it should, be sure to *disconnect the temporary ground on the CPU's R/W* line*.

18-6 Single-Step: Dtack* Blocker

The CPU demands that it receive an acknowledgment from any device it tries to read from or write to: that acknowledgment is called *dtack** ("data transfer acknowledge;" one can read this as "d-tack bar").

This scheme is intended to adjust the CPU's response time to the varying speeds of the several devices to which it is attached. Most of the time, when the computer is running, we simply hold *dtack** low, since your computer includes no slow memories or peripherals.

We do, however, take advantage of *dtack** to provide a *single-step* facility. When we choose to, we can use *dtack** to freeze the processor in the midst of each of its successive "cycles" (usually, just memory accesses). We get *dtack** to do this for us by adding some logic that locks *dtack** high (inactive) until we hit a pushbutton.

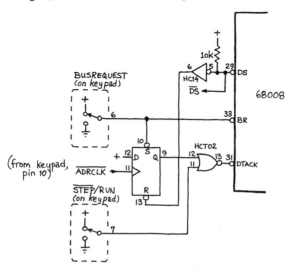

Figure L18.11: Single-step logic (*dtack**-block)

end, or rise, of ds*: at that time the signal called ds** clears the flop) *dtack** is disasserted again, hanging up the processor until we again hit the pushbutton.

By this means we provide a Single-Step function: a sort of freeze-frame slow motion. (The 'frame' here frozen is a single *cycle*, a unit smaller than a single *instruction*. For each instruction requires at least two cycles or trips to memory, and often requires four or six cycles.)

For the *Step*/*Run* switch, use the third of the slide switches on your keypad.

Because the clock labeled "StepIt" is driven by the keypad's Adr-Clk, you can speed your way through instructions at a run, using the Repeat key. Adr-Clk will not affect the address counter while you are stepping, because the counter is frozen except during *busgive**: you tied *busgive** to Cin* of the lower '469 a little earlier in this lab.

*Br** asserts *dtack** in order to allow the processor to respond to the request. (As far as the Q output is concerned, this Set* or Preset* overrides the Reset*, incidentally, in case both are asserted.)

In a few moments you will try out your single-step function. Here's a diagram to explain what's going on:

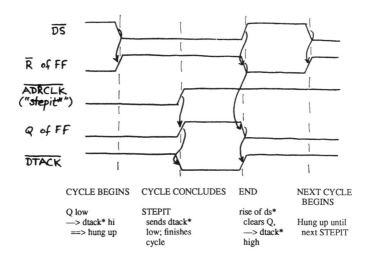

Figure L18.12: Single-Step Logic (*dtack**-block): Timing Diagram

18-7 Test Program

The following simple program will let you confirm that your computer works, if you run the machine in single-step mode.

Before writing the program itself, you should initialize the "exception vector" that tells the CPU how it should respond to a Reset. This "vector" is a pair of 32-bit values that the processor requires:

- four bytes the processor loads into the stack pointer; these you must store at addresses 0 through 3;
- four bytes the processor uses as the address of the first *instruction*; these you must store at addresses 4 through 7. The processor loads this 32-bit value into its program counter, and in effect "jumps" to this address.

Address	(hex)Code	Mnemonic	Comment
00 00	00		; stack pointer,
01	00		; to top of memory
02	20		; plus one: $2000
03	00		
04	00		; start address
05	00		; of your program
06	01		; $100
07	00		
loop: 01 00	4E	nop	; do nothing
01	71		
02	60	bra.s "loop"	; loop forever, back by
03	FC		; 4 bytes, counting
04	AA		; from 2 bytes beyond
05	BB		; start of *bra*
			; these values are meaningless, included only so that you can spot them in the instruction "pre-fetch"

Figure L18.13: First Program: tiny loop

Startup Procedure

Enter the program above.

In order to start running the program, proceed as follows:

1. Put the Run/Step* switch into the Step* position (this should be one of the slide switches on the keypad);
2. Hold reset high [1];
3. While holding reset high, release br*;
4. Release reset.

Slow Motion Checkout

Now (with the Run/Step* switch still in Step* position) push the INC key on the keypad. That should take the processor through one cycle: one trip to memory. If you get impatient, use the Repeat key.

---

1. Yes, the "reset" switch is *active high*, as this name implies. The two CPU pins that effect a reset—Reset* and Halt*—are indeed *active-low*; but an inverter stands between them and the slide switch, which thus is active *high*.

You should see the processor pick up the first eight bytes of data, then hop to address 100H. If your machine fails to do this, check that all the CPU lines that are to be disabled are disabled (in all cases, that means *tied high*): for example, IPL_1^*, $IPL_{0/2}^*$, VPA^*.

If your computer does run properly, after hopping to 100H it should begin its looping. Notice that the machine picks up the meaningless bytes AA, BB, even though these lie *outside the loop*. It does this as part of its automatic *prefetch*: the processor always picks up the next pair of bytes from memory while it is decoding the current instruction (figuring out what to do in response). Carrying out this "prefetch," the machine is simply playing the odds: most of the time, it needs that next word ("word" = 16 bits); it loses nothing by picking it up even in a case like the present one, where it does not use that word: the processor was busy anyway during the prefetch time.

When you find that your machine knows how to loop, you are entitled to feel pleased with yourself. If your machine can loop, it can run any program.

18-8 Full Speed (optional)

Switching from Step* to Run mode should set the data displays and low-order address displays into an unreadable glow. You may want to investigate what the running program looks like on the scope. See if you can find the repetition points in the loop timing picture. Look at *ds** and A_3 as starting points: these should help you get your bearings. Compare your timing diagram with the one predicted by Motorola's timing (see Chapter 11, Fig. 11.4, and 68008 timing data in this Manual's appendix).

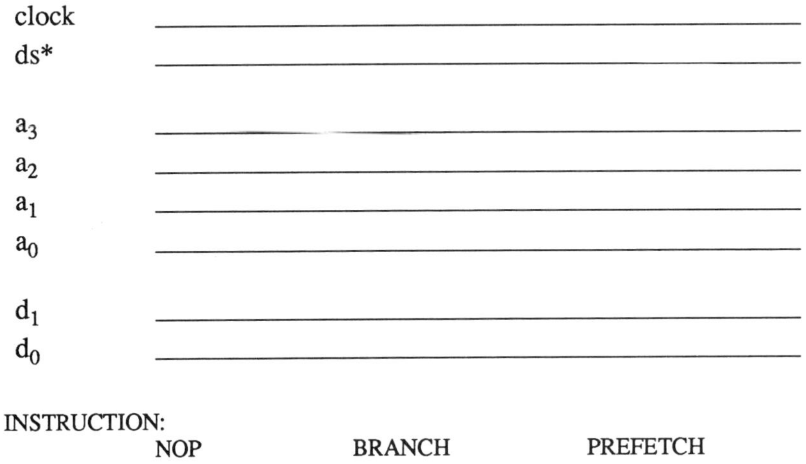

clock

ds*

a_3
a_2
a_1
a_0

d_1
d_0

INSTRUCTION:
    NOP   BRANCH   PREFETCH

Figure L18.14: Tiny Loop timing diagram

Class 19: µ2: Assembly language; Inside the CPU; I/O decoding

Topics

- *old:*

  amusing evidence that Motorola and Intel dominate the world of microprocessors: an "I/M*" pin

- *new:*

  - inside the CPU: 'watching' it execute a loop
  - *I/O decoding* with a special-purpose chip: '138 decoder
  - *applying the I/O signals:* lab computer's first Out and In ports
  - *assembly language*
    - ♦ syntax
    - ♦ the few important *addressing modes*

Old:

"Intel/Motorola-bar"

Last time, we tried to persuade you that it is not hard to adjust to the slightly different control signals used by the microprocessors of Intel (IBM PC) and Motorola (6800x). Now here is a chip to underline that point. It is a fancy peripheral designed to work with a microprocessor, and it reduces the adjustment (and thinking) process to just about nil. Look at pin 31 on this chip (a Siemens data-acquisition controller, the SDA8800):

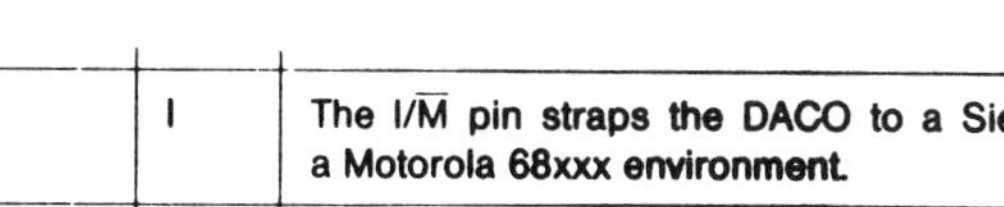

Figure N19.1: A pin called I/M*; and it means what it appears to mean. (Data used by permission of Siemens Components, Inc.)

And here you can see the consequences of controlling the I/M* pin:

I/M̄	I	The I/M̄ pin straps the DACO to a Siemens/Intel 80xxx or a Motorola 68xxx environment.
RD (R/W̄)[1]	I	Read strobe used to clock out the contents of an internal register or an HSDA memory location. (The R/W̄ line determines the direction of the data transfer).
READY (DTACK)[1]	O	Output of the internal READY or DTACK generator for all HSDA-memory accesses.

Figure N19.2: The accomodating SDA800 makes its signals fit the processor that you choose

Will you, one day soon, not need the skills **you are learning**? Will digital design be reduced to squinting at your processor to see **whether it says** "Motorola" or "Intel"?

New:

1. Inside the CPU

Most of the time we will not worry about what's going on within the CPU, except in the roughest terms. It may be useful, however, to make one effort to picture what happens within the chip as it executes a program. Hereafter, we will revert to our usual view of the CPU as another black box that works. We have made similar efforts to look within other devices: we looked into an op amp (back in Chapter 4 notes), and then into a logic gate (in Lab 13).

To keep things as simple as possible, let's consider the test program listed at the end of Lab 18. Programs don't come much more modest than this one! We'll sketch the insides of the CPU, and then try to imagine what's going on inside as it moves through the program. We'll be doing a 'freeze-frame' picture, to speak in video terms; we'll be 'single-stepping,' to speak in terms of your little machine.

First, here's a rough diagram of what's within the CPU. This is what we infer from its behavior; this is not an official portrait.

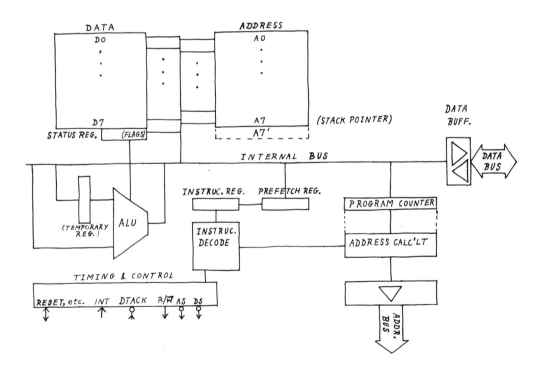

Figure N19.3: CPU innards (presumed)

The CPU In Operation

Here's that first test program, listed in Lab 19:

Address (hex)		Code	Mnemonic	Comment
00 00		00		; stack pointer,
	01	00		; to top of memory
	02	20		; plus one: $2000
	03	00		
	04	00		; start address
	05	00		; of your program
	06	01		; $100
	07	00		
loop: 01	00	4E	nop	; do nothing
	01	71		
	02	60	bra.s "loop"	; loop forever, back by 4 bytes,
	03	FC		; counting from 2 bytes beyond start of *bra*
	04	AA		; these values are meaningless,
	05	BB		; included only so that
				; you can spot them
				; in the instruction "pre-fetch"

Figure N19.4: First program: tiny loop

The program opens with 8 bytes that are not *instructions*, but instead are *constants* to be loaded into two registers, in order to set up the right initial conditions: one defines the *stack pointer*, which you will hear more of later; the other loads the *program counter*—the register that holds the address used by the CPU to pick up its current instruction. So, once the CPU has picked up that PC value—$100, here—it jumps to that location, and begins to execute whatever instructions it finds.

Let's watch this step-by-step:

First, loading constants:

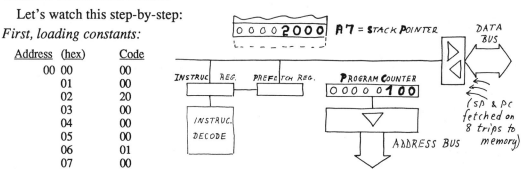

Address (hex)		Code
00 00		00
	01	00
	02	20
	03	00
	04	00
	05	00
	06	01
	07	00

Figure N19.5: CPU's first action after a Reset: pick up two 32-bit constants: Stack Pointer and Program Counter initial values

Then, Executing Instructions

Once it has picked up the constants (called the "Reset Vector"), the processor in effect jumps to the address listed: $100 ("$" means "hexadecimal," here). There it begins to do what processors normally do: picks up and executes *instructions*. It keeps picking them up in sequence, until it hits an instruction, *branch*, that tells it to depart from this usual pattern, and instead to hop back (*branch always*: bra).

Below, following the program listing, is a timing diagram showing the full program loop; below that is a scope photo showing some lines during the same loop. And to the right is a sort of frame-at-a-time picture of what the processor is doing as it goes through this loop. You can see the processor picking up a byte at a time, and using these bytes to load the *instruction register*, and then the *prefetch* register. The machine prefetches always, playing the odds. When it hits a branch, it loses, and throws away what it prefetched ("AA BB," in this case).

Address	(hex)	Code	Mnemonic
loop: 01	00	4E	nop
	01	71	
	02	60	bra.s "loop"
	03	FC	
	04	AA	junk
	05	BB	

Before the CPU can execute an instruction, it must fetch the 16-bit value and decode it.

The CPU begins by picking up 4E71—and it knows to decode it as soon as possible, not to stockpile it in the prefetch register.

Once it has picked up the 16-bit instruction, 4E71 —which turns out to mean 'Do Nothing'— the processor must figure out what the instruction means: must *decode* it. That's the job of the instruction decoder, a ROM running a little state machine. The decoding process takes quite a long time (about 0.5 μs: nearly 4 clock periods).

While it uses its brain on the decoding job, the processor keeps doing a kind of automatic, spinal-column task: it goes back to memory two more times, to fetch the next word (60FC, in the case we are considering). It puts that value into the pipeline: into the *prefetch register*.

By the time it has picked up this word it also has decoded the preceding instruction, and understands that it need do *nothing*. So it next just clocks the 60FC from the prefetch register into the instruction register, and begins the process of decoding that instruction.

While it's busy thinking out this problem, it prefetches as usual, prefetching AABB.

This time, on decoding 60FC, the processor realizes it must do something: a branch. That requires two minor computations: the processor must *sign-extend* the displacement byte (FC) to a 32-bit number which has the same value, -4_{10}: FFFF FFFC. Then it adds this negative number to the present value of the program counter (PC) (here things are a little weird: it uses $104, not $106; it ignores the prefetch, so there is a sort of "true" PC ($104) and a "pseudo" PC ($106, the value after the prefetch)).

The result is the jump destination: back four, to $100.

This addition takes a little time—and you will see this if you look at the scope photo on the next page: the processor seems to hesitate a little before asserting ds* once again. It's thinking, here.

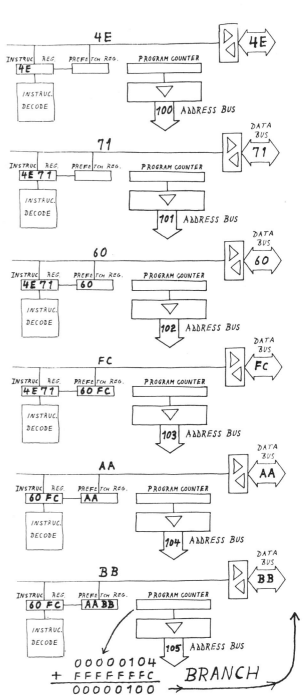

Figure N19.6: 'Freeze-frame' picture, and timing diagram: tiny loop

A snapshot of the loop

After that slow-motion account of the loop program, you can make sense of the following timing diagram. This is a scope photograph showing the tiny loop program. The eight traces are drawn by a digital multiplexer and offset-adding circuit that we put together for this purpose. (If you feel like it, you might try sketching such a circuit. All you need is a little analog circuitry and a digital mux.)

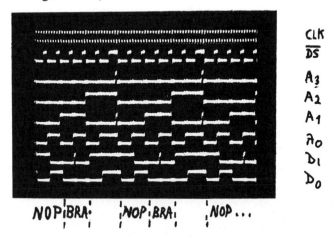

Figure N19.7: Scope photo of 68008 running 'tiny loop' program

What do you make of those brief *highs* on some of the data and address lines? Why don't they derail the processor? (*Hint*: who's driving the bus at those times? Who's reading the bus? (Trick questions.))

2. Input/Output Hardware

A memory does most of its own address decoding; a designer need do only enough address and signal decoding to turn the memory on at appropriate times.

A collection of I/O devices does not work that way. A typical I/O device occupies only one or two address locations, and one must add external address decoding in order to point to each of the I/O devices, making it respond at appropriate times.

Decoding With Gates

Suppose you want to define two input ports, two output ports, each of 8 bits. You could do this with gates. Let's use A_{15} to distinguish memory from I/O space, as in your lab computer.

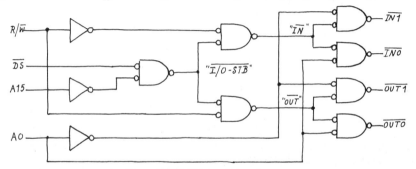

Figure N19.8: I/O decoding, using gates

But even this small example probably persuades you that you'd rather let someone integrate this function for you on a chip. Here's such a chip—a decoder: the '138. It points to one of eight outputs, and kindly provides *three* enables. Try doing the same decoding task with this chip:

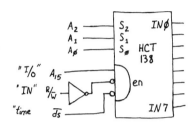

Figure N19.9: '138 decoder can put together *categorical* conditions (is it an I/O operation, etc.) and decode *device number*

Lots easier than starting with gates, isn't it?

'138 in your lab computer

Here is the lab computer's way of using the same decoder:

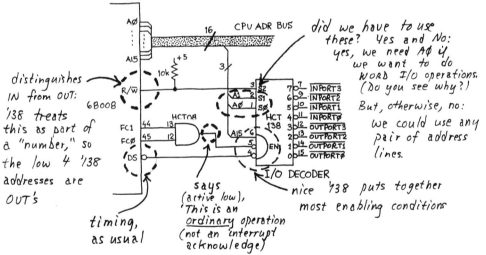

Figure N19.10: '138 as wired in lab computer: we trick it into providing both *in* and *out* ports

As soon as we have wired this decoder in the lab, we use it to drive displays and keypad buffer.

Input

Here's the keypad buffer wired as *input*:

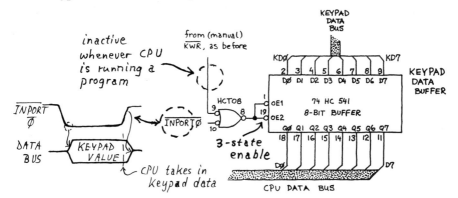

Figure N19.11: Applying I/O decoder: IN port 0

And here is a timing diagram to remind you how the CPU uses this hardware:

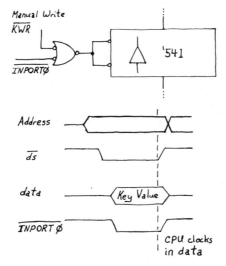

Figure N19.12: Timing for CPU's read from keypad

Output

Before we look at the lab computer's gating, let's do something simpler: use just *two* hex displays to catch a single byte output. Once we have a decoder wired, this is very easy:

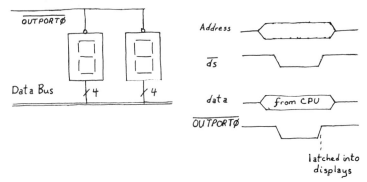

Figure N19.13: Very simple byte-display hardware: exploiting the display's built-in latch so as to make the display an output device

The pulse that comes out of the decoder updates the display; such a pulse occurs on each operation that writes to port 0 (we might call that an *out 0** operation).

And here—more ornate—is your little computer's display logic. It is complicated by the scheme that *blanks* the right-hand display except after a *word* output. Again we use the displays' built-in latches to catch outputs from the data bus; the flip-flop holds the right-hand display blanked except when its contents interest us—and that interests us only after a *word*-sized output (16 bits) from the CPU. The hardware event that signals such an output is just an output to port 1.

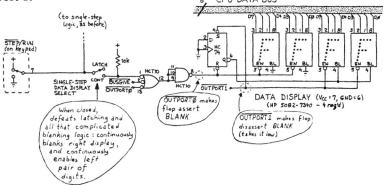

Figure N19.14: Applying I/O decoder again: display logic *complicated* by blanking of stale information on the right-hand display

If you find this hardware a little puzzling, that's not your fault: it's probably easier to design such a circuit than to follow it once it is designed.

3. Instruction Sets; Assembly Language

Addressing Modes

All processors can do roughly the same small set of operations:

- transfers of data
- arithmetic & logical operations
- jumps or branches & calls
    —the most interesting of these are conditional.

But processors differ in the orderliness and flexibility of their *addressing modes*. That is jargon for 'ways that they allow you to refer to data.' For example, you might refer to a byte stored in memory by listing its address: $1000; or by listing a register that holds its address: "A1," when you have earlier loaded $1000 into A1. Addressing modes are hard to get used to—especially this last example, which is called "indirect." Gradually, it will come clear to you. For the moment, let's just note that it is the good data-addressing scheme of the 68000/68008 that makes it especially nice to work with.

xt sec. 11.02,
.748-51

The Text lists the 12 or 14 "addressing modes" the 68000 (or 68008) makes available: a dozen ways to specify *what* an instruction will operate on. This frightening set of possibilities is enough to depress anyone new to assembly-language programming. But don't worry. The good news is that we'll find only *four* of these possible modes important:

- immediate
- direct
- absolute
- indirect

Now let's look at sample bits of code that use these four modes.

Several ways to transfer data from here to there

To get used to the syntax of assembly language, let's start with very simple examples, and work our way up to more interesting cases.

We will concentrate on the issue of *addressing modes*, and to help us do that we will use nothing but *move* instructions, in their several flavors. It turns out that the 68000 gives you great freedom in defining both source and destination for this most-used instruction: you can use most "modes" for either.

A. Immediate) addressing

Immediate data appears right in the program itself; this is in contrast to all other addressing modes, which specify data by saying where the processor should go to find it.

Example:

destination: *addressing mode:* **direct** (register 1 is an arbitrary choice among the 8 possible data registers)

move.b #100, d1

operation

size (byte)

source: "#" says 'use this value as number, not address': *immediate* mode

This is coded as follows:

```
1236
0064
```

The first *word*, the 16 bits, "1236," says what sort of operation is to be done; the second word provides the *immediate* data: notice that this data, '100_{10}'—64 in hex—appears in the program code. That's sometimes necessary, but sometimes is to be avoided, since it takes time to pick up this information from memory: two trips to get this data (requiring about 1 μs).

How the CPU decodes the MOVE instruction

Now that you have seen one example, you can understand the encoding of information in the *move* instruction's several 'fields:' sets of bits defining sub-characteristics of the instruction:

*t sec. 11.03,
'52, fig. 11.2*

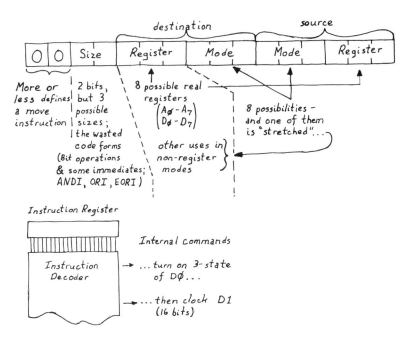

Figure N19.15: Move instruction dissected; much-simplified picture of the way the instruction decoder might implement the instruction

The lower section of the figure just above suggests the way the processor might implement a move from register to register, in hardware terms: let the source register drive the internal data bus, wait a short time, clock the destination register, whose inputs are tied to that bus. Nothing surprising here; our goal is just to feel that the processor is not mysterious, but instead a collection of familiar parts arranged to perform familiar functions.

The Text points out what a large portion of all possible codes Motorola has assigned to *moves*—nearly one quarter. That's all right, because moves are used so much. They will make up a very large percentage of all the code that you write. In the Worked Example called 'Ten Tiny Programs,' for instance, 44 of the total 86 instructions are moves (these programs, because of their extreme simplicity probably show a higher proportion of moves than is typical; but the main point stands: nearly all programs rely heavily on *moves*).

B. "Register Direct" addressing

The example above illustrates the mode Motorola calls register direct. It's easy to understand, and runs fast because the registers are right on the CPU: no trips to memory required—except two trips to "fetch" the instruction.

Other addressing modes: <u>Absolute, Indirect</u>

Suppose we want simply to transfer a byte of data from the keypad to the data display. Both these ports are at address $8000 (8000 hex). One is an input, the other an output.

C. "Absolute" addressing: specify the address explicitly

Motorola names "absolute" the mode that others call "direct" in which the instruction includes the address to be used as source or destination. Here are two examples. One uses a CPU register as a temporary storage place for the data: equivalent to a lay-over with change of planes in Chicago; the other method skips that layover. The data still is obliged to land in Chicago, but it doesn't get out or change planes.

1) Using a CPU register explicitly:

Hex Code	Mnemonic	Comment
1238	move.b $8000, d1	; get a byte from keypad
8000		

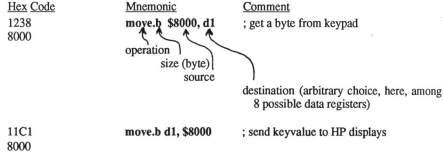

11C1	move.b d1, $8000	; send keyvalue to HP displays
8000		

Figure N19.16: "Absolute" addressing (along with data-register direct): "absolute" states source or destination address in program

The hex codes at the left remind us that the CPU has to pick up the address $8000 from the program: that requires 2 trips to memory. Soon we will see how to sidestep that tedious work.

Total trips to memory:
  to get instruction,
  including needed address info: 8
  to implement instruction: 1 to get data
                                      1 to send data
                                      _____
                                      10 trips to memory

2) Without explicit use of CPU register

Probably you can feel the silliness of using register d1 in the example we have just done. You needn't use it. How could you do the same job in a single instruction? Here's a way:

Hex Code	Mnemonic	Comment
11F8	move.b $8000, $8000	; get a byte from keypad,
8000		; send it to displays
8000		

Total trips to memory:
  to get instruction,
  including needed address info: 6
  to implement instruction: 1 to get data
                                      1 to send data
                                      _____
                                      8 trips to memory

That's progress; but we can do better.

D. "Address Register Indirect": using a pointer

movea.w #$8000, a1 ; set up a pointer (address register) to point to both keypad and data display

- operation
- size (word: required for address loads)
- source: a constant (this is the "immediate" mode that you saw just above)
- destination (arbitrary choice, here, among 7 possible address registers (8th register, A_7, is committed as stack pointer)

Now let's use the pointer:

Hex Code	Mnemonic	Comment
1291	move.b (a1), (a1)	; use pointer twice, to point to source and destination of transfer

"read from where A1 points"
"write to where A1 points"

Picking up this instruction requires just two trips to memory. Because this code is so extremely compact, it runs fast; twice as fast as the *absolute*; better than twice as fast as the very dopiest code, the first one listed above.

Total trips to memory:
- to get instruction — 2
- to implement instruction: — 2 (one input, one output)

_____
4 trips to memory

But *indirect addressing* offers another power, even greater than its superior speed: indirect addressing is not just *better*, but is *essential* where the pointer must *move*, as it must in a transfer of a block of data from one place in memory to another, or in a table search. We will look at that important case next time.

Assembly Language: Notes & Examples

Are these notes for you?

Maybe not. These pages speak at an unusually elementary level, trying to accommodate the student who has very little programming experience. We try, here, to explain a couple of basic notions that the Text treats as well: machine language versus assembly language, and addressing "modes." If these ideas are familiar to you, stop right here. These notes would bore you.

---

1. What's "assembly language"?

xt sec. 1.02

It's a weird phrase, isn't it? Regrettable, since it tends to make something simple seem arcane. *Assembly language* is the set of terms, close to English, used to describe a processor's instructions. Each instruction defines one operation: for example,

> move.b d0, d1

says, in quasi-English,

> "move a byte (really *"copy"* it) from data-register zero into data-register one."

What's this quasi-English for? It serves two purposes:

- it lets us humans talk to one another about our programs ('What's your program say?' 'It says *move.b d0, a1*.' 'Oh, that won't work....').
- more important, it lets us humans compose programs in more-or-less readable form (talking to oneself), then turn over to a computer program called an *assembler* the task of translating this quasi-English into *machine language*: the set of ones and zeros that pleases a microprocessor (and displeases a human).

Here are a couple of instructions in *assembly language*, and alongside these, their equivalents in *machine language*:

ompare Text sec. 11.03,
pecially fig. 11.2,
ssecting move; and see
ass notes 19

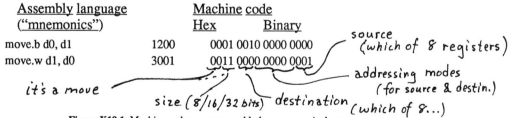

Figure X19.1: Machine code versus assembly language equivalents

Probably you need no further argument to convince you that a human *needs* assembly-language, in preference to machine code. More likely, you are yearning for a still higher level of language, like the C, Pascal or BASIC that you are more accustomed to.

The higher-level language lets you speak in terms closer to the way you are likely to want to analyze a problem. Pascal and C allow you to use terms like "If...else...." Assembly language doe not include those terms. Most operations in the higher-level language require more than one line of assembly code (if they didn't, the two languages would be equivalent; we would not need both concepts). But assembly language of course permits the "If...else" construction. You'll just need to articulate the several steps included in "If...". A function like $sin(x)$ would require many lines of assembly code. This you will discover if you make the mistake of trying to use your little computer for a purpose better suited to a big computer. You may decide to write the code that calculates $sin(x)$; but if you do you should understand that you are undertaking a project that may take you hours. In *some* cases a line of a higher-level language like C does translate to one line of assembly-language; but usually not.

2. Examples of alternative <u>addressing modes</u>?

Text sec. 10.03, pp. 680 et seq.; 11.02

Moving data: many ways to say "from here" and "to there"

<u>a) Register-to-register</u>

Let's start with this mode, which Motorola calls **register-direct**. The *registers*, you will recall, are those 32-bit-long storage places *on the CPU* (not out in memory). They behave like big banks of flip-flops, and are useful for temporary storage of information: sometimes they hold data (the so-called *data registers* usually do); sometimes they hod addresses (the so-called *address registers* usually do this job).

Text sec. 11.03, p. 752

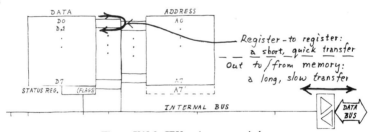

Figure X19.2: CPU registers; a reminder

The registers are handy for two reasons:

- *data registers*:
  - useful because the CPU can get at their contents (or load them) very fast: no additional trips to memory required, and trips to memory are slow (1 μs to pick up a 16-bit word).
  - a few operations are restricted to data registers. For example, an *ADD* operation must be between a data register and something else; the something else can be a memory location, but you can't add memory to memory in one operation. (See Text sec. 11.03, p.752.)

- <u>address registers:</u>
  - useful for the important technique called *indirect addressing*. We'll meet this in a few minutes, but in sum the notion is to use the address register to hold an address, then refer to that address *indirectly*: by mentioning the address register. If that sketch baffles you, hang on for a page or so. We'll return to the subject.

Now, back to the instructions: *register-to-register*. Here's an *example*:

Example:

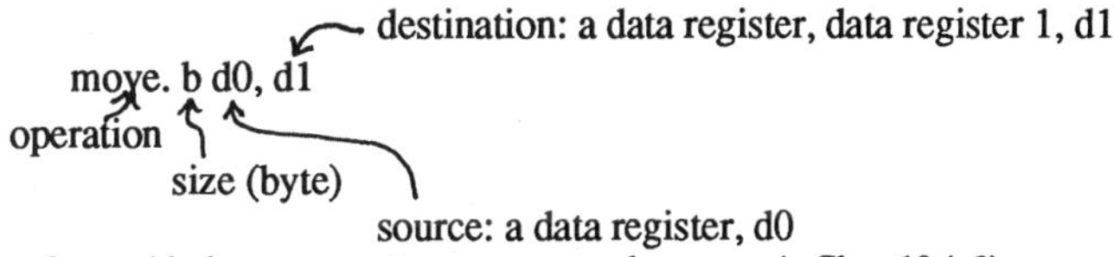

```
          destination: a data register, data register 1, d1
   move. b d0, d1
   operation
          size (byte)
                       source: a data register, d0
```

Figure X19.3: A reminder of assembly-language syntax; more examples appear in Class 19 (μ2)

In the notes for Class μ2 we look at other addressing modes; we won't repeat that here. Instead, let's make sure we agree on preliminaries: on terms, and what the listing above is trying to show.

We promised to define or explain the term, *"Addressing Modes"*
xt sec. 11.02,

This term describes the way the processor determines what to operate on.

A homely analogy

Some people hate homely and elaborate analogies. If you're one of these, or if you understand addressing pretty well, skip this silly story. But if you're struggling to get a grip on addressing modes, and not too proud to read a silly tale, read on.

Imagine that an imperious, bedridden old man—smart but unable to get things for himself—is trying to run his business from his sickroom. He keeps some records up in his room, but most are downstairs, in his numbered files. To get any information he barks orders at his faithful manservant.

Sometimes he says,

> "Get me box number D1 and box number D3, over there on the sofa. I want you to make a duplicate of what's in D1 and put it into D3 for safekeeping."

This request resembles *register-direct* addressing, by which the CPU refers to a pair of data registers, copying ("moving") data from one to another.

Sometimes he says,

> "Go downstairs to the main files; find out what's in box 195; copy what you find and bring it back to me. Put your copy into box D1 up here. Throw out the old stuff that was in box D1 when you finish."

The poor servant has to go down and up a narrow flight of stairs. This process resembles a trip to memory (or I/O device) to fetch data, and takes much longer than an operation on a box that is within the room. Motorola calls this mode *absolute* addressing. Given a choice, you—like the weary manservant—prefer to use information already present in the room (or CPU).

Sometimes the old fellow says,

> "I don't remember now what file I need, but I wrote a note yesterday showing the file number, and put it into box A1, up there on the windowsill. Take a look, and go make a copy of what you find in whatever downstairs file the note says to look at, would you, like a good man? Put your copy into box D1 up here—and, by the way, destroy whatever old stuff was in box D1."

This elaborate scheme, hard to describe, but helpful to the old curmudgeon, resembles *indirect* addressing, using an address register on the CPU to hold the address.

Okay. Enough of that. Back to the addressing modes.

Addressing Modes:

Incidentally, Motorola sometimes uses the term "effective address" to say something similar. Only examples can make these notions come clear. Let's try an example of each, and a paraphrase of what the mode means:

direct *example*: move.b d0, d1
The source or destination is specified explicitly by 'name' (in this case, both source and destination are specified in this way). Motorola reserves this term for *register* operations; conceptually, the term ought to apply equally to operations on named memory locations. Motorola calls those operations by a different term, 'absolute.'

absolute *example*: move.b $8000, $8000
Source or destination (here, both) are specified by naming its *address*. In this example, the $8000 could be a memory location or an I/O device; in your lab computer, the first $8000, used as an *input*, is the keypad port; the second $8000, used as an output, is the data display.

immediate *example*: move.w #$8000, a0
Source, here, is a constant, a value that appears in the program itself. Notice that this mode says something simpler than what *absolute* says: the processor needn't rush off to location $8000; it just puts the *value* $8000 somewhere. That 'somewhere' in this example is arbitrary: the destination, here, happens to be an address register—addressing mode is *direct*. *Immediate* mode applies only to *sources*, incidentally. You can convince yourself with a moment's thought that "destination *immediate*" makes no sense.

(address-register-) indirect:
example: move.b (a0), (a0)
If a0 has been loaded with $8000 beforehand, as in the preceding example, then this silly-looking instruction would pick up a byte from the keypad (in your lab computer) and deliver it to the displays. (This you saw in class notes µ2.)

Those are enough addressing modes for our purposes, except for the auto-increment and auto-decrement modes, which we choose not to reach until we speak of moving pointers (you will see them used in the stack operation *PUSH*, as well). We will reach that topic in a Class 21 program, and you will find an example of this mode in action in the Text program 11.1, as you may already have noticed.

Lab 19: µ2: I/O

Reading:	A second look at chapter 10 re: I/O:
	    10.06-10.08 re programmed I/O.
	    10.17 briefly, re: assembly language.
	 11.1-11.4, re 68008;
	 11.05 describing a controller much like the one you re building.
	 The following *Figures* are especially helpful:
	 Ch. 10:
	    10.7: XY output hardware
	 Ch. 11:
	    11.1-11.4 re 68008;
	    Fig. 11.5 showing parallel I/O port, again much like what you build in this lab.

Problems:	11.5-11.8 re: use of decoders and effects of incomplete address decoding

Today your circuit becomes a computer by any standard, with the addition of a little hardware that lets it do I/O operations under program control. You will find the hardware addition minimal: a second pair of data displays and some associated logic, an I/O decoder, and a couple of flip-flops with 3-states buffering their outputs. We will continue to write out the small programs for you; next time you will begin to write your own.

The changes and additions you will make today to the circuit you built last time are shown on the next page. They are indicated by shaded areas.

Big Picture II:
Changes and Additions to Lab Computer Circuit: µ1 to µ2

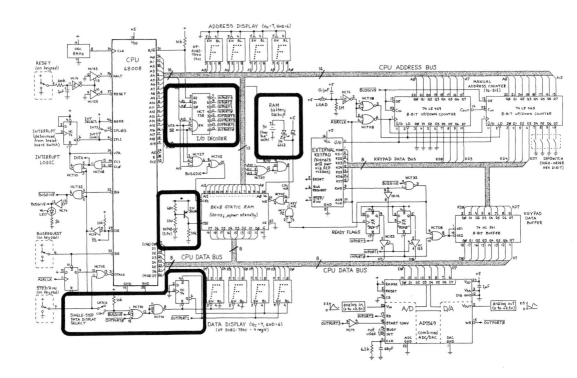

Figure L19.1: Changes and additions to circuit: µ1 to µ2

19-1 Battery Backup

The battery-backup circuit shown below will keep the RAM powered when the +5 supply is shut off: the two AA cells provide the necessary 2.0 volts (or more). Check that your battery backup does deliver this voltage to the RAM when the +5V supply is off.

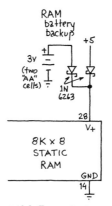

Figure L19.2: Battery backup supply

19-2 Power-Fail Detector: *Protection Against Glitches on Power Down/Up*

The circuit below senses the level of the +5 supply, and disables the RAM when the supply is low. When disabled, the RAM is immune to the spurious pulses that are likely to drive its WE* pin during the powering down and up of the gates that drive that pin.

The power-fail detector switches the RAM off at around 4.2 to 4.5 volts. You may want to bring power in to your computer close to this voltage monitor: if you do not, your power supply may be dangerously close to the 'fail' trip level under normal conditions. (The displays and processor draw about 100 mA each; these large currents, adding to about 1.5A, cause a drop along your power buses of several tenths of a volt.)

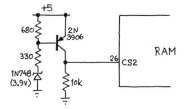

Figure L19.3: Battery backup detail: power-fail detector

Input/Output Hardware and Programming

19-3 I/O Decoder

Let's wire the decoder first. We will test it, then apply it to its first useful work: controlling the output *displays*.

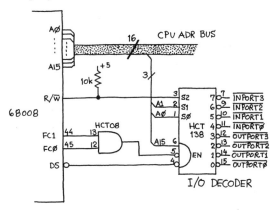

Figure L19.4: I/O decoder

As you can see, enabling of the decoder (a 74HCT138), like enabling of the RAM, is conditioned on INTA: we require that this signal *not* be asserted. The decoder is enabled by addresses at 8000H and above (up to 64K, when A_{15} rolls over to zero); your computer will use only *four* locations in this enormous I/O space (the lowest 32k of address space, you recall, is defined as memory space).

One of the 138's connections is a little curious: R/W* is connected to the 138's S_2 (MSB select). Thus the 138's eight outputs make up four INPUT ports and four OUTPUT ports. (I/O addresses asserted with R/W* high will be treated as IN, and with R/W* low will be treated as OUT.)

Testing the I/O Decoder

To give you a little reward for this wiring work, and to see if you have done it right, let's test the decoder with a very simple program. The program will do an input operation from port 0 (= address $8000), then an out to the same port. You will watch the program do this, in *single-step* mode, to confirm that the decoder asserts first its IN0* pin, then its OUT0* pin. Then we'll run a small variation on this test, and move on.

Test Program

Here's a little program to test the decoder:

```
; DECODER TEST
                    ;                   portptr = a0
                    ;                   Start Address: 110H
110    307C  MAIN:  movea.w #$8000, a0    ; point to PORT 0
       8000
114    203C         move.l #$ABCD EFBA, d0 ; set up distinctive 32-bit
       ABCD                                 ; display value (at first we will use
       EFBA                                 ; only the 'CD' byte: 8 bits of the 32)
118    1080  MORE:  move.b d0,(a0)        ; do an OUT0 operation
11A    1210         move.b (a0), d1       ; do an IN0
11C    60FA         bra.s MORE            ; keep doing it
```

Figure L19.5: Program to test I/O decoder: In, Out, at port 0

Note on word-sized pointer definitions

In using only a *word*-sized *move* to the address register, we are doing something a little unusual, but it works, and saves you some keying-in of values. The 68008 always uses 32-bit addresses, but it permits word-sized loads; it then "sign extends" the value loaded (as the process conventionally is called): fills the high 16 bits with the level of the MSB.[1]

Single-step through this. After the pre-fetch at address $11A, 11B, you should see the first novelty—which usually makes a person think something's wrong: the address bus shows $8000. '$8000?,' you may want to object, 'my whole program is down around $110!' Yes, but the address bus carries the *I/O port address* as well as memory addresses: this is your first encounter with the processor's broad-minded view that I/O operations are just like memory operations. That $8000 is, of course, the address of the I/O port we call "Port 0." On the data bus, at that time, you should see the 'BA' that you set up in data register d0. This time, it goes nowhere; ordinarily it would be clocked into a register of D flops.

During the *OUT0* operation, at the time when the address bus is showing $8000, poke the decoder's out0* pin (pin 15) with the logic probe: confirm that it sits *low*.

Now step through one more instruction and watch execution of the *IN0*: this time, the same address should appear on the address display ($8000); the data display should show $FF (how come? *Hint*: who is driving the data bus during this *IN*? (Trick question!)) Poke the decoder's in0* pin (pin 11); it should be low during the operation.

---

[1]. The process is called by this odd name because it preserves the sign of a 2'-complement number; addresses are not signed, so the concept does not fit. The word is applied here nevertheless.

Word-sized and Long-word sized operations

Now change the In and Out instructions from *byte* to *word* size: to do this, just change the 1080 and 1210 codes: the leading hex digit determines size, and should change from *1* to *3*. Now the processor should do an OUT to $8000, then $8001, then an IN from the same addresses. Confirm this, using the logic probe as you watch the address displays show $8000, $8001. Now, just to let the processor show off, you might change the size of the move to *long word*: change the 1080 and 1210 (now 3080 and 3210) to 2080 and 2210, respectively. Watch the decoder's response.

Full-speed (optional)

Now go to *full-speed*: switch from step* to run mode, and watch out0* and out1* pins of the '138 on two channels of the scope. The low pulses should last about two clock cycles (how long is that, at 8 MHz), and should be separated by four clock cycles. That is the fixed rate at which the CPU transfers data to and from a memory or peripheral.

Once you are satisfied that the decoder works properly, you can use it on its first mission: to drive the data displays.

19-4 Data Displays

The data display chips include latches, as you know; these let us achieve a stable display while a program is running, even though the data bus is rapidly and continually changing. We will use pulses from the I/O decoder to update the displays: first, byte-wide (keeping the right-hand pair of digits blanked), then word-wide (using all 4 hex displays) in the case when the CPU is putting out 16 bits, in its two 8-bit passes).

Byte versus Word Display

The left-hand pair of digits shows all byte outputs, as you know already: you have been watching this pair all along. This pair of digits also serves to show the high byte of "word" outputs (Motorola refers to 16-bit values as "words").

The right-hand pair of displays should be blanked except after the CPU has output a word. The logic shown below aims for this result.

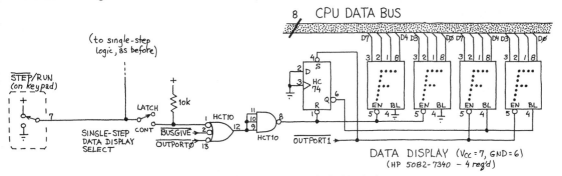

Figure L19.6: Display latching & blanking logic

Latching Outputs: Byte/Word

When the CPU sends out a *byte*, it will appear on the left display; when its sends out a *word*, it will appear on both displays.

For example, if register D0 holds the 16-bit word, ABCD:
   a byte-size write (or OUT) to 8000H
   would put out CD, on the left-hand display.
      move.b d0, 8000H

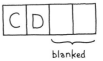

A word-size write to 8000H would put
out ABCD on the two displays.
      move.w d0, 8000H

Figure L19.7: How Byte/Word outputs appear on 16 bit display

Display Blanking

The Blanking logic shown above is designed to let the right-hand byte-display show data only when that data is helpful, rather than distracting. That means that the low-byte display should be blanked (dark) except after the CPU has done a 16-bit OUT.

Only a write to 8001H turns *on* the right-hand display. Any of three other events turns *off* the right-hand display:

1. a write to 8000H (because this might be a byte-write)
2. busgive* (because data coming from memory is always a byte wide).
3. step* because we use this for troubleshooting and usually want to see the data in each memory location as we single-step.

When you do *not* want step* have this effect, simply *open the step-with-display* switch. Then you should find that *after the first word output*, the display shows you 16 bits, not the eight that step* usually shows. This option is useful primarily for checking this display hardware; once you are satisfied that it is working you may want to leave the switch closed.

Timing Programs

The next few programs will exercise the CPU's operations of word-size and larger, and also will test your data display hardware. When you have built that hardware, try these programs.

19-5 Delay Loop: 32-bit, With Display Test

This program fragment—which will soon be demoted to "subroutine" status—is useful mostly in order to kill some time. It also tests your new output hardware.

The suggested delay value (the $1046C loaded into register D1) produces a delay of 1/4-second. If you get impatient waiting, shorten the delay.

> *Debugging note:* when single-stepping this program be sure to replace this large delay value with some tiny delay count such as 2; otherwise you will spend your afternoon in the delay loop!

The program puts out a byte, pauses, increments the display value, then does it all again.

Put this little program at address 160H and up ("160H" means "hexadecimal 160;" sometimes written as $160). You must fix the Program-counter loading vector stored down at the bottom of memory in order to run this program: replace the 0100H stored at addresses 6 and 7 with the value 0160H.

```
; LONG DELAY (32 bits)
                        ; increments display
                        ; port at about 4Hz, to demonstrate
                        ;       1) delay loop;
                        ;       2) 8 versus 16-bit-display hdwr
                        ; dsply = d0
                        ; delay = d1
                        ; portptr = a0
                        ; Start Address: 160H (not 100H))
160   307C    main:     movea.w #$8000, a0    ; point to dsply
      8000                                    ; port
164   4240              clr.w d0              ; clr dsply val.
166   1080              move.b d0,(a0)        ; show current sum
                                              ; (first pass only)
168   223C    delay:    move.l #$0001046C, d1     ; init 32-bit #
      0001                                    ; for about 1/4
      046C                                    ; second's delay
16E   5381    timekil:  subq.l #1, d1
170   66 FC             bne.s timekil         ; loop till zero
172   5240              addq.w #1, d0         ; incrmnt dsply val.
174   1080              move.b d0,(a0)        ; show current sum—
                                              ; first as byte,
                                              ; later as word
176   60 F0             bra.s delay           ; do it again
```

Figure L19.8: Delay loop, with display test

As you single-step this program, don't let the effects of *prefetch* throw you (you saw the same thing when you tested the I/O decoder, so probably you don't need this warning): the machine executes the instruction *before* the one it just picked up from memory. So, for example, you will see the CPU pick up 60 F0 from memory—an instruction that means "branch back 16 addresses"—and then the address bus will show you $8000. One might assume that the processor had jumped or branched to $8000. Not so. Again this is the *port* address, put out as the processor executes the preceding instruction,

move.b d0, (a0)

"timekil" in the program listing is a label, included for the convenience of human readers, not for the machine. You, or the assembler if you have one, must specify the relative branch distance—a 2's complement value measured from *2 bytes beyond* the branch instruction itself. If you had written this program on an assembler, the machine would have made that calculation for you. The difficulty of such computations is one of the two best arguments for assemblers (if you need any convincing). The other is, of course, that a machine should be assigned the tedious job of converting mnemonic to code.

Modify the program to put out a *word*, and confirm that your displays show you this 16-bit value. To change a *byte*-size move to *word*-size, just change the leading hex digit from a *1* to a *3*.

> *Debugging note:* for programs that output to displays
>
> While you single-step, your displays will *not* latch the output as they do during full-speed execution; instead, they will show you everything on the bus, and byte-size only. To make the displays show you latched values while you *step*, *open* the switch that links the "step*" line to pin 1 of the HCT10; pin 1 now will be pulled high, and the displays will perform as they do in full-speed operation.

When you are satisfied that the display hardware works properly, probably you will want to close the switch that links step* to pin 1 of the HCT10. Now you should find that flipping to *step** mode while the *word* output program is running at full speed makes the display revert to *byte* output.

479

Class 20: μ3: A/D<—>D/A Interfacing; Masks; Data tables

Topics:

- Stack
    - explicit use: PUSH & POP
    - implicit use: subroutines CALL

- Programming:
    - *Flags*: combinations
    - *Masking*: isolating a single bit, or group of bits

- Hardware:
    - D/A interface
    - A/D interfacing: alternative schemes

A. Programming Issues

1. The Stack
text sec. 10.03

The "stack" is just a storage table somewhere in RAM. But it works in a curious way that requires some explanation. We cannot postpone discussing it, because your programs begin to use it almost from the beginning.

The stack is a region of RAM pointed to by "the stack pointer"—which is just one of the address registers (A7). The very first action the processor takes after reset is to load its stack pointer; that is a clue to how fundamental the 68000's designers thought use of the stack was to orderly programming.

We put the stack at the top of memory (one above, in fact). Then, when we store something on the stack the pointer is decremented first, and the something is stored.

Stack Use

*Text sec. 10.03,
fig. 10.4, p. 682*

1. *Explicit "Push" of data*

   ```
                          ; stack pointer: $2000;
                          ; Let's suppose register d2 holds the value AB CD
       move.w d2, –(a7)   ; store contents of register d2
                          ; on the stack ("PUSH")
       .
       .
       .
       move.w (a7)+, d2   ; recall the saved register value ("POP")
   ```

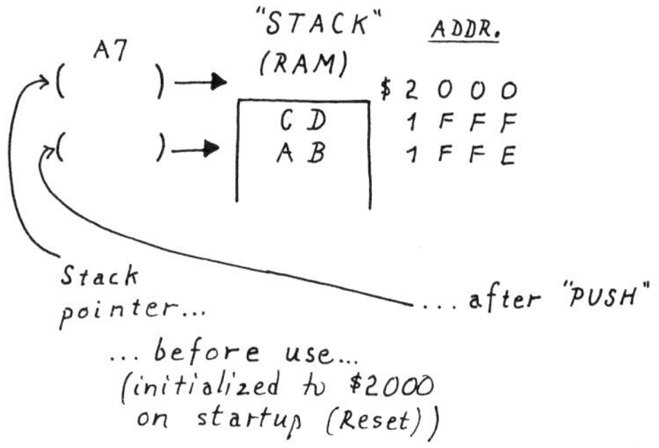

Figure N20.1: Stack used for temporary storage of register contents

2. *Automatic use of stack: "Return address"*

   The stack is used *automatically* whenever one uses an instruction called "branch to subroutine" (bsr): the processor automatically saves the "return address" on the stack, so that when it finishes executing the subroutine, it can return to the place from which it "branched to subroutine."

Here's the idea:

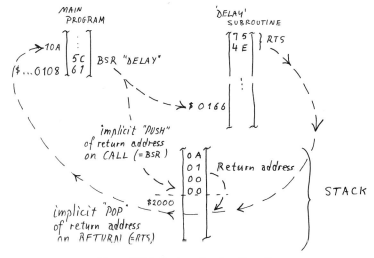

Figure N20.2: Stack use in subroutine call

Lab 20 includes a diagram showing stack use on a particular subroutine call. Here's that diagram, again:

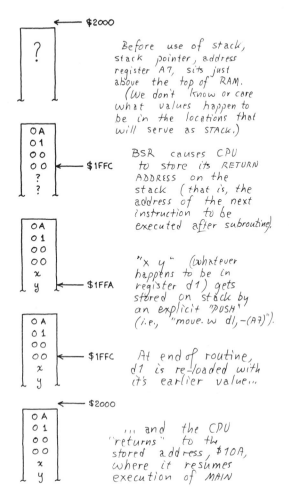

Figure N20.3: Stack use on a particular subroutine call: subroutine 'Delay' from address $108

2. Flags

Here are the *flags* of the 68000 (just flip-flops dedicated to storing these particular items of information, items reflecting results of recent CPU operations). These flops make up part of the "status word" register.

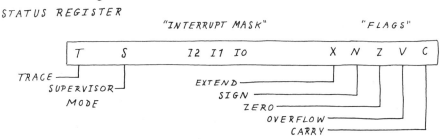

Figure N20.4: 68000 flags

We'll see in a few minutes that the 68000 allows us to use all of the *16* (!) combinations of these four flags. For the moment, however, let's think about just *one* of the flags.

The flag we use most often is the *Zero* or *Equal* flag, used, for example, in the BNE and BEQ instructions. Let's start, therefore, with a program that looks at the Z or EQ flag.

Conditional Branch:

a) *Straightforward*

Here is the first program in the second micro lab. It increments the data display every second or so. How does the processor determine whether to loop back again, or to fall through to the "addq" instruction? What does it look at in order to decide, as it executes the conditional branch instruction, *bne*?

```
            ; LONG DELAY (32 bits)
                        ; increments display
                        ; port at about 4Hz, to demonstrate
                        ;       1) delay loop;
                        ;       2) 8 versus 16-bit-display hdwr
                        ;   dsply = d0
                        ;   delay = d1
                        ;   portptr = a0
                        ;   Start Address: 160H (not 100H))
    160   307C   main:     movea.w #$8000, a0   ; point to dsply
          8000                                  ; port
    164   4240             clr.w d0             ; clr dsply val.
    166   1080             move.b d0,(a0)       ; show current sum
                                                ; (first pass only)
    168   223C   delay:    move.l #$0001046C, d1     ; init 32-bit #
          0001                                  ; for about 1/4
          046C                                  ; second's delay
    16E   5381   timekil:  subq.l #1, d1
    170   66 FC            bne.s timekil        ; loop till zero
    172   5240             addq.w #1, d0        ; incrmnt dsply val.
    174   1080             move.b d0,(a0)       ; show current sum—
                                                ; first as byte,
                                                ; later as word
    176   60 F0            bra.s delay          ; do it again
```

Figure N20.5: Straightforward use of Zero flag: decrement until a register hits zero; branch on zero flag

A plausible answer, but not quite right, would be, 'It looks at register d1.' If you look at the instruction *bne* you find this answer must be wrong: *bne* is silent concerning *what* it is that is 'not equal'—or 'not zero,' to put it more clearly.

Here's a sketch to remind you that the Zero flag is just a flip-flop that records whether the result of an operation was zero:

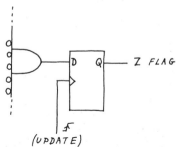

Figure N20.6: Zero flag: a flip-flop recording whether a result was zero

Bne says nothing about which data or address register to look at because the processor looks at *none*: it looks at just the single flip-flop in the *flag* register: the Zero flag. In this case, that flag does reflect the value of register d1 after the subq. But *bne* doesn't know that; *you* do—you, the programmer must make sure that the flags do describe the value that interests you. Usually this works without effort on your part. But it is possible, as you can imagine, to mess things up, if you let something intervene between the operation that interests you and the testing of a flag. You'll recognize this problem if you meet it: perhaps after an hour spent wondering why your program now and then branches wrong!

b) *Masking: a way to make flags reflect selected <u>bits</u>*

The previous program looped until a register hit zero, setting the Zero flag. Sometimes a program needs to branch on the condition of not an entire register or memory location but on the level of one or a few bits. For that case, one needs a way to ignore all the other bits.

1- Single Bit Test

Here is the lab's Ready-Check routine, which looks at the Ready Flop's Q on line d1 of the data bus, at port 1.

```
                ;   Rdycheck: leaves addend register
                ;     unchanged unless Rdy key (Wr*) has beenhit
                ;   assumes main has set up a0
                ;     to point to keyport
                ;   d2 brings in key value
                ;   a call to this would replace
                ;     "move.b (a0),d2"
Rdychk: move.w  d4, -(a7)       ; save scratch reg
        move.b  $8001,d4        ; take in Rdy & junk
        andi.b  #1, d4          ; check Rdy bit: mask all but data line d0
        beq.s   'rejoin'        ; if not ready, leave addend unchanged
        move.b  (a1),d2         ; if Rdy, take new value
                                ;   from keypad as addend
                                ;   this also clears flop
        move.w  (a7)+,d4        ; recall saved reg
rejoin: rts                     ; back to calling program
```

Figure N20.7: Lab 20 'Readycheck' routine: mask all but one bit

And here is what that *andi.b #2, d4* operation does:

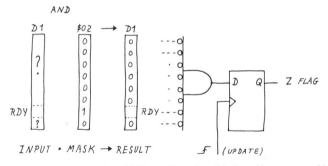

Figure N20.8: Mask operation: AND uses 0's to force bits low, 1's to preserve bits

The processor offers instructions that will work on single bits (testing, setting or clearing a single bit). So you might think you did not need to know how to apply a mask in software. Not so.

2- Testing More Than One Bit

A mask is more versatile than the bit-operation instructions. A mask can throw out, say, four bits, keeping four. Here a mask is used that way, in order to help the program discover which *single key* the user has entered at the keypad. The mask is needed in order to allow the program to concentrate on the *latest* key value, ignoring the latest-but-one.

```
                    ;KEYMATCH fragment
                    ; assumes main program has set up an address register
                    ;   to point to the keypad:
                    ;   "movea.w #keyport, a0"   ; point to keyport

200   3F04   kmatch:  move.w d4, –(a7)      ; save scratch regstr

202   1810   look:    move.b (a0), d4       ; get key value (8 bits)
204   0204            andi.b #$0F, d4       ; force high nybble to zero (this is
      000F                                  ;   the old key); keep low nybble
                                            ;   (this is the most recent key)

208   0C04            cmpi.b #$A, d4        ; is the most recent key "A"?
      000A

20C   670A            beq.s 'doA'           ; if so, do something
                                            ;   appropriate

20E   0C04            cmpi.b #$C, d4        ; is it C?
      000C

212   66EE            bne.s 'look'          ; if neither A nor C, go look
                                            ;   again (the silly human has hit
                                            ;   the wrong key!)

214   4E71            nop                   ; must be C, if we landed here
                                            ;   so this is what we do if we get C
                                            ;   (you write something more
                                            ;   interesting than "NOP"!)
```

Figure N20.9: Mask used to keep 4 bits, throw out 4

3- Other Masking Tricks: Setting, Clearing & Toggling Bits

The Boolean operations allow one to use masks to manipulate bits or sets of bits:

Figure N20.10: Masks used to force, clear or toggle selected bits

The set and clear tricks are used, for example, in a program that needs to drive a pulse on one line while holding 6 other lines constant, thus:

```
0000                    ;STARTUP subroutine: sends silence code without
0000                    ; awaiting Ready: needed on first pass.  Uses d0,
0000                    ; which is used for output code throughout
0000
0000                    ; if pointers are not already set up, define them thus:
0000
0000    367C 8003        movea.w #$8003, a3    ; point to Vox data port
0004    327C 8001        movea.w #$8001, a1    ; point to the generic Ready-bit port
0008
0008
0008    70 3E     startup moveq #$3E, d0       ; send silence code.  Strobe is low.
000A    1680             move.b d0, (a3)       ; send to talker
000C    0000 0040        ori.b #$40, d0        ; force Stb bit high, leaving other bit
0010                                           ;   unchanged
0010    1680             move.b d0, (a3)       ; send it, Stb high
0012    0200 00BF        andi.b #$BF, d0       ; force Stb low
0016    1680             move.b d0, (a3)       ; send it
0018
0018    4E75             rts                   ; the deed is done
```

Figure N20.11: Votrax talker program: sends pulse on STB Line, leaves other 7 lines untouched

4- Other Flags Besides Zero: Carry, Sign and two oddballs

The Zero flag is the one you're likely to use most often, but you sometimes will want to use others. The 68000 kindly allows you to branch on *combinations* of flags. Some of these are pretty bewildering:

CC	carry clear	C*	LS	low or same	C + Z
CS	carry set	C	LT	less than	N + V* + N*V
EQ	equal	Z	MI	minus	N
GE	greater or eq	NV + N*V*	NE	not equal	Z*
GT	greater than	NVZ* + N*V*Z*	PL	plus	N*
HI	high	C*Z*	VC	overflow clr	V*
LE	less or equal	Z + NV* + N*V	VS	overflow set	V

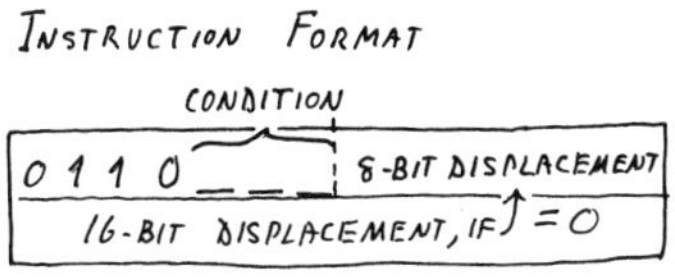

```
           INSTRUCTION FORMAT
               CONDITION
        ┌─────────┬──────────────────┐
        │ 0 1 1 0 │ 8-BIT DISPLACEMENT│
        ├─────────┴──────────────────┤
        │ 16-BIT DISPLACEMENT, IF ↑ = 0│
        └────────────────────────────┘
```

Figure N20.12: 68000 branch conditions (listed under B_{cc})

The main point to note here is that you must choose the appropriate flags depending on whether or not you are working in *signed (2's complement)* notation.

Usually we work with *unsigned* values, so we should avoid the following conditions which assume 2's comp:

>BGE
>BGT
>BLE
>BLT

Figure N20.13: *Beware* these deceptive conditions, which assume you're thinking of values as signed 2's comp usually you are not thinking of them that way—for example, when you write an *address*

Take a look at the puzzling combinations of flags these conditions use, if you need convincing.

Or consider the following case: suppose you compare a register against a constant, planning to continue until the 'moving' register (the one you are incrementing) passes the constant. Suppose you reason that you want to keep going so long as the pointer is less than or equal to the constant; that's reasonable enough, and it sounds reasonable to choose the condition, "BLT."

You might write:

```
            .
            .
            .
            movea.l #$9000, a1       ; set up end marker
doitagain:  move.b $somewhere, (A0)+ ; use the moving pointer & increment it,
                                     ; to store something in memory
            cmpa.l  A1, A0           ; moving pointer, A0, minus constant, A1
            ble doitagain            ; continue so long as A0 ≤ A1
            .
            .
            .
```

Figure N20.14: Faulty use of *signed* condition on *unsigned* value

This fails at once. *BLE* looks for a bewildering combination of flags (BLE = Z + N·V* + N*·V, where N is the *negative* condition of the sign flag, and V is the 2's comp overflow flag!) And since $9000 has the MSB high, it is treated as a negative number, less than the $400 at which the moving pointer begins. The program would fall out of its loop on the first pass. The remedy is simply to use BLS, which assumes *unsigned* values (it looks at the Z and Carry flags only).

The other conditions work fine with unsigned values. The following pair is especially easy to use—easier than the seemingly-simpler carry-set/carry-clear conditions:

>BHI
>BLS

These are used after a compare. Watch out for the fact that compare (like SUB) subtracts the first item from the second. Here's a data compare:

```
            cmpi.b #$10, D1          ; D1 minus #$10
```

And here are two possible conditions to use after an address compare:

    doit: .
               .
               .

oneway:	cmpa.l A0, A1	; A1 minus A0
	bne 'doit'	; loops until A1 = A0

versus

abetterway:	cmpa.l A0, A1	; A1 minus A0
	bls 'doit'	; loops until A1 exceeds A0 (a good way to check whether a table-read or table-fill is completed; assume A1 holds the end-of-table address). This one *works*: *BLS* does not assume 2's-comp, thus does not look at *sign* of the result.

Figure N20.15: Two conditions one might use in an end-of-table test. BNE is a little dangerous

The second condition, BLS, is a little safer than BNE: if the pointer sneaked past the end marker, BLS would not be fooled; BNE would be fooled. How could the pointer 'sneak by?' Perhaps you might put the end marker at an odd address, then change the data transfer to *word* size, for example; then your pointer would never *equal* the odd end marker. So, use BLS and BHI when you can.

B. Hardware: linking A/D & D/A to the computer

1. *Generic: separate A/D, D/A*

How would you use the I/O decoder's signals, and other hardware as needed, to link an 8-bit A/D, D/A to the computer?

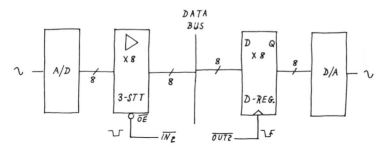

Figure N20.16: Tying separate 8-bit A/D, D/A to computer

We hope you're beginning to think it *obvious* that you need a 3-state on the input side, a register on the output.

2. Using a particular chip: one-chip A/D, D/A package: AD7569

Here our job is eased because the manufacturer has done so much of the wiring for us, on-chip:

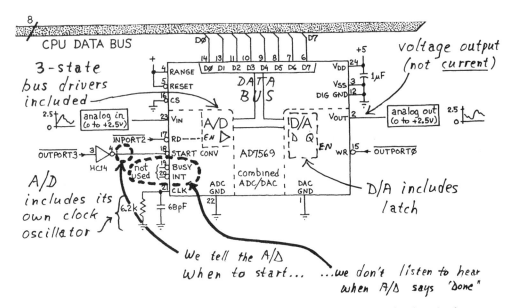

Figure N20.17: AD7569 one-chip A/D, D/A (8 bits): lab wiring: we ignore conversion-done signals

We had some choices to make in driving and listening to its control lines. We chose a very simple scheme: we do not check to see whether a conversion has been completed, as you can see from the figure above: we ignore its two signals that might give us that information (BUSY* and INT*).

This scheme leaves to the programmer the task of making sure the computer does not ask for data until the A/D has had time to complete a conversion. (That's easy, given this 2.6 us converter.)

b- Fancier: polling: Let the converter tell the computer when data is ready

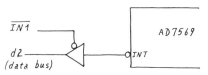

Figure N20.18: Letting converter tell CPU when data is ready: polled

This "fancy" scheme (not very fancy, really) turns out to be less clever than the simpler arrangement, because the converter is so fast: its conversion-time, 2.6 µs, does not allow the processor even enough time to ask 'Ready?'

c- Fancier still: Interrupt

The INT* line could also *interrupt* the computer, as its name makes evident. This may sound like an elegant arrangement, and sometimes it is. But not here: the interrupt response itself takes longer than the A/D takes to convert (at least 9µs versus 2.6 µs). You need to understand interrupts; but not in order to use this chip. We will postpone considering interrupts for a couple of sessions: until Lab 22.

Lab 20: µ3: Subroutines; More I/O Programming

> **Reading:** 10: 10.03 comments on *stack*; figure 10.4 helps here.
> Note how close the example in 10.08 is to the keypad plus Ready that you build in this lab (don't worry yet about interrupts, discussed as well in that example).
>
> 10 & 11: The following *Figures* are especially helpful:
> *Ch. 10:*
> 10.4 re: stack.
> Fig. 11.5, again, showing parallel I/O port; this time it is the *input* port that resembles what you add in this lab.
>
> **Problems:** 11.14 (input port); 11.13 (program to clear arrays)

Today you get a chance to put to use the hardware you wired last time: you will wire up just a little additional hardware: an 8-bit input port (this is a 45-second job: just link the decoder to a gate already in place), and a 1-bit Ready flag (this takes a little longer, both to wire and to understand). Once again the programs are very simple, though the Ready-check program introduces you to a new trick called "masking."

These notes open with a software change that appears small but is important: the *delay* routine that you wrote last time is transformed into a *subroutine*; instead of jumping or "branching" to it, you "call" it, using the "stack" to store the return address.

20-1 Invoking Delay as a *Subroutine*

The following program uses the delay routine that you entered last time, but modifies it so that it can be used as a subroutine.

What does that mean? It means that the "main" program here—the one that simply puts out the value of D0 as a byte, then as a word, increments D0 and then does it again—"calls" the delay program. The Call or "Branch to Subroutine" instruction (the latter is Motorola's name: BSR for short) is a clever sort of jump that is capable of returning to the calling program at the right address upon executing a Return from Subroutine (RTS) instruction. BSR and RTS work, as you know, by using the Stack to save the return address. This ability is just what we want in a delay routine, because we want to be able to invoke it from wherever we happen to be in some other program.

The Delay routine begins with a push to the Stack—storing the value that was in the D1 register. We do this here just to illustrate a good practice: don't let a subroutine mess up registers permanently. Such a mess-up may startle some other calling program. (Our "main" program does not use D1, so we here could safely omit this saving of D1.) Just before returning to the main program, the routine "Pop"s the old value of D1 off the stack, restoring it to D1. If the program failed to pop old D1, we might lose it, in the general case. But something much *worse* would happen, as well. Do you see what?

You will need to fill in the *branch* instructions. They take the form:

op code (8 bits) displacement (8 bits)

The displacement is a 2's-complement number, relative to (the address of the branch instruction)-plus-two. We have done one of the cases for you.

   BSR 61   (branch to subroutine; unconditional)
   BRA 60   (branch always: unconditional)
   BEQ 67   (branch if result zero)
   BNE 66   (branch if result not zero)

Labor saver: JSR

You may prefer to use Jump Subroutine, JSR, which lets you specify a 16-bit destination address rather than a displacement. (You cannot do the same lazy trick with the conditional branches.) Here is the JSR code:

4E80 0160   JSR $160

The "0160" here is arbitrary; you put here the starting address of the subroutine you want to invoke. JSR uses the stack exactly the way BSR does.

You must assign addresses, as usual: Main to 100H. The subroutine—which is just a slightly modified version of the old Delay program—can sit at 166H; if it does, you can use the code already entered to generate the delay. That will save you a little typing.

```
0000                            ;       display increment pgrm
0000
0000                            ;       dsply = d0
0000                            ;       portptr = a0
0000                            ;       Origin: Start at 100H
0000
0100   307C 8000         main:  movea.w #$8000, a0    ; point to dsplyport
0104   4240                     clr.w d0              ; clear display register
0106   1080              show:  move.b d0,(a0)        ; show current value,
                                                      ; as byte
0108   615C         (R)        bsr.s 'delay'          ; invoke delay routine
010A   3080                    move.w d0,(a0)         ; show it as word
010C   5240                    addq.w #1,d0           ; increment display
0108   xx           (R)        bsr.s 'delay'          ; pause
0110   xx           (R)        bra.s 'show'           ; do it again

                                ; DELAY subroutine: kills time, messes up no registers
                                ; place it at 166H

0166   3F01              delay: move.w d1,-(a7)       ; push to save
                                                      ; main reg
0168   223C 0001 8000          move.l #$0001046C, d1  ; init 32-bit #
                                                      ; for about
                                                      ; 1/4-second dly
0168   5381              hang:  subq.l #1, d1
0170   66 FC        (P)        bne.s hang             ; loop till zero
0172   321F                    move.w (a7)+,d1        ; pop to restore
                                                      ; main reg
0174   4E75                    rts
                                ;---
```

Figure L20.1: Byte/Word display; delay invoked as subroutine

Observing Stack Operations

This program, when single-stepped, affords you a rare opportunity to watch the CPU's use of the *stack* in slow motion. This use will look odd at first. But stick with it. It does make sense in the end.

Here is a sketch of stack use in this program. We can watch the stack and its pointer, at successive points in the program. (Compare Text figure 10.4, showing *PUSH* and *POP*.)

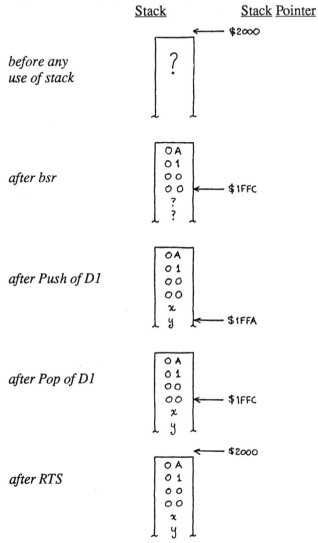

Figure L20.2: Stack use: delay & display increment program

Debugging suggestion: for programs that include call to Delay

When single-stepping this or any other program that includes a call to a *delay* subroutine, you can speed your debugging of any part of the program other than the delay call by eliminating the call to delay: temporarily replace the *BSR Delay* instruction with a NOP (4E 71).

20-2 Improved Delay Routines (*optional*)

If you are feeling pressed for time, or if you want to postpone programming complications for a while, hop on to section 20-3. But if you feel ready to try variations on a theme, you might enjoy the *Appendix* to this lab, which suggests several alternative versions of the *delay* routine you have just tried. These alternative delay routines will be useful in later labs, though none is *necessary*.

Input Hardware

20-3 Data Input Hardware

Connect the decoder's IN 0* pin (address 8000H) to the gate (74HCT08) that drives the keyboard data buffer's EN*:

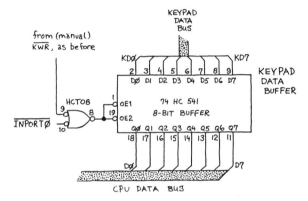

Figure L20.3: Keyboard data buffer enable logic: CPU control added

20-4 Input/Output Program

The following program allows you to input a value that is to be added to D0 while the program is running, instead of always incrementing the display value as in the earlier program. This program uses the data buffer hardware you have just connected to the IN0* line.

```
0100    307C 8000              movea.w #$8000, a0    ; point to dsplyport
0104    4240                   clr.w d0              ; clear display register
0106    4242                   clr.w d2              ; clear key-value register

0108    1080            show:  move.b d0,(a0)        ; show current value, as byte
010A    xx         (R)         bsr.s 'delay'         ; invoke delay routine
010C    1410            get:   move.b (a0),d2        ; get key value
010E    D042                   add.w d2,d0           ; form new sum
0110    xx         (R)         bsr.s 'delay'         ; pause
0112    xx         (R)         bra.s 'show'          ; go show new value
```

Figure L20.4: Input test program

Operations on Bytes/Words

When you are satisfied that this program is working properly, use a *word-sized* output ("move.w...": the first hex digit is *3* instead of *1*) and try replacing the *word-sized* add,

D0 42   add.w d2, d0

with the *byte-sized* equivalent,

D0 02   add.b d2, d0.

We hope the result will let you feel what you glimpsed last time, the tidiness of this processor's way of treating different-sized data: it can act as if it were an 8-bit processor, or 16 or 32, depending on your choice.

20-5 Ready Signal

The I/O program you just ran would work better if you could determine when to give it a new value to treat as its addition constant. As the program stands it can lead the program to take some value you are in the process of changing: you may have entered one of two new key values. Here we will add a circuit that typically is necessary on an input port: a signal that tells the computer when valid data is available.

Evolving the 'Ready' Hardware

Here is a first, poor suggestion for the Ready hardware:

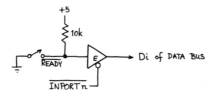

Figure L20.5: [A poor way to tell the computer data is ready

In the present application this method would work: you would need to hold the ready line down long enough to let the computer get the new data (that's no hardship at normal computer clock rates); you would find yourself giving the computer its new data thousands of times over. (The switch bounce here is harmless.)

But it is more useful to work out a design that lets us give the machine new data just once each time we hit an Enter key (which would tell the computer Ready). Usually that is the better scheme. This change requires a slight program change, and a trifle more hardware. Here is part of it:

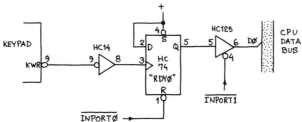

Figure L20.6: A better ready circuit: value input once each time 'enter' is hit

To finish the circuit you need to choose an appropriate line from the '138 to clear the Ready signal.

20-6 I/O Program With Enter/Ready Function Added

When you think you have your hardware correctly done, try it in this amended I/O program. In place of the instruction

move.b (a0), d2

at label 'get,' write a bsr to the following Rdychk routine:

```
0000                            ;         Rdycheck: leaves addend register
0000                            ;           unchanged unless Rdy key (Wr*) has been
0000                            ;           hit
0000
0000                            ;         assumes main has set up a0
0000                            ;           to point to keyport
0000                            ;         d2 brings in key value
0000                            ;         a call to this would replace
0000                            ;           "move.b (a0),d2"
0000
0140   3F04            Rdychk:  move.w d4, -(a7)        ; save scratch reg
0142   1810 8001               move.b $8001,d4          ; take in Rdy
                                                        ;   & junk
0146   0204 0002               andi.b #1, d4            ; check Rdy bit:
                                                        ;   mask all but
                                                        ;   data line d0
014A   xx         (R)          beq.s 'rejoin'           ; if not ready,
                                                        ;   leave addend
                                                        ;   unchanged
014C   1410                    move.b (a1),d2           ; if Rdy, take
                                                        ;   new value
                                                        ;   from keypad
                                                        ;   as addend
                                                        ;   this also
                                                        ;   clears flop
014E   381F                    move.w (a7)+,d4          ; recall saved reg

0150   4E75            rejoin: rts                      ; back to calling
                                                        ;   program
```

Figure L20.7: Ready-check routine

20-7 Decimal Arithmetic

Humans are accustomed to the decimal numbers, and accommodating processors know how to add in that base. For a byte-size add, you can simply replace the ADD with ABCD command ("add, binary-coded decimal"):

C102   abcd d2, d0

This puts the result of the addition into D0, as before. Try this in the keyboard-input program.

Decimal addition works only on inputs that are in BCD form. If you set up an input of "B" on the keypad, for example, you will find that the result of the BCD addition goes bad: it will not be restricted to BCD but will wander into Hex. ABCD is not clever enough to fix a non-BCD input; instead, the decimal addition simply amends the result of an arithmetic operation so that the result (initially in ordinary binary form) is transformed to "packed BCD": two BCD digits per byte.

For example:

Decimal	Binary	Packed BCD equivalent
8	0000 1000	same
+5	0000 0101	+ same
13	0000 1101	0001 0011
		Executing the decimal addition, the processor notes that LSD exceeds 9 after the ADD. So it cleverly adds 6 so as to fix result.

Decimal addition fails if the input is not in "packed-BCD" but ordinary binary. The following example illustrates what decimal addition cannot fix:

Decimal	Binary	Packed BCD equivalent
17	0001 0001	0001 0111
+20	0001 0100	0010 0000
37	0010 0101	0011 0111 *(This answer, correct in BCD, would have resulted only if the input had been entered in BCD form)*
		—No Change: ERROR
		This time the processor finds that neither LSD nor MSD exceeds 9 after the ADD, so it senses no need to adjust the result. But the result is *wrong* in BCD: 25.

You can confirm this limited cleverness of the processor by putting in keyboard values that exceed 9. You could write a program capable of handling such inputs, converting them to BCD form before adding them; but that task would require programming skills that we do not now assume you to have.

(End of Lab 20; Appendix listing alternative *delay* routines follows)

Appendix: Alternative Delay Routines

As we said in the notes above, you do not *need* these routines, and they may overdose you with programming issues, if you are meeting assembly-language programming for the first time. But you would find these alternative delay routines *useful*, and you may be an experienced programmer already, or just eager to look at variations on a theme—always a good device for revealing a subject.

The Delay routine listed in 20-1, above, uses a *constant* delay determined within the subroutine. To change the delay you need to go into the routine and alter the value by hand before invoking Delay. The alternative versions of Delay set out below may be more convenient as general-purpose time-killers. We assume, in suggesting that you might want to install an improved Delay, first, that you have a battery backup that will preserve any Delay you install; and, second, that you will need delays in several later programs. You need delay today in order to make the count-up process slow enough to watch. Later you will want a more versatile version, especially when you fill a table using an A/D; at that time, the delay value you want will be much smaller, by a factor of perhaps 1000; in addition, you will want to vary the delay value *easily*.

Both of these alternative Delays set out below allow one subroutine, never altered, to generate a variable delay. The method, perhaps obviously, is simply to let the *calling* program (the one that invokes the subroutine) load a delay count into a memory location; the Delay subroutine then uses that value.

Delay: Version 0

The first version, below, is just section 20-1's DELAY, listed to remind you of the scheme on which we are trying to improve.

```
            ; DELAY subroutine: kills time, messes up no registers
            ; place it at 166H
    delay:  move.w  d1,-(a7)        ; push to save
                                    ;   main reg
            move.l  #$0001046C, d1  ; init 32-bit #
                                    ; for about
                                    ;   1/4-second dly
    hang:   subq.l  #1, d1
            bne.s   hang            ; loop till zero
            move.w  (a7)+,d1        ; pop to restore
                                    ;   main reg

            rts
```

Figure L20.8: DELAY subroutine: version *0*, as listed at the start of this lab (fig. L20.1)

20-App-1: Delay: Version 1

The second version, below, loops a number of times determined by the memory value stored by the calling program. This permits very short delays, but it requires the programmer to recalculate the delay each time it is to be altered.

Notice that one needs to load the delay value only *once* somewhere in the main program. After that the value stays in place, undisturbed by its use in DELAY1.

```
0000                        ; Alternative DELAY subroutine: value stored
0000                        ;  in memory; loaded from calling program
0000                        ; place it at 166H
0000
0000                        ; assume main program includes lines like--
0000
0400                                dlycnt equ $400 ; storage location
0000                                                ; for long-word delay
0000                                                ; count
0000                        ;       move.l #$0001046C, (dlycnt) ; init 32-bit #
0000
0000                        ;-----------
0000                        ;The following delay routine would use the delay loaded
0000                        ; in the main program:
0000
0000   2F38 0400            delay   move.l (dlycnt), -(A7)  ; save val that'll
0004                                                        ;  be messed up
0004   53B8 0400            hang:   subq.l #1, (dlycnt)
0008   66 FA                        bne.s hang              ; loop till zero
000A   21DF 0400                    move.l (A7)+, (dlycnt)  ; restore val that
000E                                                        ;  was messed up
000E   4E75                         rts
```

Figure L20.9: DELAY subroutine: version *1* : calling program determines number of *loops*

20-App-2: Delay: Version 2

The third version, below, eases a programmer's work by letting him or her determine once for all what count generates a 1 ms delay; that delay value is fixed within the subroutine. Then the calling program determines how many *milliseconds* of delay it wants, using the method used in Version 1: the *number of milliseconds* count is stored in a memory location accessed by the subroutine.

This version of the routine is probably the most convenient. It does not permit very short delays. But you can, of course, go into the subroutine and alter the fixed delay by hand.

```
0000                    ;----------
0000                    ; SUBROUTINE MsDELAY
0000                    ; This one would include its own loop--say 1 ms,
0000                    ; and the calling program would determine the
0000                    ; number of iterations before calling (this time
0000                    ; a word should do: up to 65k milliseconds!
0000                    ; you need to decide what constant to use to
0000                    ; generate 1 ms delay.  Here we'll call it just
0000                    ; MSVAL
0000                              MSVAL EQU 256           ; just for example
0000
0000
0000    3F38 0400       msdelay    move.w (dlycnt), -(A7)  ; save val that'll be messed up
0004    3F01                       move.w d1, -(A7)        ; and a register
0006
0006    3238 0100       loopdeloop: move.w MSVAL, d1       ; inner loop, for 1 ms
000A    5341            msloop     subq.w #1, d1           ; count down to zero
000C    66 FC                      bne.s msloop
000E
000E    5378 0400                  subq.w #1, (dlycnt)     ; now decrement mS count
0012    66 F2                      bne.s loopdeloop        ; loop till zero
0014
0014    321F                       move.w (A7)+, d1        ; restore vals that
0016    31DF 0400                  move.w (A7)+,(dlycnt)   ;   were messed up
001A    4E75                       rts
```

Figure L20.10: DELAY2 subroutine: version 2: calling program determines number of *milliseconds* of delay

20-App-3 Delay determined during program run

Both of the revised versions of Delay—DELAY1 and DELAY2—lend themselves to a nice program device: altering the delay value while a program runs, to see quickly the effect of differing delays. This facility is very useful in the Lab 20, where the delay value will determine the A/D *sampling* rate.

The main program would need to load the delaycount memory location from the keypad. As you do this, consider some possible difficulties:

- The key value is limited to *8 bits*. If you need a delaycount value larger than that, your program will have to magnify the key value in one way or another. Here are some suggestions:

    - You might shift the key value left, one or more places. Each shift *doubles* the value, as you know.
    - You might load the key value as the *high order* half of the delaycount. (This would be equivalent to shifting left 8 times, of course.)

- The program fragment that checks the keypad will take some time, of course. So, adding this fragment will affect the total delay. That's all right if your delays are long (in the ms range); it may be annoying if the delays you want are very short.
  Probably you will want to alter the delay value only when you hit the *Ready* key. Otherwise the delay would be determined by the particular value sitting on the keypad when you began to run your program. That would annoy you.

Class 21: µ4: More Assembly-Language Programming: 12-bit port

Topics:

•*old:*

– stack

•*new:*

– Programming:

* table copy

* Getting data from A/D

– Hardware: peripherals of more than 8 bits

Old

1. The Stack

Just to confirm that you understand how the *stack* works, let's try an example; this time, let's call a subroutine from within a subroutine and watch what happens. Does the stack get confused?

Suppose the main program calls[1] a program named Dosomething, and Dosomething calls Delay. Here's how the stack history would look:

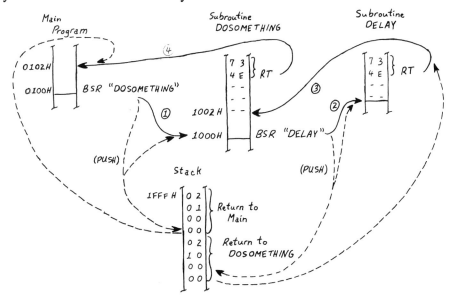

Figure N21.1: Stack use as call occurs within a subroutine

The stack automatically makes the *returns* work properly—unless we, the programmers, blunder.

Query: what would happen if the subroutine Dosomething called *itself*? (You computer science hotshots know the answer, no doubt; anyone else probably will need to stare at the diagram for a minute to settle the question.)

---

1. "call" is the generic name for the process Motorola names "branch subroutine," "bsr" or "jump subroutine," "jsr"

New:

2. The most important application of <u>indirect addressing:</u> *Moving Pointer*

We have been promoting the addressing mode called "indirect"—or "address-register indirect," as Motorola describes it. Its main virtue then appeared to be speed; it was compact, too.

Here we meet the case where indirect addressing is not just preferable but is *required*: the case where we need to work with a table in memory: to fill it with data, or read data from it, or search for some particular data. Let's look at an example.

ext sec. 11.02,
x. 11.1, p. 748

To copy $100 words from one place to another (Exercise 11.1 in Text), one can use two pointers, and another register to count transfers:

```
            movea.l #$A0000, a0    ; set up a pointer (address register), to source block
                                   ;   (in Text example, installed memory reaches much
                                   ;    higher than in our Lab machine)
            movea.l #A8000, a1     ; destination pointer
            move.w #$100, d0       ; set up count register
transfer:   move.w (a0), (a1)      ; transfer a word
            addq.w #2, a0          ; advance pointers
            addq.w #2, a1          ; (advance by two bytes: word)
            subq.w #1, d0          ; decrement count register
            bne.s 'transfer'       ; keep transferring until 16 are done
            end
```

Figure N21.2: Program to copy $100 words: uses counter register and explicit advance of each pointer

But here's a tidier way, letting the processor advance both pointers automatically after using them (so-called "postincrement register indirect" mode). Notice that the processor is smart enough to notice that it needs to advance the pointer by *two* addresses, since the data is 16 bits ("word" sized).

```
transfer:   move.w (a1)+, (a2)+    ; transfer a word, & advance pointers
            subq.w #1, d0          ; decrement count register
            bne.s 'transfer'       ; keep transferring until 16 are done
            end
```

Figure N21.3: Tidier copy program: lets CPU take care of advancing both pointers, using *auto-increment*

That's pretty neat. It can be made neater still by compressing the decrement of d0 and conditional branch into one complex instruction (see Text example program 11.1):

>   subq.b #1, d0
>   bne.s 'transfer'

compresses to

>   dbf d0 'transfer'

Figure N21.4: A further compression of code: *dbf* includes both decrement of counter and conditional branch

This compressed form is good not just because it looks neat; it also runs faster: 18 clocks, versus 26 for the pair of instructions it replaces. "dbf" is a strangely simple case of a fancy instruction. It means 'decrement a count register and branch back if not yet at –1;' the "f" stands for false, and expresses the fact that in this form the instruction assumes *false* the condition that it could, instead, be asked to test. So, we waste part of the power of the instruction, which is to look at other flags before decrementing, and then break out of the loop if the condition is true.

This instruction is so peculiar—so 68000-specific—that it is not worth teaching except as an example of the kind of processor specifics that you would be much concerned to know if you ever found yourself getting deep into assembly-language programming. That event becomes steadily less likely, as higher-level languages are applied to the programming of even microcontrollers.

3. Hardware: Peripherals of more than 8 bits

You have met this issue already: the data display that you built in Lab 19 takes a *16-bit* output. You saw that the processor is as happy to put out 16 bits as 8; from the programmer's point of view there's practically nothing to the difference: just write ".w" or ".b" to define the size; in machine code the difference is one bit (e.g., byte-sized moves begin "1...;" word-sized moves begin "3...").

Let's try another example: wiring separate *12-bit* A/D and D/A's to the computer. This job turns out to be easy. It may help to recall that the processor does not know (or care!) whether the things appearing at these two addresses are *i/o* devices or *memory* locations. It's up to you, the designer and programmer, to treat the i/o gadgets similarly: as if they were two memory locations. So, the more significant digits go into the lower address, for example.

In doing this example, assume an I/O decoder already wired, and providing as many ports as you need.

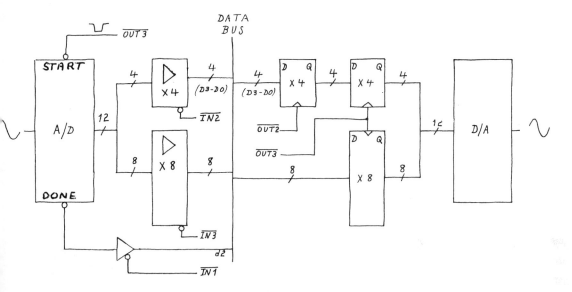

Figure N21.5: 12-bit A/D, D/A: wiring details

That extra 4-bit register on the output side makes sure that the D/A gets updated all at once; otherwise it would be fed a strange mixture of old and new samples for a short time (about 0.5 μS, if fed by our 68008).

We did not worry about the equivalent problem on the input side: making sure that all 12 incoming bits were from a single sample. (Do you recognize how a 12-bit input device of slightly different design *might* deliver such a screwy mixed sample? You'll see this issue treated in a Worked Example called *68008 peripheral of more than 8 bits*, if you're interested.)

We didn't need to worry about mixed input samples, because the processor cannot be surprised by a conversion that finishes between the two pickups (between high-4 and low-8 bits). It cannot be surprised because it is the processor that decides when a conversion is to begin, and the processor knows when a conversion has been completed. A free-running converter, in contrast, *could* cause such mischief by completing a conversion between pickups, and the Worked Example considers such an input device.

2. Programming

Getting data from the lab A/D

Here, the only difficulty comes from the rather annoying fact that the AD7569 does not begin a conversion until told to do so (it might be nicer if it made the default assumption that it was to begin a conversion when the last value was read).

So, we need to think a bit about *when* to give this *start* pulse. Here's a plausible, but foolish answer. This program is to take in A/D values endlessly, sending them to the D/A:

```
            movea.w #$8002, a2        ; point to A/D, D/A data ports
            movea.w #$8003, a3        ; point to A/D start port
  startit:  move.b d0, (a3)           ; start A/D. (query: what are
                                      ; we sending?
            move.b (a2), (a2)         ; from A/D, to D/A
            bra.s startit
```

Figure N21.6: Badly-placed A/D Start instruction

The trouble is that the A/D does not have time to convert, between Start pulse (evoked by the instruction at *startit*) and the data *read* in the next instruction. (A timing diagram in the Lab notes may help on this point.) You might think that you needed to pad the program with delays; that turns out not to be necessary, if you can find a way to let the A/D convert while the program is busy doing something it needs to do anyway.

Ch. 11: Worked Examples: Assembly Language: Ten Tiny Programs

You have seen a number of examples of small assembly-language programs, in Lab 20 and in Chapter 11. Chapter 11 concludes with a very large example, the signal-averaging instrument. In case you would like more *very small examples*, here are little programs to implement the schemes suggested in Lab 21, the lab that brings data in from an A/D and sends it out to a D/A.

0. Once for all: pointers

Let's set up some pointers, as usual, then invoke them in all the programs below. We want to use pointers because the code that uses them will run much faster than code forced to specify addresses *directly*, as you know.

```
movea.w #$8000, a0      ; point to displays (& keypad)
movea.w #$8001, a1      ; point to Ready port
movea.w #$8002, a2      ; point to converter data port
movea.w #$8003, a3      ; point to converter start pin
```

Figure X21A.1: The usual pointer initializations. We assume these in all the programs shown in these notes

By the way: sign extension

Today you probably don't care, but sometimes you will need to know, as we have said before, that a *word*-sized move to an address register gets *extended* automatically to *long* size: the CPU repeats our MSB (high, in this case), so as to preserve the sign. So, our "#$8000" gets loaded into a0 as the *32-bit* value, $FFFF8000. Strange, huh? Our circuit doesn't care. Why not?

1. Testing A/D

Your program needs to start the converter, then read it. It needs to give the converter time to finish a conversion: an attempt to read during conversion aborts the conversion, evoking bad data.

Here, to remind you, is the way the converter is wired, and a timing diagram showing what the converter wants, and when the *start* instruction delivers the start edge.

Incidentally, this example, which requires us to stare at the hardware a while before we can write code, is characteristic of the programming we want you to learn in this course: not fancy programming, but programming intimately fitted to the hardware.

Here is a reminder from the lab notes: the '7569's wiring and its timing characteristics. First, here is the way the chip is wired to your little computer:

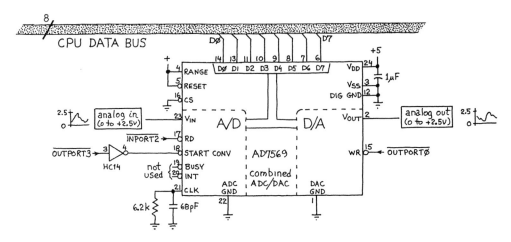

Figure X21A.2: '7569 converter as wired in Lab 21

And here is a timing diagram describing the way it behaves:

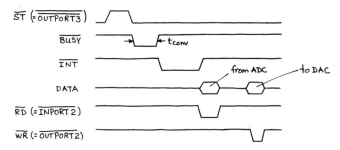

Figure X21A.3: '7569 timing

Surely the scheme we want is *roughly*

- start converter
- read A/D and display result
- do it again

Let's try writing it that way.

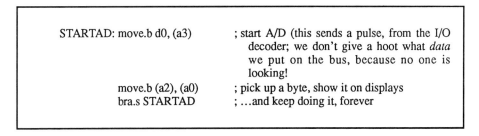

Figure X21A.4: *Defective* A/D test program: not enough time for conversion

Why use source *d0* in the first instruction? Why not send a zero, say? Because fetching a zero from memory takes time; d0 is in the CPU, so the instruction runs faster than, say, "move.b #0, (a3)."

Timing

The timing here is bad:

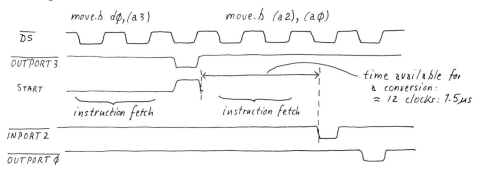

Figure X21A.5: Ill-considered placement of Start and Read instructions does not allow enough time for a conversion

The converter wants 2.6 µs to convert, worst-case; we are giving it about 1.5 µs.

Remedies

We could pad the program with a NOP or two. That's okay if necessary, but in this case it is *not* necessary. A rearrangement of the instructions solves the problem: once in the loop, let the conversion take place during the relatively-slow *branch* operation. *BRA* uses 18 clocks: 2.25 µs: almost enough in itself.

Here's the rearranged code:

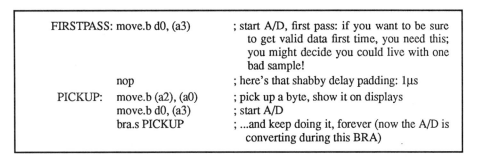

Figure X21A.6: *Corrected* A/D test program: enough time for conversion

2. Testing D/A

This is much easier. The D/A wants no start pulse. We need only send a byte to the 8-bit register that feeds the D/A. To see whether the converter is working properly, we can let the program "ramp" a digital value, and then watch to see whether the D/A output *ramps* nicely on the scope. We will watch, particularly, for strange bumps and holes in the staircase, a pattern that can reveal which bits are stuck (high or low) or interchanged with other bits.

clr.b d0	; clear the register that is to be sent to D/A: a little compulsive, but nice if you single-step the program, especially
SENDINC: move.b d0, (a2)	; send it (this time, d0 is not junk, as in the earlier example: it holds the gradually-growing value)
addq.b #1, d0	; increment the value to be sent
bra.s SENDINC	; ...forever

Figure X21A.7: D/A test program: incrementing value sent to converter

4. In & Out

After the A/D test, this is utterly easy: all we do is redefine the output port:

TRANSFR: move.b (a2), (a2)	; from A/D to D/A at a stroke!
move.b d0, (a3)	; start A/D
bra.s TRANSFR	; ...and keep doing it, forever (now the A/D is converting during this BRA)

Figure X21A.8: IN & OUT program

The line labeled "TRANSFR" looks crazy, doesn't it? Some skeptical student always protests, 'But you're not doing anything at all. You picked something up and put it back.' Someone else may object, 'The trouble is, whatever strange thing you're doing, you do it *twice*.' Do you see why these plausible objections are wrong?

That's right: those two "(a2)"s are not the same operation, and do not operate not twice on one object, but once on one object (the A/D), then on another (the D/A):

- the first use of "(a2)" is a *read*, and it takes a byte from the *A/D*
- the second "(a2)" is a *write*, and sends a byte to the *D/A*—a distinct device (though it happens to cohabit with the A/D, in a single chip)

Once you have this program working, you will want to modify it to include a call to a delay routine. That will let you bring the sampling rate down into the range of the low-pass filters suggested in Lab 21.

5. Invert

As the lab notes suggest, you need NOT, rather than NEG (2's complement), because NEG has the funny property that it leaves zero unchanged. That's good in a number system, but not good in an inverter: if the input waveform ever touched zero, the NEG'd output would stay there:

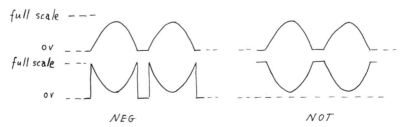

Figure X21A.9: The trouble with NEG as a waveform inverter

And here's the program. Only one instruction is changed, from the plain *In, Out* program:

```
FLIP:     not.b (a2)           ; from A/D, flipped, and back to D/A, still all
                                 at a stroke!
          move.b d0, (a3)      ; start A/D
          bra.s FLIP           ; ...and keep doing it, forever (now the A/D is
                                 converting during this BRA)
```

Figure X21A.10: INVERT program

This is a bit of a fluke (resulting from the fact that A/D and D/A sit at the same address), but isn't it magically compact? There's a lot going on *implicitly* in the instruction, "not.b (a2):"

You can figure out what's happening if you ask yourself, '*How* does the CPU flip every bit?' No, it doesn't sent little bit-flipping secret agents to wrestle with every bit out there in I/O-land (or in memory, if this happened to be an operation on memory), standing each bit on its head. No, instead, it brings the value onto the CPU, and operates on it there.

So, the instruction brings in a value, flips it, and sends it out again; implicit in this instruction, therefore, are three distinct operations:

- first, a read from (a2): from the A/D, in this case; the value is taken into some unnamed CPU register—some unmarked CIA register never mentioned in the 68008 data book: but we know it has to be there.
- then the CPU flips every bit (in this case; in another case it might add something to the value taken in, *and* it with something, or do whatever other process you wanted to specify);
- finally, the CPU writes the flipped value out to the address from which it took it. In this case, that place—though the same in *address*—is not the same as the source from which the data came: it is the D/A, not the A/D. (If this were an operation on RAM, instead, then the flipped value would go back to the same location, and would overwrite the earlier value.)

6. Full-wave Rectify

Here, for the first time, we need to do an operation *conditionally*: flip the waveform *if*.... *If* what? If the input lies below the midpoint of the waveform. How can we detect that? It's easy: the A/D output ranges from 0 to $FF; at the midpoint, the MSB goes from low to high. So, $80 will serve as midpoint (or $7F; it's a tossup: the *true* midpoint is halfway between the two!).

You might think, therefore, that we ought to compare the input to the midpoint. Something like,

```
INTAKE:   move.b (a2), d0      ; from A/D
          cmpi.b #$80, d0      ; see if input value < midpoint
```

Figure X21A.11: Clumsy way to check whether input exceeds midpoint of range

Then we'd do some branch conditional on the result of the compare.

If you're getting used to the 68000's versatile addressing, you may prefer something like

```
             move.b #$80, d1          ; load a register (for speedy compare)
FINDVAL: cmp.b (a2), d1               ; see what A/D has to offer (midpoint – value,
                                          this time)
```

Figure X21A.12: Another clumsy midpoint test

Actually this isn't such a good idea, because you need to pick up the data in any event. Still, if you thought of it, good: the 68000 can operate on memory or I/O, and sometimes you should take advantage of that power. Not this time, though.

Sneaky use of a flag

It turns out that we can skip the *compare* operation, because one of the CPU's *flags*—one of those on-chip flip-flops that records the result of recent operations, including data moves—already has just the information we need: the answer to the question whether the MSB is high or low. That flag is the *sign* flag, used in 2's complement operations. So, after picking up the data we can branch on that flag at once.

The fact that the *sign* flag always describes the MSB is worth remembering: it provides the fastest possible *single-bit test*: if you put a *Ready* key, for example, on d7 of the data bus, you need apply no mask before doing a conditional branch. You will see that done in the worked example called *68008-based frequency counter*, which follows immediately after these notes. (There, it is an overflow flag that is tested by this method.)

Now, here's the rectify program modified to use the *sign* bit:

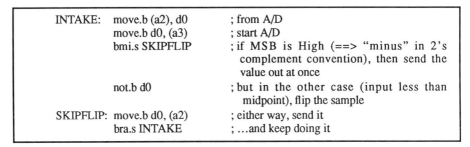

Figure X21A.13: FULL-WAVE RECTIFY program, using quickest test of input versus midpoint

This is quick, but the code includes one *dangerous* trick. Do you see it?

Here it is: *we let an instruction intervene between the operation that interests us and the branch*. Does the program work, nevertheless?

Yes. It does because we were careful (or lucky!) to use d0 again in the *start A/D* operation. Thus the sign flag still has the right information when the program reaches the *bmi* instruction.

You can see that a program could easily run awry when written this way. One might write the following code, quite plausibly:

```
INTAKE:   move.b (a2), d0      ; from A/D
          move.b d1, (a3)      ; oh, send any data register—it's only the
                                  start pulse that we want
          bmi.s SKIPFLIP
```

Figure X21A.14: Faulty variation: a *start* instruction that messes up flags

But with this plausible code the program would fail. Do you see why?

7. Half-wave Rectify

In this case, instead of flipping the input we just force it to the midpoint. So all we need do is replace the "not.b d0" instruction with

moveq #80, d0

The moveq instruction is compact (just 16 bits, including the data), but it is strange, too:

- it is always *long* in size (here we simply ignore the higher order bits, which happen to be *1*'s—see the next point:)
- the instruction sign-extends the specified constant as it loads the entire 32-bits of the register (this is the same as the effect we mentioned earlier when speaking of address registers loaded with a word).

The program now looks like this:

```
INTAKE:   move.b (a2), d0      ; from A/D
          move.b d0, (a3)      ; start A/D
          bmi.s SKIPFORCE      ; if MSB is High (==> "minus" in 2's
                                  complement convention), then send the
                                  value out at once
          moveq #$80, d0       ; but in the other case (input less than
                                  midpoint), force value to midpoint
SKIPFORCE:
          move.b d0, (a2)      ; either way, send it
          bra.s INTAKE         ; ...and keep doing it
```

Figure X21A.15: HALF-WAVE RECTIFY program

8. Low-pass Filter

Here we need to form an average, each time round the loop, between the new sample and the previous average. We want to give old and new equal weight, so we can add and divide by two—or perhaps divide by two and then add.

The general scheme is clear enough; it's the details that get sticky, as usual. Does it matter whether we add first, divide second, or vice versa? Should we use the 68000's *divide* instruction?

Let's start by doing it slightly wrong, just to appreciate what's at stake (Tom did it wrong, first try, in the lab). We don't want to be too ludicrously wrong, though: we *don't* want to use the *divide* instruction—which takes 144 clocks. Instead, we can divide by two very fast with a *shift-right* operation. Here goes:

```
           clr.b d0           ; clear old average, for clean start
SUM:       add.b (a2), d0     ; form sum of old & new
           move.b d0, (a3)    ; start A/D for next pass
           lsr.b #1, d0       ; shift right to divide by 2
           move.b d0, (a2)    ; send it to D/A
           bra.s SUM          ; ...and keep doing it
```

Figure X21A.16: FAULTY low-pass filter program

What's wrong with that? *Lsr*, "logical shift right," brings a zero into the MSB. Most of the time, that is just what we want. But what if the old average was $80, and the new value was $80. Their average *should* be $80. What is it, according to the program above? *Zero*.

Well, suppose we divide the new and old values by two before summing them. Then we can't overflow the 8-bit register. This is a little disappointing, too, though: it loses information that was carried in the LSB of each value. For example, if we average $81 with $81 by this method, after shifting each has the value $40, and their sum is $80: close, but not the best we can do.

Remedies

We need to keep the full 8 bits of each value, sum them, and then *keep* the information that a *Carry* was or was not generated, as we divide by two (that is, shift right). There are two ways to get this result:

- keep the sum as a *word*-length quantity: do an *add.w* between old average and new sample;
- or keep just the *carry* information, bringing it in as MSB when shifting right

The first method is slightly the fussier of the two. We have to be sure that the high byte of the *sample* word starts cleared; in addition, we cannot do the tidy

add.w (a2), d0

because this would bring in a high byte of junk (the A/D gives only 8 bits). We could work around both these restrictions.

The second method is easy to implement: we need only substitute, for the *logical shift*, a similar instruction called *rotate*, which (in one of its forms—the one called "rotate with extend": *ROXR*) takes in the carry bit instead of a zero at one end of the register. (We are being slightly sloppy, and generic, calling this bit "carry:" strictly, it is what Motorola calls the "extend" flag, *X*. *X* sometimes differs from carry, though rarely; it does not differ in the present case).

Here's the program, using ROXR:

```
           clr.b d0           ; clear old average, for clean start
SUM:       add.b (a2), d0     ; form sum of old & new
           move.b d0, (a3)    ; start A/D for next pass
           roxr.b #1, d0      ; form average by shifting right, but this time
                              ;   taking in Carry (or extend-) flag as MSB
           move.b d0, (a2)    ; send average to D/A
           bra.s SUM          ; ...and keep doing it: now "new" average is
                              ;   old: sic transit, once again
```

Figure X21A.17: Corrected LOW-PASS FILTER program

If you want to alter f_{3dB}, then you need to alter the relative weights given old and new samples. To do that you need to do something less crude than the divide-by-two effected by the shift or rotate instructions. Trouble is, operations that are less crude also are less fast. Let's face it: the microprocessor is not well-suited to this kind of work. It is not fast enough. Next time you want a single-pole low-pass, try an *R* and a *C*!

9. Storage Scope

compare Text sec. 11.02 , p. 753,
program 11.1; exercise 11.1, p. 748
very similar

You have been using *fixed pointers* all along: address registers used for indirect addressing. Here's a first chance to use a *moving pointer*. (If you implement this scheme in the lab, you also produce an entertaining gadget.)

Because the pointer will move, we will need to check for the end of the process: check whether the pointer has reached the end of the table. As the lab notes suggest, there are several ways to do this task:

- use a *counter register*, loaded initially with table length, and decremented each time the pointer is advanced;
- use an *address compare*, asking whether the moving pointer has hit some predefined limit address.

On a table *read* (what we call 'playback,' in the storage-scope program), the program could use one more alternative device to check for table end:

- some stored value that means 'EndOfTable.'

This method is neatest of all, but doesn't work in the storage-scope application: all of the 256 possible 8-bit combinations are needed as genuine data values. The 'EndOfTable' method *does* work for a later application: Lab 23's program that drives a talker chip. That chip does not use all possible codes, so we are able to store $FF as the last entry in each table, and let the program branch out when it reads that value.

Now, let's write the code:

```
; STORAGE SCOPE
; here is a one time only initialization; all pointers are assumed loaded as usual (see start
    of these notes):
            movea.w #$1FF0, a5      ; set up end-of-table address marker: leaves
                                        16 bytes at top of RAM, for stack
; ...and here the main program loop begins
STARTFL:    move.b (A0), d0         ; clear Ready flop with a dummy input
                                        operation (see Lab 20-10)
            move.b d0, (a3)         ; start A/D for first pass (this time we don't
                                        shrug at even one bad value, since we're
                                        saving all samples)
            movea.w #$400, a4       ; init storepointer to tablestart
STORMOR:    move.b (a2), (a4)+      ; this advances the storing pointer after using
                                        it. Notice that the source pointer does not
                                        move: it always looks to the A/D (later
                                        D/A).
            move.b d0, (a3)         ; start A/D for next pass
            cmpa.w a5, a4           ; see if we've finished the table: subtracts end
                                        marker from present value of moving
                                        pointer; does not alter a4 or a5, but makes
                                        flags reflect the result.
            bls STORMOR             ; carry on till moving pointer passes end
                                        marker
; and here begins the Playback section. We land here after the table has been filled
REINIT:     movea.w #$400, a4       ; init storepointer to tablestart
PLAYMOR:    move.b (a4)+, (a2)      ; sample from table to D/A
            move.b d0, (a3)         ; dummy start of A/D: Huh? Yes, it's a fake,
                                        designed to make the playback rate match
                                        the store rate
            cmpa.w a5, a4           ; see if we've finished the table
            bls PLAYMOR             ; carry on till moving pointer passes end
                                        marker
RDYCHK:     move.b (a1), d0         ; take in ready bit, as usual
            andi.b #$1, d0          ; mask, as usual
            beq.s REINIT            ; if not Ready, playback again
            bra.s STARTFL           ; ...but if ready, go fill the table anew
```

Figure X21A.18: STORAGE SCOPE program (delay omitted)

The playback rate matches the store rate, but a kink will appear in the playback as your program checks for Ready. This kink—lasting a handful of microseconds—is not a serious flaw, though, for there's another, nastier kink in the waveform that's played back: the end of the stored waveform will not match the start (unless you're magically-lucky).

A practical version of this program would let you control the sampling rate: between storings of successive samples, call a delay routine. (Make sure you call the same routine on playback, of course.) Once an appreciable delay pads the samples, the kink caused by RDYCHK will become negligible.

10. A program only sketched: function-select without continual scan for key value

This scheme, suggested in Lab 21-6 ("fancier...") turns out to be not very rewarding. It is also rather hard to understand, so if you are squeamish *don't read on*. If you're just beginning to get used used to assembly-language code, forget this last example.

If, however, you just long to try this program, we might point out a helpful 68000 instruction:

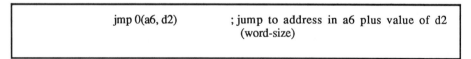

```
jmp 0(a6, d2)         ; jump to address in a6 plus value of d2
                      ;   (word-size)
```

Figure X21A.19: 68000's jump instruction can be steered by data in a register; that register could be loaded from the keypad

This is a versatile form of *jump*: one that takes as its destination a *base address* (defined in an address register) plus an *offset* defined in another register: the result is that all of your several loops (Do-A, Do-B, Do-C...) can end with a common instruction that says jump to *base plus offset*—with the wrinkle that the offset has been loaded from the keypad, and changed from time to time.

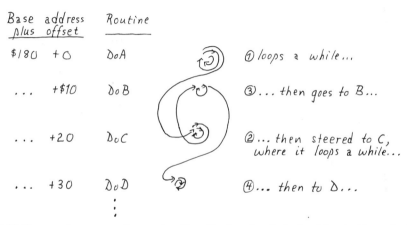

Figure X21A.20: The several response routines, each ending with a jump to a destination determined by a value loaded from the keypad upon Ready

Here's some code to exploit this neat instruction:

```
           movea.w #$180, a6      ; base address for set of small programs that
                                      DoThis, DoThat...
  Do-A    move.b (a2), (a2)      ; here you do whatever the 'A' key should
                                      evoke
           move.b d0, (a3)
           bsr STEERME            ; go check Ready flop, and change steering
                                      register only if Ready key is pressed
           jmp 0(a6, d2)          ; jump to address in a6 plus value of d2
                                      (word-size)
  Do-B    not.b (a2)             ; here you do whatever the 'B' key should
                                      evoke
           move.b d0, (a3)
           bsr STEERME
           jmp 0(a6, d2)          ; jump to patch of code selected from keypad
                                      (keypad determines offset, value in d2)
```

Figure X21A.21: Using the steered-jump

Register d2 plus a6 must point to the start of *one* of the "Do-?" program-fragments. The program loops in that selected fragment so long as d2's value is constant.

Another routine, which we will call "STEERME" checks the ready key, and alters the value of d2 only if Ready is asserted:

```
; SUBROUTINE 'STEERME': on exit, d2 holds offset to start of appropriate response
  STEERME:    move.b (a1), d4     ; the usual ready check (we're omitting the
                                      usual 'save d4 on stack.' We don't want to
                                      clutter this code—and we want the code to
                                      run fast, as well.)
              andi.b #$1, d4      ; mask
              beq.s EXIT          ; if not ready, leave steering register as it was
              move.b (a0), d2     ; but if ready, take in key value (incidentally
                                      clearing Ready flop)
              lsl.b #4, d2        ; this keeps only the most recent key value,
                                      and puts it into left nybble of byte
              subi.b #$90, d2     ; now this cuts $A0 to $10, $B0 to $20, and
                                      so on, so as to give comfortable space for
                                      each little response program
  EXIT:       rts                 ; d2 now holds the chosen offset—either
                                      because it's unchanged (Ready key not hit)
                                      or because we've just changed it
```

Figure X21A.22: Steerme routine, which alters steering register only if Ready key has been pressed

This is all pretty clever—but it doesn't work very well. The whole point was to save time, obviating the scan of all possible key codes. But this response routine is itself pretty slow: the fancy jump takes an extra microsecond to compute its destination; worse, the 'call' to STEERME—with its stacking and unstacking—takes 66 clock cycles: more than 8 microseconds. So, unless there are a great many key codes to scan for, the baby has exited with the bathwater. You could make it run a good deal faster by putting the Ready-check code *in-line*, instead of within a subroutine. But that's boring to code and write in by hand. Too bad: the scheme looked pretty smart! What we really need is an *interrupt*, and soon will meet that technique.

(——— *Program listings follow, showing assembled code* ———)

10 TINY PGMS_GLOBALS from 10 tiny pgms.Rel

```
000000:                         .Verbose
000000:                         ; Ch. 11: Worked Examples: Ten Tiny Programs
000000:
000000:                         ; init pointers
000000: 30 7C 80 00                 movea.w #$8000, a0      ; point to displays (& keypad)
000004: 32 7C 80 01                 movea.w #$8001, a1      ; point to Ready port
000008: 34 7C 80 02                 movea.w #$8002, a2      ; point to converter data port
00000C: 36 7C 80 03                 movea.w #$8003, a3      ; point to converter start pin
000010:
000010:                         ;--------------------------------
000010:                         ; TESTING A/D
000010: 16 80                   FRSTPAS    move.b d0, (a3)  ; start A/D, first pass
000012: 4E 71                       nop                     ; time killer
000014:
000014: 10 92                   PICKUP     move.b (a2), (a0) ; show A/D input value
000016: 16 80                       move.b d0, (a3)         ; start A/D for next pass
000018: 60 FA                       bra.s PICKUP
00001A:
00001A:                         ;--------------------------------
00001A:                         ; TESTING D/A
00001A: 42 00                       clr.b d0
00001C:
00001C: 14 80                   SENDINC    move.b d0, (a2)  ; send current value to D/A
00001E: 52 00                       addq.b #1, d0           ; increment value to be sent
000020: 60 FA                       bra.s SENDINC           ; ...forever
000022:
000022:                         ;--------------------------------
000022:                         ;IN & OUT
000022:
000022: 14 92                   TRANSFR    move.b (a2), (a2) ; from A/D to D/A at a stroke!
000024: 16 80                       move.b d0, (a3)         ; start A/D
000026: 60 FA                       bra.s TRANSFR           ; ...and keep doing it, forever
000028:
000028:                         ;--------------------------------
000028:                         ; INVERT
000028:
000028: 46 12                   FLIP       not.b (a2)       ; from A/D, flipped, and back to D/A,
00002A:                                                     ;   still all at a stroke!
00002A: 16 80                       move.b d0, (a3)         ; start A/D
00002C: 60 FA                       bra.s FLIP              ; ...forever
00002E:
00002E:
00002E:                         ;--------------------------------
00002E:                         ;FULL-WAVE RECTIFY
00002E:
00002E: 10 12                   GETIT      move.b (a2), d0  ; from A/D
000030: 16 80                       move.b d0, (a3)         ; start A/D
000032: 6B 02                       bmi.s NOFLIP            ; if MSB is High...then send
000034:                                                     ;   the value out at once
000034: 46 00                       not.b d0                ; but in the other case,
000036:                                                     ; ...flip the sample
000036: 14 80                   NOFLIP     move.b d0, (a2)  ; either way, send it
000038: 60 F4                       bra.s GETIT             ; ...and keep doing it
00003A:
00003A:
00003A:                         ;--------------------------------
00003A:                         ; HALF-WAVE RECTIFY
00003A:
00003A: 10 12                   INTAKE     move.b (a2), d0  ; from A/D
00003C: 16 80                       move.b d0, (a3)         ; start A/D
00003E: 6B 04                       bmi.s NOFORCE           ; if MSB is high, then send
000040:                                                     ;   the value out at once
000040: 10 3C 00 80                 move.b #$80, d0         ; but in the other case,
000044:                                                     ; ...force the value to midpoint
000044: 14 80                   NOFORCE    move.b d0, (a2)       ; either way, send it
000046: 60 F2                       bra.s INTAKE            ; ...and keep doing it
000048:
000048:
000048:                         ;--------------------------------
000048:                         ; LOW-PASS FILTER
000048:
000048: 42 00                       clr.b d0                ; clear old average, for clean start
00004A: D0 12                   SUM        add.b (a2), d0        ; form sum of old & new
00004C: 16 80                       move.b d0, (a3)         ; start A/D...
00004E: E2 10                       roxr.b #1, d0           ; form average by shifting right,
```

Figure X21A.23: Ten Tiny Programs: listing (first page of three)

```
000050:                                         ; but this time taking in Carry
000050:                                         ; (or extend-) flag as MSB
000050: 14 80             move.b d0, (a2)       ; send average to D/A
000052: 60 F6             bra.s  SUM            ; ...and keep doing it....
000054:
000054:
000054:                ;-------------------------------
000054:                ; STORAGE SCOPE
000054:
000054:                ; Initializations (other pointers assumed initialized as at start
000054:                ;   of these notes, once again)
000054: 3A 7C 1F F0       movea.w #$1FF0, a5    ; set up end-of-table address marker:
000058:                                         ;   leaves 16 bytes at top of RAM,
000058:                                         ;   for stack
000058:
000058:                ; ...and here the main program loop begins
000058: 16 80             TAKEONE  move.b d0, (a3)  ; start A/D for first pass
00005A:                                         ; ...to give A/D time to convert
00005A:
00005A: 10 10             STARTFL  move.b (a0), d0 ; clear Ready flop
00005C:                                         ;   with dummy input
00005C: 38 7C 04 00       movea.w #$400, a4     ; init storepointer to tablestart
000060:
000060: 18 D2             STORMOR  move.b (a2), (a4)+  ; pick up a sample,
000062:                                         ;   store it, and
000062:                                         ;   advance storing pointer
000062: 16 80             move.b d0, (a3)       ; start A/D
000064: B8 CD             cmpa.w a5, a4         ; see if we've finished the table...
000066: 63 F8             bls STORMOR           ; carry on till moving pointer
000068:                                         ;   passes end marker
000068:
000068:                ; and here begins the Playback section. We land here after
000068:                ;   the table has been filled
000068:
000068: 38 7C 04 00       REINIT   movea.w #$400, a4  ; init storepointer to tablestart
00006C: 14 9C             PLAYMOR  move.b (a4)+, (a2) ; sample from table to D/A
00006E: 16 80             move.b d0, (a3)       ; dummy start of A/D....
000070: B8 CD             cmpa.w a5, a4         ; see if we've finished the table...
000072: 63 F8             bls PLAYMOR           ; carry on till moving pointer
000074:                                         ;   passes end marker
000074:
000074: 10 11             RDYCHCK  move.b (a1), d0    ; take in ready bit, as usual
000076: 02 00 00 02       andi.b #$2, d0        ; mask, as usual
00007A: 67 EC             beq.s REINIT          ; if not Ready, playback again
00007C: 60 DC             bra.s STARTFL         ; ...but if Ready, go fill table anew
00007E:
00007E:
00007E:                ;-------------------------------
00007E:
00007E:                ; STEERME: lets value entered from keypad determine jump destination,
00007E:                ;   thus determining what operation program performs on input waveform
00007E:
00007E: 3C 7C 01 80       movea.w #$180, a6     ; base address for set of small
000082:                                         ;   programs that DoThis, DoThat
000082:
000082: 42 40             clr.w d0              ; startup value for Do_Offset register
000084:
000084: 14 92             DoA      move.b (a2), (a2) ; here you do whatever the 'A' key
000086:                                         ;   should evoke
000086: 16 80             move.b d0, (a3)       ; start A/D, as usual
000088: 4EBA 001C         bsr STEERME           ; go see if need to branch anywhere
00008C:                                         ;   but back to DoA
00008C: 4E F6 00 00       jmp 0(a6, d0)         ; now jump to wherever the offset directs
000090:
000090:
000090: 46 12             DoB      not.b (a2)   ; this is what, say, key B should evoke
000092: 16 80             move.b d0, (a3)       ; start A/D, as usual
000094: 4EBA 0010         bsr STEERME           ; go see if need to branch anywhere
000098:                                         ;   but back to DoB
000098: 4E F6 00 00       jmp 0(a6, d0)         ; now jump to wherever the offset directs
00009C:
00009C:
00009C:                   DoC                   ; something else...
00009C:                ; .
00009C:                ; .
00009C:                ; .                      ; but followed in all cases
00009C: 16 80             move.b d0, (a3)       ;   by these same three lines of code
00009E: 4EBA 0006         bsr STEERME           ;
0000A2: 4E F6 00 00       jmp 0(a6, d0)         ; now jump to wherever the offset directs
0000A6:                ; .
```

Figure X21A.24: Ten Tiny Programs: listing (second page of three)

```
0000A6:
0000A6:                                 ; and here's the subroutine used by all these loops: it updates the offset
0000A6:                                 ;   register, if the Ready key was pressed
0000A6:
0000A6: 18 11                   STEERME     move.b (a1), d4   ; the usual Ready check
0000A8: 02 04 00 02                         andi.b #$2, d4    ; mask
0000AC: 67 08                               beq.s EXIT        ; if  not ready, leave steering register
0000AE:                                                       ;   as it was
0000AE: 14 10                               move.b (a0), d2   ; but if ready, take in key value
0000B0:                                                       ;   incidentally clearing Ready flop)
0000B0: E9 0A                               lsl.b #4, d2      ; this keeps only the most recent key value,
0000B2:                                                       ;   and puts it into left nybble of byte
0000B2: 04 02 00 90                         subi.b #$90, d2   ; now this cuts $A0 to $10, $B0 to $20,
0000B6:                                                       ;   and so on, so as to give comfortable
0000B6:                                                       ;   space for each little response program
0000B6: 4E 75                   EXIT        rts               ; d2 now holds the chosen offset--either
0000B8:                                                       ;   because it's unchanged or because we've
0000B8:                                                       ;   just changed it
```

Figure X21A.25: Ten Tiny Programs: listing (third page of three)

Ch 11.: Worked Example: 68008-based Frequency Counter

Problem: 12-bit counter as input device

Show how to interface a 12-bit counter to a 68008 controller like your lab computer. The counter is clocked by a signal from the outside world, and is to serve as an event counter or frequency counter. The computer is to allow the counter to accumulate counts for a while, then read it when the CPU chooses to.

Assume an I/O decoder already is wired for you as in the lab computer.

Rules of the game:

- Input: counter

    - use 4-bit binary counters, synchronous, with jam clear (use a generic counter of your invention, if you like; or this could be a 74HC161);
    - make sure the CPU takes in a count that is not a mixture of new and old, as it picks up 4 bits, then 8 bits;
    - include hardware that allows the CPU to clear the counter;
    - include a flag that the CPU can read to discover whether the counter has overflowed;

- Program:

    - write code that will take frequency counts from this hardware, after first checking that no overflow has occurred.

        * The program should store *valid* counts in a data table 1K long, beginning at $400; the 12 valid bits should be stored (right-justifed), with zeros at the 4 unused bit positions);
        * In case of overflow, the program should store $FFFF in the same data table, then proceed as usual to look for the next count;
        * when the table is full, the program should branch to a label, "QUIT"

This exercise is meant to remind you of the intimate relation between code and hardware: you will not be able to write this little program without looking at the hardware that you have sketched. The exercise incidentally will show again how easily the 68008 handles transfers larger than 8 bits. Along the way, we will need to work through some operations now becoming familiar: branching on a bit, and storing with a moving pointer, for example.

The hardware details

Here's some hardware to do the job. One point requires comment: we show the *overflow* flag driving the data bus at d7. This assignment would conflict with the *second* of the lab computer's two Ready keys, one that you wire up if you use the Register-Check routine. To run this frequency counter as shown you would need to assign the Register-check Ready to another bit position. We will keep *overflow* at d7 in this example so that we can illustrate a trick we referred to earlier: this is the quick, no-mask bit test that we referred to in the notes called X21A (p. X21A-6). We use the MSB or *sign* bit to carry information; then the sign flag reflects that information after a move, and no masking or other bit-testing operation is needed.

For the rest of the circuit, we'll let our little "balloon" notes take care of explanations.

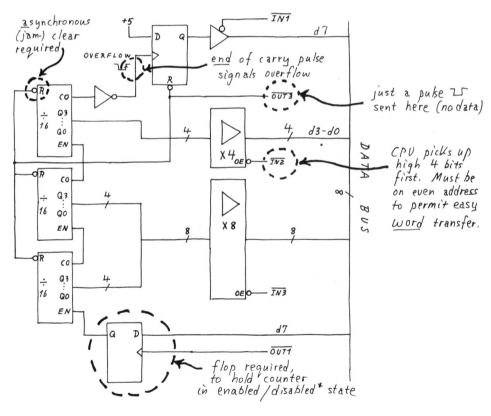

Figure X21B.1: Frequency counter hardware

Ch 11.: Worked Example: 68008-based Frequency Counter

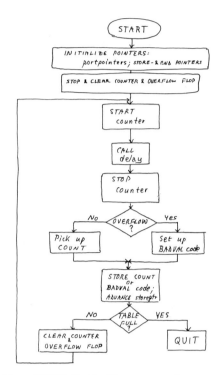

Figure X21B.2: Sketch of frequency-counter program

Now let's try the code:

```
            ; FREQUENCY COUNTER PROGRAM
            ; assumes hardware of example X21B: 12-bit counter input

; initializations
            movea.l #$8002, a2      ; point to counter data
            movea.l #$8003, a3      ; point to Clear port (counter & overflow flop)
            movea.l #$8001, a1      ; point to the usual Ready port (bit input);
                                    ;   this is also the Counter Enable port, on output
            movea.l #$400, a4       ; point to start of data table
            movea.l #$7FF, a5       ; point to end marker (= last entry address)
;-----------------
            clr.b   (a1)            ; disable counters
            move.b  d0, (a3)        ; clear counters and overflow flop (we
                                    ;   send anything at all; we need only a pulse
                                    ;   from the decoder, not a transfer of data

ANEW:       move.b  #$80, (a1)      ; start counters
            bsr.s   FIXEDDELAY      ; wait a fixed time (perhaps 0.1 s)
            clr.b   (a1)            ; disable counters

            move.b  (a1), d0        ; overflow?
            bmi.s   BADVAL          ; if overflow, go do the right thing
                                    ;   this is the sneaky way to test a single bit: assign it in hardware
                                    ;   to MSB position, then sign flag reflects its level without BTST
                                    ;   or mask.  Perhaps a little annoying to read such queer code.
                                    ;   BTST would be nice & explicit

            move.w  (a2), d0        ; but if no overflow, pick up the sample
            andi.w  #$FFF, d0       ; ...and force the top 4 bits to 0
            bra.s   STORIT          ; ...and go store this sample
BADVAL:     move.w  #$FFFF, d0      ; if overflow: set up badvalue marker, for storage

STORIT:     move.w  d0, (a4)+       ; now store it--good sample, or warning that an overflow occurred
            cmpa.l  a5, a4          ; now see if table is full
            bhi.s   QUIT            ; if moving pointer exceeds end of table, quit

            move.b  d0, (a3)        ; otherwise, clear counter (& overflow flop,
            bra.s   ANEW            ; and go do it again

QUIT:       ; ----                  ; exit from this program
```

Figure X21B.3: Frequency-counter program

Hand Assembly: 68008 Codes

Hand Assembly?

At first glance, the task of "assembling" 68000 code[1] may look hopeless: doesn't a 16-bit instruction word imply 64K possible codes? And weren't assemblers developed just to take this horrible work off human shoulders? Yes. We don't propose that you do *much* assembling of code by hand. But we have found that for tiny programs hand assembly is *faster* than the use of an assembler. Some overhead gets in the way of assembling a very small program: you have to go to the computer, type in code, invoke the assembler, print. That takes longer than writing a 6-line program by hand. And whenever someone sits down in front of a computer, the person runs the risk of getting drawn in to the entertaining but time-gobbling game of trying small improvements, revising, revising some more. That process is appropriate in a programming course, but not in a course that tries to keep your attention close to the hardware side of your own little computer.

So, we encourage students to assemble their first, very small programs by hand: little programs like the ones needed in Lab 21. The good news is that you need only a very few instructions to for nearly all these programs (the low-pass filter uses a couple of the slightly less common instructions; the other programs are extremely simple). You need *move* (lots of these), *branch* (conditional; especially *bne* and *beq*), and occasionally *andi* for the purpose of masking; you need *subq* or subtract, and you may need *compare* to check for table ends. You can get by for a while with only those instructions.

The selective list of codes that is attached to these notes helps above all with the *moves*, the most common of the instructions, the *moves*. You can read the move codes directly from the table, if you are willing to use data registers d0 and d1, and address registers a0 and a1. When you need to use another register, scan the two cases done on the table to see the pattern.

The second page of codes shows some Boolean operations (AND and OR, etc.) and a set of arithmetic operations (add, subtract, —even multiply and divide, though these are hard to code as well as slow), and *compare*. These are more painful to code, but they turn out to share a common pattern, set forth here as the "SEA" code—Size and Effective Address. Some students never dare tangle with this half of the coding sheet, but we hope you will be braver than these people. The *moves* certainly are easier, but we think *CLR.B d0*, for example, would not hold you up, either. You will be able to figure out CLR... very fast, once you have done two or three cases.

If your head begins to spin as you look at this table, you can retreat to an assembler. But give this method a try. On the following pages you will find examples.

---

[1] In case the notion of "assembly" isn't quite clear to you, it means simply finding the machine code equivalent to the mnemonic: discovering that the hex code for "move.b d0, d1" is 1200. It also includes the conversion of *labels*, convenient for human readers, into *addresses* (absolute or relative), which the computer requires.

Examples:

Moves

If you want to find the code for *move.b d1, (a0)*, you need to find the intersection between the *source* column (d1, in this case) and the *destination* row (a0, in this case):

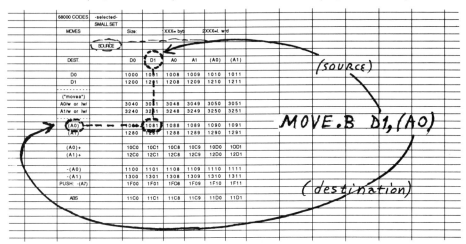

Figure Ass'y.1: A move example: the code appears at the intersection of *source* column with *destination* row

Instructions using common 'size and effective address' pattern ("SEA")

Below the code for *clr.b d1* is shown. The first byte, 42, defines the operation; the second byte says what is to be cleared. To form the second byte, we look to the SEA table to find the addressing mode, here "data register direct" (abbreviated 'data dir'). The first hex character depends on size; we want byte size so we take the "0." The instruction concludes with what the table labels "#," which means simply 'register number,' in this case *one*. So, the complete code is 42 01. The second time you do this it will seem easier; the 5th time it will seem very straightforward. When you are comfortable with codes like this one, they try, say, *subq*, which is a little harder. Don't forget that immediate instructions must include immediate *data*; this follows the first word of instruction. For example, *cmpi.b #$9, d0* is coded as 0C00 0009. Make sense?

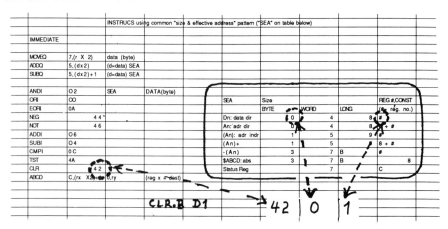

Figure Ass'y.2: CLR.B D1: the opcode is peculiar to CLR; the rest of the instruction is made up according to a formula shared by many instructions ("SEA")

The tables appear below.

68000 CODES: MOVE INSTRUCTIONS, BRANCH CONDITIONALS, ETC.

MOVE CODES

68000 CODES		Size:	1xxx = byte	2xxx = long wd	3xxx = word						
MOVES	selected										
	SOURCE →	D0	D1	A0	A1	(A0)	(A1)	(A0)+	(A1)+	(A7)+ POP:	-(A0)	-(A1)	ABS (word)	IMM (word)
DESTINATION ↓												
D0		1000	1001	1008	1009	1010	1011	1018	1019	101F	1020	1021	1038	103C
D1		1200	1201	1208	1209	1210	1211	1218	1219	121F	1220	1221	1238	123C
("movea")														
A0 (word or lw)		3040	3041	3048	3049	3050	3051	3058	3059	305F	3060	3061	3078	307C
A1 (word or lw)		3240	3241	3248	3249	3250	3251	3258	3259	325F	3260	3261	3278	327C
(A0)		1080	1081	1088	1089	1090	1091	1098	1099	109F	10A0	10A1	10B8	10BC
(A1)		1280	1281	1288	1289	1290	1291	1298	1299	129F	12A0	12A1	12B8	12BC
(A0)+		10C0	10C1	10C8	10C9	10D0	10D1	10D8	10D9	10DF	10E0	10E1	10F8	10FC
(A1)+		12C0	12C1	12C8	12C9	12D0	12D1	12D8	12D9	12DF	12E0	12E1	12F8	12FC
-(A0)		1100	1101	1108	1109	1110	1111	1118	1119	111F	1120	1121	1138	113C
-(A1)		1300	1301	1308	1309	1310	1311	1318	1319	131F	1320	1321	1338	133C
PUSH: -(A7)		1F00	1F01	1F08	1F09	1F10	1F11	1F18	1F19	1F1F	1F20	1F21	1F38	1F3C
ABS		11C0	11C1	11C8	11C9	11D0	11D1	11D8	11D9	11DF	11E0	11E1	11F8	11FC

BRANCHES — dspl (8bits, from PC + 2)

BRA	60
BSR	61
BEQ	67
BNE	66
BHI	62
BLS	63
BMI	6B
BPL	6A

SPECIAL

NOP	4E 71	
RTS	4E 75	
RTE	4E 73	
JMP	4EF8,adr	(16 bit address)
JSR	4EB8,adr	
STOP	4E72,2700	
TRAP	4E4(trap #)	
SET TRACE	00 7C 80 00	ORI.W #$8000, SR

Figure Ass'y.3: MOVE codes, BRANCH conditional, and miscellaneous other instructions

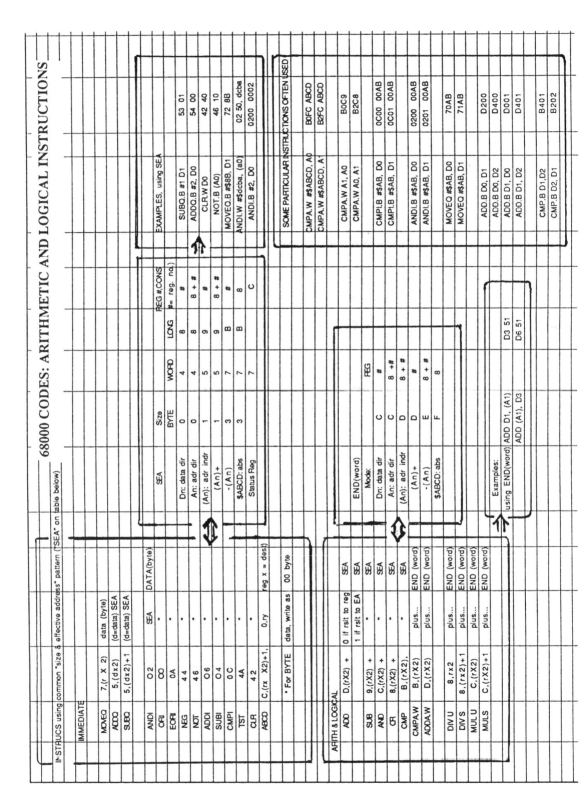

Figure Ass'y.4: Some instructions sharing a common "Size & Effective Address" code: arithmetic and logical operations

Lab 21: μ4: A/D, D/A: Data Handling

Reading:	Ch. 10: Look at programming examples in Chapter 10 re: indirect addressing (table copy, fill): program listings 10.1 (p. 682), 10.3 (p. 688);
	Ch. 11: then look at the same topic in chapter 11—where you are again invited to copy an array (ex. 11.1, p. 5).
	Then see the example of program listing 11.1. The first method should be intelligible; the second method may not be intelligible until you take a look at the little Motorola manual's description of that instruction *DBF*. It's a little obscure.
	In the detailed discussion of the *signal averager*, skim or omit most programming details at this time; concentrate on the hardware: compare the demanding A/D and D/A used in the *averager* against the pleasantly fast and easy chip we use in the present lab (AD7569).
Problems:	Embedded problems: 11.1 (*BGT*, suggested there, works; but *BHI* is simpler, and works better in the general case, with unsigned values like addresses); 11.5-11.7 you will find a relaxing reversion to combinational hardware; 11.8
	Postpone: 11.4, 11.15, 11.17. These treat interrupts, which we will discuss next time.

You will write the programs in today's lab. The computer will take data from an 8-bit A/D; it will do something to that data; then it will use its results to drive a D/A. You will be able see the effect of sampling-rate on the *reconstructed* analog signal: the D/A output will get 'steppy' as the sampling rate approaches the theoretical minimum of 2/period—but a good low-pass filter will recover the original sine, smoothing away the steps. If you are ambitious, you will use the *phase-locked loop* that you built in Lab 18 to set the filter's f_{3dB} automatically.

You will also have a chance to write some keyboard-check routines that let you use keyboard commands to steer the program into one or another action while it is running. In one of these programs you will use *indirect addressing* in the application for which it is essential: filling and reading-back tables. Throughout, we encourage you to cast your program in the form of Main and Subroutine sections (reminders on this, below).

We hope you will enjoy the challenge of writing these small programs. If you don't—and instead find yourself frustrated after a reasonable effort—then you may want to peek at our solutions which appear in a Worked Example called 'Ten Tiny Programs.'

Your programs today will be short, and can best be assembled by hand in a few minutes. We hope you will resist the inclination experienced computer users may feel to rush to a MacIntosh or PC assembler: an assembler slows you down if the program is tiny.

For hand assembly, we like the coding sheet; some people prefer to borrow patches of code from programs listed in earlier labs; still others seem to enjoy the agony of figuring the hex equivalents of the bit-by-bit code described in the Motorola manual! Take your pick. Soon we will relent and let you use a MacIntosh assembler.

Hardware Additions

Today's hardware additions are very simple:

We add an A/D on the input side, a D/A on the output side. Then we put the D/A output through a switched-capacitor low-pass filter.

We use just one chip to do both A/D and D/A conversion. The two devices share a single set of eight data lines, and they are fast enough to let us use them without the wait states required in the text's example, the *averager*.

This single-chip A/D-D/A, the AD7569, shows lots of nice features. Before we look at the details of wiring this chip to your computer, let's pause to admire the compression of functions in this one chip.

Here is the converter's block diagram:

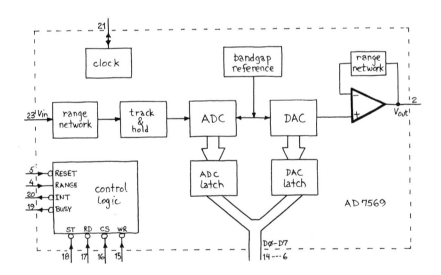

Figure L21.1: AD7569 converter: block diagram

And here are some of the characteristics of the chip that make it appealing and easy to use:

- Single, 5-volt supply possible;
- Includes a D/A with some nice features:

    - Register of D flops to feed it;
    - Short setup time, so that it does not require slowing the CPU (by delaying DTACK*);
    - Voltage output (rather than current, common on earlier D/A's);

- Includes an 8-bit A/D also showing some good features:

    - Fast: 2 to 2.6 µS;
    - On-chip voltage reference
    - Includes sample-and-hold (200nS to sample)
    - Self-clocking (includes own RC oscillator).

21-1 A/D-D/A Wiring Details

Wire the AD7569 as shown below.

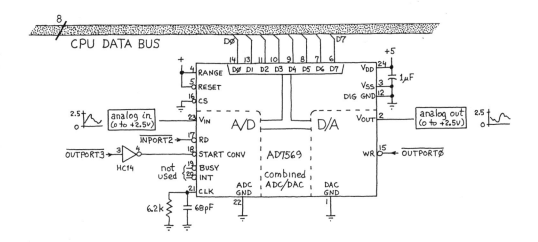

Figure L21.2: AD7569: wiring details

Some peculiarities of this circuit call for explanation:

- We *ignore* the two signals that might tell the computer whether a conversion has been completed: INT*, and BUSY*. One might expect that the computer should test one of these levels, taking a sample only when the signal had reached the proper level (so-called "programmed" or "polled" I/O), or that the processor ought to take in a sample only when *interrupted* by the appropriately-named INT* signal. Such schemes are tidy, but inefficient when the converter is as fast as this one: the processor would take about as long to *ask* whether INT* was low as the A/D takes to complete a conversion. The CPU's interrupt response is slower still.

So, we take a simpler approach: let the processor take in a sample any time—recalling, as we program, that we must not try to pick up a sample within 2.6 µS of the time when we told the converter to *start* converting. This restriction

turns out to be no hardship at all; in fact, you have to *try* in order to violate this requirement. (Compare the similar choice made in the Text design: sec. 11.05, p. 770; but their converter is slower.)
- We hold CS* constantly low.
  CS* is only a redundant gating signal, AND'd with both Rd* and Wr*: CS* does *nothing* by itself.
- We use OUT3** (the I/O decoder's OUT3*, inverted) tell the converter to start: a *falling edge* does the job. It is too bad that we need to waste an instruction on this operation; this is one of the chip's few annoying characteristics: it would be nicer if a Read could properly initiate a new conversion. (You might experiment: do conversions work properly if you drive ST* with inverted RD* instead of inverted OUT3*?)

Timing

Here is a timing diagram for D/A and A/D:

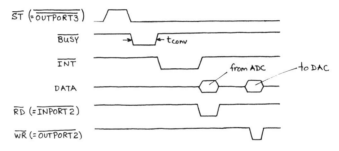

Figure L21.3: AD7569 timing as wired to 68008 (CS* held low)

PROGRAMS

Tests; In, Out; Some Processing of Input Data

<u>Preliminary: Pointers</u>

You can save yourself programming effort, and can make your programs run fast, if you set up address registers to point to the ports that you need: A2 might point to port 2, where A/D input and D/A output are located. A3 might point to port 3, where the A/D start signal is located.

Here are codes, for example, to set up two such pointers:

347C	movea.w #$8002, A2;	point to A/D, D/A data ports
8002	movea.w #$8003, A3	; point to A/D start
367C		; port (out)
8003		

21-2 Testing A/D and D/A

a) Confirming that the D/A Works

Write a program that clears a register, puts it out to the D/A, increments the register and puts it out again; and so on, endlessly. This program should produce a repeating waveform at the D/A output, ramping from zero to about 2.55 volts.

If the ramp is not smooth, you can make some quick inferences about what's gone wrong. Suppose you see one of these kinky ramps:

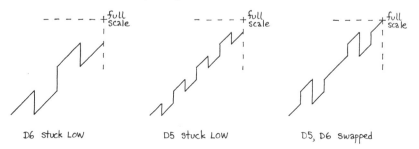

Figure L21.4: Defective ramps: count the kinks to infer which data line is screwy

You can at least tell what is the highest-order D/A data line that's not driven properly or misbehaving on the D/A: 1 kink (at midpoint of range) ==MSB is bad; 2 kinks (at 1/4 and 3/4 of range) ==> d6 is bad; and so on. This sort of information gives you a clue to what lines you should probe as you troubleshoot. It may not reveal subtler points, such as whether the line is stuck high or low or floating (floats cause the strangest waveforms).

b) Confirming that the A/D Works

Write a program that takes a value from the A/D, displays it on the data displays, then does this all again, endlessly. Feed the A/D with a DC level, from the 1k or 10k potentiometer on the breadboard:

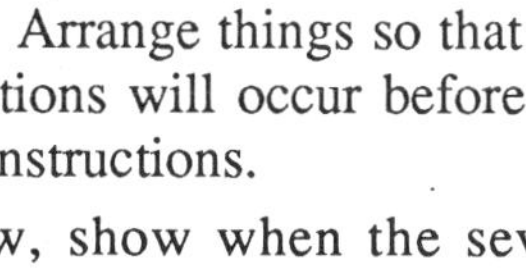

Figure L21.5: DC A/D test input

> Think about the A/D's timing requirements: make sure you leave at least 2.6 µs between the falling edge of ST* (= OUT3**) and the assertion of RD* (the A/D doesn't like to be *asked* for data before it is ready). Arrange things so that you tell the converter to Start at a time when some necessary operations will occur before the next Read, so that you don't need to kill time with do-nothing instructions.

On the timing diagram below, show when the several control signals occur, and determine whether sufficient time elapses between the start of conversion and your program's Read of the A/D output. Count clocks: the Start* edge occurs one clock into the cycle on which OUT3* occurs (and this is 8 clocks after the start of the instruction: 8 clocks are required for the instruction fetch); the Read* begins at a similar place in the IN2* instruction.

Consult the *MC68008 Instruction Execution Times*, at the end of the set of digital data sheets, for the number of clocks needed for each instruction of your program.

ST* (=OUT3**)   _____
RD* (= IN2*)   _____
WR* (=OUT2*)   _____
DATA   _____
Instruction: *Move.b Dx,(A3)*   _____   _____
Clocks: 12

Figure L21.6: AD7569 timing as driven by your program

Leave yourself 2 NOPs after the IN, OUT instruction(s), for you will shortly complicate the program a little.

Incidentally, it is good practice, when assembling by hand, to *scatter pairs of NOPs through your programs*. These NOPs can save you lots of rewriting time: if you need to add something later, you can branch out to a patch program, then branch back (2 bytes will allow a Branch).

21-2 In & Out

Write a program that will take in a single byte from the A/D, then put it out at once to the D/A. This exercise sounds exceedingly dull, but it isn't. In fact, this simple scheme provides a good setup for watching the effect of sampling rate, a topic we mentioned but did not demonstrate in Lab 17.

Effect of Sampling Rate

Gradually increase the frequency of an input sine wave, and compare the number of samples needed to make the reproduction a recognizable sine against Nyquist's theoretical minimum: at least two samples per period. As your sampling gets sparse, the 'steppiness' of the waveform gets extreme. In a few minutes we will add a low-pass filter at the output of your D/A, to make the steppy waveform prettier. But before we do that, let's get an impression of the effects of sampling rate.

See if you can find the effect called *aliasing*, by driving the converter with a sine at a frequency *above* the maximum it can convert.

If aliasing intrigues you, you may want now to build the *push-pull* speaker driver suggested at a later stage, below: sec. 21-6. Aliasing is sometimes difficult to spot on the scope, but it is very easy to *hear*.) If you *listen* for the aliased signal, you will need to slow the sampling rate by inserting a call to Lab 20's *delay* routine within the program loop. You will need to plant a *small* delay value in that routine—less than a millisecond, whereas the value used in Lab 20-5 and 20-6 produced a delay of 100's of milliseconds.

21-3 Low-pass filter; PLL control

Here's a chance to make Nyquist's surprising claim believable, by cleaning up the D/A's steppy output. If you're feeling energetic, this also a chance to put to work the phase-locked loop you built in Lab 17. We will begin by filtering the D/A output *without* the use of the PLL; you can then add the PLL if you like.

21-3-1 Applying the filter to D/A output

Find the MF4 low-pass filter that you built back in Lab 12. Add a blocking capacitor and bias divider, so that the filter can run where it's happy (centered on $V_{cc}/2$) while the D/A output sits where it must (in the range 0 to 2.55V: that is, centered on about 1.28V). Use a second external function generator to clock the filter.

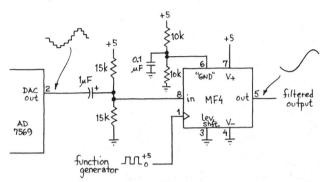

Figure L21.7: D/A driving filter

To use your filter, you will have to slow down your computer's sampling rate a great deal: the filter's top f_{3dB} is under 10kHz, whereas your computer and A/D-D/A can handle much higher frequencies. (How much higher? How long does your program take to bring in a sample and put one out again?) You can slow your program with any of the *delay* routines suggested in Lab 20; the easiest scheme is surely just to use the *delay* already in place, as we suggested above in 21-2. The most fun scheme, however, would allow you to alter the sampling rate from the keypad (see appendix to Lab 20: delay, version 3). Take your pick; but don't spend your afternoon on the delay routine!

When you have brought the sampling rate down to a few kHz, apply the D/A output to the filter input, and watch the filtered and unfiltered D/A output on the scope. See how close you can come to 2 samples/period while retaining a coherent sine out of the filter. Does your output filter solve the aliasing problem?

Anti-aliasing filter (optional, again: this is for perfectionists)

A complete A/D-D/A circuit would include an anti-aliasing filter. It is easy to provide such a circuit here: add an MF4 ahead of the A/D input, clocked with the same signal that clocks the output MF4.

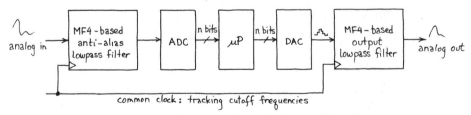

Figure L21.8: MF4 filter again: this time used as anti-aliasing filter

Note the *AC-coupling* used on the anti-aliasing filter, once more: again the goal is to keep both filter and A/D happy.

This filter, properly clocked (that's up to you), should keep out most of the signals that are too high to handle at a given sampling rate. You should now find that aliased signals appear briefly, at rather low amplitude as you move just past the minimum adequate sampling rate (Nyquist's); then the aliased output should disappear altogether at frequencies substantially higher.

Footnote: the anti-aliasing filter, you may recall from Lab 12, can *alias*! (It does if f_{in} exceeds $0.5 f_{clock}$! Strictly, therefore, one should precede the MF4 with a conventional low-pass (one that does not sample). We'll skip this refinement. Things are complicated enough, and about to get more complicated with the resurrection of your phase-locked loop!

21-3-2 Using Phase-Locked Loop to adjust f_{3dB} to sampling rate (optional: skip if you feel pressed for time)

Now here's the grand scheme that lets us put the PLL to work. Find the PLL that you built in Lab 17. Add a flip-flop to feed the *signal* input of its phase detector (the *sample* pulses that come out of the '138 I/O decoder are too narrow to work satisfactorily):

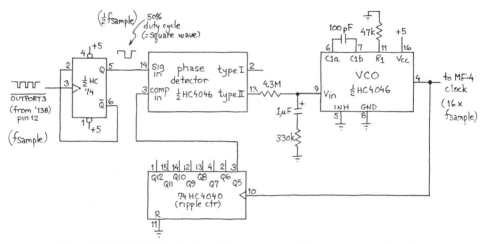

Figure L21.9: Lab 18 PLL: flip-flop added to input, to provide square wave to phase detector

Use the output of the VCO to drive the *clock* terminal of the MF4 (remove the line that has been driving that terminal until now, of course). Now see if the filter can follow your computer as the computer varies the sampling rate and thus the analog input range the converter can handle.

Back to the main stream:

21-4 Invert; Rectify: Half-Wave and Full

a. Invert

Replace the simple In & Out instruction with one that will put out an inverted version of the signal that comes in from the A/D. (Do you want NEG or NOT?: 2's comp or 1's comp inversion? Consider how you want a value of *zero* treated.)

b. Rectify

1- Half-Wave

Let a sine input generate a 1/2-wave rectified output (rectified about the midpoint of the voltage range of A/D and D/A; that range, in case you care, is 0 to about 2.55 v.).

Figure L21.10: Half-Wave rectified output (about midpoint)

Suggestions:

- Your program should branch on the MSB of the input. The "sign" flag (called MI or PL in the conditional instructions) reflects this bit: MI == MSB high; PL == MSB low.
- Note that when you want to simulate "zero out" you should force the output not to a digital zero but to the midpoint of the output range.

2- Full-Wave

Let a sine input generate a full-wave rectified output:

Figure L21.11: Full-wave rectified output (about midpoint)

21-5 Filter: Low-Pass

It turns out to be very easy to write a program to simulate a low-pass filter: let it average the current sample with the previous average, and output the result. (Give the most recent sample a weight equal to the previous average).

Test your routine by feeding it a square wave of low frequency (100 Hz or lower). Does the shape look roughly like

$$(\text{amplitude-in}) \times (1 - e^{-(t/RC)})?$$

If you're not sure: does it climb fast at first, then slowly? Does the waveform travel in each step just halfway to its destination (about like Xeno's hare)? The answer to all these questions should be Yes.

Test the circuit's treatment of a sine wave. Do you find the usual low-pass filter phase-shift effects? (You will see a constant *delay*; this is an artifact of the digital processing, different from the phase shift that characterizes the analog low-pass filter.)

Here are two codes you may find helpful:

    E20[8 + x]   LSR.B #1, Dx
                           ; ("x" = register number); (*E3…* for shift *left*)
                           ; shifts register Dx right by 1 bit; brings 0 into MSB,
                           ; puts old LSB into Carry and Extend flags
    E21[x]   ROXR.B #1, Dx
                           ; rotates right, bringing in Cy or Extend
                           ; flag as MSB

Sample coding:

E209	= LSR.B #1, D1
E308	= LSL.B #1, D0
E211	= ROXR.B #1 D1

Wrinkles

How could you change the frequency-response of your filter—the filter's f_{3dB}? A clumsy way: slow the program with a delay loop, so as to slow the averaging process. This produces a steppy output. Can you invent another way?

Next time you will have a chance to use programs you have written today, if you choose to improve your signal processor by allowing keyboard control. And you will have at least two more chances to use the A/D, D/A that you have installed today. Your machine now has grown as far as it must. From here on, it will be up to you to choose and devise hardware improvements, and to write the code that lets your little machine do something you find rewarding.

Class 22: μ5:: Interrupts & Other 'Exceptions'

Topics:

- *old:*
    - Pointers to read and fill tables
    - A/D, D/A hardware

- *new:*
    - Exceptions:
      A peculiar Motorola term, encompassing software and hardware events that can break into normal program execution
    - Interrupt hardware
      Alternative schemes: autovectored versus fully-vectored

Old

1. Pointers to read and fill tables

We talked about ways to copy a table from one location in memory to another, last time; the Text shows two methods, in program 11.1.

The main choices here concern how to check for table's end. So far, we've seen two: use a counter register, and do an address compare. What's to be said in favor of each?

If you use address compare, what flags/conditions should your program look at?

For example, after

    cmpa.l #$1F00, a1   ; a1 is moving pointer, $1F00 is end marker

Should you use BNE? BLS? BLT? Does it matter? (Yes!)

2. Hardware: A/D, D/A; Integrated Version (Lab 21)

Should you wire the chip so that the computer checks the signal that says 'Conversion Done'? No. Remember why? Should you use an interrupt? No. Today you will begin to understand better why not, and also when one might be useful.

New:

Why interrupts?

You will meet some cases where interrupts would be useful, in Lab 22: both the 'storage scope' and 'function select' programs could benefit. And here is a simpler case, taken from back in Chapter 10: a keyboard input:

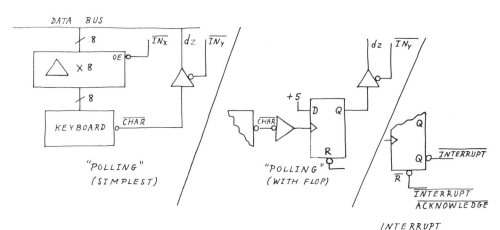

Figure N22.1: Keyboard input: how should the buffer tell the computer that it has a character ready?

We have discussed the method called *polling*: the processor asks, from time to time, 'character ready?' That method can be inefficient, or worse: the processor could be too slow, and could miss a character; the human could be too slow, and oblige the machine to ask, over and over, 'Ready?,' 'Ready?'—before getting anything new. Interrupts solve both problems.

Let's generalize what we have just described for the keyboard: the conditions that call for interrupts:

a peripheral needs quick attention
> An melodramatic example is a power-failure sensor: it wants to warn the computer that the sky is falling, so the computer can save some information while it still is able to. A more typical case, because it requires much faster response, is a disk or tape drive that needs to load or unload data—in quick, periodic bursts.

a peripheral needs *infrequent* attention
> Extreme example: a "real time clock" that wants to say, 'It's a new day. Update your calendar.' It would be a shame to ask the computer to *poll* the calendar continually.

In these notes we will speak, first, of the general category that Motorola calls "exceptions;" then we will concentrate on the most important member of this category, the *interrupt*.

68000 "Exception Processing"

The 68000 treats interrupts as one of a class of many "exceptional" conditions that call for a specialized response.

Some of the exceptions are *software events*:

- "illegal" instructions (more on this, below);
- an attempt to divide by zero;
- a "trap" instruction—an instruction that deliberately invokes exception-processing (compare the IBM PC's 'software interrupt' discussed in Text sec. 10.11, p. 701).

- "trace:" a facility designed to ease debugging: the processor behaves as if it had interrupted itself, after each instruction (this we use in the *Register Check* routine, included in this Manual)

Other exceptions are *hardware events*, signaled to the processor by assertion of one or more of its input pins:

- reset (this you have been using, of course);
- bus error (device fails to respond);
- interrupt.

The chip responds to all of these "exceptions" in a standard way:

- the processor saves minimal information that will let it resume what it was doing, when it has finished responding to the exception:
    - on the stack, it saves a copy of—
        - the program counter; and
        - the "status register," which includes the flags.

Then—

- the processor looks to a particular address low in memory, to find the address where it ought to begin executing an exception-response routine or program.
- at the end of the response routine, the processor executes a RTE instruction (return from exception); RTE pops the old PC and SR values from the stack, restoring them to the PC and SR registers, thus allowing the machine to resume execution of the program that the exception broke into.

Lab 22 includes a diagram (fig. 22.4) spelling out this standard response to an exception. The process is a bit convoluted; it may help to take a look at that figure.

B. A Hardware Exception: *Interrupt*

An *Interrupt*, the most important of hardware 'exceptions,' evokes a response almost identical to what you saw for *illegal*. The important difference is simply that an interrupt is initiated not by a software event, but by the pulling of a pin (1 or more): an interrupt pin, called IPL2/0 or IPL1 on the 68008.

The interesting difficulty—as for the software event—is how to steer the processor to the appropriate response. Pulling a pin cannot, in itself, tell the processor *where* it ought to branch in order to respond appropriately. Microprocessor designers use two methods to solve this problem:

How the processor finds the interrupt response routine
Two alternative methods:

> **autovector**   the processor automatically looks to a particular location (in the 'vector table'—low memory) and loads the 32-bit value there into its program counter. Thus it does an indirect jump. This is *identical* to the response to a software exception (except for the effect on the *interrupt mask*; we'll reach this topic in a moment).
>
> **(fully-) vectored**   the processor reads a value from the data bus (the value is fed by the interrupting device), and uses that value to steer it to the address of the response routine; in the 68000/8 the value read from the data bus (8 bits) steers the processor to an entry in the vector table; so, this vectored jump might be called double-indirect.

In Lab 22 we suggest that you use the simpler of these two schemes, *autovector*. Fully vectored interrupts are a little faster, and permit the use of more interrupting devices.

Hardware to Implement Interrupts
Autovector Hardware

Here is the Text's example of hardware to implement autovectored interrupts with the 68008. The scheme is similar to the one we use in the lab.

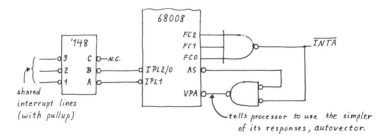

Figure N22.2: Autovectored interrupts

The '148, shown above, is a *priority encoder*—a chip that puts out the ("encoded-") binary number of the highest-numbered of its inputs that is asserted. So, if one asserts input 2, the chip puts out *010*. (In the lab, we omit this chip, because we have only one interrupting device).

The 68008 uses more than one interrupt pin so as to let it determine *priority* among interrupts, on chip[1]. (Most earlier processors—but not all (see, e.g., Intel's 8085)—used just one interrupt pin.) Asserting the *VPA* pin tells the processor to use this *autovector* response.

We will return, in a minute, to the question how the processor implements this priority scheme.

---
[1]. Two pins, of the three available on the 68000, are tied together internally (IPL$_0$* and IPL$_2$*) to save pins, so the 68008 provides three interrupt "levels": 2, 5 and 7, rather than the 6 provided by the 68000 (level *zero*, for both processors simply means *no* interrupt).

Vectored Interrupt

Here, in contrast, is a hardware to implement a vectored scheme:

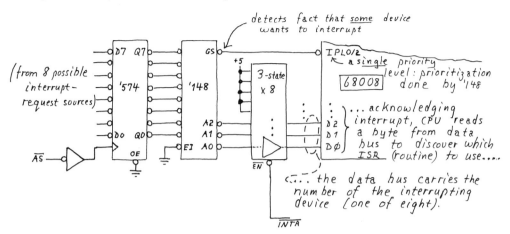

Figure N22.3: Vectored interrupt

Questions:

- How does this hardware implement prioritization—while using just one of the CPU's IPL* pins? (Nothing subtle, here.)
- Why is the '574 (octal D register) included? (This is a harder question; one posed in the Text.)

Priority among interrupts

The 68008, with its multiple interrupt pins, can be programmed to allow interrupts of high urgency (or "priority"), while ignoring interrupts of lower urgency. In addition, when it responds to an interrupt, the processor *automatically* makes itself ignore interrupts of the same priority or lower. That makes sense: finish the urgent job first. (The highest priority, 7, does not work this way: it is usually called *non-maskable (NMI)*; so, level 7 can interrupt level 7.)

How does the processor implement this *priority* scheme? The methods turns out to be straightforward: the processor uses a 3-bit *mask* in one of its registers (the "Status Register") to define the level of interrupt that will be allowed to break in.

Interrupt Mask

Figure N22.4: Interrupt mask, in Status Register

The programmer sets these 3 bits as he or she chooses, just as one sets the contents of any other register. The *default* value—to which the mask is set on Reset—is 7, as one might expect: no interrupts are permitted, except NMI.

Here is a homelier image of what the mask does: it's a sort of adjustable fence; behind it sits the CPU, trying to get some work done. To keep the neighborhood safe, however, he needs to pay attention to neighborhood creatures (maybe he has to feed them?). He adjusts the fence according to how busy he is:

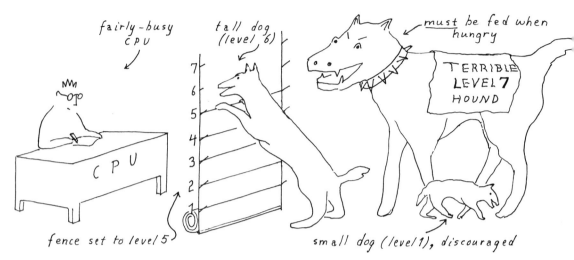

Figure N22.5: CPU's interrupt mask: adjustable to keep out small dogs, allow in tall dogs

When an interrupt occurs, the interrupt mask is saved on the stack (as part of the SR, in the normal exception response procedure). Then the mask in use is *raised* to the level of the present interrupt. On exit from the exception response, the popping of the old SR automatically restores the old mask level.

Autovector hardware again: Variation on Text's Hardware: Lab 22 Flop

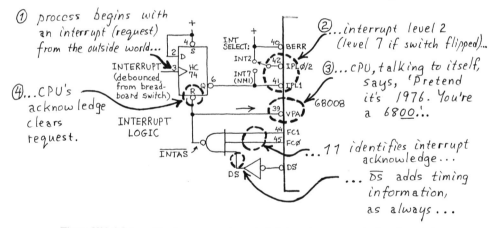

Figure N22.6: Interrupt hardware: generates a single interrupt request on closing of a switch

Probably you recognize this use of the edge-triggered flop as similar to the trick used on Lab 20's Ready key. Again the motive is to make sure that the processor can clear the request as soon as it begins to service it. This is not always necessary. The Text examples omit this feature, assuming that the request will not persist after the response is completed.

A debugging aid: Register-Check: two versions

Two versions of register-check: <u>Baby Check</u> and <u>Register Check</u>

These two programs can help you debug a program. But you should feel free to ignore these notes, if you like. If you are feeling pressed for time and extremely eager to get on with your own programming, rush on; don't feel obliged even to *read* this handout!

<u>Baby Check</u> is a primitive version, <u>Register Check</u> a fancier version, but they do essentially the same thing: they allow you to discover the values of all the CPU *registers*, at any point in your program.

<u>Baby Check</u> is very easy to enter: it's a three-line program; <u>Register Check</u> takes perhaps half an hour to enter and check: it's about 100 lines long. It also requires that you wire the second of the two *Ready* keys shown in the circuit Big Picture.

<u>Register Check</u> can reward you for the harder work, however: this longer program lets you discover register values while the program runs, using the keypad: type D2 to discover the word-length value of D2; type A7 to get the value of the stack pointer; and so on. It lets you do this as you move through a program as slowly as you like, one instruction at a time.

<u>Baby Check</u>, in contrast, lets you discover register values by taking the bus and walking through the stack, looking at the stored values. You'll need a pencil and paper to help you keep track of what register you're looking at.

BABY CHECK: Poor Man's (or woman's) version of Register-Check

The fancy register-check routine works by pushing the contents of all registers onto the stack, then letting you inspect those stored stack values by pressing particular keys on the keypad.

The simple register-check routine suggested below also pushes all registers onto the stack. It then, however, stops. You inspect those values by taking the bus and looking at the stored values. Before quitting, the program puts out the address of the stack pointer on the displays, to tell you where to begin looking for the stored values.

To invoke this **Baby Check** routine, you should plant a *breakpoint* somewhere in the program that you are checking; that breakpoint is a *TRAP* instruction—one of the software exceptions:

'Breakpoint' command
BKPT: 4E40   ; TRAP 0

Vector
As for all other *exceptions*, this Trap stores the return address (next instruction) on the stack, then looks to the *vector table* for the Trap_0 vector, a 32-bit address stored at location $80.

So, if your Baby Check routine sits at $300, you plant that address in the vector table:

<u>Vector</u> <u>Table</u>:

address	stored 'vector'	
$80	00 00 03 00	; TRAP 0 vector: start address of your response routine

Routine:

Baby Check:

31CE 8000	move.w a7, $8000	; show stack pointer on displays
48A7 FFFF	movem.w d0-d7/a0-a7, –(a7)	; push all registers onto stack
4E72 2700	stop	; that's all, folks!

Figure RGCHK.1: Baby Check listing (tiny!)

If you go and inspect the stack, you will find that it holds, first, the address where the breakpoint occurred, then the status register, and then the 16 stored registers, in the order shown below:

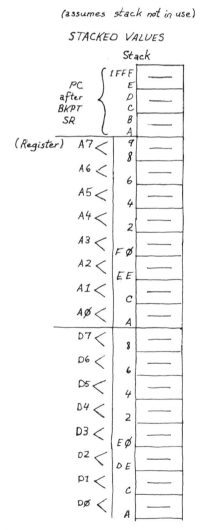

Figure RGCHK.2: The stack after Baby Check does its job; assumes stack not in use when Baby Check invoked

The fancier REGISTER CHECK

Preliminary: should you take time to install this fancier debug routine in your machine?

Maybe not. Entering it and debugging it will probably take a half hour or so. Until now we have studiously avoided asking you to type in a heap of code that is hard to understand, because that's depressing work.

Two arguments run in favor of entering this Register-Check routine:

– First, if you are about to undertake a project that requires a program longer than those we have put in so far, you may be grateful for a debugging aid;
– Second, this routine will give you a taste of a facility normally available to an assembly-language programmer.

Software Single-Step, & Breakpoint

The debugging aid described here is a supplement to the dtack*-blocking *single-step* function built into your computer. That hardware stepper is good for checking hardware, and adequate to let you watch programs in slow motion. The *software single-step* described here gives you additional information about the operation of a program in action: it lets you watch the program's effect on all the CPU's registers.

The 68000 provides a *"Trace"* mode of operation for just this purpose, and we will take advantage of that mode. We will also suggest a way to plant a "breakpoint" instruction that invokes the register-check routine only when the program hits the breakpoint.

TRACE

We tell the processor to *Trace* by setting a single bit in the "Status Register" ("SR"): we set the *Trace* bit by OR'ing in a 1 at that bit position in the SR.

So, to turn on *Trace* function, insert, just before the section of program that you want to trace, the following instruction:

    00 7C 80 00 ORI.W #$80 00, SR

As it excutes the *Trace* exception, the processor saves the program counter and status register, just as it does in response to any other exception (see Lab 22, if you have forgotten this pattern; fig. L22.4). It looks to the exception vector stored at 24H; there it picks up the start address of the routine that responds to the *Trace* exception: here, that response is our Reg-check routine.

Incidentally, the processor prudently turns off the *Trace* function while executing a *Trace* (a good idea!) Exiting the Reg-Check routine, the processor turns on the *Trace* bit once again, and resumes execution of the program being *Traced*, at the next instruction. So, the processor interrupts itself after executing a single instruction and lets us inspect its registers.

Neat?

You will need to load the *Trace* vector with the starting address of your response routine (which will be the longer Register-Check program):

Vector Table:		
address	stored 'vector'	
$24	00 00 03 00	; TRACE vector: start address of your response routine

Figure RGCHK.3: *Trace* Vector

Register-Check Routine

How Reg-Check Behaves

The Register-Check routine shows, on the Data Displays, the 16-bit value of any of the processor's registers; the user selects which register by typing, for example "D3" on the keypad: this evokes the contents of register D3.

You can look at as many registers as you like, or as few. When you want to resume execution—carrying out the next instruction in the program you are Tracing—you hit the keypad's INC key. The processor executes one more instruction, and again shows you register contents.

The routine will continue to *Trace*, until you break in to clear the *Trace* bit. To stop tracing, you must take the bus and change the Reset vector so as to start execution just *after* the instruction that turns On *Trace*. (Reset turns *Trace* off.)

How Reg-Check Does What it Does

The routine first saves all registers on the stack—including the stack pointer, a7. It then fixes the stored value of a7, adding a constant to restore a7 to the value it held *before* the *Trace* exception began. (It also keeps a copy of the modified a7, which will be needed on exit from Reg-Check.)

Then the program looks at the keypad to find out which register the user would like to see revealed on the data display. It checks the high nybble to find whether the user asks for A (address), D (data), F (flags & the rest of the Status Register). Any other key evokes the Program Counter (PC). The routine uses the low nybble coming in from the keypad to point to a particular one of the registers, if Data or Address registers are requested.

To watch the progress of a program, press the C key (or almost any other, as noted above) the data displays will show the PC. Hit INC on the keypad; incidentally, this is the key used for *hardware* single-step, as well.[1]

The program counter will advance, one instruction at a time; when you get to a point here you are curious to check register values, evoke the values of the registers that interest you, using the other keypad values.

---

1. A wrinkle, in case you use the hardware single-step to debug the Reg-Check routine: because both hardware and software single-steps use the same Adr-Clk signal, note that Reg-Check will always find its Ready flop in the high state—clocked by the hardware-single-step's use of Adr-Clk.

Hardware

To run the Reg-Check routine, you need to wire up a *second* "Ready" signal, just like the one you installed in Lab Micro-2, except that this one uses a flip-flop clocked by the keypad's Adr-Clk signal rather than the keypad's Kwr*.

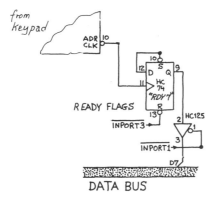

Figure RGCHK.4: Second "Ready" key

We ask you to install this *second* Ready for two reasons:

- This second Ready allows you to debug a program that uses the other Ready key;
- Use of Adr-Clk to clock this Ready flop lets you step rapidly through a program, taking advantage of the keypad's Repeat function.

Breakpoint

Instead of *Tracing* every instruction, you can plant a Trap instruction in your program, so as to let the program run full-speed up to the Trap instruction; the Trap then can invoke Reg-Check just once.

Such a *breakpoint* is useful in two ways:

- It lets you determine whether the processor ever reaches that particular point in the program;
- it lets you inspect register values at a point where they particularly interest you. (For example, you might check a value input from the A/D, to see whether it seems to vary on successive passes through the program, reflecting changes in the analog input.

The code and vector you need for a breakpoint ("TRAP 0," in the microprocessor's own terms) are shown above under *Baby Check*. A listing of the full Register-Check program appears below.

REGCHECK 6/89_GLOBALS from regcheck 6/89.Rel

```
000000:                         .verbose
000000:                         .ListToFile
000000:                         ; REGCHECK 7/16/89
000000:                         ; This routine evokes any of registers plus PC & SR,
000000:                         ; using keypad.
000000:                         ; Ax evokes contents of that address register
000000:                         ; Dx gets that data reg
000000:                         ; Fx gets flags ("status register")
000000:                         ; anything else gets program counter
000000:                         ; Values limited to word-size, to fit displays
000000:
000000:                         ;  As written, expects to be reached through
000000:                         ;  an exception vector; PC & CR evocation
000000:                         ;  depend on that assumption.
000000:
000000:                         ; To invoke this as response to TRACE,
000000:                         ;   turn on Trace by oring 8000H with SR,
000000:                         ;       in pgm to be traced
000000:
000000:                         ; register use:    a0      points to Kbd & Dsply
000000:                         ;                  a1      points to Rdy port
000000:                         ;                  a2      points to Clear port
000000:
000000:                         ;                  d0      holds offset from key,
000000:                         ;                          to pick value from tbl
000000:                         ;                  d1      holds high nybble, to
000000:                         ;                          determine Adr/Data/Other
000000:
000000:                         ;                  a3      temporary copies of stack ptr
000000:                         ;                  a6
000000:
000000: 48 A7 FF FF     main        movem.w d0-d7/a0-a7,-(a7)  ; push D and A regs
000004:                                                        ;   onto syst stack
000004: 2C 4F                       movea.l a7,a6       ; temp copies of stack pointer
000006: 26 4F                       movea.l a7,a3       ;  so we can mess up original
000008: 42 40                       clr.w   d0          ; clear top byte of register
00000A:                                                 ;  later used as index & later
00000A:                                                 ;  fixed only in low byte
00000A:
00000A: 30 7C 80 00                 movea.w #$8000, a0  ; point to keypad & dsply
00000E: 32 7C 80 01                 movea.w #$8001, a1  ; point to Readyport
000012: 34 7C 80 03                 movea.w #$8003, a2  ; point to Clearport
000016:
000016:
000016: D6 FC 00 26                 adda.w  #38,a3      ; this should amend copy of SP
00001A:                                                 ;  to value it held before
00001A:                                                 ;  exception began
00001A: 3F 4B 00 1E                 move.w  a3, 30(a7)  ; this overwrites stored
00001E:                                                 ;  a7 (=SP) with value it had
00001E:                                                 ;  before the exception
00001E:
00001E: 10 12                       move.b  (a2), d0    ; clear  Ready flop: dummy
000020:                                                 ;  input, uses I/O decoder
000020:                                                 ;  to clear Ad-Clk Rdy flop
000020: 10 10           get         move.b  (a0), d0    ; get keyvalue, to dtrmn
000022:                                                 ;  which of saved regs to show
000022:
000022: 12 00                       move.b  d0,d1       ; and save a copy, to work
000024:                                                 ;  on, dscvrx high nybble
000024:
000024: 02 00 00 0F                 andi.b  #$0f,d0     ; mask to get reg #
000028:                                                 ;  (low nybble)
000028: E3 08                       lsl.b   #1,d0       ; dbl the low-nybble bec
00002A:                                                 ;  two bytes/reg
00002A:
00002A: 02 01 00 F0                 andi.b  #$f0,d1     ; Now look at high nybble:
00002E:                                                 ;  mask to dtrmn whether
00002E:                                                 ;  data, address or control
```

Figure RGCHK.5: Register-check program listing: first page of two

A debugging aid: Register-Check

```
00002E:                                                  ;   (PC or SR)
00002E:  0C 01 00 D0         cmpi.b  #$d0,d1             ; D?
000032:  67 12               beq.s   rejoin              ; if so, go use reg. # to form
000034:                                                  ;   pointer
000034:
000034:  0C 01 00 A0         cmpi.b  #$a0,d1             ; A?
000038:  67 08               beq.s   afix                ; if so, go fix up reg-# pointer
00003A:
00003A:  0C 01 00 F0         cmpi.b  #$f0,d1             ; F?
00003E:  67 0C               beq.s   flags               ; if so, fix pointer to get at
000040:                                                  ;   status reg
000040:
000040:  60 10               bra.s   catchall            ; ... show PC
000042:
000042:  00 00 00 10    afix    ori.b   #$10,d0          ; for an A reg, need to offset
000046:                                                  ;   by 16 to get past data regs
000046:
000046:  30 B7 00 00    rejoin  move.w  0(a7,d0),(a0)    ; this uses low nybble X 2
00004A:                                                  ;   (+ 16 if an adrs reg)
00004A:                                                  ;   as index added to stack
00004A:                                                  ;   ptr, to get register
00004A:                                                  ;   contents (2 bytes) &
00004A:                                                  ;   send them to display
00004A:
00004A:  60 0A               bra.s   check               ; see if Rdy (back to
00004C:                                                  ;   traced pgm?)
00004C:
00004C:  30 AF 00 20    flags   move.w  32(a7),(a0)      ; display saved status reg
000050:  60 04               bra.s   check               ; see if Rdy
000052:
000052:  30 AF 00 24    catchall move.w 36(a7),(a0)      ; show pre-exception PC, if
000056:                                                  ;   nothing else requested from
000056:                                                  ;   keypad
000056:
000056:  08 11 00 07    check:  btst.b  #7,(a1)          ; see if Rdy key hit
00005A:  67 C4               beq.s   get                 ; if not Rdy, go look again at
00005C:                                                  ;   keyvalue
00005C:                                                  ; But if Rdy, back to main pgm;
00005C:                                                  ;   terminate register dsply
00005C:
00005C:
00005C:  10 12          resume: move.b  (a2), d0         ; clr Rdy flop (dummy input)
00005E:  3F 4E 00 1E         move.w  a6, 30(a7)          ; restore a7's post-
000062:                                                  ;   exception value
000062:  4C 9F FF FF         movem.w (a7)+, d0-d7/a0-a7  ; restore the registers
000066:  4E 73               rte                         ; and return
000068:
000068:
000068:
```

Figure RGCHK.6: Register-check program listing: second page of two

Lab 22: µ5: 'Storage Scope;' Interrupts & Other 'Exceptions'

> **Reading:** Ch. 10:
> 10.09-10.11 <u>re:</u> interrupts, in general and for the IBM PC: especially the section on *autovectored...* response (pp. 698-99), which describes the scheme you will use in Lab 22.
> Ch. 11:
> 11.4 <u>re:</u> interrupt hardware & autovectored vs fully-vectored response;
> 11.07 <u>re:</u> programming response to interrupt: p. 793, in general; but note that much of this is heavy going: much more complicated than what we ask you to understand in the present lab.
> Lab 22:
> Try to make sure that you understand the interrupt hardware, here: why we use a flip-flop; what happens to RAM and I/O decoder during an interrupt response; what the assertion of VPA* does.
>
> **Problems:** 11.4 (interrupt hardware),11.12 (p. 780), 11.15 (p. 794).

More assembly-language programming: storage scope; Exceptions

The first two exercises in this lab ask you to continue using the A/D, D/A that you installed last time. The first exercise, **the** 'storage scope,' is fun and will give you a workout on some important programming issues: use of moving pointers; table-end tests; matching program delays. The second exercise is less important. There you use the keypad to select one of the functions you implemented last time: invert, rectify, and so on. This program obliges you for the first time to use a mask that keeps more than one bit; and the program will make you feel the need for program speed, because the processing that you need to do in order to discover which key has been hit will put a noticeable dead (or *constant*) region into the D/A output.

The lab then shifts to a new topic: *exceptions*—Motorola's general term for unusual events that are allowed to break into program execution. Here we ask you to do an exercise that requires *no* new hardware at all, and entails writing fewer than a dozen lines of code. This exercise—in which you teach the processor what to do if it encounters an "illegal" instruction—gives you an efficient introduction to the topic of "exceptions" while skirting the work required to demonstrate *interrupts*.

The last part of the lab, treating interrupts, we call optional because it takes a fair amount of time (perhaps an hour), and probably is not as satisfying as getting on to devising some micro application of your own—the task that awaits you in "Lab 23," a lab that really is just an open invitation to do something fun with your little machine. We would not be disappointed if it turned out that only students who planned to use interrupts in Lab 23 paused to do the interrupt exercise.

We are as anxious as you are to make sure that you have time to do some project of your own, now that you have invested considerable effort in getting your little computer to work, and in learning something about assembly-language programming.

Two Debugging Aids: <u>Register-check routines</u>

Following these lab notes you will find two versions of a debugging aid: a program fragment that allows you to break into a running program and discover the contents of all registers. You have done enough troubleshooting by now to know that this information often is useful. The simpler version is very easy to write (it's just three lines), but relatively hard to use (you need to take the bus, then go look at values the program stored in a data table). The complex version is about 100 lines long, and requires a second *Ready* key; so, it takes longer to get going. It pays you back, however, by being extremely easy to use: you just poke keys on the keypad indicating what registers you want to inspect: type *D1*, for example, and the program displays the value of that register on the data displays (word-size). Take your pick of these two routines—or ignore them, if you like, and keep on debugging as before, with the hardware single-step. From here on, you should do with your computer whatever you find most satisfying.

22-1 Using a Moving Pointer: A "Storage Scope"

Rather than put out data as soon as it comes in, the computer can of course store it. With the data safely stored, the machine then can process away at its leisure. In our programs so far, even our modest rectify and keyboard-check operations made us feel pressed for time: we had to worry about getting the data out promptly (in "real time," is the strange jargon for this). The next program, which stores data in a table, could allow the computer to take its time processing the data, then output its results when it was ready. At first, however, we will ask the computer to do no processing: simply to "play back" the saved values.

To store and recall the data we need a movable pointer to memory. The 68000 provides 8 such **address registers** (one, A7, is dedicated to use as system stack pointer). It also takes care of advancing the pointer, in the "post-increment" or "pre-decrement" mode. Your task is reduced to initializing the pointer, and then checking for the end of your table.

Here is a flow chart describing what the program ought to do. In this scheme, the computer fills a table with data (remember, the CPU can use the full 8K of memory, less the protected 1K at the bottom, and a bit of space for the stack at the top). The program then plays back the stored data to the scope—endlessly, or until you hit the Enter key, signaling the machine that it should fill the table anew. The flow-diagram shows this Enter check occurring only at the end of a playback cycle. You may prefer to check more often.

If you care that the waveform played back should be at just the same frequency as the original, you will have to make sure the playback loop runs at the same speed as the store loop. How can you make sure that Ready-Check, needed in the playback portion, does not upset the timing match?

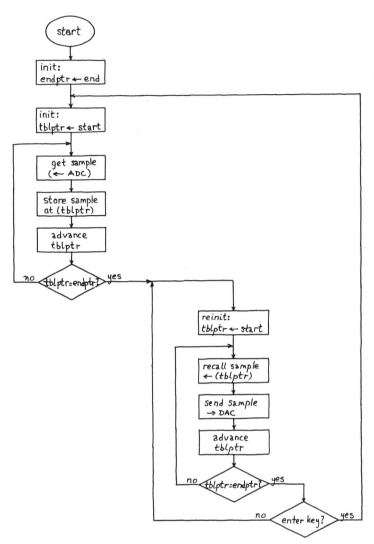

Figure L22.1: Storage scope: flow chart

Setting sampling-rate

The flow chart is over-simple in assuming that you want to sample at full speed. Probably you do not want to. It would be nice—once again—to be able to control the rate of sampling and playback from the keypad; but at least you probably should plant a delay routine, then take the bus to tinker with the delay value. (Make sure both fill and play call the same routine, so that you needn't alter *two* values!)

End-of-Table Tests

There are several ways to determine when the pointer has reached the end of a table.

- The neatest method uses and End-of-Table code. We cannot use this method in this case, because there is no data value that cannot be a valid input; thus there is no code available to us.
- Less neat, but easy: use a counter register; decrement it each time the pointer is advanced; test for zero. This method provides *position-independent code*, unlike the next method—
- Address-compare: watch for the end of table by comparing pointer with some boundary value. In your little computer, this may be the best scheme, since you want to make sure you do not overwrite the Stack, which sits at the very top of memory.

Optional (Fun): Use 'Storage Scope' to Record Sound

Many weeks ago you built a microphone amplifier with an output in the range 0 to +2.5 volts. Now you meet the reason for that choice of biasing point (about 1.25 V quiescent): the amplifier output is centered in the range of your A/D. If you saved that circuit, you can use it with the "storage scope" to look at voice or other waveforms generated by the mike. If you listen, you will find yourself obliged to learn about the effect called *aliasing*—a conversion error that results from an inadequate sampling rate, as you may have had a chance to demonstrate last time. Your *ears* are extremely sensitive to this effect—though the effect sometimes is hard to *see* on a scope screen.

After looking at the waveforms, you may want to do more:

- If you want to listen to the output, build a push-pull follower (with op amp, powered from ±15V) to drive the 8-ohm speaker. Remember to AC-couple the push-pull's input, since the D/A output is centered at about 1.2 volts.
- You may want to do some processing to the stored signal, rather than simply play it out. You could easily play back the samples at a rate different from the rate at which you recorded them, for example. So, you could turn yourself into chipmunk or whale.
- It would be harder to make a machine that changes the playback rate without changing frequency. (Such devices are used by blind people to speed up recorded readings.)
- It is also harder to make a reverb or echo machine. (Sum present sample with attenuated version of old sample.)

You may want to undertake such a substantial programming job as part of a final project in this course, rather than now. But if you are eager, here are some suggestions for some processing control you might do now.

22-2 Keyboard Control *(optional)*

If we imagine that you are building a signal-processing gadget, then it's reasonable to ask that you make it controllable from the keyboard. Here you will add function keys that will let you choose among the above operations: straight versus invert versus 1/2-wave versus full-wave rectification. No additional hardware is required. *Note*: Do no more of this than you find interesting. You should be sure to reach the last section of this lab, the 'storage scope,' though it is not important that you do so on the day when you *begin* this lab.

Here is the idea:

    MAIN
    Subroutine: CMND

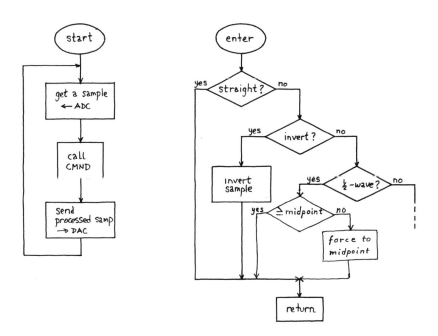

Figure L22.2: Keyboard control program: suggested flowchart

And here are some more specific suggestions:

1. Try writing the program as a very simple MAIN program, and one subroutine that checks the keyboard and alters the input value according to what the keyboard commands require.
2. Mask out the bits that do not interest you (these will be the 4 MSB's, which are set not by the latest key-pressing but by the one before).
                                then—
    do a series of "compare immediate's" with the codes for each of the relevant keys; on finding a match, the program should alter the input value appropriately.
    If no match is found, let the program put out the data as it came in.

Here, to illustrate the method, is a patch of code that determines whether the key A or C has been hit, branching to a particular place in response to each value, or back to the start of the loop if neither A nor C has been pressed.

```
                        ;KEYMATCH fragment
                        ; assumes main program has set up an address register
                        ;   to point to the keypad:
                        ;   "movea.w #keyport, a0"  ; point to keyport

200   3F04   kmatch:   move.w d4, –(a7)       ; save scratch regstr

202   1810   look:     move.b (a0), d4        ; get key value (8 bits)
204   0204             andi.b #$0F, d4        ; force high nybble to zero (this is
      000F                                    ;  the old key); keep low nybble
                                              ;  (this is the most recent key)
208   0C04             cmpi.b #$A, d4         ; is the most recent key "A"?
      000A
20C   670A             beq.s 'doA'            ; if so, do something
                                              ; appropriate
20E   0C04             cmpi.b #$C, d4         ; is it C?
      000C
212   66EE             bne.s 'look'           ; if neither A nor C, go look
                                              ;  again (the silly human has hit
                                              ;  the wrong key!)
214   4E71             nop                    ; must be C, if we landed here
                                              ;  so this is what we do if we get C
                                              ;  (you write something more
                                              ;  interesting than "NOP"!)
216   6002             bra.s 'back'           ; then go back to calling
                                              ;  program
218   4E71   doA:      nop                    ; this is what we should do
                                              ;  if we get A
21A   381F   back:     move.w (a7)+, d4       ; recall saved register
21C   4E75             rts                    ; back to calling pgm: task now
                                              ;  is done
```

Figure L22.3: Sample key-match program fragment

A Fancier Version

The scheme we have suggested requires the program to spend a good deal of time checking commands, before processing each byte of input data. You will recognize this in the output waveform: the horizontal plateaus in the output waveform last longer for the operations that result from a branch later in the key-check routine.

It would be neater to do this keyboard check only in the event that we want to *change* the operation. An Interrupt function is perfectly suited to this task, but we will save that technique for the next lab. The second best method might be to check only the *Ready* key regularly, and go into the slow key-check routine only if Ready is asserted.

If you want to hurry on to storing waveforms, don't bother to make that improvement; instead, just note that you have met an application where speed counts, and where even this fast processor does not seem fast enough. (You might also reasonably conclude that we are using the processor in a perverse way: it performs less well than 25-cents'-worth of diode and resistor, or R and C.)

Exceptions

For a very quick sampling of the 68008's *exception response*, try only the *software* exception that opens this section of the lab notes. That requires no additional hardware, and should take you no more than ten minutes.

If you are a little more ambitious, try *interrupts* minimally by installing the one required flip-flop and then entering the *test* program shown. That work should take about an hour.

After that you may want to devise a way to put interrupts to practical use; but you should not feel obliged to do so. Instead—as we have suggested already—we invite you to hurry on to applying your computer to some task that you consider challenging and fun. If that task includes interrupts, then carry on here.

22-3 A Software Exception: *"Illegal" Instruction*

This exercise looks at an exception that requires *no* hardware additions to the computer. You will watch the CPU's response to an invalid instruction code. We propose here that you write a response routine that simply shows "BADI"—or some other distinctive message—on the data display. Later, if you install the register-check routine, you will want to let the *Illegal* response invoke that more versatile routine.

Here is a diagram showing what the processor ought to do in response to an illegal instruction planted at address 102H. The Illegal "vector"—holding the start-address of the service routine—is stored at address 10H. In this example, the service routine begins at address 1E0H.

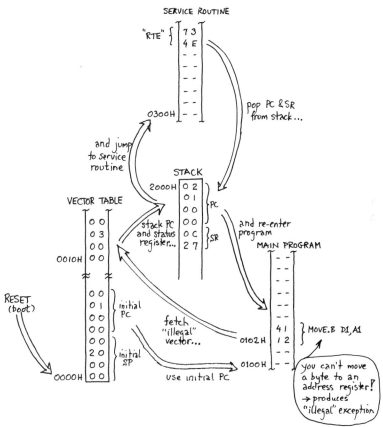

Figure L22.4: A particular exception response: illegal instruction

(A detail: the processor's response to *illegal* is atypical in one respect: the program counter value stored is not the address of the *next* instruction, as usual, but instead is the address of the illegal instruction itself. An interrupt or a subroutine call, as you know, would store the address of the *next* instruction. Perhaps you can see why the *Illegal* process *cannot* follow this usual pattern.)

Write a routine that sends out "BAD!" or some other distinctive display when the machine meets an illegal instruction. Try a main program that is simply a NOP followed by a bad instruction. For example,

                       1241 move.b d1, a1

will do. The processor can see that this instruction is improper. Can you?

22-3 A Hardware Exception: *Interrupt* (optional: for zealots)

Before you use the interrupt line to tell the computer to do anything interesting, such as taking samples of switch bounce (see below), we suggest that you try the little interrupt-test routine provided first. That should let you confirm that your interrupt hardware is working, and should also give you some insight into the way interrupts operate. The notes on interrupts, just below, reiterate some elementary points made at greater length in the Text and some 68008 details that may be new even to someone experienced with other machines. Skip the first page or so if you already understand interrupts in general.

A. *Preliminary: possible applications*

You have met, already, some cases where use of an interrupt would have served you better than the alternative method, polling. The function-select program of section 22-2 would have benefited; so might the storage-scope, 22-1.

In each case, the interrupt permits the main program to run at full speed. The program need not take time to ask whether some control signal from the outside world has been asserted. A pretty good argument for interrupts appears if one imagines using the storage-scope program to capture an image of *switch bounce*: the CPU takes quite a long time (several milliseconds) to display a table of values. If a new switch-bounce event occurs during that display, the CPU must drop everything and hurry to sample that new event before it's too late. The CPU must not finish its displaying and then go poll the input to see if the switch is bouncing. Perhaps frequent polling could do the job in this case, but the use of an interrupt allows faster sampling and playback.

Once you have interrupts working properly you may want to try using them to modify one of those two earlier programs (22-1 or 22-2).

B. *Hardware*

Let's talk about a particular example, rather than speak in general terms. Let's suppose that we mean to use the storage-scope program to catch and display an image of switch bounce.

We need to interrupt the computer by asserting one of the CPU's two IPL* pins. Two of those available on the 68000 are tied together (IPL_0* and IPL_2*) to save pins, so the 68008 provides three interrupt "levels": 2, 5 and 7, rather than the 6 provided by the 68000 (level *zero*, for both processors simply means *no* interrupt).

These "levels" describe an internal interrupt prioritization scheme—something often accomplished by a chip external to the CPU. An internal interrupt "*mask*" is set under program control to determine what interrupt priority qualifies to initiate an interrupt. Then any interrupt request higher than that mask level is permitted to generate an interrupt.[1]

That mask is simply three bits in the Status Register (D_{10}-D_8 in the 16-bit word), and can be established with a *MOVE to SR* instruction. The level 7 interrupt request is always honored (it is called "non-maskable" on other processors).

In this lab we will start by using IPL_1*, a pin that generates a Level 2 interrupt request when asserted.

---

[1] Level 7 interrupt does not follow this rule: it is recognized even though the mask is set to 7.

Hardware to Request Interrupt

We need, first, to build a little hardware that asserts $IPL_1{*}$ at an appropriate time; for our present purposes, that time will be the time when the voltage at our pushbutton switch goes low. At first we will use that simply as a "test interrupts" key. Once we have interrupts working properly we can use this hardware to start the sampling of switch bounce in the interrupt-triggered 'storage scope.' What hardware is required? Your first thought might be to do this:

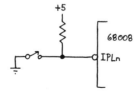

Figure L22.5: Interrupt hardware: poor

That's a start, but a flawed start. You don't want to interrupt over and over as the switch bounces, but once per 'set of bounces.' Recognizing that, you might use a debounced signal to interrupt, thus:

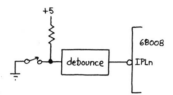

Figure L22.6: Interrupt hardware: a little better

That is better, but still not right. (No doubt you have recognized that we are rehearsing the reasoning that led to the design of the "Ready" key hardware in Lab 19.) The problem is that the CPU's $IPL_x{*}$ inputs respond not to a falling *edge* but to a low *level*. So, to make the CPU execute your "interrupt service routine" just once each time you hit the interrupt switch, you need to add some logic, just like the Lab 19 Ready key's, to make sure that $IPL_x{*}$ goes high as soon as the CPU "acknowledges" the interrupt and begins to "service" it (sorry for the jargon, but we might as well get used to these conventional terms). In effect, it is up to you to transform a level-sensitive input into an edge-sensitive input. (You may recall the Text's observations on the IBM PC's edge-sensitive interrupt lines. See p. 701.)

Here, once more, is a flip-flop to do that:

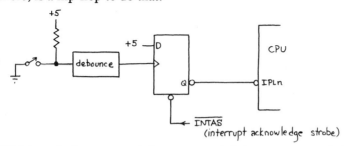

Figure L22.7: Interrupt hardware: a proper design: generates a single interrupt request on closing of a switch

Interrupt Response

a) Processor Saves present Condition

This you saw in the response to an Illegal instruction.

b) Processor Determines What to Do When Interrupted: finds the interrupt service routine (ISR)

We will use the slower and simpler of the 68000's two ways of finding what to do in response to an interrupt. We will use the chip's *autovectoring* (which mimics the response of a 6800, Motorola's earlier 8-bit processor). The fancier method would require a little more hardware: an additional octal 3-state, and we want to spare you that work unnecessary in this application.

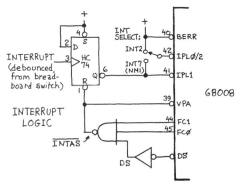

Figure L22.8: Interrupt hardware: for autovector response

Our logic asserts VPA* during INTAS*[2] The 68000 responds by looking to a particular vector address (in the manner of a 6800[3] (The toggle switch that feeds $IPL_{0/2}$* will allow us to transform our interrupt request to Level 7, if we choose to do so later.)

Note a curious side-effect of VPA* on your computer: your hardware single-step will fail during interrupt response: the computer will run through the ISR at full speed, even when you put the computer into STEP* mode, because VPA* tells the processor to pay no attention to the level on DTACK*. When the program pops out of the ISR, the single step will work normally again.

For our "Level 2" interrupt request, the processor looks to address 68H for the 32-bit address of the ISR.

---

[2]. We had to synthesize "INTAS*" ("interrupt acknowledge strobe") because, as usual, the chip designers ran short of pins: they provide no such signal on a single pin. Highs on both fc_1 and fc_0 indicate **INTAS*** and are used to disable both memory and the I/O decoder, the '138 you wired in Lab 19.

[3]. Asserting VPA* causes the CPU to autovector, and also makes the 68000 emulate 6800 timing; this feature eases use of the new processor with old 6800 peripheral chips, which expect both a slow clock rate (1 MHz, for the garden variety peripheral) and timing that does not depend on DTACK*).

22-4 Programs: Main & Service Routine

Reminder: Interrupt Mask

Note that the CPU ignores interrupts unless you tell it to pay attention to them. (This statement does not apply to the Non-Maskable Interrupt, of course.) You tell it to pay attention to a level 2 interrupt by writing an instruction that sets the interrupt mask to a level lower than 2. We will set it to level 1 (we also keep off a function called "Trace"; you will meet that in the *register-check* routine, if you are interested).

```
46 FC 71 00      move #$7100 to SR    ; Trace off, level 1 int mask
```

A *reset* of the processor sets the mask to maximum, blocking all but level 7 interrupts. So, you must lower the mask at the start of your program—or whenever your program is ready to receive an interrupt. The interrupt response itself also raises the interrupt mask to the interrupting level; thus interrupts are disabled during execution of the ISR, unless they have higher priority. The return from exception (RTE) that terminates the ISR then enables Level-2 interrupts again, by restoring the SR mask to the value it held before the interrupt response began. Tidy scheme, isn't it?

Reminder: Interrupt Vector

You will need to tell the processor where to look for its interrupt service routine. In the example below, we have placed the routine at address 00D0—a bit below our main program. The ISR, you will notice, is tiny, so it fits handily in this niche.

Vectors

Adr (hex) where vector is stored		Vector value	
0	Stack Pointer	00 00 20 00H; (as usual)	
3	Reset: PC main	00 00 01 00	; this is just the delay-and-increment ; program
68	Interrupt 2	00 00 00 D0	; ISR start address (ISR just displays ABCD, pauses, and returns)

Interrupt Test Programs

Here is a simple pair of programs to let you test your interrupt hardware. The 'Main' program just increments the display. The ISR puts out a distinctive display—"ABCD"—pauses for a few seconds, and then returns to the main program. Pressing the Interrupt button should evoke the ABCD display.

```
                    ; VECTORS:
                    ;       address     function        comment
                    ;       00          stack ptr       as usual
                    ;       04          PC, main        as in lab 20
                    ;       68          int level 2     int srvc rtn: (its start address).
                    ;                                   We have suggested $D0, but
                    ;                                   any value will do
;_____
; MAIN program
307C 8000           MAIN:   movea.w #$80000, a0         ; the usual display pointer
027C F1FF                   andi.w #$F1FF, SR           ; mask, to force interrupt mask
                                                            down to level 1, without
                                                            disturbing the remainder of
                                                            the Status Register

4240                        clr.w d0                    ; clear display value
3080                DISP:   move.w d0, (a0)             ; show current display value
61 ??                       bsr DELAY                   ; call delay—we hope it's still
                                                            around, so you needn't rewrite
                                                            it. In Lab 20, it sat at $166

5240                        addq.w #1, d0               ; increment display value
60F8                        bra.s DISP                  ; ...and keep doing it

;_____
; DELAY Routine (assumed to sit at $166)
2F01                DELAY:  move.l d1, –(a7)            ; save main register
223C 0001                   move.l #$0001046C, d1       ; 1/4 second delay
5381                LINGER: subq.l #1, d1               ; decrement 66FC
                            bne.s LINGER                ; ...till you hit zero
221F                        move.l (a7)+, d1            ; recall main register
4E75                        rts                         ; back to calling program

;_____
; INTERRUPT SERVICE Routine
48A7 C000                   movem.l d0-d1, –(a7)        ; save contents of a register we'll
                                                            mess up—plus one we won't
                                                            use (d0). That's a silly thing
                                                            to do, but we want to
                                                            introduce you to this neat
                                                            instruction
                                                            ("MOVEMultiple"), which
                                                            stores as many registers as you
                                                            need, at a stroke; often handy
                                                            for interrupt service.
30BC ABCD                   move.w #$ABCD, (a0)         ; set up distinctive display
                                                            (assumes a0 points to display,
                                                            as it does, thanks to main
                                                            program; a bit odd to use that
                                                            assumption within ISR)
720A                        moveq #$0A, d1              ; set up 10X delay counter
4EB8 0166  LOOPLOOP:        jsr.w $0166                 ; call delay, assumed to sit at old
                                                            location, $166
5381                        subq.l #1, d1               ; call dly 10X
66F8                        bne.s LOOPLOOP
4CDF 0003                   movem.l (a7)+, d0-d1        ; restore registers
4E73                        rte                         ; and back to main (*NOT RTS*!)
```

Figure L22.9: Interrupt test program: main program & interrupt service routine (isr)

22-4 NMI: Level 7 Interrupt

Now we will try a *non-maskable* interrupt: level 7. To use it, we must install a level 7 interrupt vector:

Vectors

Adr (hex) where vector is stored		Vector value	
7C	Interrupt 7	00 00 00 B0	; NMI ISR

And you must write an ISR at address $B0. Perhaps your program could put out another distinctive display, such as "DCBA."

Remove, from the main program, the instruction that lowers interrupt mask to level 1; lacking this instruction, the CPU will keep the interrupt mask at 7. If you try to interrupt at Level 2, the program should ignore you.

If you flip the toggle switch that drives $IPL_{0/2}{}^*$, your interrupt switch now will drive a *level 7* interrupt request (see fig. L22.8, p. L22-11). You should find that now the processor recognizes an interrupt from your interrupt switch, regardless of the level of the interrupt mask. Later, we may make use of this NMI to allow breaking in to a running program and checking register contents.

Using Interrupts

When you are satisfied that your interrupt hardware works satisfactorily, you should apply it to a problem that requires the quick response interrupts allow. You might try the switch-bounce capture: the storage scope loaded promptly upon the beginning of bounce. The 7K memory space will allow you to capture more samples than you need; your A/D gives you a new sample about every 3 µs. If you take these samples in at full speed, filling the 'storage scope' table, a 7K-byte table will let you get around 20 ms of data, much more than you need for normal switch bounce. But there is no harm in that; the important point is that you should get the *early* switch bounce. For a clean scope display, your program should put out a Trigger pulse to the scope just before starting a playback cycle. For this purpose, use any I/O decoder line not otherwise busy in your program.

Class 23: μ6: Wrap-up: Buying and Building

The bag of tricks Horowitz & Hill promised you at the start of the Text now is pretty well loaded (though it can, of course, hold a great deal more). Here's a set of alternatives you have seen, presented partly just to let you feel pleased with yourself, but partly also to remind you that you may face choices when you need to produce some electronic device. You may have to choose analog versus digital, hard-wired versus programmable, integrated versus discrete.

An Inventory of Your Bag of Tricks: a tree

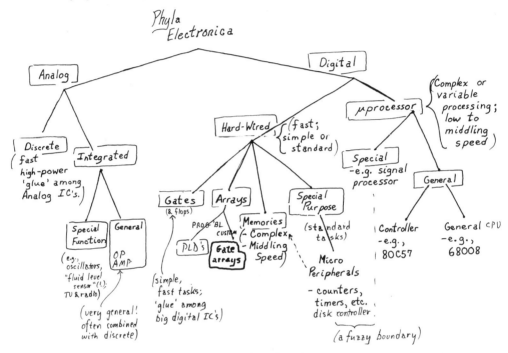

Figure N23.1: Alternatives

The *tree* should not be taken to suggest that when you design something you end up on one twig and stay there. When you land on a branch like *op amp* or *microprocessor* you are very likely to reach across to take some *discrete* transistors or gates or flip-flops. So, if the structure is taken to be a tree, then a designer is not like a caterpillar who creeps out to the end of one twig; the designer is more like a monkey: nimble, with long arms, not shy about leaping or reaching about among the branches to pick up whatever fruit might go nicely with the main meal.

For some applications, the right general choice is obvious. For example:

amplify a sine wave × 2:
    don't use an A/D, computer (left shift) and D/A!

count photons (suppose that each generates a unit pulse). The information is digital; don't use an op amp to integrate the pulses!

store a page of typed words, allowing changes
    don't convert the key codes to voltages and store them in analog form!

Sometimes the right answer is not obvious. A long-term integrator or sample-and-hold, for example, could be done with analog or digital methods. At some long period the digital method becomes clearly preferable; but it's not obvious where.

Build versus Buy

Sometimes you should buy the whole gadget: a digital thermometer; a power supply. The more interesting questions come when you decide to build, but wonder at what level to begin—in the range from transistor up to fancy large-scale IC or even circuit card.

Use an IC that does the job, if it's available

This is obvious, but worth recalling. If you need a small temperature sensor, for example, it would be a shame not to take advantage of the nifty linearized sensors & amplifiers squeezed onto a tiny chip by Analog Devices or National Semiconductor. You should build your own only if the IC version won't do (perhaps you need a still tinier sensor, for example).

A less obvious example: you want to make a hand-held remote controller, or sense the level of fluid in a tank. Would you have *guessed* that the following devices were available?

```
  LM903  Fluid Level Detector .....................................
  LM1812 Ultrasonic Transceiver ...................................
  LM1851 Ground Fault Interrupter .................................
* LM1893/LM2893 Biline Carrier Current Transceiver ................
  LM1871 RC Encoder/Transmitter ...................................
  LM1872 Radio Control Receiver/Decoder ...........................
  LM1884 TV Stereo Decoder ........................................
```

Figure N23.2: Some special-purpose IC's (reproduced with permission of National Semiconductor Corp.)

You might well want to use one of these rather than an equivalent circuit built with op amps and cleverness. Same answer for, say, a logarithmic converter.

This observation leads to an obvious question: ***Query:***How is one to discover whether an IC exists to do the job?

Answer: Use a book called IC Master, a big list of IC's, published annually (with updates); it will steer you to the next source: manufacturers' data books. Here's what you would see if you were looking for telephone dialers, for example:

IC MASTER

LINEAR—Telecommunication Circuits (Cont'd)

Function	Device	Source	Line	Function	Device	Source	Line	Function	Device	Source	Line
Dialer, Pulse/Tone (with last number redial)				Dialer, Tone with Last Number Redial and Speech Network	MA530	Marconi		Direct Access Adapter	UM9161	UMC	
		(Cont'd)									
	MK53721	SGS-Thomson		Dialer, Tone with Speech Network and DC Line Voltage Regulator	LB1008AE	AT&T		DMI (Digital Multiplexed Interface) Chip Set, Consists Of:	DS-1 Chip Set	AT&T	
	MK53762	SGS-Thomson			PBL3780	Ericsson					
	TEL5413A	STC			MC34010A	Motorola (2974)		DMI (Digital Multiplexed Interface) Framer	N229CG	AT&T	
	UM91230	UMC			MC34011A	Motorola (2974, 2990)	55		N229GB	AT&T	
Dialer, Pulse/Tone (with LCD driver)	LR48106	Sharp	5					DMI (Digital Multiplexed Interface) Maintenance Buffer	N229FB	AT&T	95
Dialer, Pulse/Tone with Redial	MC145410	Motorola (2974)		Dialer, One Key	UM95080	UMC					

Figure N23.3: IC Master excerpt (reprinted with permission of IC Master)

Another query: How does one get all those data books?
Answer: it's surprisingly hard, unless you're working for a company that buys a fair amount of stuff. Nobody seems to want to send a data book to a plain old academic. Go visit the distributors, or go to a trade show. If you have to, you can even *buy* data books from some manufacturers: Texas Instruments and National will sell you any of their data books. That may not sound like doing you a favor, but really it is.

You can get by with just a few books:

- *analog ("linear"):* National is essential; Burr-Brown, Analog Devices and Linear Technology are nice to have as well.
- *digital:* National or Motorola: CMOS, FAST (National) bipolar digital
- *conversion:* Analog Devices

To get at the many other data books that you need now and again, befriend some fanatic who collects them, or turn yourself into a little company, and become a fanatic yourself.

Microcontrollers:

Full PC versus little controller

Usually it's easiest in a lab to use a PC, and to buy the board to do a standard job (like A/D, D/A). Save your energy for odd peripherals.

You may, however find or invent applications for little controllers: circuits that are many (Paul's "META" SETI machine—a big, fast multi-channel analyzer—uses 144 identical computer/controllers running in parallel), fast (the SETI machine offers a good example again), remote or low-power (or both, like many of Winfield Hill's gizmos, some of which sit deep under water; or like the data loggers a graduate of this course leaves out in the woods to monitor this and that environmental condition).

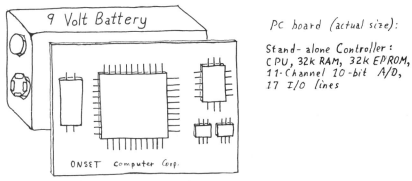

Figure N23.4: A Really Tiny Controller (after data sheet of Onset Computer Corp., N. Falmouth, MA. Shown with permission of Onset Corp.)

Build versus Buy

It's hard to dream up applications where you will *not* want to buy someone else's controller board (available for a few hundred dollars), unless you're manufacturing something yourself. In the latter case, you should consider a highly integrated *controller* chip (see opening of Ch. 11): Intel's 8051, or heaps of fancier gadgets. Some include A/D, D/A; most include serial ports, timers, ROM & RAM. A few even let you talk to them in high-level languages. But beware the investment in *time* required by any chip that is new to you, with its new assembly-language, if you are obliged to program it at the level.

Generally, you can program at a higher level, if you prefer: perhaps in C or FORTH, or even BASIC.

Ways to Build Things: (getting off breadboards)

You probably know that the way you have built circuit in this course is not the way you will want to build anything that should survive for a while. Consult the Text's Chapter 12 when it's time to convert a design to permanent form. What you learn in Chapter 12 will include a good deal about the three most important ways to put together a circuit:

solder breadboards
    good for one-of-a-kind circuits, analog or digital

wire-wrap
    good for one-of-a-kind circuits; convenient for IC's, less convenient for discrete parts, and not so good for high-frequency analog

custom printed-circuit
    good for all kinds of circuits when you need more than one

Wire wrap is odd enough so that a diagram may be worthwhile:

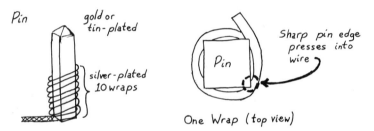

Figure N23.5: Wire wrap connection: about 40 gas-tight connections link wire to post

You can see how the wire makes its 40-odd connections on each wrap. These connections are gas-tight, so they remain good long after the exposed surfaces have oxidized. On an old wire wrap board, the silver-plated wire itself may turn nearly black, while the board continues to work just fine.

What next?

You now know pretty well how to teach yourself more electronics. Students often tell us that when they look again at the Text *after* finishing the course they find they understand passages that puzzled them on their first reading. We expect you will feel the same effect, too, and that you will look again at familiar circuits and continue to discover characteristics that you had not noticed on your first meeting. (This happens to *teachers* of electronics, too, and helps keep the process of teaching fresh and intriguing.) We hope that you are feeling more or less literate, now, in this field: we hope that you now are able to *read* circuit diagrams, and to understand designs.

That skill helps to answer another question that we often hear: what next? We cannot guide you to a next course, or even tell you whether you should keep studying or now go get a job that will give you some experience. At least you will want to keep reading so, that you can follow the rapid changes in the field. This course has given you only a start, and its teachings will not remain up-to-date for long. There are good trade magazines in the field that explain new technology, show circuit designs, and (maybe most fun) advertise what's new: <u>Electronics</u>, <u>EDN</u> and <u>Electronic Design</u> all serve this purpose. A weekly newspaper called <u>Electronic Engineering Times</u> gives a more summary form of the same kind of information, and it's full of speculation, industry gossip and business news.

You will keep learning best, of course, if you continue to design circuits. Next best is to look at the work of others. You know enough to be able to understand explanations of issues subtler than what we have attempted in this Manual. You might try Chapter 13 on high-frequency techniques, for example, or Chapter 14 on low-power circuits; then you might also find yourself inspired to go take an engineering course that treats some of the fundamental issues we have dodged! And you have seen enough data sheets so that you know how to get at the world's amazing variety of semiconductors—especially integrated circuits. You should be able to read even a thick data sheet without fear, now that you have wrestled with a 48-pin monster of a chip, considered its timing requirements, and *tamed* the thing: made it run loops for you.

When you meet a new design problem you will ask yourself, 'What does this problem resemble?'—hoping to apply what you know to the new problem. That's when your bag of tricks is supposed to serve you. If the problem is interesting, nothing in your bag will fit *exactly*. You will begin to call up images of this and that circuit, and will begin to try variations, try splicing this and that fragment together, amending this and that standard circuit. So, you will begin to cross the boundary from work that can be satisfying but not exciting—building standard circuits, adjusting them until they work—into the headier region where you make something at least slightly *new*. Enjoy yourself.

Lab 23: µ6: Applying Your Microcomputer (*Toy Catalog*)

> **Reading:** Nope.
> **Problems:** Only to come up with a fun and instructive use for your little computer.

This lab simply invites you to play with the available peripherals; we provide only a little guidance. Scan these suggestions, then choose or invent a task. The gadgets and their possible applications described here are meant to inspire you, not restrict you. The best project is a feasible enterprise that you find challenging and fun.

The projects described here can be simple, if done in their most rudimentary form, or complex, if done with refinement.

1. *X-Y scope displays*: Dead Easy in its simplest form; much more difficult in its refined forms—particularly the true-vector scheme. You may want to try a simple version of this exercise even if you then plan to do something else: some people get a big kick out of seeing their initials, for example, on a scope screen.

2. *Light pen*: A phototransistor detects the CRT beam; the position of the beam is controlled by the computer, and the computer thus can tell where the pen sits. This can be straightforward to program, but will require some fussy analog work; for that reason it's a good wrap-up exercise.

3. *Voice output:* Not quite so easy as the scope display, even in its simplest form. It offers some challenging programming if you choose one of the neater ways to feed the voice chip. A suggested application—a talking logic probe or talking voltmeter—would give pretty flashy results for an effort of middling magnitude.

4. *Driving a stepper motor:* This can be very easy: you first write a program that sends the sequences of 2-bit codes that will make the motor turn. If you do more than simply spin the motor, you will have a chance for smarter programming.

   You can, of course, invent some new stepper-controlled gadget, like a primitive crane, to lift and move a load.

5. *Little computer meets PC*: Some of you are about to go to work in labs, where you are more likely to find an IBM PC than a 68000-based computer (though this may not remain true). You may want to take this occasion to get used to a PC interface. Presumably you will want to let your little computer do some task that suits it (it runs faster than the PC; but you want to keep its *programs* very simple, since you have to key them in by hand), and pass the results to the PC, which is better suited to fancy operations (since you can program it at a higher level).

6. *Games:* Here the X-Y display that begins these notes probably is essential. You can use a *joystick* to input position data to your game. The joystick would give you a chance to build the very useful device, a 2:1 analog multiplexer, in order to let you use a single A/D (already in place) to accept both X and Y information. The analog mux is not hard to build, but it is a circuit worth meeting, and one that would oblige you to consider some timing issues.

7. *Sound sampling/generation:* Again, this can start out simple and get as fancy as you like. By now you have stored a waveform, using your A/D (back in Lab Micro 3) You can do more by processing the stored information: you might, for example, provide echo or "reverberation;" you could also splice samples together under keypad control to make a sort of synthesizer. You could even make a talker, this way.

8. *Any other project that intrigues you:* If you dream up a project that you find challenging, we will be well satisfied. Aim to keep hardware and program simple. Above all, try to take advantage of the powers of this lab computer, and to dodge its weaknesses: it lets you build and control hardware; it makes programming laborious. So, don't do a job you can do much better on an ordinary PC or on a big computer. Don't write a word-processing program.

23-1 X-Y SCOPE DISPLAY

This is the quickest and easiest way to draw a picture on a CRT screen. If you use a capacitor to join the points your program puts out, you achieve a quick and easy 'vector' display, good at drawing lines, including diagonals, whereas the more usual raster-scan (TV-like) display does lines laboriously with a succession of dots—and incidentally makes ugly lines whenever the slope approaches vertical or horizontal (an effect called "aliasing"). More generally, since most TV and monitor displays are coarse enough to show you the individual pixels, a raster scan display usually is pretty ugly.

Hardware

You will need to give your computer a second D/A (*two* more if you implement the hardware "Zoom": see (d), below). The complicating wrinkles suggested in the later subsections require a bit more hardware. Most of the work, however, lies in the programming, and in building the data tables your program is to send out. The payoff comes when you get to put whatever you choose on the scope screen.

Preliminary note: Two D/A Types

In the lab we have two sorts of D/A:

"microprocessor compatible," single supply: the AD7569 (A/D, D/A in one package) and the AD558, an 8-bit single-supply part (you used this D/A in Lab 17, where it provided feedback for the successive-approximation A/D). Avoid the AD558 in any application where it is fed directly from the data bus, because of its unacceptably-long setup time. (See sec. 11.05, discussing this characteristic of the '558.)

"multiplying:" this type offers an output that is the *product* of its digital input and an (analog-) input current. We exploit that characteristic in the Zoom feature (23-1-b-3, below). The multiplying converter we have in the lab, the MC1408, requires three power supply voltages. It also lacks internal latches. So you should use the 1408 only if you want to use its distinctive multiplying feature.

The 68000's multiply instructions should let you achieve the same results in software. But if the hardware method appeals to you, you should decide at the outset whether you mean to Zoom, so as to choose the appropriate D/A.

23-1-a 16 X 16 Dot Array Hardware

For a 16 X 16 array of dots, two 4-bit D/A's are sufficient. It is convenient to drive these from a single (8-bit-) port, and to take advantage of the D/A's internal register. Here is the laziest scheme, which does not require disconnecting lines from your D/A-A/D. If you can tolerate an error of about 1 LSB in one of the outputs, this lazy scheme is good enough:

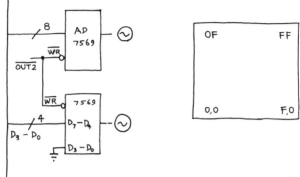

Figure L23.1: 16 X 16 X-Y display hardware; scope screen map

The program should fetch a byte at a time from successive locations in a table of x-y display values, and put out those bytes. If you include a call to your delay subroutine, you will be able to tinker with the drawing rate.

Complications/Improvements

23-1-b-1 Connect the Dots

As you learned in elementary school, connecting a few dots can give a coherent picture. So, with relatively little programming effort (fewer coordinates to list in your data table), you can draw straight-line pictures. To connect the dots, add a low-pass filter at the output of each D/A. Try _RC_ of a few microseconds (e.g., $R = 2.2k$, $C = 0.01$ µF). But you will have to experiment; the visual effect will vary with the drawing-update frequency, as well as with the filter's *RC*. The filter slows the movement of the output voltages, of course, so that the movement of the scope trace becomes visible. You will notice that this new scheme obliges you to pay attention to the sequence in which your program puts out these dots.

23-1-b-2 256 X 256 Dot Array

You can use two output ports, of course, to feed the full 8 bits to each D/A. If you do this, you should arrange things so that the new information reaches the two D/A's simultaneously. That requires use of an 8-bit register of D flip-flops. Here a 558, fed by the output of the D register, may be easier to wire than the 7569. We will leave to you the details of this hardware. (Use a 74HCT574 8-bit D register; enable the '574 3-states continuously.)

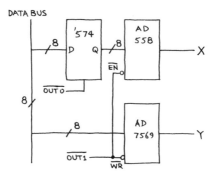

Figure L23.2: 256 × 256 X-Y display hardware: register added for simultaneous updates

If you enter a large number of data points, you will begin to see flicker in the display, even if your program includes no delay routine. The screen needs to be refreshed about 30 times/second in order not to flicker annoyingly. You can calculate the number of dots you can get away with—or you can just try a big table and examine the resulting display for flicker.

23-1-b-3 Size Control; Zoom

As we have suggested already, you can *zoom* in software. But if you want to make your programming task a little easier, and like the neat feature of a picture symmetrical about zero as its size changes, then consider the zoom hardware described just below:

Zooming, you will recall, is the operation that can exploit the multiplying, current-output D/A's (MC1408) rather than the AD7569. The '1408 takes longer to wire because it does not include an input latch and because it obliges you to take care of an annoying number of pins: for frequency compensation, and multiple supplies.

'1408 Details

Here are wiring details for the '1408:

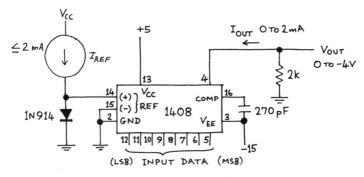

Figure L23.3: '1408 multiplying DAC: pinout and wiring

Incidentally, in case you choose to modify your circuit, the '1408's reference current flows into the summing junction of an internal op amp; therefore, you can provide a *voltage* rather than *current* reference, if you prefer. In that case you simply feed the '1408's pin 14 through an appropriate series resistor (for example, 7.5k from a +15V supply).

You may prefer a newer single-supply multiplying D/A such as the AD7524; if you use one, you will need to modify the circuits we have shown below, which assume a '1408.

One more output port (4 to 8 D flops) and D/A, plus a current mirror, will allow you to control the size of the *X-Y* pattern put out by your first two D/A's if you used multiplying D/A's for the *X-Y* output.

In the circuit below, we feed a constant current into just one of the three D/A's—the "*Size*" D/A. The other two D/A's are fed not a constant I_{ref} but, instead, the *Size* D/A's output current (mirrored by duplicate mirrors). Thus the computer can use the *Size* D/A to scale the *X-Y* outputs.

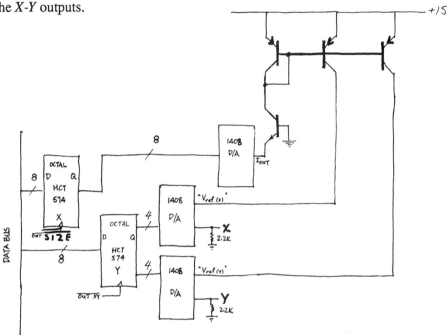

Figure L23.4: *X-Y* size control, using third d/a and current mirror; *receding square*, drawn with this hardware

Below is a suprisingly-subtle image drawn with this hardware. This pyramid appears to show cleverly-gradated shading. In fact it is only a diminishing square. The square was drawn as usual by defining four points, then 'linking' them by slowing the D/A output with an *RC*. The apparent shading results from the CRT beam's gradual slowing (exponential) as it moves from source to destination.

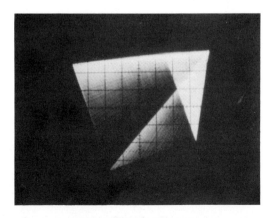

Figure L23.5: Receding square

A circuit refinement is proposed below to center the image on the CRT regardless of *size*. Without this circuit addition, a change of size also moves or translates the image: since *X* and Y are always negative, the visual effect will be as if a figure that grows were moving down and to the left on the screen, as well as toward you. If you prefer to make your figures "approach" straight-on, then you should make the modification shown below: an op-amp is added to the output of each of the D/A's (X and Y); the op-amp allows you to center the coordinate output at zero volts.

In the circuit below, half the D/A input current is applied to the output op amp's summing junction, and thus is subtracted from the output, centering that voltage as the scaling current varies.[1]

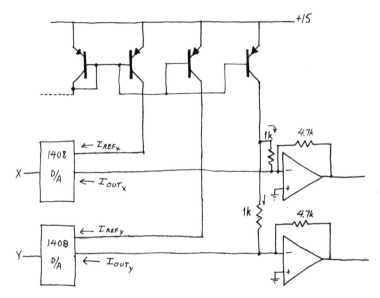

Figure L23.6: Op amp added to D/A to give V_{out} symmetrical about zero volts

23-1-b-4 True Vector Drawing: Position-Relative [2]

Your drawing table need not store absolute screen locations. Instead, it can store vectors: direction and length relative to present screen location. The relative vectors suggested below are listed in the manner of compass headings: ESE = East-South East. This scheme would work with a 4-bit direction specification 16 directions). The remaining four bits could define magnitude; that magnitude could be used with the multiply instruction to stretch the unit-length direction vector (approximate unit lengths will do!).

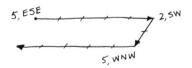

Figure L23.7: Drawing with true vectors: *relative* movement

This way of defining a figure allows one to rotate the figure without much difficulty. The programming is a good deal harder than for an absolute X-Y figure.

---

1. Thanks to Extension student D. Durlach for this nifty amendment (1985).
2. Thanks to Summer School student Scott Lee for proposing and demonstrating this arrangement (1985).

23-1-b-5 Animation

If you enter the coordinates for two or more similar but different pictures in memory, and then draw these pictures in quick succession (changing the picture, say, every 10th of a second) then you can 'animate' a drawing. Evidently, you will need a large table to animate even a simple figure, so start modestly. You will want to use the follow-the-dots scheme to minimize the umber of data points required to draw a single image. Here's a crude example:

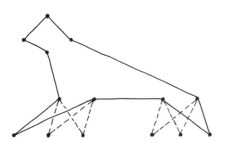

Figure L23.8: Rudimentary animation

23-1-b-6 Computed Drawing

The tedious loading of values into memory demanded by the table-reading schemes we have suggested probably have made you yearn to to turn over to the machine the job of determining what points it should draw. This you can do, of course. You could write a program that would draw a rectangle, say, by incrementing the X register for some steps while holding Y fixed, then incrementing the Y register while holding X fixed, then decrementing the X register..., and so on.

Two excessively-experienced programmers have used the <u>X-Y</u> hardware to draw cubes, which they then could rotate about any of three axes.[3]

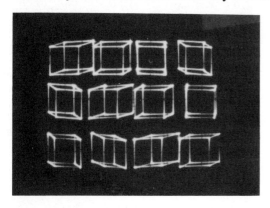
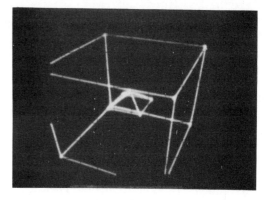

Figure L23.9: Computed drawing: two versions of cube that can be rotated about its axes. The photo showing multiple cubes is a multiple exposure showing the cube rotating over time

This task requires too much programming sophistication—and too much code—for ordinary mortals.

---

[3]. Grant Shumaker did this first (1986), and did it entirely in assembly-language, entered by hand. This is a feat roughly comparable to climbing one of the middle-sized Himalayas without oxygen. David Gingold did it again, this time putting shapes within cubes; David used oxygen, as noted below under 23-6 in a discussion of David's remarkable *asteroids* game.

23-2 LIGHT PEN

This task combines software and hardware, analog and digital in a satisfying way. The analog part is ticklish: you are likely to spend a good deal of effort fighting ambient light, for example; and you will get best results if you can assemble a little light wand, using fiber optics and some tape or heat-shrink tubing.

Hardware

Output: You need the *X-Y* output hardware; 4 bits per D/A is plenty.

Input: You need a photosensor circuit, mounted on a flexible cable. You can use the photo-transistor plus op amp current-to-voltage converter that you built back in Lab exercise 8-6-b. You may want to put a potentiometer in the feedback loop, so that you can adjust gain (this setting turns out to be delicate). In addition, to fend off the effect of ambient light, including light from the scope beam before it arrives at the point the light pen means to watch, you will find a light shield helpful. We used a fiber optic bundle or fiber attached with opaque heat-shrink tubing to the phototransistor:

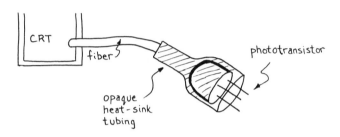

Figure L23.10: Light pen improvement: light shield and fiber optics

No special treatment of the fiber ends is required; just try to make clean cuts.

A comparator can watch the output of the light-sensing circuit; its threshold should be adjustable. The comparator will tell the computer when the CRT beam has arrived.

Suggestions: if your pen has trouble detecting the beam, which is present only briefly, you may want to modify the sensor circuit, adding a capacitor in the feedback loop so as to let the circuit integrate the effect of successive appearances of the beam. This should not cause errors, since the 'last straw' of charge that tips the integrator output past the comparator's threshold will occur while the CRT beam is at the sensor's position.

Program

The foundation of the program is just an *X-Y* raster-scan (that is, TV-like scanning). You should include a delay, to give the light-sensor circuit time to respond. When the program detects a response, the register holding the current *X-Y* data suddenly becomes significant, and can be passed to whatever program you devise to use the light pen input.

Suggestions: ways to use the light pen

Draw on the scope screen

Probably it is most fun to let the pen do something to the display: it could brighten an *X-Y* dot, for example, allowing the pen to draw on the screen. For such an application you can use the scope's so-called *Z axis* input: a scope input that dims the scope a good deal when a positive voltage is applied: +5V usually is sufficient; some scopes want +15V to show an appreciable effect.

One difficulty with this scheme, perhaps evidently, is that the light-pen operation requires a beam bright enough for the pen; the selected dots cannot be much brighter.

Select among several functions: music?

Light pens are used, as you know, to make a computer do this or that operation—and the operation need not be one that alters the screen display. You could use the pen to alter the pitch of an output square wave, for example, so that the pen formed the input of a strange musical instrument (the X value might control pitch, for example, while Y determined amplitude; X could be passed to a count-down delay loop that would give a kick to the speaker at the end of each count-down). You would have to tinker with delay *constants*, of course, to get the frequency down into audible range.

23-3 VOICE OUTPUT

Two nearly-equivalent chips will put out humanoid sounds in response to a 6-bit code defining a "phoneme" or "allophone"—a fragment of sound such as the "b" and "oo" in "boo." You may find a talker chip that you prefer; these are not state of the art, but are easy to use. The chips are—

- SC-02
    (developed by Votrax International, Troy Michigan; now sourced only by Artic Technology)
- SC-01
    (this is an old Votrax chip, hard to find; it is very similar to the SC-02, but simpler, lacking its programmable attack, duration and pitch)

You will need to check the data sheet for the talker that you have in the lab, of course. The notes below describe use of the SC-01, and will serve as only an approximate model if you use another chip.

A. Hardware

1- What the Synthesizer Requires

A VOTRAX voice synthesizer chip, or equivalent, does all the hard work for you: generates one waveform in response to each digital "phoneme" code. The computer's job is only to provide a new code promptly, when the talker chip says that it has finished putting out its current sound.
These are the signals used by the VOTRAX SC-01:

- a 6-bit phoneme code ($D_0 - D_5$)
- a Strobe signal from the computer to the chip (D_6).
    This signal tells the chip that the phoneme code is valid; the rising edge clocks the code into the chip's internal register.
- A Ready signal from the chip to the computer (D_7). This signal tells the computer that the chip has finished generating its last phoneme, and needs a new phoneme code. (Strobe clears Ready, incidentally.)

The Strobe signal must not occur until at least a microsecond after the phoneme code is presented to the chip:

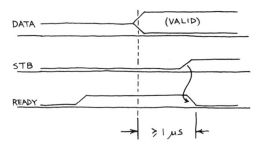

Figure L23.11: VOTRAX timing

This requirement results from the long setup time of the chip's D register (the voice chip is made of slow CMOS).

2- Hardware That You Must Provide

You will need one output port (7 bits), and one input bit:

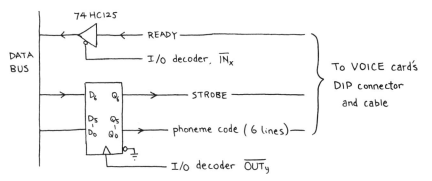

Figure L23.12: Hardware to drive voice chip

B. Software

1- Two useful talker routines

The interesting part of this task comes when you choose a way to get data from a table of phonemes, in order to send it to the chip. There are some duller chores to take care of, as well. Specifically, you need patches of program (which we have written out for you) to do the following jobs:

- give the chip's STB an initial kick, and tell the chip to be silent;
- check the Ready line, then send the 6-bit phoneme code when Ready comes High, then assert STB after the Data's long setup time.

Two utility routines to drive the talker chip

We have written these as subroutines. Your more interesting Main program can call these at its pleasure.

Subroutine: STARTUP

```
0000                    ;STARTUP subroutine: sends silence code without
0000                                       ; awaiting Ready: needed on first pass.  Uses d0,
0000                                       ; which is used for output code throughout
0000
0000                                       ; if pointers are not already set up, define them thus:
0000
0000    367C 8003            movea.w #$8003, a3      ; point to Vox data port
0004    327C 8001            movea.w #$8001, a1      ; point to the generic Ready-bit port
0008
0008
0008    70 3E         startup moveq #$3E, d0         ; send silence code.  Strobe is low.
000A    1680                  move.b d0, (a3)        ; send to talker
000C    0000 0040             ori.b #$40, d0         ; force Stb bit high, leaving other bit
0010                                                 ;   unchanged
0010    1680                  move.b d0, (a3)        ; send it, Stb high
0012    0200 00BF             andi.b #$BF, d0        ; force Stb low
0016    1680                  move.b d0, (a3)        ; send it
0018
0018    4E75                  rts                    ; the deed is done
```

Subroutine: SENDOUT

```
                        ;SENDOUT subroutine: sends whatever is in d0,
                        ;    but first awaits Ready from Votrax, and
                        ;    then sends STB pulse as required

0000    1F01          sendout move.b d1, -(a7)       ; save register which
0000                                                 ; we'll mess up
0002    1213          test:   move.b (a3), d1        ; Ready? (check bit 7)
0004    0201 0080             andi.b #$80, d1        ; mask all but bit 7
0008    67 F8                 beq.s test             ; loop till ready

000A    0200 00BF             andi.b #$bf, d0        ; force STB low
                                                     ; in phoneme code
000E    1680                  move.b d0, (a3) ;        send it, STB low

0010    0000 0040             ori.b #$40, d0         ; force STB high
0014    1680                  move.b d0, (a3) ;        send it

0016    0200 00BF             andi.b #$bf, d0        ; force STB low again
001A    1680                  move.b d0, (a3) ;        send it

001C    121F                  move.b (a7)+, d1       ; restore saved register
001E    4E75                  rts                    ; back to calling pgm
```

Figure L23.13: Two utility routines

Before attaching your computer to the Votrax, check that your hardware and program appear to work. You can do this by single-stepping with the Repeat key and watching your program's output on your data displays (insert an OUT to Displays at the point where the program later will drive the voice chip).

You can also check your program's response to the READY signal that later will come from the chip (see below). For a continuous output of phonemes, hold Ready high. When you are satisfied that your circuit works, attach your computer to the VOTRAX, and try sending this table of data to the voice chip:

Message Table

Address (you define start address)	Data
XX X0	0B
1	09
2	2A
3	03
4	2D
5	3A
6	2B
7	19
8	1F
9	3E
A	3E
B	FF; this is one possible STOP or EOWord code

Figure L23.14: A table of phonemes for testing your program & hardware

When you are ready to compose messages of your own, consult the talker chip's data sheet for the list of its phoneme codes, and a 'dictionary' of words defined as sequences of phonemes. If you are using an assembler (as we hope you are, by this point), you can define, for the assembler, the equivalence between phonemes and their binary codes. 'UH1 = $19,' for example, where UH1 is the sound of the 'o' in *love*. Then you can write out words by listing their phoneme names, not hex codes: you can write *love* as 'L1, UH1, V.'

2- Neater ways to get codes

Typing in those few phoneme codes probably convinced you that if you were using the Votrax to say more than a few words it would be useful to leave some standard words stored permanently in memory, then pluck out whole words rather than single phonemes. Such a scheme would pay when applied to any word used more than once. The 68000 is eager to help you with its versatile addressing modes!

3- An Application: Talking Voltmeter or Logic Probe

A DC input to your computer's A/D can serve to select a message to be played back. You will want a tidy way to tell the computer to state its message anew. One reasonable possibility is to let the 'meter' speak whenever it has a new value to report. That would be especially appropriate for a logic probe, which would speak only when it found High or Low after float; it could respond to High and Low in immediate succession by saying something like "switching."

23-4 STEPPER MOTOR

23-4-a Generally

A stepper motor contains two coils surrounding a permanent-magnet rotor; the rotor looks like a gear, and its 'teeth' like to line up with one or the other of the slightly-offset coils, depending on how the current flows in the coils. A DC current through these coils holds the rotor fixed. A reversal of current in either coil moves the rotor one 'step' clockwise or counter-clockwise (a coarse stepper may move tens of degrees in a step; a moderately fine stepper may move 1.875 degrees: 200 steps/revolution). The sequence in which the currents are reversed (a gray code) determines the direction of rotation.

For example:

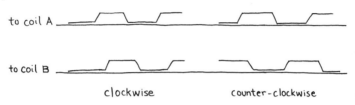

Figure L23.15: Sample waveforms to drive stepper motor's two coils

Integrated stepper-motor driver chips can make driving a stepper extremely easy: the chip usually is just a bidirectional counter/shift-register, capable of sinking and sourcing more current than an ordinary logic gate. But since a computer can easily do most of the work that the chip does, it makes sense to wire up some power transistors and then write a little routine to do the job. Specifically, the routine should take care of these tasks:

- store the pattern most recently sent to the motor;
- generate the next pattern, either clockwise or counter-clockwise;
- send the new pattern after a delay that (added to others) will determine the rate of rotation.

Hardware

The motors can be driven from power MOSFETs (VN01 is marginal but possible; the RFP4N05 could do the job easily; see Lab 11). Here is the scheme:

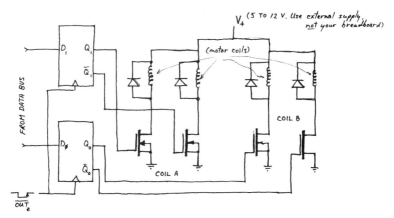

Figure L23.16: Stepper driver hardware

One way to generate such successive patterns is to load a value into a register and rotate that value, feeding two adjacent bits to the motor's two coil-drivers:

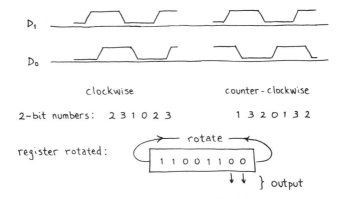

Figure L23.17: Rotated register can produce the 2-bit drive pattern for stepper

23-4-b Using the motor

You will dream up your own ways to use a motor, but here are a few uses students have tried.

Crane

Two motors, each with a spool on its shaft to wind cord, can rest on a tabletop, with the load suspended from both cords. The two motors can work together to lift, move, then lower the load. This is easy to do crudely, but would be challenging if one wanted to let the load move vertically, then horizontally, then vertically again. A small electromagnet could let the machine pick up and drop an iron load (a washer, perhaps). (A small spool of wire-wrap wire is handy for an instant home-made coil; put an iron bolt through its core.) (One student built such an electromagnet, hung it from a stepper-driven 'winch,' and mounted the whole thing on the tractor described below.[4])

---
4. This was Dylan Jones' ingenious work. (1987)

Drawing machine

Two motors can drive the X and Y knobs of a child's sketching toy (the kind with a glass cover, coated with aluminum dust on its sealed underside; a sharp stylus etches a dark trace in this material). Such a machine makes a curious form of semi-permanent drawing. The scheme resembles 'turtle graphics': any new vector begins where the last one terminated. This application requires motors of exceptionally high torque, and a little machine-shop skill in order to link the motors to the two shafts.

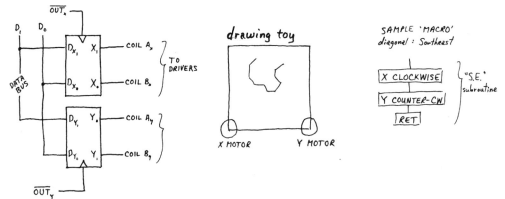

Figure L23.18: Two stepper motors can drive an X-Y drawing toy

Little tractor

Two motors mounted end to end, and each with a rubber-tired wheel on its shaft form a tractor that can be steered (in the manner of a bulldozer or tank) by driving the two wheels independently.

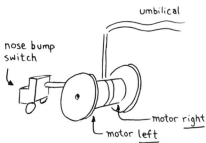

Figure L23.19: Tractor/tank: steered by independent drive of its two rear wheels

A couple of ambitious students taught such a gizmo—which was equipped with a switch on its nose to detect collisions—to blunder its way through a maze. It then cruised calmly out, avoiding all the blind alleys it had found on the way in[5].

That was a challenging project, and putting together the tractor itself requires some machine-shop skills. Because the motors draw a lot of current, the tractor was fed through an umbilical cable, which carried logic signals as well as power for the motors.

Brilliant & Harebrained Projects

We have not seen the following done, but it would be nice: to reproduce with microcomputer control what was perhaps the most spectacular project ever: a pencil balancer. The original project was done with entirely analog methods. This machine was

---

[5]. Mike Pahre and Danny Vanderryn performed this feat (1988). They even had their little machine do a victory dance (in the manner of a football player who has just scored a touchdown) when it had completed its graceful exit from the maze.

able to keep a pencil balanced on its point for a few seconds, by moving a little platform about on a smooth base.

The analog design shone an LED on the upright pencil while a pair of photodetectors watched to see if the pencil leaned away from the vertical. The photodetectors fed a differential amplifier, which drove a DC motor whose response tended to force the pencil upright. The motor did this by spinning a pulley on which a string was wound, and the string would draw the balancing-block one way or the other on the base platform.

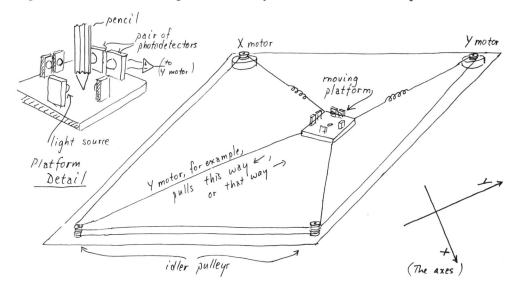

Figure L23.20: Pencil balancer mechanism[6]

Two such LED-photodetector-pair units took care of watching for X and Y lean. The negative feedback in this arrangement made the mechanism work quite well even though the construction was crude in the extreme: sewing thread, wooden pulleys, a composition-board base, and odds and ends of adhesive tape were its materials.

The mechanism worked pretty well until the little platform reached one of the edges of the base. There, the pencil would fall. What the whole design lacked was any awareness of the position of the moving block upon the base. A computer-driven version of this machine might fill this gap. If the computer used stepper motors, it could easily keep track of the block's absolute position even as the motors carried out balancing corrections according to the scheme used in Paul's analog machine. As the block approached any limit, the computer could exaggerate corrections in a sense tending to force the pencil to lean back toward the center of the base. Easier said than done, we admit, but perhaps someone will feel like using the microcomputer to improve on Paul's already-impressive design.

---

6. This balancer was the work of the inspired tinkerer *extraordinaire*, Paul Titcomb (ca. 1986). Another of Paul's amazing projects deserves mention here; it may inspire someone to a similar effort. Paul made a drawing-mimicker using his lab computer. Paul rigged a 2-jointed arm that held a pen. He placed a potentiometer at each joint, and as he guided the arm, drawing a picture by hand, he had the computer read the joint-potentiometers at regular time-intervals. Then, when he had finished a drawing, Paul would ask the computer to 'play back' the motions it had sensed, by driving a pair of stepper motors on the joints of a similar arm also holding a pencil. The machine never quite worked: its drawings showed a rather severe palsy—probably an effect of its mechanical crudity and the fact that (unlike the pencil balancer) it operated *open loop*—without the error-forgiving magic of feedback. It was an impressive gadget, nevertheless, and the arm and its last, shaky drawings languish at the back of our lab. They await someone who will perfect the scheme.

23-5 Little Computer Meets PC

Here is a hardware suggestion for an interface between your little computer and an IBM PC. The hardware assumes that the PC is going to run full tilt, unaware of what the little computer is doing. The little computer is given the chores of checking whether the PC has given it a value (*Flag To68*) or has picked up a value and thus can be fed another (*Flag From68*). If the PC is running a BASIC program, there's no question the little machine will outrun it, keeping the big machine satisfied—picking up all it can send, or feeding it all it can eat.

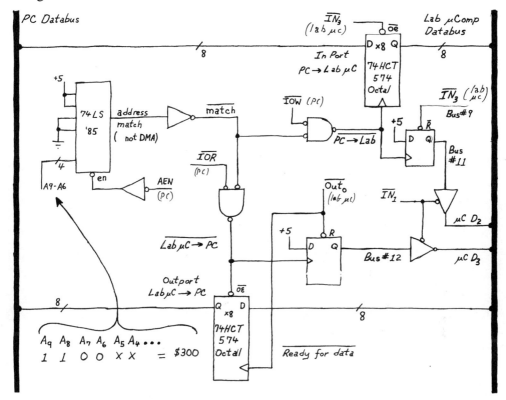

Figure L23.21: A PC-to-little-computer interface

A pioneering pair of students[7] used this hardware to let the little machine fill a table with data from the A/D, then pass the data to the PC, which did a Fourier transform. This was an impressive exercise, but a bit heavy on programming. We encourage you to choose a project that stresses the hardware issues closer to the core material of this course. We suggest—once more—that you make all your programs *very* modest.

Notes:

1. You can build the PC interface on a breadboard mounted on a PC plug-in card (e.g., Global Specialties PB-88). The little computer can connect to this card through a ribbon cable terminated in an RS-232-style connector (DB-25) that plugs into the PC card.

2. If you program the PC in BASIC, note that you should use *INP(address)* not *Peek*, since these ports are *I/O* ports, not memory locations.

---

7. Wolf Baum and Tom Killian demonstrated this nice exercise (1988).

23-6 GAMES

Using the computer to drive the scope in its *X-Y* display mode one can, of course, draw anything on the scope screen. A few years ago, every kid was fascinated with video games. Maybe you can work up some enthusiasm for a game that you have fashioned yourself.

The first video games, as you know, resembled ping-pong or squash. You can, if you like, use a potentiometer driving your A/D to determine the position of a "paddle" (a line drawn on the scope screen). Then the ball—a traveling dot or figure—can rebound or not, depending on whether it "hits" the paddle. One ambitious student made the angle of rebound depend on whether the hit was near the center of the paddle or near either end. A ball that hits the wall presumably should "bounce"; what would that imply for the slope of its travel on rebound?

If, like most video game players these days, you find batting a ball too peaceable, you can redefine 'ball hits paddle' to mean something more alarming, like 'ABM hits ICBM,' and you can make exciting things happen in consequence: let parts scatter on the scope screen, or let a sound output change, as one student did: (beeps change to whoops?)

 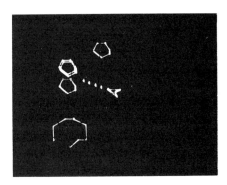

Figure L23.22: Some CRT games; one a mad *tour de force* on the little machine: asteroids

The 'tour de force' shown above suggests how far you can go, given time and skill: this zealot[8] wrote an assembly-language program for his little machine that let it take in serial data and load it into specified sections of memory. Then he wrote his 68000 code on a full-scale computer, in C. He also expanded his little computer's RAM to 32K and built some nice vector graphics hardware that drew straight lines of uniform brightness, regardless of length. This was a project wildly excessive for our course. This student happened to be a very good programmer who was able to combine work for two courses on this project. So don't think you're falling short of some standard because you decide not to write a game of asteroids.

Joystick

A joystick—two pots linked—can provide *X-Y* information to your game, so that you are not restricted to sliding a paddle up and down on the screen, but instead can make a figure move anywhere on the scope screen (fleeing aggressive blobs, if your game is conventional, as one recent student's was). The analog switch DG403 makes the 2:1 multiplexer needed for this *X-Y* input easy to implement; you will need a flip-flop to hold the switch in either mode until the computer changes it. Check the switch specifications to be sure the switch output has settled before you start converting the value.

---

8. David Gingold did this project, and did it solo (1988).

23-7 SOUND

23-7-a: Processing Stored Inputs

It's fun to alter stored sound waveforms, drawn from your A/D. An echo machine, suggested in Lab 22, lets you pretend to be down in a cavern, well or culvert, depending on the echo-delay and echo-attenuation that you choose. Your program can take in new samples as it plays out old, or it can work on a stored table of values. Probably the *live input* version is the more fun of the two. You may want to let yourself vary the echo dimensions from the keypad, as the program runs. (This would be a primitive version of a machine used by recording studios to put acoustically-bland studio music into fictitious settings: small nightclub, big concert hall, cathedral.)

It's also possible to alter the playback rate without altering frequency, with some effort. This trick is used to let the blind listen efficiently to slow-talking teachers (like Tom). Another version of the machine could be used to take the stored speech of a quick talker (like Paul), and then play back this speech closer to the information-input rate of the ordinary human brain.

23-7-b Making Noises ("music"?)

You can make some pretty ugly noises by using your computer to drive a speaker—especially if the amplitude is a little big. (Like oxymorons?). This task offers a virtually-unlimited range of difficulty. In three minutes you will be able to produce a tone. But whoops are a little harder; melodies are harder than that; so is keypad control; and you can surely imagine lots of grandiose possibilities, like a scheme that lets the computer memorize a melody entered at the keypad, or a drum machine (need a cymbal? Perhaps a software pseudo-random noise generator would help: see Text sec. 9.32 et seq.).

An easy way to let the computer output a chosen frequency is to let an output from the I/O decoder drive a MOSFET that drives a speaker. (This you have done before, back in Lab 15, as you may recall. Remember the *programmable* $\div$ *n* counter?) For higher volume, *square* the waveform with a toggling flip-flop before driving the transistor. On second thought, maybe we shouldn't be telling you how to make *louder* noises!

Whatever you do with your little machine, don't forget to *have fun*.

Chs. 10, 11 Review: Important Topics

Microcomputers, Microprocessors

1. Hardware

    a. Buses: 3-states to drive bus, flops to take information off it
    b. address decoding; using control lines as well: e.g., I/O decoding; *partial* address decoding
    c. word/byte transfers
    d. bit input, pulse output:

        i. an occasional external event is best saved in a flip-flop, usually edge-triggered with constant input
        ii. pulse output: simple way, if pulse can be short, just use decoder output; for longer pulse, need an output flop

2. Programming

    a. branch on condition

        i. loop iteration (e.g., delay)
        ii. check bit
        iii. compare, looking for pattern (e.g., keycheck in Lab 22)

    b. pointers

        i. for speed
        ii. for table reads and fills; need way(s) to tell whether table done

    c. stack: it works so automatically that you seldom need think about it; you do need to appreciate the difference between a subroutine CALL (useful for multiple use of one patch of code) and garden-variety branch or jump
    d. exception response: this is much less important, for our purposes; but it is not hard to understand, once you understand the use of the stack in an ordinary subroutine call.

Chapter 10, 11: Microcomputer/Microprocessor Jargon and Terms

These two chapters are especially rich (especially annoying) in their use of *acronyms*. It's hard not to imagine that the manufacturers sometimes *start* with a clever acronym, and then demand that the research department invent something to fit it ('RIOT', for example, an old chip that included Ram, I/O and a Timer. Which came first: the riot or the chip?) So, this list of terms includes some substantial notions and some terms that are no more than names.

absolute addressing	contrasted with *relative* addressing: address specified as a number rather than as a distance from the present location of the computer's *program counter*.
access time	time a memory takes to deliver data after a stable address is applied. Typical values 25 to 200 ns. Sec. 11.12.
asynchronous bus	a misnomer, really; the Text's term, *default wait*, is better. Means only that the CPU proceeds only when told to do so by the things it talks to and listens to (the 68000 uses dtack* to achieve this result). Sec. 10.14.
bus	a set of lines connected to many points in a circuit: address and data buses are the most important; "control bus" is a term used less widely; power buses are used in all of electronics, not just digital.
dtack*	68000 *input* signal, 'DataTransferAcknowledge.' (This signal is a 68000 special, so may deserve a definition.) It tells the processor to proceed. It is roughly equivalent to an active-high signal that one might call 'Wait.'
EEPROM	ElectricallyErasablePROM.
EPROM	ErasablePROM; means more than that: the name is reserved for the device that is erasable by ultra-violet light.
exception	Motorola's peculiar term for all sorts of events (hardware and software) that cause a break in ordinary program execution and diversion to a response routine.
interrupt	process by which a processor allows a hardware signal to cause a break in ordinary program execution and diversion to execution of an 'interrupt response routine.'
PROM	the sort of ROM that a user can load. 'Programmable ROM.'
RAM	a memory one can write to easily. Contrasted with ROM, a memory that can be loaded only with difficulty. RAM is an acronym for 'Random Access Memory.' But that's unfortunate, since *all* semiconductor memories are Random Access Memories (except bubble memories and FIFOs). (Re: memories of all sorts, see Sec. 11.12.)
ROM	Read Only Memory: this acronym fits. It is a memory that one write to only rarely and with considerable effort; true ROM, when contrasted with PROM, can be loaded only at the manufacturing stage. Holds data when power is removed, so useful for *bootstrap* program of computer.
vector(s)	the noun refers to a stored value that steers program execution to a particular destination. The 68000 uses nearly 1K of low memory space to store such values. The 'Reset vector,' stored at locations 0-7, loads stack pointer and program counter, for example.

APPENDIXES

A. Equipment and Parts

B. Selected Data Sheets
2N5485 JFET
74HC74 dual D flip-flop
AD7569 8-bit A/D, D/A
68008 execution times & timing diagrams
25120 write-only-memory (*Note*: add salt)

C. Schematic of lab computer: 'Big Picture'

D. Pinouts: parts used in lab exercises

APPENDIX A: EQUIPMENT AND PARTS LIST

EQUIPMENT
Suggested for each setup

Device	Description
Scope	
Tektronix 2225	dual channel, 50 MHz
Func. Gen.	
Krohn-Hite 1400	sine, triangle, square, logic level; sweep; to 3 MHz
DVM	
Beckman DM25L	3 1/2 digit
VOM	
Simpson 260-8P	multimeter with overload protection (essential!)
Power Supply	
B & K Precision 1630	0-30V, 3A, linear

Breadboard (Powered)
Global Specialties PB503
   3 supplies (+5V, ±15V), function generator including logic level; debounced swithes, 8 LED indicators

Breadboards, not powered
Global PB-105   set of six strips, mounted
Global UBS-100   single strip
(*Beware* of cheap imitations. We've tried them! Make sure that any *mounted* strips are screwed down to a base, not taped or glued.)

Logic Probe
OK PRB50   O.K Industries: 50 ns glitch capture
or
HP545A   (Hewlett Packard: 50ns glitch capture; easier to use; $$$!)

Resistor Substitution Box
RD111 (Contact East)
   1Ω to $11M\Omega$ in 1Ω steps; ±1%
or
Ohmite Ohm-Ranger 3420

Capacitor Substitution Box
Ohmite 3430A Cap-Ranger
   100 pF to 11.111 µF in 100pF steps; ±2% accuracy

Resistor Kit
Ohmite CAB-54
"little devil"
   1/4-watt carbon composition; 10Ω to $10M\Omega$

Equipment and Parts List

> **Keypad**
> *CUSTOM: Note:* the schematic is included in this Student Guide (it follows Lab 15). The authors have made manufacturing arrangements, and can supply complete keyboard units. Write or call Paul Horowitz, Harvard Department of Physics, Lyman Laboratory, Cambridge, MA 02138: tel. (617)495-3265.

Tools
- pliers: chain nose
  - C.K 3772-1H
- wire stripper
  - e.g., Contact East cat. no. 100

Odds and Ends
- patch cords
  - "stacking banana": black, red E.g., Pomona B-18
- BNC cables
- BNC tees
- BNC-to-minigrabber converters

Equipment: *one* for the laboratory
- Palladin PA100
  - ribbon cable stripper (Contact East): hugely helpful in the microprocessor labs
- (your choice)
  - a frequency counter: occasionally very helpful

PARTS

Suppliers

We have a couple of favorites to recommend: for tools and supplies, Contact East, N. Andover, MA ((508) 682-2000); for many other parts, Digikey, Thief River Falls, MN ((800) 344-4539). Both are fast and efficient: take orders quickly and ship promptly.

Passive

Diodes

PART NO.	DESCRIPTION
1N914	silicon diode
1N748A	zener, 3.9V
1N749A	zener, 4.3v.
1N751A	zener, 5.1v.
1N5817	Schottky, 1 A.

LEDs
- HLMP-4700 — Red, low current
- HLMP-4719 — Yellow, low current
- HLMP-3950 — Green, high efficiency

Lamps
- #47 lamp
- #344 lamp

Inductor
- 10mH: (E.g., Mouser 43LJ310)

Transformer
- 6.3 v, C.T. transformer, 1A, with power cord, and banana jacks on secondary side

Switches
- pushbutton
  - Panasonic: EVQ-PXR04K (Digikey Cat. # P9950)
- toggle
  - spdt (may substitute DIP toggle, below)
- DIP:
  - 4-position toggle spdt: Grayhill 76STC04 (alternative to single toggle suggested above)
  - 4-position slide

Microphone
- 25LM045
  - electret: Panasonic (Digikey #P9931 or Mouser #25LM045)

> **Cable**
> NFBC 5022-1
> **custom order:** ribbon cable, #22 solid conductor, 50 conductors; 8 colors repeat, black through violet. 100-foot minimum. National Wire & Cable, Los Angeles

Heat Sink

HS130   TO-92: Aavid

Battery Holder

BH2AA-SF
    battery holder, 2 AA: Memory Protection Devices

12BC254   battery snap leads, 24 AWG solid (Mouser)

Miscellaneous

hookup wire
    #22 AWG hookup wire, solid; 10 colors, black through violet (e.g., Belden #8530, 100' roll)

IC Sockets:

ICA·326·S·TG
    32–pin screw machine socket, to protect HP data displays, whose pins are exceedingly fragile. (Robinson-Nugent)

or

532AG11D Augat.
    40–pin sockets also do the job, but must be cut to 32 pins with a razor saw. These would be R-N: ICA·406·S·TG, or Augat: 540AG11D.

Data Books

National Semiconductor
    Linear Data Book (multiple volumes)
    Logic Data Books (including HCMOS)

Motorola:

M68000RG/AD
    pocket reference guide

M68008/AC
    programming card

AD1939R2 68008 hardware manual

COMPONENTS

Capacitors

Ceramic

| Z5U | 0.01 |
| | 0.1 |

CK05BX-	
–100K	10 pF
–470K	47 pF
–101K	100 pF
–221K	220 pF
–331K	330 pF
–471K	470 pF
–561K	560 pF
–681K	680 pF
–102K	1000 pF

Mylar

0.001	axial-lead 50V or 100V (= polyester)
0.0033	for example Cornell-Dublier WMF
0.01	
0.1	
0.33	

Tantalum

CS13B-type (axial lead):

1 µF		35V
4.7 µF	35V	
15 µF		20V
68 µF		15V

Electrolytic

500 µF   aluminum electrolytic, 25V

Potentiometer

100k   single-turn 1/2-W (e.g., Bourns 3323)
1M

Resistors

These are in addition to the standard 5% selection suggested under <u>equipment</u>. Few are required in the laboratory exercises; we suggest these values as occasionally useful.

1%

    1/4-watt metal film (e.g., RN55D type)

    100 ohms, 1K, 2K, 10K, 49.9K, 100K, 1.0M, 10.0M

ACTIVE DEVICES
Transistors
bipolar

2N4400	NPN, power
2N3904	NPN, small-signal
2N3906	PNP, small-signal
2N5962	NPN superbeta
LPT100	phototransistor (Siemens)

FET

2N5485	JFET
2N3958	JFET, matched pair
RFP4N05L	power MOSFET, low threshold: IRL 510 (IR) is equivalent
VN0104N3	power MOSFET, n-channel (Supertex)
VP0104N3	power MOSFET, p-channel (Supertex)

Miscellaneous

MCR218-3 silicon-controlled rectifier (SCR), 8A, TO-220 package (Motorola)

INTEGRATED CIRCUITS

Linear

CA3096CE bipolar transistor array: npn, pnp
CD4007A or CD4007UB MOSFET array (RCA CA3600 is the same part)

LM311N	comparator
TLC372CP	comparator, single supply
LM723CN	voltage regulator
LM317H	voltage reg, adjustable
LM358N	dual op amp, single supply
LM385-2.5	voltage reference
LF411CN	op amp, FET input
LM741CN	op amp
LM78L05CZ	voltage reg, 3-term, TO-92
ICM7555IPA	timer/oscillator (CMOS)
MF4CN-50	switched-capacitor 4-pole Butterworth low-pass filter
DG403CJ	analog switch (Siliconix)

Digital

TTL

74LS00	
74LS469	8-bit up/down counter: MMI only
74LS502	8-bit successive-approximation register: National only

CMOS: HC

74HC-

–00	quad NAND
–04	hex inverter
–14	hex Schmitt Trigger inverter
–74	dual D flip-flop
–112	dual J-K flip-flop
–125	quad 3-state buffer
–161	4-bit binary counter, jam clear
–175	quad D flops
–541	octal 3-state buffer
–4040	12-stage ripple counter
–4046	phase-locked loop

74HCT-

–02	quad NOR
–08	quad AND
–10	triple 3-in NAND
–27	triple 3-in NOR
–32	quad OR
–138	1-of-8 decoder
–574	octal D register, 3-state

NMOS

68008-P8 CPU, 8 MHz (Motorola)

Miscellaneous

HM6264-LP-12	8K × 8 SRAM, 120ns, low-power
AD558JN	D/A, single supply, with register
AD7569JN	A/D, D/A, microprocessor-compatible
HP5082-7340	hexadecimal display with decoder & latch
F5C-8 MHz	Fox 8 MHz oscillator, 45/55% duty cycle

Parts not generally required, but useful for projects

MC1408 or AD5724
    8-bit multiplying D/A
SC-02 or SC-01
    speech synthesizer (Votrax: SC-01 discontinued; SC-02 from Artic Technology, Rochester, Michigan)

APPENDIX B: SELECTED DATA SHEETS
2N5485 JFET

n-channel JFETs designed for . . .

- **VHF/UHF Amplifiers**
- **Mixers**
- **Oscillators**
- **Analog Switches**

Siliconix

Performance Curves NH
See Section 4

BENEFITS
- Low Cost
- Completely Specified for 400 MHz Operation
- Low Error Analog Switch
  Very Little Charge Coupling
  $C_{rss} < 1.0$ pF

TO-92
See Section 6

Plastic

Bottom View
Source & Drain Interchangeable

*ABSOLUTE MAXIMUM RATINGS (25°C)

Drain-Gate Voltage	25 V
Source Gate Voltage	25 V
Drain Current	30 mA
Forward Gate Current	10 mA
Total Device Dissipation @ 25°C	360 mW
Derate above 25°C	3.27 mW/°C
Operating Junction Temperature Range	−65 to +135°C
Storage Temperature Range	−65 to +150°C
Lead Temperature (1/16" from case for 10 seconds)	240°C

*ELECTRICAL CHARACTERISTICS (25°C unless otherwise noted)

#		Characteristic		2N5484 Min	2N5484 Max	2N5485 Min	2N5485 Max	2N5486 Min	2N5486 Max	Unit	Test Conditions	
1	S	I_{GSS}	Gate Reverse Current		−1.0		−1.0		−1.0	nA	$V_{GS} = -20$ V, $V_{DS} = 0$	
2	T				−200		−200		−200			$T_A = +100°C$
3	A	BV_{GSS}	Gate-Source Breakdown Voltage	−25		−25		−25		V	$I_G = -1 \mu A$, $V_{DS} = 0$	
4	T I	$V_{GS(off)}$	Gate-Source Cutoff Voltage	−0.3	−3.0	−0.5	−4.0	−2.0	−6.0		$V_{DS} = 15$ V, $I_D = 10$ nA	
5	C	I_{DSS}	Saturation Drain Current	1.0	5.0	4.0	10	8.0	20	mA	$V_{DS} = 15$ V, $V_{GS} = 0$ (Note 1)	
6		g_{fs}	Common-Source Forward Transconductance	3,000	6,000	3,500	7,000	4,000	8,000			f = 1 kHz
7		g_{os}	Common-Source Output Conductance		50		60		75			
8		$Re(y_{fs})$	Common-Source Forward Transconductance	2,500								f = 100 MHz
9						3,000		3,500		μmhos		f = 400 MHz
10		$Re(y_{os})$	Common-Source Output Conductance		75							f = 100 MHz
11							100		100		$V_{DS} = 15$ V, $V_{GS} = 0$	f = 400 MHz
12	D	$Re(y_{is})$	Common-Source Input Conductance		100							f = 100 MHz
13	Y						1,000		1,000			f = 400 MHz
14	N A	C_{iss}	Common-Source Input Capacitance		5.0		5.0		5.0			
15	M I	C_{rss}	Common-Source Reverse Transfer Capacitance		1.0		1.0		1.0	pF		f = 1 MHz
16	C	C_{oss}	Common-Source Output Capacitance		2.0		2.0		2.0			
17		NF	Noise Figure		2.5		2.5		2.5		$V_{DS} = 15$ V, $V_{GS} = 0$, $R_G = 1$ MΩ	f = 1 kHz
18					3.0						$V_{DS} = 15$ V, $I_D = 1$ mA, $R_G = $	
19							2.0		2.0			f = 100 MHz
20							4.0		4.0	dB	$V_{DS} = 15$ V, $I_D = 4$ mA, $R_G = 1$ kΩ	f = 400 MHz
21		G_{ps}	Common-Source Power Gain	16	25						$V_{DS} = 15$ V, $I_D = 1$ mA	f = 100 MHz
22						18	30	18	30		$V_{DS} = 15$ V, $I_D = 4$ mA	
23						10	20	10	20			f = 400 MHz

* JEDEC registered data

NOTE:
1 Pulse Test PW 300 μs, duty cycle ≤ 3%

NH

2N5485 n-channel JFET
reproduced with persmission of Siliconix, Inc.

DG403 ANALOG SWITCH

DG400-405
Low-Power – High-Speed CMOS Analog Switches

FEATURES
- ±15 V Input Range
- ON Resistance < 35 Ω
- Fast Switching Action
  t_{ON} < 150 ns
  t_{OFF} < 100 ns
- Ultra Low Power Requirements (P_D < 35 μW)
- TTL, CMOS Compatible

BENEFITS
- Wide Dynamic Range
- Low Signal Errors and Distortion
- Break-Before-Make Switching Action
- Simple Interfacing

APPLICATIONS
- High Performance Audio and Video Switching
- Sample and Hold Circuits
- Battery Operation

DESCRIPTION

The DG400 family of monolithic analog switches were designed to provide precision, high performance switching of analog signals. Combining low power (< 35 μW) with high speed (t_{ON} < 150 ns), the DG400 series is ideally suited for portable and battery powered industrial and military applications.

Built on the Siliconix proprietary high voltage silicon gate process to achieve high voltage rating and superior switch ON/OFF performance, break-before-make is guaranteed for the SPDT configurations. An epitaxial layer prevents latchup.

Each switch conducts equally well in both directions when ON, and blocks up to 30 volts peak-to-peak when OFF. ON resistance is very flat over the full ±15 V analog range, rivaling JFET performance without the inherent dynamic range limitations.

The six devices in this series are differentiated by the type of switch action as shown in the functional block diagrams. Package options include the 16-pin plastic, CerDIP and LCC package. Performance grades include industrial, D suffix (–40 to 85°C), and military, A suffix (–55 to 125°C). Additionally, the DG403 and DG405 are available in the narrow body surface mount package, SO-16.

FUNCTIONAL BLOCK DIAGRAM, PIN CONFIGURATION AND TRUTH TABLE

Dual-In-Line Package
Order Numbers:
CerDIP: DG403AK
        DG403AK/883
Plastic: DG403DJ

Leadless Chip Carrier
Order Number: DG403AZ

DG403 Two SPDT Switches per Package Truth Table

LOGIC	SWITCH 1 SWITCH 2	SWITCH 3 SWITCH 4
0	OFF	ON
1	ON	OFF

Logic "0" = V_{IN} ≤ 0.8 V
Logic "1" = V_{IN} ≥ 2.4 V

SO Package
(Same pinout as DIP)
Order Number: DG403DY

ELECTRICAL CHARACTERISTICS [a]

PARAMETER	SYMBOL	Test Conditions Unless Otherwise Specified: $V+ = 15$ V, $V- = -15$ V, $V_L = 5$ V, GND = 0 V, $V_{IN} = 2.4$ V, 0.8 V [e]	TEMP [d]	1=25°C TYP	2=125,85°C 3=–55,–40°C A SUFFIX –55 to 125°C MIN [b]	MAX [b]	D SUFFIX –40 to 85°C MIN [b]	MAX [b]	UNIT
SWITCH									
Analog Signal Range [c]	V_{ANALOG}				–15	15	–15	15	V
Drain-Source ON Resistance	$r_{DS(ON)}$	$V+ = 13.5$ V, $V- = -13.5$ V, $I_S = -10$ mA, $V_D = \pm 10$ V	1 2,3	20		35 45		45 55	Ω
Delta Drain-Source ON Resistance	$\Delta r_{DS(ON)}$	$V+ = 16.5$ V, $V- = -16.5$ V, $I_S = -10$ mA, $V_D = 5, 0, -5$ V	1 2,3	3.0		3.0 5.0		3.0 5.0	Ω
Switch OFF Leakage Current	$I_{S(OFF)}$ $I_{D(OFF)}$	$V+ = 16.5$ V, $V- = -16.5$ V, $V_D = -15.5$ V, $V_S = 15.5$ V	1 2	–.01	–0.25 –20	0.25 20	–0.50 –20	0.50 20	nA
Channel ON Leakage Current	$I_{D(ON)} +$ $I_{S(ON)}$	$V+ = 16.5$ V, $V- = -16.5$ V, $V_D = V_S = \pm 15.5$ V	1 2	–0.04	–0.4 –40	0.4 40	–1.0 –40	1.0 40	nA
INPUT									
Input Current with V_{IN} LOW	I_{IL}	V_{IN} Under Test = 0.8 V All Other = 2.4 V	1,2	.005	–1.0	1.0	–1.0	1.0	μA
Input Current with V_{IN} HIGH	I_{IH}	V_{IN} Under Test = 2.4 V All Other = 0.8 V	1,2	.005	–1.0	1.0	–1.0	1.0	μA
DYNAMIC									
Turn-ON Time	t_{ON}	$R_L = 300$ Ω, $C_L = 35$ pF See Figure 1A	1	100		150		150	ns
Turn-OFF Time	t_{OFF}		1	60		100		100	ns
Break-Before-Make Time Delay	t_D	$R_L = 300$ Ω, $C_L = 35$ pF DG402/DG403	1	20	10.0		10.0		ns
Charge Injection	Q	$C_L = 10,000$ pF, $V_{gen} = 0$ V, $R_{gen} = 0$ Ω	1	60		100		100	pC
Off Isolation		$R_L = 100$ Ω, $C_L = 5$ pF f = 1 MHz	1	72					dB
Crosstalk [f] (Channel-to-Channel)		Any Other Channel Switches $R_L = 100$ Ω, $C_L = 5$ pF f = 1 MHz	1	90					dB
Source-OFF Capacitance	$C_{S(OFF)}$	$V_S = 0$ V f = 1 MHz	1	12					pF
Drain-OFF Capacitance	$C_{D(OFF)}$		1	12					pF
Drain and Source ON Capacitance	$C_{D(ON)} +$ $C_{S(ON)}$		1	39					pF

DG403 dual spdt analog switch reproduced with persmission of Siliconix, Inc.

DG403 ANALOG SWITCH

DG400-405

ELECTRICAL CHARACTERISTICS [a]

Test Conditions Unless Otherwise Specified:
V+ = 15 V
V− = −15 V
V_L = 5 V
GND = 0 V
V_{IN} = 2.4V, 0.8 V [e]

PARAMETER	SYMBOL	Test Conditions	TEMP [d] 1=25°C 2=125,85°C 3=−55,−40°C	TYP	LIMITS A SUFFIX −55 to 125°C MIN [b]	MAX [b]	LIMITS D SUFFIX −40 to 85°C MIN [b]	MAX [b]	UNIT
SUPPLY									
Positive Supply Current	I+	V+ = 16.5 V, V− = −16.5 V V_{IN} = 0.0 or 5.0 V	1	0.01		1		1	μA
			2,3			5		5	
Negative Supply Current	I−		1	−0.01	−1		−1		
			2,3		−5		−5		
Logic Supply Current	I_L		1	0.01		1		1	
			2,3			5		5	
Ground Current	I_{GND}		1	−0.01	−1		−1		
			2,3		−5		−5		

NOTES:
a. Refer to PROCESS OPTION FLOWCHART for additional information.
b. The algebraic convention whereby the most negative value is a minimum and the most positive a maximum, is used in this data sheet.
c. Guaranteed by design, not subject to production test.
d. Typical values are for DESIGN AID ONLY, not guaranteed nor subject to production testing.
e. V_{IN} = Input voltage to perform proper function.
f. Crosstalk performance is improved to 110 dB (typ.) with LCC package.

ABSOLUTE MAXIMUM RATINGS

V+ to V−	44V
GND to V−	25V
V_L to V−	(GND − 0.3 V) to 44 V
Digital Inputs[1] V_S, V_D	(V−) −2 V to (V+ plus 2 V) or 30 mA, whichever occurs first
Current (Any Terminal) Continuous	30 mA
Current, S or D (Pulsed 1 ms 10% duty)	100 mA
Storage Temperature (A Suffix)	−65 to 150°C
(D Suffix)	−65 to 125°C
Operating Temperature (A Suffix)	−55 to 125°C
(D Suffix)	−40 to 85°C
Power Dissipation (Package)*	
16-Pin Plastic DIP**	450 mW
16-Pin CerDIP***	900 mW
20-Pin LCC****	750 mW
16-Pin SO*****	600 mW

* All leads welded or soldered to PC board.
** Derate 6 mW/°C above 75°C
*** Derate 12 mW/°C above 75°C
**** Derate 10 mW/°C above 75°C
***** Derate 7.6 mW/°C above 75°C

[1] Signals on Sx, Dx or INx exceeding V+ or V− will be clamped by internal diodes. Limit forward diode current to maximum current ratings.

74HC74 DUAL D FLIP-FLOP

National Semiconductor — microCMOS

MM54HC74/MM74HC74 Dual D Flip-Flop with Preset and Clear

General Description

The MM54HC74/MM74HC74 utilizes microCMOS Technology, 3 micron silicon gate P-well CMOS, to achieve operating speeds similar to the equivalent LS-TTL part. It possesses the high noise immunity and low power consumption of standard CMOS integrated circuits, along with the ability to drive 10 LS-TTL loads.

This flip-flop has independent data, preset, clear, and clock inputs and Q and $\overline{Q}$ outputs. The logic level present at the data input is transferred to the output during the positive-going transition of the clock pulse. Preset and clear are independent of the clock and accomplished by a low level at the appropriate input.

The 54HC/74HC logic family is functionally and pinout compatible with the standard 54LS/74LS logic family. All inputs are protected from damage due to static discharge by internal diode clamps to V_{CC} and ground.

Features
- Typical propagation delay: 20 ns
- Wide power supply range: 2–6V
- Low quiescent current: 40 μA maximum (74HC series)
- Low input current: 1 μA maximum
- Fanout of 10 LS-TTL loads

Truth Table

Inputs				Outputs	
PR	CLR	CLK	D	Q	$\overline{Q}$
L	H	X	X	H	L*
H	L	X	X	L*	H
L	L	X	X	H*	H*
H	H	↑	H	H	L
H	H	↑	L	L	H
H	H	L	X	Q0	$\overline{Q0}$

Note: Q0 = the level of Q before the indicated input conditions were established.
* This configuration is nonstable; that is, it will not persist when preset and clear inputs return to their inactive (high) level.

Connection Diagram
Dual-In-Line Package

MM54HC74 (J) 74HC74 (J,N)

Logic Diagram

					Supply Voltage (V_{CC})	−0.5 to +7.0V
					DC Input Voltage (V_{IN})	−1.5 to V_{CC} +1.5V
					DC Output Voltage (V_{OUT})	−0.5 to V_{CC} +0.5V
					Clamp Diode Current (I_{IK}, I_{OK})	±20 mA
					DC Output Current, per pin (I_{OUT})	±25 mA
					DC V_{CC} or GND Current, per pin (I_{CC})	±50 mA
					Storage Temperature Range (T_{STG})	−65°C to +150°C
					Power Dissipation (P_D) (Note 3)	500 mW
					Lead Temperature (T_L) (Soldering 10 seconds)	260°C

	Supply Voltage (V_{CC})	Min	Max	Units
		2	6	V
	DC Input or Output Voltage (V_{IN}, V_{OUT})	0	V_{CC}	V
	Operating Temperature Range (T_A)			°C
	MM74HC	−40	+85	°C
	MM54HC	−55	+125	°C
	Input Rise or Fall Times (t_r, t_f) $V_{CC}=2.0V$		1000	ns
	$V_{CC}=4.5V$		500	ns
	$V_{CC}=6.0V$		400	ns

DC Electrical Characteristics (Note 4)

Symbol	Parameter	Conditions	V_{CC}	$T_A=25°C$		$74HC$ $T_A=−40$ to $85°C$	$54HC$ $T_A=−55$ to $125°C$	Units
				Typ	Guaranteed Limits			
V_{IH}	Minimum High Level Input Voltage		2.0V		1.5	1.5	1.5	V
			4.5V		3.15	3.15	3.15	V
			6.0V		4.2	4.2	4.2	V
V_{IL}	Maximum Low Level Input Voltage		2.0V		0.3	0.3	0.3	V
			4.5V		0.9	0.9	0.9	V
			6.0V		1.2	1.2	1.2	V
V_{OH}	Minimum High Level Output Voltage	$V_{IN}=V_{IH}$ or V_{IL} $\|I_{OUT}\|≤20$ μA	2.0V	2.0	1.9	1.9	1.9	V
			4.5V	4.5	4.4	4.4	4.4	V
			6.0V	6.0	5.9	5.9	5.9	V
		$V_{IN}=V_{IH}$ or V_{IL} $\|I_{OUT}\|≤4.0$ mA	4.5V	4.3	3.98	3.84	3.7	V
		$\|I_{OUT}\|≤5.2$ mA	6.0V	5.2	5.48	5.34	5.2	V
V_{OL}	Maximum Low Level Output Voltage	$V_{IN}=V_{IH}$ or V_{IL} $\|I_{OUT}\|≤20$ μA	2.0V	0	0.1	0.1	0.1	V
			4.5V	0	0.1	0.1	0.1	V
			6.0V	0	0.1	0.1	0.1	V
		$V_{IN}=V_{IH}$ or V_{IL} $\|I_{OUT}\|≤4.0$ mA	4.5V	0.2	0.26	0.33	0.4	V
		$\|I_{OUT}\|≤5.2$ mA	6.0V	0.2	0.26	0.33	0.4	V
I_{IN}	Maximum Input Current	$V_{IN}=V_{CC}$ or GND	6.0V		±0.1	±1.0	±1.0	μA
I_{CC}	Maximum Quiescent Supply Current	$V_{IN}=V_{CC}$ or GND $I_{OUT}=0$ μA	6.0V		4.0	40	80	μA

Note 1: Absolute Maximum Ratings are those values beyond which damage to the device may occur.
Note 2: Unless otherwise specified all voltages are referenced to ground.
Note 3: Power Dissipation temperature derating — plastic "N" package: −12 mW/°C from 65°C to 85°C; ceramic "J" package: −12 mW/°C from 100°C to 125°C.
Note 4: For a power supply of 5V ±10% the worst case output voltages (V_{OH} and V_{OL}) occur for HC at 4.5V. Thus the 4.5V values should be used when designing with this supply. Worst case V_{IH} and V_{IL} occur at $V_{CC}=5.5V$ and 4.5V respectively. (The V_{IH} value at 5.5V is 3.85V.) The worst case leakage current (I_{IN}, I_{CC}, and I_{OZ}) occur for CMOS at the higher voltage and so the 6.0V values should be used.

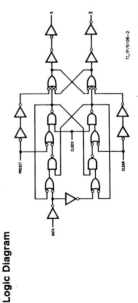

74HC74 dual D flip-flop, edge-triggered
reproduced with permission of National Semiconductor Corp.

74HC74 DUAL D FLIP-FLOP

AC Electrical Characteristics $V_{CC} = 5V$, $T_A = 25°C$, $C_L = 15$ pF, $t_r = t_f = 6$ ns

Symbol	Parameter	Conditions	Typ	Guaranteed Limit	Units
f_{MAX}	Maximum Operating Frequency		50	30	MHz
t_{PHL}, t_{PLH}	Maximum Propagation Delay Clock to Q or $\bar{Q}$		16	30	ns
t_{PHL}, t_{PLH}	Maximum Propagation Delay Preset or Clear to Q or $\bar{Q}$		25	40	ns
t_{REM}	Minimum Removal Time, Preset or Clear to Clock			5	ns
t_S	Minimum Set Up Time Data to Clock			20	ns
t_H	Minimum Hold Time Clock to Data			0	ns
t_W	Minimum Pulse Width Clock, Preset or Clear			16	ns

AC Electrical Characteristics $C_L = 50$ pF, $t_r = t_f = 6$ ns (unless otherwise specified)

Symbol	Parameter	Conditions	V_{CC}	$T_A = 25°C$ Typ	74HC $T_A = -40$ to $85°C$ Guaranteed Limits	54HC $T_A = -55$ to $125°C$	Units
f_{MAX}	Maximum Operating Frequency		2.0V	5	4	4	MHz
			4.5V	27	21	18	MHz
			6.0V	32	25	21	MHz
t_{PHL}, t_{PLH}	Maximum Propagation Delay Clock to Q or $\bar{Q}$		2.0V	88	175	221	261	ns
			4.5V	18	35	44	52	ns
			6.0V	15	30	37	44	ns
t_{PHL}, t_{PLH}	Maximum Propagation Delay Preset or Clear To Q or $\bar{Q}$		2.0V	98	230	290	343	ns
			4.5V	30	46	58	69	ns
			6.0V	28	39	49	58	ns
t_{REM}	Minimum Removal Time Preset or Clear To Clock		2.0V		25	32	37	ns
			4.5V		5	6	7	ns
			6.0V		4	5	6	ns
t_S	Minimum Set Up Time Data to Clock		2.0V		100	126	149	ns
			4.5V		20	25	30	ns
			6.0V		17	21	25	ns
t_H	Minimum Hold Time Clock to Data		2.0V		0	0	0	ns
			4.5V		0	0	0	ns
			6.0V		0	0	0	ns
t_W	Minimum, Pulse Width Clock, Preset or Clear		2.0V	30	80	101	119	ns
			4.5V	9	16	20	24	ns
			6.0V	8	14	17	20	ns
t_{TLH}, t_{THL}	Maximum Output Rise and Fall Time		2.0V	25	75	95	110	ns
			4.5V	7	15	19	22	ns
			6.0V	6	13	16	19	ns
t_r, t_f	Maximum Input Rise and Fall Time		2.0V		1000	1000	1000	ns
			4.5V		500	500	500	ns
			6.0V		400	400	400	ns
C_{PD}	Power Dissipation Capacitance (Note 5)	(per flip-flop)		80			pF
C_{IN}	Maximum Input Capacitance			5	10	10	pF

Note 5: C_{PD} determines the no load dynamic power consumption. $P_D = C_{PD} V_{CC}^2 f + I_{CC} V_{CC}$, and the no load dynamic current consumption, $I_S = C_{PD} V_{CC} f + I_{CC}$.

Note 6: Refer to back of this section for Typical MM54/74HC AC Switching Waveforms and Test Circuits.

AD7569 8-BIT A/D, D/A

FEATURES
- 2µs ADC with Track/Hold
- 1µs DAC with Output Amplifier
- On-Chip Bandgap Reference
- Fast Bus Interface
- Single or Dual 5V Supplies

GENERAL DESCRIPTION

The AD7569 is a complete, 8-bit, analog I/O system on a single monolithic chip. It contains a high-speed successive approximation ADC with 2µs conversion time, a track/hold with 200kHz bandwidth, a DAC and output buffer amplifier with 1µs settling time. A temperature-compensated 1.25V bandgap reference provides a precision reference voltage for the ADC and the DAC.

A choice of analog input/output ranges is available. Using a supply voltage of +5V, input and output ranges of zero to 1.25V and zero to 2.5 volts may be programmed using the RANGE input pin. Using a ±5V supply, bipolar ranges of ±1.25V or ±2.5V may be programmed.

Digital interfacing is via an 8-bit I/O port and standard microprocessor control lines. Bus interface timing is extremely fast, allowing easy connection to all popular 8-bit microprocessors. A separate start convert line controls the track/hold and ADC to give precise control of the sampling period.

The AD7569 is fabricated in Linear-Compatible CMOS (LC²MOS), an advanced, mixed technology process combining precision bipolar circuits with low-power CMOS logic. The part is packaged in a 24-pin, 0.3" wide "skinny" DIP and is also available in plastic leaded chip carrier (PLCC) and ceramic leadless chip carrier (LCCC).

PRODUCT HIGHLIGHTS

1. **Complete Analog I/O on a Single Chip.**
   The AD7569 provides everything necessary to interface a microprocessor to the analog world. No external components or user trims are required, and the overall accuracy of the system is tightly specified, eliminating the need to calculate error budgets from individual component specifications.

2. **Dynamic Specifications for DSP Users.**
   In addition to the traditional ADC and DAC specifications the AD7569 is specified for AC parameters, including signal-to-noise ratio, distortion and input bandwidth.

3. **Fast Microprocessor Interface.**
   The AD7569 has bus interface timing compatible with all modern microprocessors, with bus access and relinquish times less than 75ns and Write pulse width less than 80ns.

4. **Low Power.**
   Thanks to the combination of high-speed linear circuits with low-power CMOS logic, the AD7569 offers power consumption less than 60mW — considerably lower than any system of comparable performance.

DAC SPECIFICATIONS

($V_{DD} = +5V \pm 5\%$, $V_{SS}^1 = $ RANGE $= $ AGND$_{DAC} = $ AGND$_{ADC} = $ DGND $= 0V$; $R_L = 2k\Omega$, $C_L = 100pF$ unless otherwise stated). (All specifications T_{min} to T_{max} unless otherwise stated.)

Parameter	AD7569J[2]/AD7569A	AD7569K/AD7569B	AD7569S	AD7569T	Units	Conditions/Comments
STATIC PERFORMANCE						
Resolution	8	8	8	8	Bits	
Total Unadjusted Error[4]	±2	±2	±3	±3	LSB typ	
Relative Accuracy[4]	±1	±1/2	±1	±1/2	LSB max	Guaranteed Monotonic
Differential Nonlinearity[4]	±1	±3/4	±1	±3/4	LSB max	
Unipolar Offset Error						
@ 25°C	±2	±1.5	±2	±1.5	LSB max	
T_{min} to T_{max}	±2.5	±2	±2.5	±2	LSB max	
Bipolar/Zero Offset Error						
@ 25°C	±2	±1.5	±2	±1.5	LSB max	
T_{min} to T_{max}	±2.5	±2	±2.5	±2	LSB max	
Full-Scale Error[4]						
@ 25°C	±1	±1	±1	±1	LSB max	Typical tempco is 10µV/°C for ±1.25V range
T_{min} to T_{max}	±3	±2	±3	±2	LSB max	Typical tempco is 20µV/°C for ±1.25V range
ΔFull Scale/ΔV_{DD}, $T_A = 25°C$	0.5	0.5	0.5	0.5	LSB max	$V_{OUT} = \pm 2.5V$; $\Delta V_{DD} = \pm 5\%$
ΔFull Scale/ΔV_{SS}, $T_A = 25°C$	0.5	0.5	0.5	0.5	LSB max	$V_{OUT} = \pm 2.5V$; $\Delta V_{SS} = \pm 5\%$
Load Regulation at Full Scale	0.2	0.2	0.2	0.2	LSB max	$R_L = 2k\Omega$ to 0V
DYNAMIC PERFORMANCE						
Signal-to-Noise Ratio[4] (SNR)	44	44	44	46	dB min	$V_{IN} = 20kHz$ full-scale sine wave with $f_{SAMPLING} = 400kHz$
Total Harmonic Distortion[4] (THD)	48	48	48	48	dB max	$V_{IN} = 20kHz$ full-scale sine wave with $f_{SAMPLING} = 400kHz$
Intermodulation Distortion[4] (IMD)	55	55	55	55	dB typ	$f_a = 18.4kHz$, $f_b = 14.5kHz$ with $f_{SAMPLING} = 400kHz$
ANALOG OUTPUT						
Output Voltage Ranges						
Unipolar	0 to +1.25/2.5				Volts	$V_{DD} = +5V$, $V_{SS} = 0V$
Bipolar	±1.25/±2.5				Volts	$V_{DD} = +5V$, $V_{SS} = -5V$
LOGIC INPUTS						
$\overline{CS}$, $\overline{WR}$, RANGE, $\overline{RESET}$, DB0-DB7						
Input Low Voltage, V_{INL}	0.8	0.8	0.8	0.8	V max	
Input High Voltage, V_{INH}	2.4	2.4	2.4	2.4	V min	
Input Leakage Current	10	10	10	10	µA max	
Input Capacitance[4]	10	10	10	10	pF max	
DB0-DB7						
Input Coding (Single Supply)	Binary					
Input Coding (Dual Supply)	2s Complement					
AC CHARACTERISTICS[4]						
Voltage Output Settling Time						Settling time to within ±1/2LSB of final value
Positive Full-Scale Change	2	2	2	2	µs max	Typically 1µs
Negative Full-Scale Change (Single Supply)	4	4	4	4	µs max	Typically 2µs
Negative Full-Scale Change (Dual Supply)	2	2	2	2	µs max	Typically 1µs
Digital-to-Analog Glitch Impulse[4]	15	15	15	15	nV secs typ	
Digital Feedthrough[4]	15	15	15	15	nV secs typ	
V_{IN} to V_{OUT} Isolation	60	60	60	60	dB typ	$V_{IN} = \pm 2.5V$, 50kHz Sine Wave
POWER REQUIREMENTS						
V_{DD} Range	4.75/5.25	4.75/5.25	4.75/5.25	4.75/5.25	Vmin/Vmax	For Specified Performance
V_{SS} Range (Dual Supplies)	-4.75/-5.25	-4.75/-5.25	-4.75/-5.25	-4.75/-5.25	Vmin/Vmax	Specified Performance also applies to $V_{SS} = 0V$ for Unipolar output ranges.
I_{DD}						
@ 25°C	12	12	12	12	mA max	$V_{OUT} = V_{IN} = 2.5V$; Logic Inputs = 2.4V; CLK = 0.8V
T_{min} to T_{max}	13	13	13	13	mA max	Output Unloaded
I_{SS} (Dual Supplies)						
@ 25°C	4	4	4	4	mA max	$V_{OUT} = V_{IN} = -2.5V$; Logic Inputs = 2.4V; CLK = 0.8V
T_{min} to T_{max}	4	4	4	4	mA max	Output Unloaded
DAC/ADC MATCHING						
Gain Matching[4]						
@ 25°C	1	1	1	1	% typ	V_{IN} to V_{OUT} match with $V_{IN} = \pm 2.5V$,
T_{min} to T_{max}	1	1	1	1	% typ	20kHz sine wave

NOTES
Except where noted, specifications apply for all output ranges including bipolar ranges with dual supply operation.
[2]Temperature Ranges are as follows: AD7569J, K: 0 to +70°C; AD7569A, B: -25°C to +85°C; AD7569S, T: -55°C to +125°C.
[1]1LSB = 4.88mV for 0 to +1.25V output range; 9.76mV for 0 to +2.5V and ±1.25V ranges and 19.5mV for ±2.5V range.
[4]Includes internal voltage reference error and is calculated after offset error has been adjusted out. Ideal unipolar full-scale voltage is (FS - 1LSB); ideal bipolar positive full-scale voltage is (FS/2 - 1LSB) and ideal bipolar negative full-scale voltage is -FS/2.
[3]FSR is 1.25V for the 0 to +1.25V range; 2.5V for the 0 to +2.5V and ±1.25V ranges and 5V for the ±2.5V range.
*See Terminology.
Sample tested at 25°C to ensure compliance.
Specifications subject to change without notice.

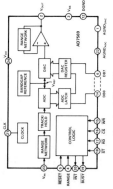

Functional Block Diagram

One Technology Way; P. O. Box 9106; Norwood, MA 02062-9106 U.S.A.
Tel: 617/329-4700 Twx: 710/394-6577
Telex: 924491 Cables: ANALOG NORWOODMASS

Information furnished by Analog Devices, Inc. is believed to be accurate and reliable. However, no responsibility is assumed by Analog Devices for its use; nor for any infringements of patents or other rights of third parties which may result from its use. No license is granted by implication or otherwise under any patent or patent rights of Analog Devices.

AD7569 8-bit A/D, D/A
reproduced with permission of Analog Devices, Inc.

AD7569 8-BIT A/D, D/A

ADC SPECIFICATIONS

($V_{DD} = +5V \pm 5\%$, $V_{SS} = -5V \pm 5\%$ = RANGE = AGND$_{DAC}$ = AGND$_{ADC}$ = DGND = 0V, f_{CLK} = 5MHz external unless otherwise stated). (All specifications T_{min} to T_{max} unless otherwise stated.) Specifications apply to Mode 1 interface.

Parameter	AD7569J/AD7569A	AD7569K/AD7569B	AD7569S	AD7569T	Units	Conditions/Comments
DC ACCURACY						
Resolution	8	8	8	8	Bits	
Total Unadjusted Error[4]	±3	±3	±4	±4	LSB typ	
Relative Accuracy[4]	±1	±1/2	±1	±1/2	LSB max	
Differential Nonlinearity[4]	±1	±3/4	±1	±3/4	LSB max	No Missing Codes
Unipolar Offset Error @25°C	±2	±1.5	±2	±1.5	LSB max	Typical tempco is 10μV/°C for +1.25V range, V_{SS} = 0V
T_{min} to T_{max}	±3	±2.5	±3	±2.5	LSB max	
Bipolar Zero Offset Error @25°C	±3	±2.5	±3	±2.5	LSB max	Typical tempco is 20μV/°C for +1.25V range, V_{SS} = -5V
T_{min} to T_{max}	±3.5	±3	±3.5	±3	LSB max	
Full-Scale Error @25°C	-4,+0	-4,+0	-4,+0	-4,+0	LSB max	V_{IN} = 5V
ΔFull-Scale/ΔV_{DD}, T_A = 25°C	-5.5,+1.5	-5.5,+1.5	-6.5,+2	-6.5,+2	LSB max	Typical tempco is ±25ppm of FSR/°C
ΔFull-Scale/ΔV_{SS}, T_A = 25°C	0.5	0.5	0.5	0.5	LSB max	V_{IN} = ±2.5V; ΔV_{DD} = ±5%
	0.5	0.5	0.5	0.5	LSB max	V_{IN} = -2.5V; ΔV_{SS} = ±5%
DYNAMIC PERFORMANCE						
Signal-to-Noise Ratio[5] SNR	44	44	46	46	dB min	V_{IN} = 100kHz full-scale sine wave with $f_{SAMPLING}$ = 400kHz[2]
Total Harmonic Distortion[5] (THD)	48	48	48	48	dB max	V_{IN} = 100kHz full-scale sine wave with $f_{SAMPLING}$ = 400kHz
Intermodulation Distortion[5] (IMD)	60	60	60	60	dB typ	fa = 99kHz, fb = 96.7kHz with $f_{SAMPLING}$ = 400kHz
Frequency Response	0.1	0.1	0.1	0.1	dB typ	D.C. to 200kHz sine wave
Track/Hold Acquisition Time[5]	200	200	300	300	ns typ	
ANALOG INPUT						
Input Voltage Ranges		0 to +1.25V ±2.5			Volts	V_{DD} = +5V; V_{SS} = 0V
Unipolar		±1.25 to ±2.5			Volts	V_{DD} = +5V; V_{SS} = -5V
Bipolar						See equivalent circuit Fig. 5
Input Current	±300	±300	±300	±300	μA max	
Input Capacitance	10	10	10	10	pF typ	
LOGIC INPUTS						
$\overline{CS}$, $\overline{RD}$, ST, CLK, RESET, ...ANGE						
Input Low Voltage, V_{INL}	0.8	0.8	0.8	0.8	V max	
Input High Voltage, V_{INH}	2.4	2.4	2.4	2.4	V min	
Input Capacitance[5]	10	10	10	10	pF max	
$\overline{CS}$, $\overline{RD}$, ST, RANGE, $\overline{RESET}$						
Input Current	10	10	10	10	μA max	V_{IN} = 0V; V_{IN} = V_{DD}
CLK						
Input Current						
I_{SNK}	-1.6	-1.6	-1.6	-1.6	mA max	I_{SINK} = 1.6mA
I_{SRC}						I_{SOURCE} = 200μA
LOGIC OUTPUTS						
DB0-DB7, $\overline{INT}$, BUSY						
V_{OL}, Output Low Voltage	0.4	0.4	0.4	0.4	V max	
V_{OH}, Output High Voltage	4.0	4.0	4.0	4.0	V min	
DB0-DB7						
Floating State Leakage Current	10	10	10	10	μA max	
Floating State Output Capacitance[5]	10	10	10	10	pF max	
Output Coding (Single Supply)		Binary				
Output Coding (Dual Supply)		2s Complement				
CONVERSION TIME						
With External Clock	2	2	2	2	μs max	f_{CLK} = 5MHz
With Internal Clock, T_A = 25°C	1.6	1.6	1.6	1.6	μs min	Using recommended clock components shown in Figure 20. Clock frequency can be adjusted by varying R_{CLK}.
	2.6	2.6	2.6	2.6	μs max	
POWER REQUIREMENTS						
	As per DAC Specifications					

NOTES
[1] Except where noted, specifications apply for all output ranges including bipolar ranges with dual supply operation.
[2] Temperature Ranges are as follows: AD7569J, K: 0 to +70°C; AD7569A, B: -25°C to +85°C; AD7569S: -55°C to +125°C; AD7569T: -55°C to +125°C.
[3] 1LSB = 4.88mV for 0 to +1.25V output range, 9.76mV for 0 to +2.5V and ±1.25V ranges and 19.5mV for ±2.5V range.
[4] Includes internal voltage reference error and is calculated after offset error has been adjusted out. Ideal unipolar last code transition occurs at (FS2-3/2LSB). Ideal bipolar last offset error occurs at (FS-3/2LSB).
[5] FSR is 1.25V for the 0 to +1.25V range, 2.5V for the 0 to +2.5V and ±1.25V ranges and 5V for the ±2.5V range. Exact frequencies are 101kHz sinewave, 384kHz to avoid harmonics coinciding with sampling frequency.
[6] Rising edge of BUSY to falling edge of ST. The time given refers to the acquisition time which gives a 3dB degradation in SNR from the nominal figure.
[7] Sample tested at 25°C to ensure compliance.
Specifications subject to change without notice.

TIMING CHARACTERISTICS[1]

(See Figures 8, 9, 11); V_{DD} = 5V ±5%; V_{SS} = ±5V, or -5V ±5%)

Parameter	Limit at 25°C (All Grades)	Limit at T_{min}, T_{max} (J, K, A, B Grades)	Limit at T_{min}, T_{max} (S, T Grades)	Units	Test Conditions/Comments
DAC Timing					
t_1	80	80	90	ns min	$\overline{WR}$ Pulse Width
t_2	0	0	0	ns min	$\overline{CS}$ to $\overline{WR}$ Setup Time
t_3	0	0	0	ns min	$\overline{CS}$ to $\overline{WR}$ Hold Time
t_4	60	70	80	ns min	Data Valid to $\overline{WR}$ Setup Time
t_5	10	10	10	ns min	Data Valid to $\overline{WR}$ Hold Time
ADC Timing					
t_6	50	50	50	ns max	ST Pulse Width
t_7	110	130	150	ns max	ST to BUSY Delay
t_8	20	30	30	ns max	BUSY to $\overline{INT}$ Delay
t_9	0	0	0	ns max	BUSY to $\overline{CS}$ Delay
t_{10}	60	75	90	ns max	$\overline{CS}$ to $\overline{RD}$ Setup Time
t_{11}	0	0	0	ns max	$\overline{CS}$ to $\overline{RD}$ Hold Time
t_{12}	60	75	90	ns max	$\overline{RD}$ Pulse Width. Determined by t_{13}
t_{13}	95	120	135	ns max	Data Access Time after $\overline{RD}$, C_L = 20pF
t_{14}[4]	10	10	10	ns max	Data Access Time after $\overline{RD}$, C_L = 100pF
t_{15}	60	75	85	ns max	Bus Relinquish Time after $\overline{RD}$
t_{16}	65	85	85	ns max	$\overline{RD}$ to $\overline{INT}$ Delay
t_{17}	120	140	160	ns max	$\overline{RD}$ to BUSY Delay
	60	75	90	ns max	Data Valid Time after BUSY, C_L = 20pF
	90	115	135	ns max	Data Valid Time after BUSY, C_L = 100pF

NOTES
[1] Sample tested at 25°C to ensure compliance. All input control signals are specified with $t_R = t_F$ = 5ns (10% to 90% of +5V) and timed from a voltage level of 1.6V.
[2] t_{13} and t_{14} are measured with the load circuits of Figure 1 and defined as the time required for an output to cross either 0.8V or 2.4V.
[3] t_{15} is defined as the time required for the data line to change 0.5V when loaded with the circuit of Figure 2.
Specifications subject to change without notice.

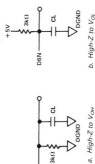

Figure 1. Load Circuits for Data Access Time Test

Figure 2. Load Circuits for Bus Relinquish Time Test

ABSOLUTE MAXIMUM RATINGS

V_{DD} to AGND$_{DAC}$ or AGND$_{ADC}$	-0.3V, +7V
V_{DD} to DGND	-0.3V, +7V
V_{DD} to V_{SS}	-0.3V, +14V
AGND$_{DAC}$ or AGND$_{ADC}$ to DGND	-0.3V, V_{DD} +0.3V
AGND$_{DAC}$ to AGND$_{ADC}$	-0.3V, V_{DD} +0.3V
Logic Voltage to DGND	-0.3V, V_{DD} +0.3V
CLK Input Voltage to DGND	-0.3V, V_{DD} +0.3V
V_{OUT} to AGND$_{DAC}$	V_{SS} -0.3V, V_{DD} +0.3V
V_{IN} to AGND$_{ADC}$	V_{SS} -0.3V, V_{DD} +0.3V
Power Dissipation (Any Package) to +75°C	450mW
Derates above 75°C by	6mW/°C
Operating Temperature Range	
Commercial (J, K)	0 to +70°C
Industrial (A, B)	-25°C to +85°C
Extended (S, T)	-55°C to +125°C
Storage Temperature Range	-65°C to +150°C
Lead Temperature (Soldering, 10 Secs)	+300°C

*Stresses above those listed under "Absolute Maximum Ratings" may cause permanent damage to the device. This is a stress rating only and functional operation of the device at these or any other conditions above those indicated in the operational sections of this specification is not implied. Exposure to absolute maximum rating conditions for extended periods may affect device reliability.

NOTE
Output may be shorted to any voltage in the range V_{SS} to V_{DD} provided that the power dissipation of the package is not exceeded. Typical short circuit current for a short to AGND or V_{SS} is 50mA.

CAUTION
ESD (Electro-Static-Discharge) sensitive device. The digital control inputs are zener protected; however, permanent damage may occur on unconnected devices subject to high energy electrostatic fields. Unused devices must be stored in conductive foam or shunts. The protective foam should be discharged to the destination socket before devices are removed.

WARNING! ESD SENSITIVE DEVICE

AD7569 8-BIT A/D, D/A

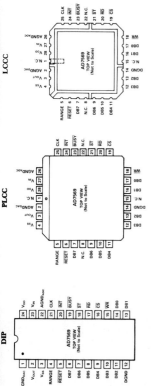

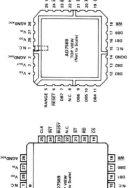

PIN FUNCTION DESCRIPTION
(Pin Nos as Per DIP Pin Configuration)

PIN	MNEMONIC	DESCRIPTION
1	$AGND_{DAC}$	Analog Ground for the DAC. Separate ground return paths are provided for the DAC and ADC to minimize crosstalk.
2	V_{OUT}	Output Voltage. This is the buffered output voltage from the DAC. Four different output voltage ranges can be achieved (see Table I).
3	V_{SS}	Negative Supply Voltage (-5V for dual supply or 0V for single supply). This pin is also used with the RANGE pin to select the different input/output ranges and changes the data format from binary ($V_{SS} = 0$V) to 2s complement ($V_{SS} = -5$V) (see Table I).

Table I. Input/Output Ranges

Range	V_{SS}	Input/Output Voltage Range	Data Format
0	0V	0 to $+1.25$V	Binary
1	0V	0 to $+2.5$V	Binary
0	-5V	± 1.25V	2s Complement
1	-5V	± 2.5V	2s Complement

PIN	MNEMONIC	DESCRIPTION
4	RANGE	Range Selection Input. This is used with the V_{SS} input to select the different ranges as per Table I above. The range selected applies to both the analog input voltage of the ADC and the output voltage from the DAC.
5	RESET	Reset Input (Active Low). This is an asynchronous system reset which clears the DAC register to all 0s and clears the INT line of the ADC (i.e., makes the ADC ready for new conversion). In unipolar operation this input sets the output voltage to 0V; in bipolar operation it sets the output to negative full scale.
6	DB7	Data Bit 7. Most Significant Bit (MSB).
7-11	DB6-DB2	Data Bit 6 to Data Bit 2.
12	DGND	Digital Ground.
13	DB1	Data Bit 1.
14	DB0	Data Bit 0. Least Significant Bit (LSB).
15	$\overline{WR}$	Write Input (Edge triggered). This is used in conjunction with $\overline{CS}$ to write data into the DAC register. Data is transferred on the rising edge of $\overline{WR}$.
16	$\overline{CS}$	Chip Select Input (Active Low). The device is selected when this input is active.
17	$\overline{RD}$	READ Input (Active Low). This input must be active to access data from the part. In the Mode 2 interface, $\overline{RD}$ going low starts conversion. It is used in conjunction with the $\overline{CS}$ input (see Digital Interface Section).
18	$\overline{ST}$	Start Conversion (Edge triggered). This is used when precise sampling is required. The falling edge of $\overline{ST}$ starts conversion and drives $\overline{BUSY}$ low. The $\overline{ST}$ signal is not gated with $\overline{CS}$.
19	$\overline{BUSY}$	BUSY Status Output (Active Low). When this pin is active the ADC is performing a conversion. The input signal is held prior to the falling edge of $\overline{BUSY}$ (see Digital Interface Section).
20	$\overline{INT}$	INTERRUPT Output (Active Low). $\overline{INT}$ going low indicates that the conversion is complete. $\overline{INT}$ goes high on the rising edge of $\overline{CS}$ or $\overline{RD}$ and is also set high by a low pulse on RESET (see Digital Interface Section).
21	CLK	A TTL compatible clock signal may be used to determine the ADC conversion time. Internal clock operation is achieved by connecting a resistor and capacitor to ground.
22	$AGND_{ADC}$	Analog Ground for the ADC.
23	V_{IN}	Analog Input. Various input ranges can be selected (see Table I).
24	V_{DD}	Positive Supply Voltage ($+5$V).

68008 EXECUTION TIMES & TIMING DIAGRAMS

Table 7-3. Effective Address Calculation Times

Addressing Mode		Byte	Word	Long
Register				
Dn	Data Register Direct	0(0/0)	0(0/0)	0(0/0)
An	Address Register Direct	0(0/0)	0(0/0)	0(0/0)
Memory				
(An)	Address Register Indirect	4(1/0)	8(2/0)	16(4/0)
(An)+	Address Register Indirect with Postincrement	4(1/0)	8(2/0)	16(4/0)
−(An)	Address Register Indirect with Predecrement	6(1/0)	10(2/0)	18(4/0)
d(An)	Address Register Indirect with Displacement	12(3/0)	16(4/0)	24(6/0)
d(An, ix)*	Address Register Indirect with Index	14(3/0)	18(4/0)	26(6/0)
xxx.W	Absolute Short	12(3/0)	16(4/0)	24(6/0)
xxx.L	Absolute Long	20(5/0)	24(6/0)	32(8/0)
d(PC)	Program Counter with Displacement	12(3/0)	16(4/0)	24(6/0)
d(PC, ix)*	Program Counter with Index	14(3/0)	18(4/0)	26(6/0)
#xxx	Immediate	8(2/0)	8(2/0)	16(4/0)

*The size of the index register (ix) does not affect execution time.

Table 7-4. Move Byte Instruction Execution Times

					Destination				
Source	Dn	An	(An)	(An)+	−(An)	d(An)	d(An, ix)*	xxx.W	xxx.L
Dn	8(2/1)	8(2/1)	12(2/1)	12(2/1)	12(2/1)	20(4/1)	22(4/1)	20(4/1)	28(6/1)
An	8(2/1)	8(2/1)	12(2/1)	12(2/1)	12(2/1)	20(4/1)	22(4/1)	20(4/1)	28(6/1)
(An)	12(3/0)	12(3/0)	16(3/1)	16(3/1)	16(3/1)	24(5/1)	26(5/1)	24(5/1)	32(7/1)
(An)+	12(3/0)	12(3/0)	16(3/1)	16(3/1)	16(3/1)	24(5/1)	26(5/1)	24(5/1)	32(7/1)
−(An)	14(3/0)	14(3/0)	18(3/1)	18(3/1)	18(3/1)	26(5/1)	28(5/1)	26(5/1)	34(7/1)
d(An)	20(5/0)	20(5/0)	24(5/1)	24(5/1)	24(5/1)	32(7/1)	34(7/1)	32(7/1)	40(9/1)
d(An, ix)*	22(5/0)	22(5/0)	26(5/1)	26(5/1)	26(5/1)	34(7/1)	36(7/1)	34(7/1)	42(9/1)
xxx.W	20(5/0)	20(5/0)	24(5/1)	24(5/1)	24(5/1)	32(7/1)	34(7/1)	32(7/1)	40(9/1)
xxx.L	28(7/0)	28(7/0)	32(7/1)	32(7/1)	32(7/1)	40(9/1)	42(9/1)	40(9/1)	48(11/1)
d(PC)	20(5/0)	20(5/0)	24(5/1)	24(5/1)	24(5/1)	32(7/1)	34(7/1)	32(7/1)	40(9/1)
d(PC, ix)*	22(5/0)	22(5/0)	26(5/1)	26(5/1)	26(5/1)	34(7/1)	36(7/1)	34(7/1)	42(9/1)
#xxx	16(4/0)	—	20(4/1)	20(4/1)	20(4/1)	28(6/1)	30(6/1)	28(6/1)	36(8/1)

*The size of the index register (ix) does not affect execution time.

Table 7-5. Move Word Instruction Execution Times

					Destination				
Source	Dn	An	(An)	(An)+	−(An)	d(An)	d(An, ix)*	xxx.W	xxx.L
Dn	8(2/1)	8(2/1)	16(2/2)	16(2/2)	16(2/2)	24(4/2)	26(4/2)	24(4/2)	32(6/2)
An	8(2/1)	8(2/1)	16(2/2)	16(2/2)	16(2/2)	24(4/2)	26(4/2)	24(4/2)	32(6/2)
(An)	16(4/0)	16(4/0)	24(4/2)	24(4/2)	24(4/2)	32(6/2)	34(6/2)	32(6/2)	40(8/2)
(An)+	16(4/0)	16(4/0)	24(4/2)	24(4/2)	24(4/2)	32(6/2)	34(6/2)	32(6/2)	40(8/2)
−(An)	18(4/0)	18(4/0)	26(4/2)	26(4/2)	26(4/2)	34(6/2)	36(6/2)	34(6/2)	42(8/2)
d(An)	24(6/0)	24(6/0)	32(6/2)	32(6/2)	32(6/2)	40(8/2)	42(8/2)	40(8/2)	48(10/2)
d(An, ix)*	26(6/0)	26(6/0)	34(6/2)	34(6/2)	34(6/2)	42(8/2)	44(8/2)	42(8/2)	50(10/2)
xxx.W	24(6/0)	24(6/0)	32(6/2)	32(6/2)	32(6/2)	40(8/2)	42(8/2)	40(8/2)	48(10/2)
xxx.L	32(8/0)	32(8/0)	40(8/2)	40(8/2)	40(8/2)	48(10/2)	50(10/2)	48(10/2)	56(12/2)
d(PC)	24(6/0)	24(6/0)	32(6/2)	32(6/2)	32(6/2)	40(8/2)	42(8/2)	40(8/2)	48(10/2)
d(PC, ix)*	26(6/0)	26(6/0)	34(6/2)	34(6/2)	34(6/2)	42(8/2)	44(8/2)	42(8/2)	50(10/2)
#xxx	16(4/0)	16(4/0)	24(4/2)	24(4/2)	24(4/2)	32(6/2)	34(6/2)	32(6/2)	40(8/2)

*The size of the index register (ix) does not affect execution time.

Table 7-6. Move Long Instruction Execution Times

					Destination				
Source	Dn	An	(An)	(An)+	−(An)	d(An)	d(An, ix)*	xxx.W	xxx.L
Dn	8(2/0)	8(2/0)	24(2/4)	24(2/4)	24(2/4)	32(4/4)	34(4/4)	32(4/4)	40(6/4)
An	8(2/0)	8(2/0)	24(2/4)	24(2/4)	24(2/4)	32(4/4)	34(4/4)	32(4/4)	40(6/4)
(An)	24(6/0)	24(6/0)	40(6/4)	40(6/4)	40(6/4)	48(8/4)	50(8/4)	48(8/4)	56(10/4)
(An)+	24(6/0)	24(6/0)	40(6/4)	40(6/4)	40(6/4)	48(8/4)	50(8/4)	48(8/4)	56(10/4)
−(An)	26(6/0)	26(6/0)	42(6/4)	42(6/4)	42(6/4)	50(8/4)	52(8/4)	50(8/4)	58(10/4)
d(An)	32(8/0)	32(8/0)	48(8/4)	48(8/4)	48(8/4)	56(10/4)	58(10/4)	56(10/4)	64(12/4)
d(An, ix)*	34(8/0)	34(8/0)	50(8/4)	50(8/4)	50(8/4)	58(10/4)	60(10/4)	58(10/4)	66(12/4)
xxx.W	32(8/0)	32(8/0)	48(8/4)	48(8/4)	48(8/4)	56(10/4)	58(10/4)	56(10/4)	64(12/4)
xxx.L	40(10/0)	40(10/0)	56(10/4)	56(10/4)	56(10/4)	64(12/4)	66(12/4)	64(12/4)	72(14/4)
d(PC)	32(8/0)	32(8/0)	48(8/4)	48(8/4)	48(8/4)	56(10/4)	58(10/4)	56(10/4)	64(12/4)
d(PC, ix)*	34(8/0)	34(8/0)	50(8/4)	50(8/4)	50(8/4)	58(10/4)	60(10/4)	58(10/4)	66(12/4)
#xxx	24(6/0)	24(6/0)	40(6/4)	40(6/4)	40(6/4)	48(8/4)	50(8/4)	48(8/4)	56(10/4)

*The size of the index register (ix) does not affect execution time.

Table 7-7. Standard Instruction Execution Times

Instruction	Size	op <ea>, An	op <ea>, Dn	op Dn, <M>
ADD	Byte	—	8(2/0)+	12(2/1)+
	Word	12(2/0)+	8(2/0)+	16(2/2)+
	Long	10(2/0)+**	10(2/0)+**	24(2/4)+
AND	Byte	—	8(2/0)+	12(2/1)+
	Word	—	8(2/0)+	16(2/2)+
	Long	—	10(2/0)+**	24(2/4)+
CMP	Byte	—	8(2/0)+	—
	Word	10(2/0)+	8(2/0)+	—
	Long	—	10(2/0)+	—
DIVS	—	—	162(2/0)+*	—
DIVU	—	—	144(2/0)+*	—
EOR	Byte	—	8(2/0)+***	12(2/1)+
	Word	—	8(2/0)+***	16(2/2)+
	Long	—	12(2/0)+***	24(2/4)+
MULS	—	—	74(2/0)+*	—
MULU	—	—	74(2/0)+*	—
OR	Byte	—	8(2/0)+	12(2/1)+
	Word	—	8(2/0)+	16(2/2)+
	Long	—	10(2/0)+**	24(2/4)+
SUB	Byte	—	8(2/0)+	12(2/1)+
	Word	12(2/0)+	8(2/0)+	16(2/2)+
	Long	10(2/0)+**	10(2/0)+**	24(2/4)+

NOTES:
+ Add effective address calculation time
* Indicates maximum value
** The base time of 10 clock periods is increased to 12 if the effective address mode is register direct or immediate (effective add ress time should also be added).
*** Only available effective address mode is data register direct

DIVS, DIVU — The divide algorithm used by the MC68008 provides less than 10% difference between the best and worst case timings.

MULS, MULU — The multiply algorithm requires 42 + 2n clocks where n is defined as:
  MULS: n = tag the <ea> with a zero as the MSB; n is the resultant number of 10 or 01 patterns in the 17-bit source, i.e., worst case happens when the source is $5555.
  MULU: n = the number of ones in the <ea>

68008 timing data: 8/16/32-bit microprocessor reproduced with permission of Motorola Semiconductor Products, Inc.

68008 TIMING

Table 7-8. Immediate Instruction Clock Periods

Instruction	Size	op<f, Dn	op<f, An	op<f, M
ADDI	Byte	8(2/0)	—	12(2/1)+
	Word	16(4/0)	—	16(2/2)+
	Long	28(6/0)	—	24(2/4)+
ADDQ	Byte	8(2/0)	—	12(2/1)+
	Word	8(2/0)	12(2/0)	16(2/2)+
	Long	12(2/0)	12(2/0)	24(2/4)+
ANDI	Byte	16(4/0)	—	20(4/1)+
	Word	16(4/0)	—	24(4/2)+
	Long	28(6/0)	—	40(6/4)+
CMPI	Byte	16(4/0)	—	16(4/0)+
	Word	16(4/0)	—	16(4/0)+
	Long	26(6/0)	—	24(6/0)+
EORI	Byte	16(4/0)	—	20(4/1)+
	Word	16(4/0)	—	24(4/2)+
	Long	28(6/0)	—	40(6/4)+
MOVEQ	Byte	8(2/0)	—	—
ORI	Byte	16(4/0)	—	20(4/1)+
	Word	16(4/0)	—	24(4/2)+
	Long	28(6/0)	—	40(6/4)+
SUBI	Byte	16(4/0)	—	12(2/1)+
	Word	16(4/0)	—	16(2/2)+
	Long	28(6/0)	—	24(2/4)+
SUBQ	Byte	8(2/0)	—	20(4/1)+
	Word	8(2/0)	12(2/0)	24(4/2)+
	Long	12(2/0)	12(2/0)	40(6/4)+

+ add effective address calculation time

Table 7-9. Single Operand Instruction Execution Times

Instruction	Size	Register	Memory
CLR	Byte	8(2/0)	12(2/1)+
	Word	8(2/0)	16(2/2)+
	Long	10(2/0)	24(2/4)+
NBCD	Byte	8(2/0)	12(2/1)+
NEG	Byte	8(2/0)	12(2/1)+
	Word	8(2/0)	16(2/2)+
	Long	10(2/0)	24(2/4)+
NEGX	Byte	8(2/0)	12(2/1)+
	Word	8(2/0)	16(2/2)+
	Long	10(2/0)	24(2/4)+
NOT	Byte	8(2/0)	12(2/1)+
	Word	8(2/0)	16(2/2)+
	Long	10(2/0)	24(2/4)+
Scc	Byte, False	8(2/0)	12(2/1)+
	Byte, True	10(2/0)	12(2/1)+
TAS	Byte	14(2/1)	14(2/1)+
TST	Byte	8(2/0)	8(2/0)+
	Word	8(2/0)	8(2/0)+
	Long	8(2/0)	8(2/0)+

+ add effective address calculation time.

Table 7-10. Shift/Rotate Instruction Clock Periods

Instruction	Size	Register	Memory
ASR, ASL	Byte	10 + 2n(2/0)	—
	Word	10 + 2n(2/0)	16(2/2)+
	Long	12 + 2n(2/0)	—
LSR, LSL	Byte	10 + 2n(2/0)	—
	Word	10 + 2n(2/0)	16(2/2)+
	Long	12 + 2n(2/0)	—
ROR, ROL	Byte	10 + 2n(2/0)	—
	Word	10 + 2n(2/0)	16(2/2)+
	Long	12 + 2n(2/0)	—
ROXR, ROXL	Byte	10 + 2n(2/0)	—
	Word	10 + 2n(2/0)	16(2/2)+
	Long	12 + 2n(2/0)	—

+ add effective address calculation time
n is the shift count

Table 7-11. Bit Manipulation Instruction Execution Times

Instruction	Size	Dynamic Register	Dynamic Memory	Static Register	Static Memory
BCHG	Byte	—	12(2/1)+	—	20(4/1)+
	Long	12(2/0)*	—	20(4/0)*	—
BCLR	Byte	—	12(2/1)+	—	20(4/1)+
	Long	14(2/0)*	—	22(4/0)*	—
BSET	Byte	—	12(2/1)+	—	20(4/1)+
	Long	12(2/0)*	—	20(4/0)*	—
BTST	Byte	—	8(2/0)+	—	16(4/0)+
	Long	10(2/0)	—	18(4/0)	—

+ add effective address calculation time
*indicates maximum value

Table 7-12. Conditional Instruction Execution Times

Instruction	Displacement	Trap or Branch Taken	Trap or Branch Not Taken
Bcc	Byte	18(4/0)	12(2/0)
	Word	18(4/0)	20(4/0)
BRA	Byte	18(4/0)	—
	Word	18(4/0)	—
BSR	Byte	34(4/4)	—
	Word	34(4/4)	—
DBcc	CC True	18(4/0)	20(4/0)
	CC False	—	26(6/0)
CHK	—	68(8/6)+*	14(2/0)+
TRAP	—	62(8/6)	—
TRAPV	—	66(10/6)	8(2/0)

+ add effective address calculation time
*indicates maximum value

68008 TIMING

Table 7-13. JMP, JSR, LEA, PEA, and MOVEM Instruction Execution Times

Instruction	Size	(An)	(An)+	−(An)	d(An)	d(An, ix)*	xxx.W	xxx.L	d(PC)	d(PC, ix)*
JMP	—	16(4/0)	—	—	18(4/0)	22(4/0)	18(4/0)	24(6/0)	18(4/0)	22(4/0)
JSR	—	32(4/4)	—	—	34(4/4)	38(4/4)	34(4/4)	40(6/4)	34(4/4)	38(4/4)
LEA	—	8(2/0)	—	—	16(4/0)	20(4/0)	16(4/0)	24(6/0)	16(4/0)	20(4/0)
PEA	—	24(2/4)	—	—	32(4/4)	36(4/4)	32(4/4)	40(6/4)	32(4/4)	36(4/4)
MOVEM	Word	24+8n (6+2n/0)	24+8n (6+2n/0)	—	32+8n (8+2n/0)	34+8n (8+2n/0)	32+8n (8+2n/0)	40+8n (10+n/0)	32+8n (8+2n/0)	34+8n (8+2n/0)
M → R	Long	24+16n (6+4n/0)	24+16n (6+4n/0)	—	32+16n (8+4n/0)	34+16n (8+4n/0)	32+16n (8+4n/0)	40+16n (8+4n/0)	32+16n (8+4n/0)	34+16n (8+4n/0)
MOVEM	Word	16+8n (4/2n)	—	16+8n (4/2n)	24+8n (6/2n)	26+8n (6/2n)	24+8n (6/2n)	32+8n (8/2n)	—	—
R → M	Long	16+16n (4/4n)	—	16+16n (4/4n)	24+16n (6/4n)	26+16n (8/4n)	24+16n (6/4n)	32+16n (6/4n)	—	—

n is the number of registers to move
* is the size of the index register (ix) does not affect the instruction's execution time

Table 7-14. Multi-Precision Instruction Execution Times

Instruction	Size	op Dn, Dn	op M, M
ADDX	Byte	8(2/0)	22(4/1)
	Word	8(2/0)	50(6/2)
	Long	12(2/0)	58(10/4)
CMPM	Byte	—	16(4/0)
	Word	—	24(6/0)
	Long	—	40(10/0)
SUBX	Byte	8(2/0)	22(4/1)
	Word	8(2/0)	50(6/2)
	Long	12(2/0)	58(10/4)
ABCD	Byte	10(2/0)	20(4/1)
SBCD	Byte	10(2/0)	20(4/1)

Table 7-15. Miscellaneous Instruction Execution Times

Instruction	Register	Memory
ANDI to CCR	32(6/0)	—
ANDI to SR	32(6/0)	—
EORI to CCR	32(6/0)	—
EORI to SR	32(6/0)	—
EXG	10(2/0)	—
EXT	8(2/0)	—
LINK	32(4/4)	—
MOVE to CCR	18(4/0)	18(4/0)+
MOVE to SR	18(4/0)	18(4/0)+
MOVE from SR	10(2/0)	16(2/2)+
MOVE to USP	8(2/0)	—
MOVE from USP	8(2/0)	—
NOP	8(2/0)	—
ORI to CCR	32(6/0)	—
ORI to SR	32(6/0)	—
RESET	136(2/0)	—
RTE	40(10/0)	—
RTR	40(10/0)	—
RTS	32(8/0)	—
STOP	4(0/0)	—
SWAP	8(2/0)	—
UNLK	24(6/0)	—

+ add effective address calculation time

Table 7-16. Move Peripheral Instruction Execution Times

Instruction	Size	Register → Memory	Memory → Register
MOVEP	Word	24(4/2)	24(6/0)
	Long	32(4/4)	32(8/0)

+ add effective address calculation time

Table 7-17. Exception Processing Execution Times

Exception	Periods
Address Error	94(8/14)
Bus Error	94(8/14)
CHK Instruction	68(8/6) +
Interrupt	72(9/16)*
Illegal Instruction	62(8/6)
Privileged Instruction	62(8/6)
Trace	62(8/6)
TRAP Instruction	62(8/6)
TRAPV Instruction	66(10/6)
Divide by Zero	66(8/6) +
RESET**	64(12/0)

+ add effective address calculation time
* The interrupt acknowledge bus cycle is assumed to take four external clock periods
** Indicates the time from when RESET and HALT are first sampled as negated to when instruction execution starts.

68008 TIMING

8.6 AC ELECTRICAL SPECIFICATIONS — READ AND WRITE CYCLES

(VCC = 5.0 Vdc ± 5%; GND = 0 Vdc; $T_A = T_L$ to T_H; see Figures 8-6 and 8-7)

Num.	Characteristic	Symbol	8 MHz Min	8 MHz Max	10 MHz Min	10 MHz Max	Unit
1	Clock Period	t_{cyc}	125	500	100	500	ns
2	Clock Width Low	t_{CL}	55	250	45	250	ns
3	Clock Width High	t_{CH}	55	250	45	250	ns
4	Clock Fall Time	t_{Cf}	—	10	—	10	ns
5	Clock Rise Time	t_{Cr}	—	10	—	10	ns
6	Clock Low to Address Valid	t_{CLAV}	—	70	—	60	ns
6A	Clock High to FC Valid	t_{CHFCV}	—	70	—	60	ns
7	Clock High to Address, Data Bus High Impedance (Maximum)	t_{CHADZ}	—	80	—	70	ns
8	Clock High to Address, FC Invalid (Minimum)	t_{CHAFI}	0	—	0	—	ns
9[1]	Clock High to $\overline{AS}$, $\overline{DS}$ Low	t_{CHSL}	0	60	0	55	ns
11A2,7	Address Valid to $\overline{AS}$, $\overline{DS}$ Low (Read)/$\overline{AS}$ Low (Write)	t_{AVSL}	30	—	20	—	ns
11[2]	FC Valid to $\overline{AS}$, $\overline{DS}$ Low (Read)/$\overline{AS}$ Low (Write)	t_{FCVSL}	60	—	50	—	ns
12[1]	Clock Low to $\overline{AS}$, $\overline{DS}$ High	t_{CLSH}	—	70	—	60	ns
13[2]	$\overline{AS}$, $\overline{DS}$ High to Address/FC Invalid	t_{SHARI}	30	—	20	—	ns
14A2,5	$\overline{AS}$, $\overline{DS}$ Width Low (Read) / $\overline{AS}$ Low (Write)	t_{SL}	270	—	195	—	ns
15[2]	$\overline{AS}$, $\overline{DS}$ Width High	t_{SH}	150	—	105	—	ns
16	Clock High to Control Bus High Impedance	t_{CHCZ}	—	80	—	70	ns
17[2]	$\overline{AS}$, $\overline{DS}$ High to R/$\overline{W}$ High (Read)	t_{SHRH}	40	—	20	—	ns
18[1]	Clock High to R/$\overline{W}$ High	t_{CHRH}	0	40	0	40	ns
20[1]	Clock High to R/$\overline{W}$ Low	t_{CHRL}	—	40	—	40	ns
20A[6]	$\overline{AS}$ Low to R/$\overline{W}$ Valid (Write)	t_{ASRV}	—	20	—	20	ns
21[2]	Address Valid to R/$\overline{W}$ Low (Write)	t_{AVRL}	20	—	10	—	ns
21A2,7	FC Valid to R/$\overline{W}$ Low (Write)	t_{FCVRL}	60	—	50	—	ns
22[2]	R/$\overline{W}$ Low to $\overline{DS}$ Low (Write)	t_{RLSL}	80	—	50	—	ns
23	Clock Low to Data Out Valid (Write)	t_{CLDO}	—	70	—	55	ns
25[2]	$\overline{AS}$, $\overline{DS}$ High to Data Out Invalid (Write)	t_{SHDOI}	50	—	20	—	ns
26[2]	Data Out Valid to $\overline{DS}$ Low (Write)	t_{DOSL}	35	—	20	—	ns
27[5]	Data In to Clock Low (Setup Time on Read)	t_{DICL}	15	—	10	—	ns
28[2,6]	$\overline{AS}$, $\overline{DS}$ High to $\overline{DTACK}$ High	t_{SHDAH}	0	245	0	190	ns
29	$\overline{AS}$, $\overline{DS}$ High to Data In Invalid (Hold Time on Read)	t_{SHDII}	0	—	0	—	ns
30	$\overline{AS}$, $\overline{DS}$ High to BERR High	t_{SHBEH}	0	—	0	—	ns
31[2,5]	$\overline{DTACK}$ Low to Data Valid (Asynchronous Setup Time on Read)	t_{DALDI}	—	90	—	65	ns
32	HALT and $\overline{RESET}$ Input Transition Time	$t_{RHr, f}$	0	200	0	200	ns
33	Clock High to $\overline{BG}$ Low	t_{CHGL}	—	40	—	40	ns
34	Clock High to $\overline{BG}$ High	t_{CHGH}	—	40	—	40	ns
35	$\overline{BR}$ Low to $\overline{BG}$ Low	t_{BRLGL}	1.5	90 ns +3.5	1.5	80 ns +3.5	Clk. Per.
36[8]	$\overline{BR}$ High to $\overline{BG}$ High	t_{BRHGH}	1.5	90 ns +3.5	1.5	80 ns +3.5	Clk. Per.
37	$\overline{BGACK}$ Low to $\overline{BG}$ High (52-Pin Version Only)	t_{GALGH}	1.5	90 ns +3.5	1.5	80 ns +3.5	Clk. Per.
37A[9]	$\overline{BGACK}$ Low to $\overline{BR}$ High (52-Pin Version Only)	t_{GALBRH}	20	1.5 Clocks	—	1.5 Clocks	ns
38	$\overline{BG}$ Low to Control, Address, Data Bus High Impedance ($\overline{AS}$ High)	t_{GLZ}	—	80	—	70	ns
39	$\overline{BG}$ Width High	t_{GH}	1.5	—	1.5	—	Clk. Per.
41	Clock Low to E Transition	t_{CLET}	—	50	—	50	ns
42	E Output Rise and Fall Time	$t_{Er, f}$	—	15	—	15	ns
44	$\overline{AS}$, $\overline{DS}$ High to $\overline{VPA}$ High	t_{SHVPH}	0	120	0	90	ns

— Continued

8.6 AC ELECTRICAL SPECIFICATIONS — READ AND WRITE CYCLES (Continued)

(VCC = 5.0 Vdc ± 5%; GND = 0 Vdc; $T_A = T_L$ to T_H; see Figures 8-6 and 8-7)

Num.	Characteristic	Symbol	8 MHz Min	8 MHz Max	10 MHz Min	10 MHz Max	Unit
45	E Low to Control, Address Bus Invalid (Address Hold Time)	t_{ELCAI}	30	—	10	—	ns
46	$\overline{BGACK}$ Width Low (52-Pin Version Only)	t_{GAL}	1.5	—	1.5	—	Clk. Per.
47[5]	Asynchronous Input Setup Time	t_{ASI}	10	—	10	—	ns
48[3]	BERR Low to $\overline{DTACK}$ Low	t_{BELDAL}	20	—	20	—	ns
49[10]	$\overline{AS}$, $\overline{DS}$ High to E Low	t_{SHEL}	−80	80	−80	80	ns
50	E Width High	t_{EH}	450	—	360	—	ns
51	E Width Low	t_{EL}	700	—	550	—	ns
53	Clock High to Data Out Invalid	t_{CHDOI}	0	—	0	—	ns
54	E Low to Data Out Invalid	t_{ELDOI}	30	—	20	—	ns
55	R/$\overline{W}$ to Data Bus Impedance Driven	—	—	—	—	—	ns
56[4]	HALT/$\overline{RESET}$ Pulse Width	t_{HRPW}	10	—	10	—	Clk. Per.
57	$\overline{BGACK}$ High to Control Bus Driven (52-Pin Version Only)	t_{GABD}	1.5	—	1.5	—	Clk. Per.
58[8]	$\overline{BG}$ High to Control Bus Driven	t_{GHBD}	1.5	—	1.5	—	Clk. Per.

NOTES:
1. For a loading capacitance of less than or equal to 50 picofarads, subtract 5 nanoseconds from the values given in these columns.
2. Actual value depends on clock period.
3. If #47 is satisfied for both $\overline{DTACK}$ and $\overline{BERR}$, #48 may be 0 nanoseconds.
4. For power up, the MPU must be held in RESET state for 100 milliseconds to allow stabilization of on-chip circuitry. After the system is powered up, #56 refers to the minimum pulse width required to reset the system.
5. If the asynchronous setup time (#47) requirements are satisfied, the $\overline{DTACK}$ low-to-data setup time (#31) requirement can be ignored. The data must only satisfy the #27 data-in to clock low setup time for the following cycle.
6. When $\overline{AS}$ and R/$\overline{W}$ are equally loaded (±20%), subtract 10 nanoseconds from the values in these columns.
7. Setup time to guarantee recognition on next falling edge of clock.
8. The processor will negate $\overline{BG}$ and begin driving the bus again if external arbitration logic negates $\overline{BR}$ before asserting $\overline{BGACK}$.
9. The minimum value must be met to guarantee proper operation. If the maximum value is exceeded, $\overline{BG}$ may be reasserted.
10. The falling edge of S6 triggers both the negation of the strobes ($\overline{AS}$ and $\overline{DS}$) and the falling edge of E. Either of these events can occur first, depending upon the loading on each signal. Specification #49 indicates the absolute maximum skew that will occur between the rising edge of the strobes and the falling edge of the E clock.

68008 TIMING

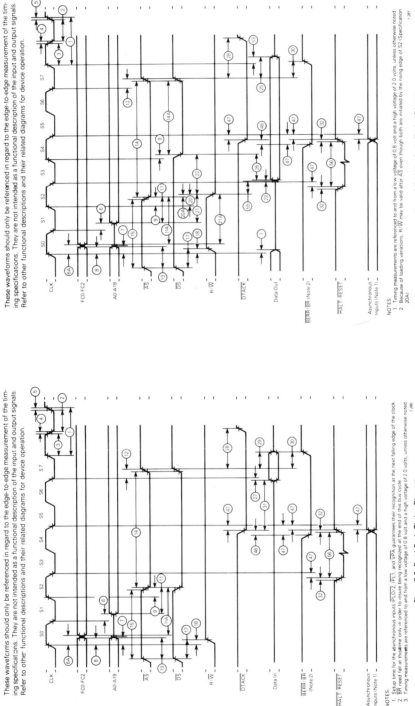

Figure 8-7. Write Cycle Timing Diagram

Figure 8-6. Read Cycle Timing Diagram

25120 WRITE-ONLY-MEMORY

25120 FULLY ENCODED, 9046×N, RANDOM ACCESS WRITE-ONLY-MEMORY

TYPICAL CHARACTERISTIC CURVES

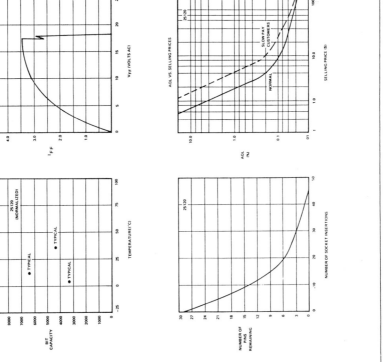

DESCRIPTION

The Signetics 25000 Series 9046XN Random Access Write-Only-Memory employs both enhancement and depletion mode P-Channel, N-Channel, and neuI(1) channel MOS devices. Although a static device, a single TTL level clock phase is required to drive the on-board multi-port clock generator. Data refresh is accomplished during CB and LH periods(11). Quadri-state outputs (when applicable) allow expansion in many directions, depending on organization.

The static memory cells are operated dynamically to yield extremely low power dissipation. All inputs and outputs are directly TTL compatible when proper interfacing circuitry is employed.

Device construction is more or less S.O.S.(2).

FEATURES

- FULLY ENCODED MULTI-PORT ADDRESSING
- WRITE CYCLE TIME 80nS (MAX. TYPICAL)
- WRITE ACCESS TIME(3)
- POWER DISSIPATION 10uW/BIT TYPICAL
- CELL REFRESH TIME 2mS (MIN. TYPICAL)
- TTL/DTL COMPATIBLE INPUTS(4)
- AVAILABLE OUTPUTS "n"
- CLOCK LINE CAPACITANCE 2pF MAX.(5)
- $V_{CC} = +10V$
- $V_{DD} = 0V \pm 2\%$
- $V_{FF} = 6.3V_{ac}$ (6)

APPLICATIONS

- DON'T CARE BUFFER STORES
- LEAST SIGNIFICANT CONTROL MEMORIES
- POST MORTEM MEMORIES (WEAPON SYSTEMS)
- ARTIFICIAL MEMORY SYSTEMS
- NON-INTELLIGENT MICRO CONTROLLERS
- FIRST-IN NEVER-OUT (FINO) ASYNCHRONOUS BUFFERS
- OVERFLOW REGISTER (BIT BUCKET)

PROCESS TECHNOLOGY

The use of Signetics unique SEX(7) process yields Vth (var.) and allows the design(8) and production(9) of higher performance MOS circuits than can be obtained by competitor's techniques.

---

1. "Neu," channel devices enhance or deplete regardless of gate polarity, either simultaneously or randomly. Sometimes not at all.
2. "S.O.S." copyrighted U.S. Army Commissary, 1940.
3. Not applicable
4. You can somehow drive these inputs from TTL, the method is obvious.
5. Measure at 1MHz, 25mVac, 1.9pF in series.
6. For the filaments, what else!

FINAL SPECIFICATION(10)

BIPOLAR COMPATIBILITY

All data and clock inputs plus applicable outputs will interface directly or nearly directly with bipolar circuits of suitable characteristics. In any event use 1 amp fuses in all power supply and data lines.

INPUT PROTECTION

All terminals are provided with slip-on latex protectors for the prevention of Voltage Destruction. (PILL packaged devices do not require protection).

SILICON PACKAGING

Low cost silicon DIP packaging is implemented and reliability is assured by the use of a non-hermetic sealing technique which prevents the entrapment of harmful ions, but which allows the free exchange of friendly ions.

SPECIAL FEATURES

Because of the employment of the Signetics' proprietary Sanderson-Rabbet Channel the 25120 will provide 50% higher speed than you will obtain.

COOLING

The 25120 is easily cooled by employment of a six-foot fan, 1/2" from the package. If the device fails, you have exceeded the ratings. In such cases, more air is recommended.

BLOCK DIAGRAM

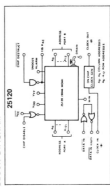

PART IDENTIFICATION

TYPE	"n"	TEMP. RANGE	PACKAGE
25120	0	0 to -70°C	Whatever's Right

---

7. "You have a dirty mind. S.E.X. is Signetics EXtra Secret process. "One Shovel Full to One Shovel Full", patented by Yagura, Kashkooli, Converse and Al. Circa 1921.
8. J. Kane calls it design (we humor him).
9. See "Modern Production Techniques" by T. Arrieta (not yet written).
10. Final until we got a look at some actual parts.
11. Coffee breaks and lunch hours.
12. Due credit to EIMAC for inspiration.

25120 data sheet reproduced with permission of Signetics, Inc.

APPENDIX C: Big Picture: Schematic of Lab Microcomputer

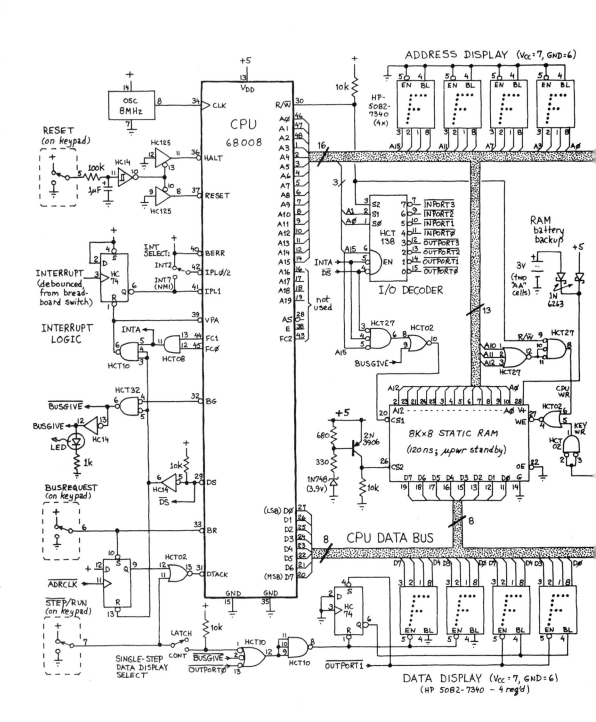

Figure N18.1: Lab Microcomputer schematic: left side

Big Picture: Schematic of Lab Microcomputer

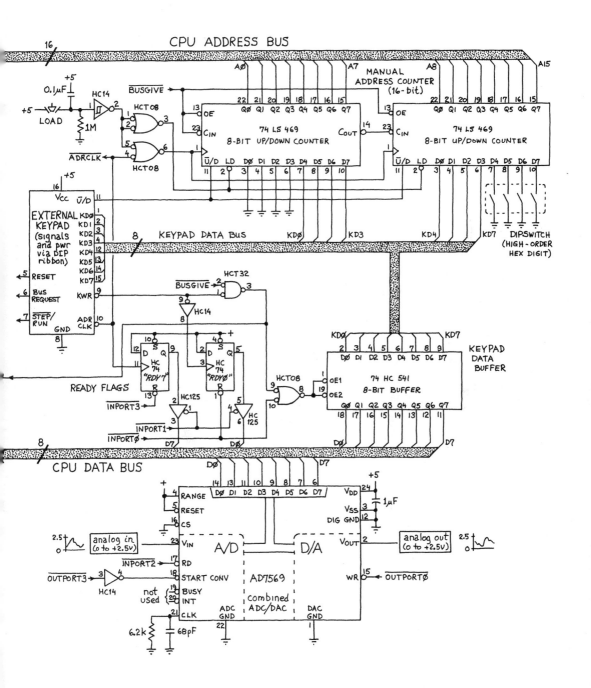

Figure N18.2: Lab Microcomputer schematic: right side

APPENDIX D: PINOUTS

TRANSISTORS

TO-108: LPT-100, FPT-100 (C, B, E)

TO-220: IRF 511, IRL 510, RFI 4N05, MTP 5N06 (G, D, S)

TO-92: 2N3904, 2N3906, 2N5962, 2N4400 (E, B, C)

TO-220: SCR (K, A, G)

TO-18: 2N3958 (S_1, D_1, G_1, G_2, D_2, S_2)

TO-92: JFET 2N5485 (D, S, G); MOS VN01, VP01 (D, G, S)

LINEAR IC's

miniDIP: 311
- GND (1), V+ (8)
- IN+ (2), OUT (7)
- IN− (3), BAL/STR (6)
- V− (4), BAL (5)

TO-39: 317
- out, adj, in

miniDIP: 411 / 741
- BAL (1), NC (8)
- IN− (2), V+ (7)
- IN+ (3), OUT (6)
- V− (4), BAL (5)

miniDIP: 358
- (1), V+ (8)
- (2), (7)
- (3), (6)
- (4) V−, (5)

miniDIP: 7555
- GND (1), V+ (8)
- TRIG (2), DIS (7)
- OUT (3), THRESH (6)
- RESET (4), CONTROL (5)

miniDIP: MF4
- CLK (1), IN (8)
- CLK R (2), V+ (7)
- level shift (3), AGND (6)
- V− (4), OUT (5)

DIP: 723
- 5 NI, 4 INV, 6 VR REF, 7 GND, 12 V+, 11 Vc, Vo 10, Vz 9, CL 2, COMP 13, CS 3

CA3096 — 16-pin DIP (substrate = pin 16): 3 NPN's, 2 PNP's

TO-92: 78L05 (in, gnd, out)

TLC 372 Dual Comparator (pins 1–8, GND)

 TO-92: LM385 (N.C.)

DG403 Analog Switch (16-pin):
16 S_1, 15 IN_1, 14 $V_−$, 13 GND, 12 V_L, 11 V_+, 10 IN_2, 9 S_2
1 D_1, 2 NC, 3 D_3, 4 S_3, 5 S_4, 6 D_4, 7 NC, 8 D_2

PINOUTS

DIGITAL IC's: GATES

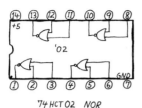

74 HCT 02   NOR

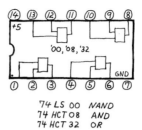

74 LS 00   NAND
74 HCT 08   AND
74 HCT 32   OR

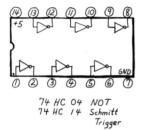

74 HC 04   NOT
74 HC 14   Schmitt
                Trigger

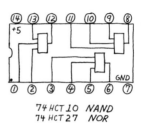

74 HCT 10   NAND
74 HCT 27   NOR

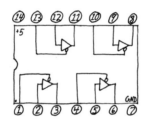

74 HC 125
3-State Buffer

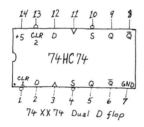

74 XX 74   Dual D flop

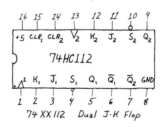

74 XX 112   Dual J-K Flop

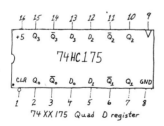

74 XX 175   Quad D register

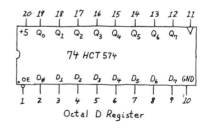

Octal D Register

PINOUTS

DIGITAL (cont.) & INTERFACE

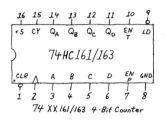

74 XX 161/163 4-Bit Counter

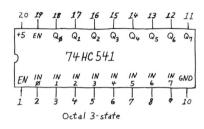

Octal 3-state

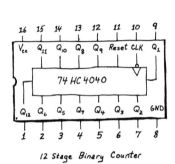

12 Stage Binary Counter

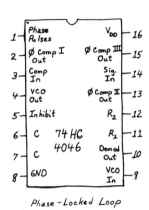

Phase-Locked Loop

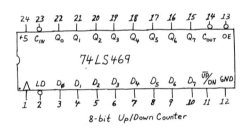

8-bit Up/Down Counter

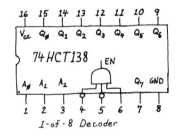

1-of-8 Decoder

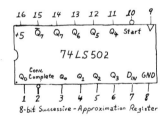

8-bit Successive-Approximation Register

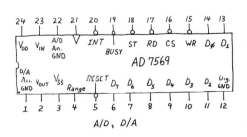

A/D, D/A

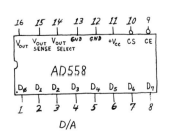

D/A

PINOUTS

MISCELLANEOUS IC's

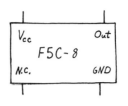

Crystal Oscillator

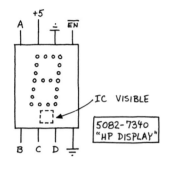

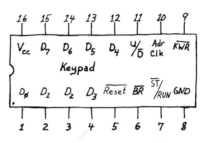

Keypad Cable DIP connector

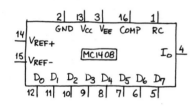

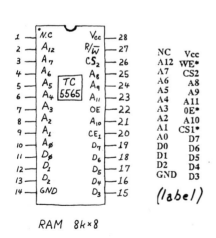

RAM 8k × 8

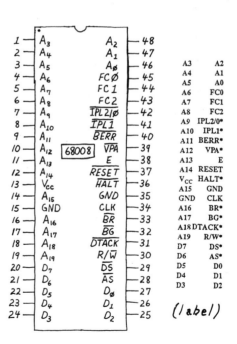

μprocessor

INDEX

This index directly covers only *class notes*. It reaches Labs and Worked Examples indirectly: topics treated in labs and worked examples are discussed in the class notes for the same day. So, to find the lab that treats the bipolar current source, for example, you would follow the index to Class 4, then look to Lab 4. There you would find the current source circuit that you build in the lab.

A

A ("open-loop gain") 208
AB ("loop gain") 219
absolute addressing................ 465
AC amplifier 193
access time 379
acquisition time (S & H) 249
"active low" 288
active pullup 292, 320
active rectifier 188
address 434
    – decoding.................... 459
addressing modes................... 462
    – immediate 463
    – absolute..................... 465
    – indirect..................... 466
    – moving pointer.......... 499
aliasing.................................. 408
amplitude resolution 402
analog switch 245
assertion-level symbols.......... 288
autovector (interrupt) 538

B

B ("fraction fed back")............ 208
balancing resistive paths 191
BG* (busgrant*)
    (68008 signal) 438
bias current (op amp) 190
biasing.................................... 86
binary.................................... 283
    – digit ("bit") 285
    – number 285
binary search.................. 411, 413
bit.. 285
bit test 482
BJT 148
BNC cable 62
Bode plot 40
Boolean algebra 287
BRANCH (68008
    instruction)....................... 457

bus 375
Butterworth low-pass.............. 44

C

CALL operation............. 480, 498
capacitance (FET: stray)........ 245
    – meter 350
capacitor 32
cascading (counters) 346
cascode 129
ceramic capacitor.................... 32
charge injection 249
clear 329
    – sync vs async 329
CMOS
    – analog switch 246
    – digital gate 290
combinational logic 287
common-emitter amplifier 88
common-mode gain 126
comparator
    – analog 210
    – digital........................ 435
complex plane......................... 42
conditional branch 482
conductances 3
counters
    – generally 330
    – ripple................. 330, 344
    – synchronous.............. 330
"crowbar" overvoltage
    protection.......................... 271
current amplifier 85
current hogging..................... 245
current limit
    – voltage regulator 268
    – op amp 194
current mirror........................ 103
current source 33, **87**
    – bipolar transistor......... 87
    – FET............................ 145
    – op amp 171
current-to-voltage converter .. 172

D

dB: (decibels)
    3dB point...................... 39
    6dB/octave slope 43
DBF (instruction) 500
decoder 460
    – I/O decoder 461
deMorgan's theorem.............. 287
depletion mode .... 14?
differential amplifier............. 124
differential gain 126
differentiator.................... 32, 3?
    – op amp version .... 18?
digital (vs analog) .... 28?
diode
    – circuits 65f
    –I vs V curve 6
direct memory access
    (DMA)............................... 43
distortion................................ 10?
droop
    – generally ....
    – sample & hold .... 24?
dropout voltage...................... 26?
DS* (data strobe*)
    (68008 signal)..................... 4?
DTACK* (data transfer
    acknowledge...................... 4?
dual-slope A/D .... 411, 4?
dynamic RAM .... 3?
dynamic resistance .... 7, ?

DVM ....

E

Ebers-Moll equation .... 1?
Early effect .... 1?
edge trigger............................ 3?
    – vs level-sensitive .... ?
emitter follower ....
emitter resistor as feedback ... ?
encode..................................... ?
enhancement mode .... ?

EPROM 378
error size:
   A-D conversion.................. 407
exceptions (68000) 536

F

555 oscillator 214
feedback
       –emitter resistor as 105
FET... 141
field effect transistor 141
filters 32, 38
finite state machine: see "state machine"
flags 481
function codes 438
"flash" A/D 412
flip-flop 325
floating input (logic gate)........ 292
follower
       – emitter follower .. 85, 87
       – as rose-colored lens..... 85
       – FET source follower . 146
       – op amp 171
Fourier
       – analysis 63
       – analyzer 64
       –components of
              square wave 64
frequency compensation 220
frequency domain 39
frequency multiplier................ 418
full-wave bridge rectifier 65
fuse rating 69

.... 148
...m.................................. 150
Golden Rules (op amp) 168
ground 10
grounded-emitter amplifier 104

hexadecimal display 348
high-pass.................................. 39
half-wave rectifier.................... 66
H (sample & hold) 248
"hold step" 249
hysteresis 212

I

IBM PC................................... 434
I_{DSS} 144
illegal exception 536
immediate addressing 463
impedance................................. 5
       – "looking" in a
            specified direction.... 87
indirect addressing:
   see *addressing*
inductor.................................... 37
input port 435
instruction register 457
integrator 32, 35
       – op amp version 185
interrupt 488, 535
       – mask......................... 539
intrinsic emitter resistance
(r_e) 101
inverting amplifier (op amp)... 169

J

"jam" clear............................ 329
JFET (junction FET).............. 142
J-K flip-flop 330

K

Karnaugh map 323
Kirchhoff's Laws...................... 2
keypad: explanation............... 372
keypad buffer........................ 461

L

latch 325
LC.. 43
       –resonant circuit 64
load function (counter) 346
loading (generally).................... 5
log plot 39
loop gain (AB)....................... 219
low-pass................................. 39

M

masking 483
master-slave flip-flop 327
memory................................. 377
memory vs I/O....................... 434

meter movement 9
Miller effect........................... 127
minimizing logic.................... 322
MOSFET 142
mylar capacitor........................ 32

N

NAND 287
       – NAND latch.............. 325
negative feedback 166
negative impedance converter
(NIC) 212
n-channel FETs..................... 143
noise immunity 284
noise margin 291
non-inverting amplifier (op
amp)...................................... 169
NMI: non-maskable
   interrupt 539
non-volatile memory 378
NOR 287
NOT...................................... 287
Nyquist sampling rate............ 407

O

offset current.......................... 192
offset voltage 189
Ohm's law 1
ohms per volt........................... 9
one-shot 331
open-loop circuits 168
open collector output 292, 321
open-loop gain (A)................. 208
oscillators 32
output impedance 5
       – of current source....... 112
output current limit
       – voltage regulator........ 268
       – op amp 194

P

parallel circuits 2
passband 44
passive pullup........................ 292
PC (two senses!)
       – IBM PC 434
       – program counter........ 457
phase detectors 416
phase-locked loop................. 416
       – stability 419

phase shift 40
phasor diagrams 41
polling 488
ports (μcomputer) 435
positive feedback 210
prefetch 457
primary (transformer) 69
printed circuit 565
priority among
    interrupts 539
probe compensation 63
program counter 457
programmable counter 347
PROM 378
PUSH (CPU operation) 480
push-pull 88

R

RAM 378
ramp 33
RC oscillator
    – op amp
        "relaxation osc." 212
    – IC version: '555 214
r_e 101
    – deriving 113
reactance 37
rectifier 65
relaxation oscillator 212
reset 329
resistance 3
return address 480
ripple (power supply) 68
ripple counter 330, 344
roll-off of gain (op amp) 194
ROM 378
R_{ON} (FET) 245
rms voltage 68
roll-off 43
R-2R ladder 410
R/W* (68008 signal) 434

S

sample
    – sample & hold ... 247, 248
sampling rate 407

saturation
    – for FET 145
    – for bipolar trans. 245
Schmitt trigger 212
scope probe 62
secondary (transformer) 69
secondary breakdown 245
sequential circuits 325
series circuit 2
setup time 344
shift register 331
signed condition
    (branch instruction) 486
68008 434
slew rate 194
slow-blow 70
solder breadboard 564
stack 479
    – pointer 479
state machine 331, 380
status register 537
stiff (voltage source) 7
"stopwatch" 349
subroutine 480
successive approximation 413
summing circuit (op amp) 171
sweep (function generator) 64
switch
    – FET vs bipolar 245
    – bounce 326
switching delays
    – analog switch 249
switching regulator 272
synchronous
    – circuits generally 343
    – counters 330, 345

T

T resistor configuration 186
temperature compensation 107
temperature effects 106
thermal shutdown 270
three-state output 292, 320
threshold
    – in Schmitt trigger 212
    – in logic gate 291
time-constant 35
time domain 39

timeout 436
times-ten rule of thumb 61
trace 537
tracking A/D 413
trade magazines 565
transconductance (bipolar
    transistor) 106
    —FET 148
transistor (bipolar):
    first, simple views .... 8?
transparent latch 327
trap .... 53(
triangle waveform .... 3∢
tri-state (trademark of NSC) .. 29²
TTL .... 29(
two's complement 28²

U

universal gate 28

V

vector (for 6800x
    'exceptions') .... 540, 5∢
V_{GS} .... 1∢
voltage regulators
    – generally 2
    – linear 2
    – switching 2
    – three-terminal 2
VOM ....
VPA* (68000 signal) .... ⁵
$V_{pinchoff}$ ....

W

Wien bridge oscillator ....
Wilson mirror ....
wire-wrap ....
word-sized output ....
worst-case analysis ....

Z

Z_{in}, Z_{out} ....
zener diode ....